THE PRINCIPLES
OF MATHEMATICS

BY

BERTRAND RUSSELL

W · W · NORTON & COMPANY
New York · London

Norton paperback edition reissued 1996

ISBN 0-393-31404-9

W. W. Norton & Company, Inc., 500 Fifth Avenue, New York, NY 10110
W. W. Norton & Company Ltd., 10 Coptic Street, London WC1A 1PU

PRINTED IN THE UNITED STATES OF AMERICA

8 9 0

INTRODUCTION TO THE
SECOND EDITION

" THE Principles of Mathematics " was published in 1903, and most of it was written in 1900. In the subsequent years the subjects of which it treats have been widely discussed, and the technique of mathematical logic has been greatly improved ; while some new problems have arisen, some old ones have been solved, and others, though they remain in a controversial condition, have taken on completely new forms. In these circumstances, it seemed useless to attempt to amend this or that, in the book, which no longer expresses my present views. Such interest as the book now possesses is historical, and consists in the fact that it represents a certain stage in the development of its subject. I have therefore altered nothing, but shall endeavour, in this Introduction, to say in what respects I adhere to the opinions which it expresses, and in what other respects subsequent research seems to me to have shown them to be erroneous.

The fundamental thesis of the following pages, that mathematics and logic are identical, is one which I have never since seen any reason to modify. This thesis was, at first, unpopular, because logic is traditionally associated with philosophy and Aristotle, so that mathematicians felt it to be none of their business, and those who considered themselves logicians resented being asked to master a new and rather difficult mathematical technique. But such feelings would have had no lasting influence if they had been unable to find support in more serious reasons for doubt. These reasons are, broadly speaking, of two opposite kinds : first, that there are certain unsolved difficulties in mathematical logic, which make it appear less certain than mathematics is believed to be ; and secondly that, if the logical basis of mathematics is accepted, it justifies, or tends to justify, much work, such as that of Georg Cantor, which is viewed with suspicion by many mathematicians on account of the unsolved paradoxes which it shares with logic. These two opposite lines of criticism are represented by the formalists, led by Hilbert, and the intuitionists, led by Brouwer.

The formalist interpretation of mathematics is by no means new, but for our purposes we may ignore its older forms. As presented by Hilbert, for example in the sphere of number, it consists in leaving the integers undefined, but asserting concerning them such axioms as shall make possible the deduction of the usual arithmetical propositions. That is to say, we do not assign any meaning to our symbols 0, 1, 2, . . except that they are to have certain properties enumerated in the axioms. These symbols are, therefore, to be regarded as variables. The later integers may be defined when 0 is given, but 0 is to be merely something having the assigned characteristics. Accordingly the symbols 0, 1, 2, . . . do not represent one definite series, but any progression whatever. The formalists have forgotten that numbers are needed, not only for doing sums, but for counting. Such propositions as " There were 12 Apostles " or " London has 6,000,000 inhabitants " cannot be interpreted in their system. For the symbol " 0 " may be taken to mean any finite integer, without thereby making any of Hilbert's axioms false ; and thus every number-symbol becomes infinitely ambiguous. The formalists are like a watchmaker who is so absorbed in making his watches look pretty that he has forgotten their purpose of telling the time, and has therefore omitted to insert any works.

There is another difficulty in the formalist position, and that is as regards existence. Hilbert assumes that if a set of axioms does not lead to a contradiction, there must be some set of objects which satisfies the axioms ; accordingly, in place of seeking to establish existence theorems by producing an instance, he devotes himself to methods of proving the self-consistency of his axioms. For him, " existence," as usually understood, is an unnecessarily metaphysical concept, which should be replaced by the precise concept of non-contradiction. Here, again, he has forgotten that arithmetic has practical uses. There is no limit to the systems of non-contradictory axioms that might be invented. Our reasons for being specially interested in the axioms that lead to ordinary arithmetic lie outside arithmetic, and have to do with the application of number to empirical material. This application itself forms no part of either logic or arithmetic ; but a theory which makes it *a priori* impossible cannot be right. The logical definition of numbers makes their connection with the actual world of countable objects intelligible ; the formalist theory does not.

The intuitionist theory, represented first by Brouwer and later by Weyl, is a more serious matter. There is a philosophy associated with the theory, which, for our purposes, we may ignore ; it is only its bearing on logic and mathematics that concerns us. The essential point here is the refusal to regard a proposition as either true or false unless some method exists of deciding the alternative. Brouwer denies the law of excluded middle where no such method exists. This destroys, for example, the proof that there are more real numbers than rational numbers, and that, in the series of real numbers, every progression has a limit. Consequently large parts of analysis, which for centuries have been thought well established, are rendered doubtful.

Associated with this theory is the doctrine called finitism, which calls in question propositions involving infinite collections or infinite series, on the ground that such propositions are unverifiable. This doctrine is an aspect of thorough-going empiricism, and must, if taken seriously, have consequences even more destructive than those that are recognized by its advocates. Men, for example, though they form a finite class, are, practically and empirically, just as impossible to enumerate as if their number were infinite. If the finitist's principle is admitted, we must not make *any* general statement—such as " All men are mortal " —about a collection defined by its properties, not by actual mention of all its members. This would make a clean sweep of all science and of all mathematics, not only of the parts which the intuitionists consider questionable. Disastrous consequences, however, cannot be regarded as proving that a doctrine is false ; and the finitist doctrine, if it is to be disproved, can only be met by a complete theory of knowledge. I do not believe it to be true, but I think no short and easy refutation of it is possible.

An excellent and very full discussion of the question whether mathematics and logic are identical will be found in Vol. III. of Jörgensen's " Treatise of Formal Logic," pp. 57-200, where the reader will find a dispassionate examination of the arguments that have been adduced against this thesis, with a conclusion which is, broadly speaking, the same as mine, namely that, while quite new grounds have been given in recent years for refusing to reduce mathematics to logic, none of these grounds is in any degree conclusive.

This brings me to the definition of mathematics which forms the first sentence of the " Principles." In this definition various changes are necessary. To begin with, the form " p implies q " is only one of many logical forms that mathematical propositions may take. I was originally led to emphasise this form by the consideration of Geometry. It was clear that Euclidean and non-Euclidean systems alike must be included in pure mathematics, and must not be regarded as mutually inconsistent ; we must, therefore, only assert that the axioms imply the propositions, not that the axioms are true and therefore the propositions are true. Such instances led me to lay undue stress on implication, which is only one among truth-functions, and no more important than the others. Next : when it is said that " p and q are propositions containing one or more variables," it would, of course, be more correct to say that they are propositional functions ; what is said, however, may be excused on the ground that propositional functions had not yet been defined, and were not yet familiar to logicians or mathematicians.

I come next to a more serious matter, namely the statement that " neither p nor q contains any constants except logical constants." I postpone, for the moment, the discussion as to what logical constants are. Assuming this known, my present point is that the absence of non-logical constants, though a necessary condition for the mathematical character of a proposition, is not a sufficient condition. Of this, perhaps, the best examples are statements concerning the number of things in the world. Take, say : " There are at least three things in the world." This is equivalent to : " There exist objects x, y, z, and properties φ, ψ, χ, such that

x but not y has the property φ, x but not z has the property ψ, and y but not z has the property χ." This statement can be enunciated in purely logical terms, and it can be logically proved to be true of classes of classes of classes : of these there must, in fact, be at least 4, even if the universe did not exist. For in that case there would be one class, the null-class ; two classes of classes, namely, the class of no classes and the class whose only member is the null class ; and four classes of classes of classes, namely the one which is null, the one whose only member is the null class of classes, the one whose only member is the class whose only member is the null class, and the one which is the sum of the two last. But in the lower types, that of individuals, that of classes, and that of classes of classes, we cannot logically prove that there are at least three members. From the very nature of logic, something of this sort is to be expected ; for logic aims at independence of empirical fact, and the existence of the universe is an empirical fact. It is true that if the world did not exist, logic-books would not exist ; but the existence of logic-books is not one of the premisses of logic, nor can it be inferred from any proposition that has a right to be in a logic-book.

In practice, a great deal of mathematics is possible without assuming the existence of anything. All the elementary arithmetic of finite integers and rational fractions can be constructed ; but whatever involves infinite classes of integers becomes impossible. This excludes real numbers and the whole of analysis. To include them, we need the " axiom of infinity," which states that, if n is any finite number, there is at least one class having n members. At the time when I wrote the " Principles," I supposed that this could be proved, but by the time that Dr. Whitehead and I published " Principia Mathematica," we had become convinced that the supposed proof was fallacious.

The above argument depends upon the doctrine of types, which, although it occurs in a crude form in Appendix B of the " Principles," had not yet reached the stage of development at which it showed that the existence of infinite classes cannot be demonstrated logically. What is said as to existence-theorems in the last paragraph of the last chapter of the " Principles " (pp. 497-8) no longer appears to me to be valid : such existence-theorems, with certain exceptions, are, I should now say, examples of propositions which can be *enunciated* in logical terms, but can only be proved or disproved by empirical evidence.

Another example is the multiplicative axiom, or its equivalent, Zermelo's axiom of selection. This asserts that, given a set of mutually exclusive classes, none of which is null, there is at least one class consisting of one representative from each class of the set. Whether this is true or not, no one knows. It is easy to imagine universes in which it would be true, and it is impossible to prove that there are possible universes in which it would be false ; but it is also impossible (at least, so I believe) to prove that there are no possible universes in which it would be false. I did not become aware of the necessity for this axiom until a year after the " Principles " was published. This book contains, in consequence, certain errors, for example the assertion, in §119 (p. 123), that the two definitions of infinity are equivalent, which can only be proved if the multiplicative axiom is assumed.

Such examples—which might be multiplied indefinitely—show that a proposition may satisfy the definition with which the " Principles " opens, and yet may be incapable of logical or mathematical proof or disproof. All mathematical propositions are included under the definition (with certain minor emendations), but not all propositions that are included are mathematical. In order that a proposition may belong to mathematics it must have a further property: according to some it must be " tautological," and according to Carnap it must be " analytic." It is by no means easy to get an exact definition of this characteristic ; moreover, Carnap has shown that it is necessary to distinguish between " analytic " and " demonstrable," the latter being a somewhat narrower concept. And the question whether a proposition is or is not " analytic," or " demonstrable " depends upon the apparatus of premisses with which we begin. Unless, therefore, we have some criterion as to admissible logical premisses, the whole question as to what are logical propositions becomes to a very considerable extent arbitrary. This is a very unsatisfactory conclusion, and I do not accept it as final. But before anything more can be said on this subject, it is necessary to discuss the question of " logical constants," which play an essential part in the definition of mathematics in the first sentence of the " Principles."

There are three questions in regard to logical constants : First, are there such things ? Second, how are they defined ? Third, do they occur in the propositions of logic ? Of these questions, the first and third are highly ambiguous, but their various meanings can be made clearer by a little discussion.

First : Are there logical constants ? There is one sense of this question in which we can give a perfectly definite affirmative answer : in the linguistic or symbolic expression of logical propositions, there are words or symbols which play a constant part, *i.e.*, make the same contribution to the significance of propositions wherever they occur. Such are, for example, " or," " and," " not," " if-then," " the null-class," " 0," " 1," " 2," . . . The difficulty is that, when we analyse the propositions in the written expression of which such symbols occur, we find that they have no constituents corresponding to the expressions in question. In some cases this is fairly obvious : not even the most ardent Platonist would suppose that the perfect " or " is laid up in heaven, and that the " or's " here on earth are imperfect copies of the celestial archetype. But in the case of numbers this is far less obvious. The doctrines of Pythagoras, which began with arithmetical mysticism, influenced all subsequent philosophy and mathematics more profoundly than is generally realized. Numbers were immutable and eternal, like the heavenly bodies; numbers were intelligible : the science of numbers was the key to the universe. The last of these beliefs has misled mathematicians and the Board of Education down to the present day. Consequently, to say that numbers are symbols which mean nothing appears as a horrible form of atheism. At the time when I wrote the " Principles," I shared with Frege a belief in the Platonic reality of numbers, which, in my imagination, peopled the timeless realm of Being. It was a comforting faith, which I later abandoned with regret. Something must now be said of the steps by which I was led to abandon it.

In Chapter IV of the " Principles " it is said that " every word occurring in a sentence must have *some* meaning " ; and again " Whatever may be an object of thought, or may occur in any true or false proposition, or can be counted as *one*, I call a *term*. . . . A man, a moment, a number, a class, a relation, a chimæra, or anything else that can be mentioned, is sure to be a term ; and to deny that such and such a thing is a term must always be false." This way of understanding language turned out to be mistaken. That a word " must have *some* meaning "—the word, of course, being not gibberish, but one which has an intelligible use— is not always true if taken as applying to the word in isolation. What is true is that the word contributes to the meaning of the sentence in which it occurs ; but that is a very different matter.

The first step in the process was the theory of descriptions. According to this theory, in the proposition " Scott is the author of Waverley," there is no constituent corresponding to " the author of Waverley " : the analysis of the proposition is, roughly : " Scott wrote Waverley, and whoever wrote Waverley was Scott " ; or, more accurately : "The propositional function ' x *wrote Waverley* is equivalent to x *is Scott* ' is true for all values of x." This theory swept away the contention—advanced, for instance, by Meinong—that there must, in the realm of Being, be such objects as the golden mountain and the round square, since we can talk about them. " The round square does not exist " had always been a difficult proposition ; for it was natural to ask " What is it that does not exist ? " and any possible answer had seemed to imply that, in some sense, there is such an object as the round square, though this object has the odd property of not existing. The theory of descriptions avoided this and other difficulties.

The next step was the abolition of classes. This step was taken in " Principia Mathematica," where it is said : " The symbols for classes, like those for descriptions, are, in our system, incomplete symbols ; their *uses* are defined, but they themselves are not assumed to mean anything at all. . . . Thus classes, so far as we introduce them, are merely symbolic or linguistic conveniences, not genuine objects " (Vol. I, pp. 71-2). Seeing that cardinal numbers had been defined as classes of classes, they also became " merely symbolic or linguistic conveniences." Thus, for example, the proposition " $1+1=2$," somewhat simplified, becomes the following : " Form the propositional function ' a is not b, and whatever x may be, x *is a* γ is always equivalent to x *is a or x is b* ' ; form also the propositional function ' a is a γ, and, whatever x may be, x *is a* γ *but is not a* is always equivalent to x *is b*." Then, whatever γ may be, the assertion that one of these propositional functions is not always false (for different values of a and b) is equivalent to the assertion that the other is not always false." Here the numbers 1 and 2 have entirely disappeared. and a similar analysis can be applied to any arithmetical proposition.

Dr. Whitehead, at this stage, persuaded me to abandon points of space, instants of time, and particles of matter, substituting for them logical constructions composed of events. In the end, it seemed to result that none of the raw material of the world has smooth logical properties, but that whatever appears to have such properties is constructed artificially in order to have them. I do not mean that statements apparently about points or instants or numbers, or any of the other entities which Occam's razor abolishes, are false, but only that they need interpretation which shows that their linguistic form is misleading, and that, when they are rightly analysed, the pseudo-entities in question are found to be not mentioned in them. "Time consists of instants," for example, may or may not be a true statement, but in either case it mentions neither time nor instants. It may, roughly, be interpreted as follows : Given any event x, let us define as its "contemporaries" those which end after it begins, but begin before it ends ; and among these let us define as "initial contemporaries" of x those which are not wholly later than any other contemporaries of x. Then the statement "time consists of instants" is true if, given any event x, every event which is wholly later than some contemporary of x is wholly later than some initial contemporary of x. A similar process of interpretation is necessary in regard to most, if not all, purely logical constants.

Thus the question whether logical constants occur in the propositions of logic becomes more difficult than it seemed at first sight. It is, in fact, a question to which, as things stand, no definite answer can be given, because there is no exact definition of "occurring in" a proposition. But something can be said. In the first place, no proposition of logic can mention any particular object. The statement "If Socrates is a man and all men are mortal, then Socrates is mortal" is not a proposition of logic ; the logical proposition of which the above is a particular case is : "If x has the property of φ, and whatever has the property φ has the property ψ, then x has the property ψ, whatever x, φ, ψ may be." The word "property," which occurs here, disappears from the correct symbolic statement of the proposition ; but "if-then," or something serving the same purpose, remains. After the utmost efforts to reduce the number of undefined elements in the logical calculus, we shall find ourselves left with two (at least) which seem indispensable : one is incompatibility ; the other is the truth of all values of a propositional function. (By the "incompatibility" of two propositions is meant that they are not both true.) Neither of these looks very substantial. What was said earlier about "or" applies equally to incompatibility ; and it would seem absurd to say that generality is a constituent of a general proposition.

Logical constants, therefore, if we are to be able to say anything definite about them, must be treated as part of the language, not as part of what the language speaks about. In this way, logic becomes much more linguistic than I believed it to be at the time when I wrote the "Principles." It will still be true that no constants except logical constants occur in the verbal or symbolic expression of logical propositions, but it will not be true that these logical constants are names of objects, as " Socrates " is intended to be.

To define logic, or mathematics, is therefore by no means easy except in relation to some given set of premisses. A logical premiss must have *certain* characteristics which can be defined : it must have complete generality, in the sense that it mentions no particular thing or quality ; and it must be true in virtue of its form. Given a definite set of logical premisses, we can define logic, *in relation to them*, as whatever they enable us to demonstrate. But (1) it is hard to say what makes a proposition true in virtue of its form ; (2) it is difficult to see any way of proving that the system resulting from a given set of premisses is complete, in the sense of embracing everything that we should wish to include among logical propositions. As regards this second point, it has been customary to accept current logic and mathematics as a datum, and seek the fewest premisses from which this datum can be reconstructed. But when doubts arise—as they have arisen—concerning the validity of certain parts of mathematics, this method leaves us in the lurch.

It seems clear that there must be some way of defining logic otherwise than in relation to a particular logical language. The fundamental characteristic of logic, obviously, is that which is indicated when we say that logical propositions are true in virtue of their form. The question of demonstrability cannot enter in, since every proposition which, in one system, is deduced from the premisses, might, in another system, be itself taken as a premiss. If the proposition is complicated, this is inconvenient, but it cannot be impossible. All the propositions that are demonstrable in any admissible logical system must share with the premisses the property of being true in virtue of their form ; and all propositions which are true in virtue of their form ought to be included in any adequate logic. Some writers, for example Carnap in his " Logical Syntax of Language," treat the whole problem as being more a matter of liguistic choice than I can believe it to be. In the above-mentioned work, Carnap has two logical languages, one of which admits the multiplicative axiom and the axiom of infinity, while the other does not. I cannot myself regard such a matter as one to be decided by our arbitrary choice. It seems to me that these axioms either do, or do not, have the characteristic of formal truth which characterizes logic, and that in the former event every logic must include them, while in the latter every logic must exclude them. I confess, however, that I am unable to give any clear account of what is meant by saying that a proposition is " true in virtue of its form." But this phrase, inadequate as it is, points, I think, to the problem which must be solved if an adequate definition of logic is to be found.

I come finally to the question of the contradictions and the doctrine of types. Henri Poincaré, who considered mathematical logic to be no help in discovery, and therefore sterile, rejoiced in the contradictions : " La logistique n'est plus stérile ; elle engendre la contradiction ! "

All that mathematical logic did, however, was to make it evident that contradictions follow from premisses previously accepted by all logicians, however innocent of mathematics. Nor were the contradictions all new ; some dated from Greek times.

In the "Principles," only three contradictions are mentioned: Burali Forti's concerning the greatest ordinal, the contradiction concerning the greatest cardinal, and mine concerning the classes that are not members of themselves (pp. 323, 366, and 101). What is said as to possible solutions may be ignored, except Appendix B, on the theory of types ; and this itself is only a rough sketch. The literature on the contradictions is vast, and the subject is still controversial. The most complete treatment of the subject known to me is to be found in Carnap's "Logical Syntax of Language" (Kegan Paul, 1937). What he says on the subject seems to me either right or so difficult to refute that a refutation could not possibly be attempted in a short space. I shall, therefore, confine myself to a few general remarks.

At first sight, the contradictions seem to be of three sorts : those that are mathematical, those that are logical, and those that may be suspected of being due to some more or less trivial linguistic trick. Of the definitely mathematical contradictions, those concerning the greatest ordinal and the greatest cardinal may be taken as typical.

The first of these, Burali Forti's, is as follows : Let us arrange all ordinal numbers in order of magnitude ; then the last of these, which we will call N, is the greatest of ordinals. But the number of all ordinals from 0 up to N is $N+1$, which is greater than N. We cannot escape by suggesting that the series of ordinal numbers has no last term ; for in that case equally this series itself has an ordinal number greater than any term of the series, *i.e.*, greater than any ordinal number.

The second contradiction, that concerning the greatest cardinal, has the merit of making peculiarly evident the need for some doctrine of types. We know from elementary arithmetic that the number of combinations of n things any number at a time is 2^n, *i.e.*, that a class of n terms has 2^n sub-classes. We can prove that this proposition remains true when n is infinite. And Cantor proved that 2^n is always greater than n. Hence there can be no greatest cardinal. Yet one would have supposed that the class containing everything would have the greatest possible number of terms. Since, however, the number of classes of things exceeds the number of things, clearly classes of things are not things. (I will explain shortly what this statement can mean.)

Of the obviously logical contradictions, one is discussed in Chapter X ; in the linguistic group, the most famous, that of the liar, was invented by the Greeks. It is as follows : Suppose a man says "I am lying." If he is lying, his statement is true, and therefore he is not lying ; if he is not lying, then, when he says he is lying, he is lying. Thus either hypothesis implies its contradictory.

The logical and mathematical contradictions, as might be expected, are not really distinguishable ; but the linguistic group, according to

Ramsey*, can be solved by what may be called, in a broad sense, linguistic considerations. They are distinguished from the logical group by the fact that they introduce empirical notions, such as what somebody asserts or means ; and since these notions are not logical, it is possible to find solutions which depend upon other than logical considerations. This renders possible a great simplification of the theory of types, which, as it emerges from Ramsey's discussion, ceases wholly to appear unplausible or artificial or a mere *ad hoc* hypothesis designed to avoid the contradictions.

The technical essence of the theory of types is merely this : Given a propositional function " φx " of which all values are true, there are expressions which it is not legitimate to substitute for " x ." For example: All values of " if x is a man x is a mortal " are true, and we can infer "if Socrates is a man, Socrates is a mortal" ; but we cannot infer "if the law of contradiction is a man, the law of contradiction is a mortal." The theory of types declares this latter set of words to be nonsense, and gives rules as to permissible values of " x " in " φx ." In the detail there are difficulties and complications, but the general principle is merely a more precise form of one that has always been recognized. In the older conventional logic, it was customary to point out that such a form of words as " virtue is triangular " is neither true nor false, but no attempt was made to arrive at a definite set of rules for deciding whether a given series of words was or was not significant. This the theory of types achieves. Thus, for example I stated above that " classes of things are not things." This will mean : " If 'x is a member of the class a' is a proposition, and 'φx' is a proposition, then 'φa' is not a proposition, but a meaningless collection of symbols."

There are still many controversial questions in mathematical logic, which, in the above pages, I have made no attempt to solve. I have mentioned only those matters as to which, in my opinion, there has been some fairly definite advance since the time when the " Principles " was written. Broadly speaking, I still think this book is in the right where it disagrees with what had been previously held, but where it agrees with older theories it is apt to be wrong. The changes in philosophy which seem to me to be called for are partly due to the technical advances of mathematical logic in the intervening thirty-four years, which have simplified the apparatus of primitive ideas and propositions, and have swept away many apparent entities, such as classes, points, and instants. Broadly, the result is an outlook which is less Platonic, or less realist in the mediæval sense of the word. How far it is possible to go in the direction of nominalism remains, to my mind, an unsolved question, but one which, whether completely soluble or not, can only be adequately investigated by means of mathematical logic.

---

* Foundations of Mathematics, Kegan Paul, 1931, p.20 ff.

PREFACE.

THE present work has two main objects. One of these, the proof that all pure mathematics deals exclusively with concepts definable in terms of a very small number of fundamental logical concepts, and that all its propositions are deducible from a very small number of fundamental logical principles, is undertaken in Parts II.—VII. of this Volume, and will be established by strict symbolic reasoning in Volume II. The demonstration of this thesis has, if I am not mistaken, all the certainty and precision of which mathematical demonstrations are capable. As the thesis is very recent among mathematicians, and is almost universally denied by philosophers, I have undertaken, in this volume, to defend its various parts, as occasion arose, against such adverse theories as appeared most widely held or most difficult to disprove. I have also endeavoured to present, in language as untechnical as possible, the more important stages in the deductions by which the thesis is established.

The other object of this work, which occupies Part I., is the explanation of the fundamental concepts which mathematics accepts as indefinable. This is a purely philosophical task, and I cannot flatter myself that I have done more than indicate a vast field of inquiry, and give a sample of the methods by which the inquiry may be conducted. The discussion of indefinables—which forms the chief part of philosophical logic—is the endeavour to see clearly, and to make others see clearly, the entities concerned, in order that the mind may have that kind of acquaintance with them which it has with redness or the taste of a pineapple. Where, as in the present case, the indefinables are obtained primarily as the necessary residue in a process of analysis, it is often easier to know that there must be such entities than actually to perceive them ; there is a process analogous to that which resulted in the discovery of Neptune, with the difference that the final stage—the search with a mental telescope for the entity which has been inferred—is often the most difficult part of the undertaking. In the case of classes, I must confess, I have failed to perceive any concept fulfilling the conditions requisite for the notion of *class*. And the contradiction discussed in Chapter x. proves that something is amiss, but what this is I have hitherto failed to discover.

The second volume, in which I have had the great good fortune to secure the collaboration of Mr A. N. Whitehead, will be addressed exclusively to mathematicians; it will contain chains of deductions, from the premises of symbolic logic through Arithmetic, finite and infinite, to Geometry, in an order similar to that adopted in the present volume; it will also contain various original developments, in which the method of Professor Peano, as supplemented by the Logic of Relations, has shown itself a powerful instrument of mathematical investigation.

The present volume, which may be regarded either as a commentary upon, or as an introduction to, the second volume, is addressed in equal measure to the philosopher and to the mathematician; but some parts will be more interesting to the one, others to the other. I should advise mathematicians, unless they are specially interested in Symbolic Logic, to begin with Part IV., and only refer to earlier parts as occasion arises. The following portions are more specially philosophical: Part I. (omitting Chapter II.); Part II., Chapters XI., XV., XVI., XVII.; Part III.; Part IV., § 207, Chapters XXVI., XXVII., XXXI.; Part V., Chapters XLI., XLII., XLIII.; Part VI., Chapters L., LI., LII.; Part VII., Chapters LIII., LIV., LV., LVII., LVIII.; and the two Appendices, which belong to Part I., and should be read in connection with it. Professor Frege's work, which largely anticipates my own, was for the most part unknown to me when the printing of the present work began; I had seen his *Grundgesetze der Arithmetik*, but, owing to the great difficulty of his symbolism, I had failed to grasp its importance or to understand its contents. The only method, at so late a stage, of doing justice to his work, was to devote an Appendix to it; and in some points the views contained in the Appendix differ from those in Chapter VI., especially in §§ 71, 73, 74. On questions discussed in these sections, I discovered errors after passing the sheets for the press; these errors, of which the chief are the denial of the null-class, and the identification of a term with the class whose only member it is, are rectified in the Appendices. The subjects treated are so difficult that I feel little confidence in my present opinions, and regard any conclusions which may be advocated as essentially hypotheses.

A few words as to the origin of the present work may serve to show the importance of the questions discussed. About six years ago, I began an investigation into the philosophy of Dynamics. I was met by the difficulty that, when a particle is subject to several forces, no one of the component accelerations actually occurs, but only the resultant acceleration, of which they are not parts; this fact rendered illusory such causation of particulars by particulars as is affirmed, at first sight, by the law of gravitation. It appeared also that the difficulty in regard to absolute motion is insoluble on a relational theory of space. From these two questions I was led to a re-examination of the principles of Geometry, thence to the philosophy of continuity and infinity, and thence, with a view to discovering the meaning of the word *any*, to Symbolic Logic. The final outcome, as regards the philosophy of Dynamics, is perhaps rather slender; the reason of this is, that almost all the problems of Dynamics appear to me empirical, and therefore outside the scope of such a work as the present. Many very interesting questions have had to be omitted, especially in Parts VI. and VII., as not relevant to my purpose, which, for fear of misunderstandings, it may be well to explain at this stage.

When actual objects are counted, or when Geometry and Dynamics are applied to actual space or actual matter, or when, in any other way, mathematical reasoning is applied to what exists, the reasoning employed has a form not dependent upon the objects to which it is applied being just those objects that they are, but only upon their having certain general properties. In pure mathematics, actual objects in the world of existence will never be in question, but only hypothetical objects having those general properties upon which depends whatever deduction is being considered; and these general properties will always be expressible in terms of the fundamental concepts which I have called logical constants. Thus when space or motion is spoken of in pure mathematics, it is not actual space or actual motion, as we know them in experience, that are spoken of, but any entity possessing those abstract general properties of space or motion that are employed in the reasonings of geometry or dynamics. The question whether these properties belong, as a matter of fact, to actual space or actual motion, is irrelevant to pure mathematics, and therefore to the present work, being, in my opinion, a purely empirical question, to be investigated in the laboratory or the observatory. Indirectly, it is true, the discussions connected with pure mathematics have a very important bearing upon such empirical questions, since mathematical space and motion are held by many, perhaps most, philosophers to be self-contradictory, and therefore necessarily different from actual space and motion, whereas, if the views advocated in the following pages be valid, no such self-contradictions are to be found in mathematical space and motion. But extra-mathematical considerations of this kind have been almost wholly excluded from the present work.

On fundamental questions of philosophy, my position, in all its chief features, is derived from Mr G. E. Moore. I have accepted from him the non-existential nature of propositions (except such as happen to assert existence) and their independence of any knowing mind; also the pluralism which regards the world, both that of existents and that of entities, as composed of an infinite number of mutually independent entities, with relations which are ultimate, and not reducible to adjectives of their terms or of the whole which these compose. Before learning these views from him, I found myself completely unable to construct any philosophy of arithmetic, whereas their acceptance brought about an immediate liberation from a large number of difficulties which I believe to be otherwise insuperable. The doctrines just mentioned are, in my opinion, quite indispensable to any even tolerably satisfactory philosophy of mathematics, as I hope the following pages will show. But I must leave it to my readers to judge how far the reasoning assumes these doctrines, and how far it supports them. Formally, my premises are simply assumed; but the fact that they allow mathematics to be true, which most current philosophies do not, is surely a powerful argument in their favour.

In Mathematics, my chief obligations, as is indeed evident, are to Georg Cantor and Professor Peano. If I had become acquainted sooner with the work of Professor Frege, I should have owed a great deal to him, but as it is I arrived independently at many results which he had already established. At every stage of my work, I have been assisted more than I can express by the suggestions, the criticisms, and the generous encouragement of Mr A. N. Whitehead; he also has kindly read my proofs, and greatly improved the final expression of a very large number of passages. Many useful hints I owe also to Mr W. E. Johnson; and in the more philosophical parts of the book I owe much to Mr G. E. Moore besides the general position which underlies the whole.

In the endeavour to cover so wide a field, it has been impossible to acquire an exhaustive knowledge of the literature. There are doubtless many important works with which I am unacquainted; but where the labour of thinking and writing necessarily absorbs so much time, such ignorance, however regrettable, seems not wholly avoidable.

Many words will be found, in the course of discussion, to be defined in senses apparently departing widely from common usage. Such departures, I must ask the reader to believe, are never wanton, but have been made with great reluctance. In philosophical matters, they have been necessitated mainly by two causes. First, it often happens that two cognate notions are both to be considered, and that language has two names for the one, but none for the other. It is then highly convenient to distinguish between the two names commonly used as synonyms, keeping one for the usual, the other for the hitherto nameless sense. The other cause arises from philosophical disagreement with received views. Where two qualities are commonly supposed inseparably conjoined, but are here regarded as separable, the name which has applied to their combination will usually have to be restricted to one or other. For example, propositions are commonly regarded as (1) true or false, (2) mental. Holding, as I do, that what is true or false is not in general mental, I require a name for the true or false as such, and this name can scarcely be other than *proposition*. In such a case, the departure from usage is in no degree arbitrary. As regards mathematical terms, the necessity for establishing the existence-theorem in each case—*i.e.* the proof that there are entities of the kind in question—has led to many definitions which appear widely different from the notions usually attached to the terms in question. Instances of this are the definitions of cardinal, ordinal and complex numbers. In the two former of these, and in many other cases, the definition as a class, derived from the principle of abstraction, is mainly recommended by the fact that it leaves no doubt as to the existence-theorem. But in many instances of such apparent departure from usage, it may be doubted whether more has been done than to give precision to a notion which had hitherto been more or less vague.

For publishing a work containing so many unsolved difficulties, my apology is, that investigation revealed no near prospect of adequately resolving the contradiction discussed in Chapter x., or of acquiring a better insight into the nature of classes. The repeated discovery of errors in solutions which for a time had satisfied me caused these problems to appear such as would have been only concealed by any seemingly satisfactory theories which a slightly longer reflection might have produced ; it seemed better, therefore, merely to state the difficulties, than to wait until I had become persuaded of the truth of some almost certainly erroneous doctrine.

My thanks are due to the Syndics of the University Press, and to their Secretary, Mr R. T. Wright, for their kindness and courtesy in regard to the present volume.

LONDON,
December, 1902.

TABLE OF CONTENTS

CHAPTER III.

IMPLICATION AND FORMAL IMPLICATION.

CHAPTER IV.

PROPER NAMES, ADJECTIVES AND VERBS.

CHAPTER V.

DENOTING.

CHAPTER VI.

CLASSES.

CHAPTER VII.

PROPOSITIONAL FUNCTIONS.

CHAPTER VIII.

THE VARIABLE.

CHAPTER IX.

RELATIONS.

CHAPTER X.

THE CONTRADICTION.

PART II.

NUMBER.

CHAPTER XI.

DEFINITION OF CARDINAL NUMBERS.

CHAPTER XII.

ADDITION AND MULTIPLICATION.

CHAPTER XIII.

FINITE AND INFINITE.

CHAPTER XIV.

THEORY OF FINITE NUMBERS.

CHAPTER XV.

ADDITION OF TERMS AND ADDITION OF CLASSES.

CHAPTER XVI.

WHOLE AND PART.

CHAPTER XVII.

INFINITE WHOLES.

CHAPTER XVIII.

RATIOS AND FRACTIONS.

PART III.

QUANTITY.

CHAPTER XIX.

THE MEANING OF MAGNITUDE.

CHAPTER XX.

THE RANGE OF QUANTITY.

CHAPTER XXI.

NUMBERS AS EXPRESSING MAGNITUDES : MEASUREMENT.

CHAPTER XXII.

ZERO.

CHAPTER XXIII.

INFINITY, THE INFINITESIMAL, AND CONTINUITY.

PART IV.

ORDER.

CHAPTER XXIV.

THE GENESIS OF SERIES.

CHAPTER XXV.

THE MEANING OF ORDER.

CHAPTER XXVI.

ASYMMETRICAL RELATIONS.

CHAPTER XXVII.

DIFFERENCE OF SENSE AND DIFFERENCE OF SIGN.

CHAPTER XXVIII.

ON THE DIFFERENCE BETWEEN OPEN AND CLOSED SERIES.

CHAPTER XXIX.

PROGRESSIONS AND ORDINAL NUMBERS.

CHAPTER XXX.

DEDEKIND'S THEORY OF NUMBER.

CHAPTER XXXI.

DISTANCE.

PART V.

INFINITY AND CONTINUITY.

CHAPTER XXXII.

THE CORRELATION OF SERIES.

CHAPTER XXXIII.

REAL NUMBERS.

CHAPTER XXXIV.

LIMITS AND IRRATIONAL NUMBERS.

CHAPTER XXXV.

CANTOR'S FIRST DEFINITION OF CONTINUITY.

CHAPTER XXXVI.

ORDINAL CONTINUITY.

CHAPTER XXXVII.

TRANSFINITE CARDINALS.

CHAPTER XXXVIII.

TRANSFINITE ORDINALS.

CHAPTER XXXIX.

THE INFINITESIMAL CALCULUS.

CHAPTER XL.

THE INFINITESIMAL AND THE IMPROPER INFINITE.

CHAPTER XLI.

PHILOSOPHICAL ARGUMENTS CONCERNING THE INFINITESIMAL.

CHAPTER XLII.

THE PHILOSOPHY OF THE CONTINUUM.

CHAPTER XLIII.

THE PHILOSOPHY OF THE INFINITE.

PART VI.

SPACE.

CHAPTER XLIV.

DIMENSIONS AND COMPLEX NUMBERS.

CHAPTER XLV.

PROJECTIVE GEOMETRY.

CHAPTER XLVI.

DESCRIPTIVE GEOMETRY.

CHAPTER XLVII.

METRICAL GEOMETRY.

CHAPTER XLVIII.

RELATION OF METRICAL TO PROJECTIVE AND DESCRIPTIVE GEOMETRY.

CHAPTER XLIX.

DEFINITIONS OF VARIOUS SPACES.

CHAPTER L.

THE CONTINUITY OF SPACE.

CHAPTER LI.

LOGICAL ARGUMENTS AGAINST POINTS.

CHAPTER LII.

KANT'S THEORY OF SPACE.

PART VII.

MATTER AND MOTION.

CHAPTER LIII.

MATTER.

CHAPTER LIV.

MOTION.

CHAPTER LV.

CAUSALITY.

CHAPTER LVI.

DEFINITION OF A DYNAMICAL WORLD.

CHAPTER LVII.

NEWTON'S LAWS OF MOTION.

CHAPTER LVIII.

ABSOLUTE AND RELATIVE MOTION.

CHAPTER LIX.

HERTZ'S DYNAMICS.

APPENDIX A.

THE LOGICAL AND ARITHMETICAL DOCTRINES OF FREGE.

Table of Contents

APPENDIX B.

THE DOCTRINE OF TYPES.

PART I.

THE INDEFINABLES OF MATHEMATICS.

CHAPTER I.

DEFINITION OF PURE MATHEMATICS.

1. PURE Mathematics is the class of all propositions of the form
" p implies q," where p and q are propositions containing one or more
variables, the same in the two propositions, and neither p nor q contains
any constants except logical constants. And logical constants are all
notions definable in terms of the following: Implication, the relation
of a term to a class of which it is a member, the notion of *such that*,
the notion of relation, and such further notions as may be involved
in the general notion of propositions of the above form. In addition
to these, mathematics *uses* a notion which is not a constituent of the
propositions which it considers, namely the notion of truth.

2. The above definition of pure mathematics is, no doubt, somewhat unusual. Its various parts, nevertheless, appear to be capable of
exact justification—a justification which it will be the object of the
present work to provide. It will be shown that whatever has, in the
past, been regarded as pure mathematics, is included in our definition,
and that whatever else is included possesses those marks by which
mathematics is commonly though vaguely distinguished from other
studies. The definition professes to be, not an arbitrary decision to
use a common word in an uncommon signification, but rather a precise
analysis of the ideas which, more or less unconsciously, are implied in
the ordinary employment of the term. Our method will therefore be
one of analysis, and our problem may be called philosophical—in the
sense, that is to say, that we seek to pass from the complex to the
simple, from the demonstrable to its indemonstrable premisses. But
in one respect not a few of our discussions will differ from those that
are usually called philosophical. We shall be able, thanks to the labours
of the mathematicians themselves, to arrive at certainty in regard to
most of the questions with which we shall be concerned; and among
those capable of an exact solution we shall find many of the problems
which, in the past, have been involved in all the traditional uncertainty
of philosophical strife. The nature of number, of infinity, of space,
time and motion, and of mathematical inference itself, are all questions

to which, in the present work, an answer professing itself demonstrable with mathematical certainty will be given—an answer which, however, consists in reducing the above problems to problems in pure logic, which last will not be found satisfactorily solved in what follows.

3. The Philosophy of Mathematics has been hitherto as controversial, obscure and unprogressive as the other branches of philosophy. Although it was generally agreed that mathematics is in some sense true, philosophers disputed as to what mathematical propositions really meant: although something was true, no two people were agreed as to what it was that was true, and if something was known, no one knew what it was that was known. So long, however, as this was doubtful, it could hardly be said that any certain and exact knowledge was to be obtained in mathematics. We find, accordingly, that idealists have tended more and more to regard all mathematics as dealing with mere appearance, while empiricists have held everything mathematical to be approximation to some exact truth about which they had nothing to tell us. This state of things, it must be confessed, was thoroughly unsatisfactory. Philosophy asks of Mathematics: What does it mean? Mathematics in the past was unable to answer, and Philosophy answered by introducing the totally irrelevant notion of mind. But now Mathematics is able to answer, so far at least as to reduce the whole of its propositions to certain fundamental notions of logic. At this point, the discussion must be resumed by Philosophy. I shall endeavour to indicate what are the fundamental notions involved, to prove at length that no others occur in mathematics, and to point out briefly the philosophical difficulties involved in the analysis of these notions. A complete treatment of these difficulties would involve a treatise on Logic, which will not be found in the following pages.

4. There was, until very lately, a special difficulty in the principles of mathematics. It seemed plain that mathematics consists of deductions, and yet the orthodox accounts of deduction were largely or wholly inapplicable to existing mathematics. Not only the Aristotelian syllogistic theory, but also the modern doctrines of Symbolic Logic, were either theoretically inadequate to mathematical reasoning, or at any rate required such artificial forms of statement that they could not be practically applied. In this fact lay the strength of the Kantian view, which asserted that mathematical reasoning is not strictly formal, but always uses intuitions, *i.e.* the *à priori* knowledge of space and time. Thanks to the progress of Symbolic Logic, especially as treated by Professor Peano, this part of the Kantian philosophy is now capable of a final and irrevocable refutation. By the help of ten principles of deduction and ten other premisses of a general logical nature (*e.g.* "implication is a relation"), all mathematics can be strictly and formally deduced; and all the entities that occur in mathematics can be defined in terms of those that occur in the above twenty premisses.

In this statement, Mathematics includes not only Arithmetic and Analysis, but also Geometry, Euclidean and non-Euclidean, rational Dynamics, and an indefinite number of other studies still unborn or in their infancy. The fact that all Mathematics is Symbolic Logic is one of the greatest discoveries of our age; and when this fact has been established, the remainder of the principles of mathematics consists in the analysis of Symbolic Logic itself.

5. The general doctrine that all mathematics is deduction by logical principles from logical principles was strongly advocated by Leibniz, who urged constantly that axioms ought to be proved and that all except a few fundamental notions ought to be defined. But owing partly to a faulty logic, partly to belief in the logical necessity of Euclidean Geometry, he was led into hopeless errors in the endeavour to carry out in detail a view which, in its general outline, is now known to be correct*. The actual propositions of Euclid, for example, do not follow from the principles of logic alone; and the perception of this fact led Kant to his innovations in the theory of knowledge. But since the growth of non-Euclidean Geometry, it has appeared that pure mathematics has no concern with the question whether the axioms and propositions of Euclid hold of actual space or not: this is a question for applied mathematics, to be decided, so far as any decision is possible, by experiment and observation. What pure mathematics asserts is merely that the Euclidean propositions follow from the Euclidean axioms—*i.e.* it asserts an implication: any space which has such and such properties has also such and such other properties. Thus, as dealt with in pure mathematics, the Euclidean and non-Euclidean Geometries are equally true: in each nothing is affirmed except implications. All propositions as to what actually exists, like the space we live in, belong to experimental or empirical science, not to mathematics; when they belong to applied mathematics, they arise from giving to one or more of the variables in a proposition of pure mathematics some constant value satisfying the hypothesis, and thus enabling us, for that value of the variable, actually to assert both hypothesis and consequent instead of asserting merely the implication. We assert always in mathematics that if a certain assertion p is true of any entity x, or of any set of entities $x, y, z, \ldots$, then some other assertion q is true of those entities; but we do not assert either p or q separately of our entities. We assert a relation between the assertions p and q, which I shall call *formal implication*.

6. Mathematical propositions are not only characterized by the fact that they assert implications, but also by the fact that they contain *variables*. The notion of the variable is one of the most difficult with which Logic has to deal, and in the present work a satisfactory theory

---

* On this subject, cf. Couturat, *La Logique de Leibniz*, Paris, 1901.

as to its nature, in spite of much discussion, will hardly be found. For the present, I only wish to make it plain that there are variables in all mathematical propositions, even where at first sight they might seem to be absent. Elementary Arithmetic might be thought to form an exception: $1 + 1 = 2$ appears neither to contain variables nor to assert an implication. But as a matter of fact, as will be shown in Part II, the true meaning of this proposition is: "If x is one and y is one, and x differs from y, then x and y are two." And this proposition both contains variables and asserts an implication. We shall find always, in all mathematical propositions, that the words *any* or *some* occur; and these words are the marks of a variable and a formal implication. Thus the above proposition may be expressed in the form: "Any unit and any other unit are two units." The typical proposition of mathematics is of the form "$\phi(x, y, z, \ldots)$ implies $\psi(x, y, z, \ldots)$, whatever values $x, y, z, \ldots$ may have"; where $\phi(x, y, z, \ldots)$ and $\psi(x, y, z, \ldots)$, for every set of values of $x, y, z, \ldots$, are propositions. It is not asserted that ϕ is always true, nor yet that ψ is always true, but merely that, in all cases, when ϕ is false as much as when ϕ is true, ψ follows from it.

The distinction between a variable and a constant is somewhat obscured by mathematical usage. It is customary, for example, to speak of parameters as in some sense constants, but this is a usage which we shall have to reject. A constant is to be something absolutely definite, concerning which there is no ambiguity whatever. Thus 1, 2, 3, e, π, Socrates, are constants; and so are *man*, and the human race, past, present and future, considered collectively. Proposition, implication, class, etc. are constants; but a proposition, any proposition, some proposition, are not constants, for these phrases do not denote one definite object. And thus what are called parameters are simply variables. Take, for example, the equation $ax + by + c = 0$, considered as the equation to a straight line in a plane. Here we say that x and y are variables, while a, b, c are constants. But unless we are dealing with one absolutely particular line, say the line from a particular point in London to a particular point in Cambridge, our a, b, c are not definite numbers, but stand for *any* numbers, and are thus also variables. And in Geometry nobody does deal with actual particular lines; we always discuss *any* line. The point is that we collect the various couples x, y into classes of classes, each class being defined as those couples that have a certain fixed relation to one triad (a, b, c). But from class to class, a, b, c also vary, and are therefore properly variables.

7. It is customary in mathematics to regard our variables as restricted to certain classes: in Arithmetic, for instance, they are supposed to stand for numbers. But this only means that *if* they stand for numbers, they satisfy some formula, *i.e.* the hypothesis that they are numbers implies the formula. This, then, is what is really

asserted, and in this proposition it is no longer necessary that our variables should be numbers: the implication holds equally when they are not so. Thus, for example, the proposition "x and y are numbers implies $(x + y)^2 = x^2 + 2xy + y^2$" holds equally if for x and y we substitute Socrates and Plato* : both hypothesis and consequent, in this case, will be false, but the implication will still be true. Thus in every proposition of pure mathematics, when fully stated, the variables have an absolutely unrestricted field: any conceivable entity may be substituted for any one of our variables without impairing the truth of our proposition.

8. We can now understand why the constants in mathematics are to be restricted to logical constants in the sense defined above. The process of transforming constants in a proposition into variables leads to what is called generalization, and gives us, as it were, the formal essence of a proposition. Mathematics is interested exclusively in *types* of propositions; if a proposition p containing only constants be proposed, and for a certain one of its terms we imagine others to be successively substituted, the result will in general be sometimes true and sometimes false. Thus, for example, we have "Socrates is a man"; here we turn Socrates into a variable, and consider "x is a man." Some hypotheses as to x, for example, "x is a Greek," insure the truth of "x is a man"; thus "x is a Greek" implies "x is a man," and this holds for all values of x. But the statement is not one of pure mathematics, because it depends upon the particular nature of *Greek* and *man*. We may, however, vary these too, and obtain: If a and b are classes, and a is contained in b, then "x is an a" implies "x is a b." Here at last we have a proposition of pure mathematics, containing three variables and the constants *class*, *contained in*, and those involved in the notion of formal implications with variables. So long as any term in our proposition can be turned into a variable, our proposition can be generalized; and so long as this is possible, it is the business of mathematics to do it. If there are several chains of deduction which differ only as to the meaning of the symbols, so that propositions symbolically identical become capable of several interpretations, the proper course, mathematically, is to form the class of meanings which may attach to the symbols, and to assert that the formula in question follows from the hypothesis that the symbols belong to the class in question. In this way, symbols which stood for constants become transformed into variables, and new constants are substituted, consisting of classes to which the old constants belong. Cases of such generalization are so frequent that many will occur at once to every mathematician, and innumerable instances will be given in the present work. Whenever two sets of terms have mutual relations of the same

---

* It is necessary to suppose arithmetical addition and multiplication defined (as may be easily done) so that the above formula remains significant when x and y are not numbers.

type, the same form of deduction will apply to both. For example, the mutual relations of points in a Euclidean plane are of the same type as those of the complex numbers; hence plane geometry, considered as a branch of pure mathematics, ought not to decide whether its variables are points or complex numbers or some other set of entities having the same type of mutual relations. Speaking generally, we ought to deal, in every branch of mathematics, with any class of entities whose mutual relations are of a specified type; thus the class, as well as the particular term considered, becomes a variable, and the only true constants are the types of relations and what they involve. Now a *type* of relation is to mean, in this discussion, a class of relations characterized by the above formal identity of the deductions possible in regard to the various members of the class; and hence a type of relations, as will appear more fully hereafter, if not already evident, is always a class definable in terms of logical constants*. We may therefore define a type of relations as a class of relations defined by some property definable in terms of logical constants alone.

9. Thus pure mathematics must contain no indefinables except logical constants, and consequently no premisses, or indemonstrable propositions, but such as are concerned exclusively with logical constants and with variables. It is precisely this that distinguishes pure from applied mathematics. In applied mathematics, results which have been shown by pure mathematics to follow from some hypothesis as to the variable are actually asserted of some constant satisfying the hypothesis in question. Thus terms which were variables become constant, and a new premiss is always required, namely: this particular entity satisfies the hypothesis in question. Thus for example Euclidean Geometry, as a branch of pure mathematics, consists wholly of propositions having the hypothesis "*S* is a Euclidean space." If we go on to: "The space that exists is Euclidean," this enables us to assert of the space that exists the consequents of all the hypotheticals constituting Euclidean Geometry, where now the variable *S* is replaced by the constant *actual space*. But by this step we pass from pure to applied mathematics.

10. The connection of mathematics with logic, according to the above account, is exceedingly close. The fact that all mathematical constants are logical constants, and that all the premisses of mathematics are concerned with these, gives, I believe, the precise statement of what philosophers have meant in asserting that mathematics is *à priori*. The fact is that, when once the apparatus of logic has been accepted, all mathematics necessarily follows. The logical constants themselves are to be defined only by enumeration, for they are so fundamental that all the properties by which the class of them might be defined presuppose

---

* One-one, many-one, transitive, symmetrical, are instances of types of relations with which we shall be often concerned.

some terms of the class. But practically, the method of discovering the logical constants is the analysis of symbolic logic, which will be the business of the following chapters. The distinction of mathematics from logic is very arbitrary, but if a distinction is desired, it may be made as follows. Logic consists of the premisses of mathematics, together with all other propositions which are concerned exclusively with logical constants and with variables but do not fulfil the above definition of mathematics (§ 1). Mathematics consists of all the consequences of the above premisses which assert formal implications containing variables, together with such of the premisses themselves as have these marks. Thus some of the premisses of mathematics, *e.g.* the principle of the syllogism, "if *p* implies *q* and *q* implies *r*, then *p* implies *r*," will belong to mathematics, while others, such as "implication is a relation," will belong to logic but not to mathematics. But for the desire to adhere to usage, we might identify mathematics and logic, and define either as the class of propositions containing only variables and logical constants; but respect for tradition leads me rather to adhere to the above distinction, while recognizing that certain propositions belong to both sciences.

From what has now been said, the reader will perceive that the present work has to fulfil two objects, first, to show that all mathematics follows from symbolic logic, and secondly to discover, as far as possible, what are the principles of symbolic logic itself. The first of these objects will be pursued in the following Parts, while the second belongs to Part I. And first of all, as a preliminary to a critical analysis, it will be necessary to give an outline of Symbolic Logic considered simply as a branch of mathematics. This will occupy the following chapter.

'CHAPTER II.

SYMBOLIC LOGIC.

11. SYMBOLIC or Formal Logic—I shall use these terms as synonyms—is the study of the various general types of deduction. The word *symbolic* designates the subject by an accidental characteristic, for the employment of mathematical symbols, here as elsewhere, is merely a theoretically irrelevant convenience. The syllogism in all its figures belongs to Symbolic Logic, and would be the whole subject if all deduction were syllogistic, as the scholastic tradition supposed. It is from the recognition of asyllogistic inferences that modern Symbolic Logic, from Leibniz onward, has derived the motive to progress. Since the publication of Boole's *Laws of Thought* (1854), the subject has been pursued with a certain vigour, and has attained to a very considerable technical development*. Nevertheless, the subject achieved almost nothing of utility either to philosophy or to other branches of mathematics, until it was transformed by the new methods of Professor Peano†. Symbolic Logic has now become not only absolutely essential to every philosophical logician, but also necessary for the comprehension of mathematics generally, and even for the successful practice of certain branches of mathematics. How useful it is in practice can only be judged by those who have experienced the increase of power derived from acquiring it; its theoretical functions must be briefly set forth in the present chapter‡.

---

* By far the most complete account of the non-Peanesque methods will be found in the three volumes of Schröder, *Vorlesungen über die Algebra der Logik*, Leipzig, 1890, 1891, 1895.

† See *Formulaire de Mathématiques*, Turin, 1895, with subsequent editions in later years; also *Revue de Mathématiques*, Vol. VII, No. 1 (1900). The editions of the *Formulaire* will be quoted as *F.* 1895 and so on. The *Revue de Mathématiques*, which was originally the *Rivista di Matematica*, will be referred to as *R. d. M.*

‡ In what follows the main outlines are due to Professor Peano, except as regards relations; even in those cases where I depart from his views, the problems considered have been suggested to me by his works.

12. Symbolic Logic is essentially concerned with inference in general*, and is distinguished from various special branches of mathematics mainly by its generality. Neither mathematics nor symbolic logic will study such special relations as (say) temporal priority, but mathematics will deal explicitly with the class of relations possessing the formal properties of temporal priority—properties which are summed up in the notion of continuity†. And the formal properties of a relation may be defined as those that can be expressed in terms of logical constants, or again as those which, while they are preserved, permit our relation to be varied without invalidating any inference in which the said relation is regarded in the light of a variable. But symbolic logic, in the narrower sense which is convenient, will not investigate what inferences are possible in respect of continuous relations (*i.e.* relations generating continuous series); this investigation belongs to mathematics, but is still too special for symbolic logic. What symbolic logic does investigate is the general rules by which inferences are made, and it requires a classification of relations or propositions only in so far as these general rules introduce particular notions. The particular notions which appear in the propositions of symbolic logic, and all others definable in terms of these notions, are the logical constants. The number of indefinable logical constants is not great: it appears, in fact, to be eight or nine. These notions alone form the subject-matter of the whole of mathematics: no others, except such as are definable in terms of the original eight or nine, occur anywhere in Arithmetic, Geometry, or rational Dynamics. For the technical study of Symbolic Logic, it is convenient to take as a single indefinable the notion of a formal implication, *i.e.* of such propositions as " x is a man implies x is a mortal, for all values of x "—propositions whose general type is: " $\phi(x)$ implies $\psi(x)$ for all values of x," where $\phi(x)$, $\psi(x)$, for all values of x, are propositions. The analysis of this notion of formal implication belongs to the principles of the subject, but is not required for its formal development. In addition to this notion, we require as indefinables the following: Implication between propositions not containing variables, the relation of a term to a class of which it is a member, the notion of *such that*, the notion of relation, and truth. By means of these notions, all the propositions of symbolic logic can be stated.

13. The subject of Symbolic Logic consists of three parts, the calculus of propositions, the calculus of classes, and the calculus of relations. Between the first two, there is, within limits, a certain parallelism, which arises as follows: In any symbolic expression, the

---

* I may as well say at once that I do not distinguish between inference and deduction. What is called induction appears to me to be either disguised deduction or a mere method of making plausible guesses

† See below, Part V, Chap. xxxvi.

letters may be interpreted as classes or as propositions, and the relation of inclusion in the one case may be replaced by that of formal implication in the other. Thus, for example, in the principle of the syllogism, if a, b, c be classes, and a is contained in b, b in c, then a is contained in c; but if a, b, c be propositions, and a implies b, b implies c, then a implies c. A great deal has been made of this duality, and in the later editions of the *Formulaire*, Peano appears to have sacrificed logical precision to its preservation *. But, as a matter of fact, there are many ways in which the calculus of propositions differs from that of classes. Consider, for example, the following: " If p, q, r are propositions, and p implies q or r, then p implies q or p implies r." This proposition is true; but its correlative is false, namely: " If a, b, c are classes, and a is contained in b or c, then a is contained in b or a is contained in c." For example, English people are all either men or women, but are not all men nor yet all women. The fact is that the duality holds for propositions asserting of a variable term that it belongs to a class, *i.e.* such propositions as " x is a man," provided that the implication involved be formal, *i.e.* one which holds for all values of x. But " x is a man " is itself not a proposition at all, being neither true nor false; and it is not with such entities that we are concerned in the propositional calculus, but with genuine propositions. To continue the above illustration: It is true that, for all values of x, " x is a man or a woman " either implies " x is a man " or implies " x is a woman." But it is false that " x is a man or woman " either implies " x is a man " for all values of x, or implies " x is a woman " for all values of x. Thus the implication involved, which is always one of the two, is not formal, since it does not hold for all values of x, being not always the same one of the two. The symbolic affinity of the propositional and the class logic is, in fact, something of a snare, and we have to decide which of the two we are to make fundamental. Mr McColl, in an important series of papers†, has contended for the view that implication and propositions are more fundamental than inclusion and classes; and in this opinion I agree with him. But he does not appear to me to realize adequately the distinction between genuine propositions and such as contain a real variable: thus he is led to speak of propositions as sometimes true and sometimes false, which of course is impossible with a genuine proposition. As the distinction involved is of very great importance, I shall dwell on it before proceeding further. A proposition, we may say, is anything that is true or that is

---

* On the points where the duality breaks down, cf. Schröder, *op. cit.*, Vol. ii, Lecture 21.

† Cf. "The Calculus of Equivalent Statements," *Proceedings of the London Mathematical Society*, Vol. ix and subsequent volumes; "Symbolic Reasoning," *Mind*, Jan. 1880, Oct. 1897, and Jan. 1900; "La Logique Symbolique et ses Applications," *Bibliothèque du Congrès International de Philosophie*, Vol. iii (Paris, 1901). I shall in future quote the proceedings of the above Congress by the title *Congrès*.

false. An expression such as " x is a man " is therefore not a proposition, for it is neither true nor false. If we give to x any constant value
whatever, the expression becomes a proposition : it is thus as it were a
schematic form standing for any one of a whole class of propositions.
And when we say " x is a man implies x is a mortal for all values of x,"
we are not asserting a single implication, but a class of implications ;
we have now a genuine proposition, in which, though the letter x appears,
there is no real variable: the variable is absorbed in the same kind of
way as the x under the integral sign in a definite integral, so that the
result is no longer a function of x. Peano distinguishes a variable which
appears in this way as *apparent*, since the proposition does not depend
upon the variable ; whereas in " x is a man " there are different propositions for different values of the variable, and the variable is what Peano
calls *real**. I shall speak of propositions exclusively where there is no
real variable: where there are one or more real variables, and for all
values of the variables the expression involved is a proposition, I shall
call the expression a *propositional function*. The study of genuine
propositions is, in my opinion, more fundamental than that of classes ;
but the study of propositional functions appears to be strictly on a
par with that of classes, and indeed scarcely distinguishable therefrom.
Peano, like McColl, at first regarded propositions as more fundamental
than classes, but he, even more definitely, considered propositional functions rather than propositions. From this criticism, Schröder is exempt:
his second volume deals with genuine propositions, and points out their
formal differences from classes.

A. *The Propositional Calculus.*

14. The propositional calculus is characterized by the fact that
all its propositions have as hypothesis and as consequent the assertion of
a material implication. Usually, the hypothesis is of the form " p implies p," etc., which (§ 16) is equivalent to the assertion that the letters
which occur in the consequent are propositions. Thus the consequents
consist of propositional functions which are true of all propositions.
It is important to observe that, though the letters employed are symbols
for variables, and the consequents are true when the variables are given
values which are propositions, these values must be genuine propositions,
not propositional functions. The hypothesis " p is a proposition " is
not satisfied if for p we put " x is a man," but it is satisfied if we put
" Socrates is a man " or if we put " x is a man implies x is a mortal for
all values of x." Shortly, we may say that the propositions represented
by single letters in this calculus are variables, but do not contain
variables—in the case, that is to say, where the hypotheses of the
propositions which the calculus asserts are satisfied.

* *F.* 1901, p. 2.

15. Our calculus studies the relation of *implication* between propositions. This relation must be distinguished from the relation of *formal* implication, which holds between propositional functions when the one implies the other for all values of the variable. Formal implication is also involved in this calculus, but is not explicitly studied: we do not consider propositional functions in general, but only certain definite propositional functions which occur in the propositions of our calculus. How far formal implication is definable in terms of implication simply, or material implication as it may be called, is a difficult question, which will be discussed in Chapter III. What the difference is between the two, an illustration will explain. The fifth proposition of Euclid follows from the fourth: if the fourth is true, so is the fifth, while if the fifth is false, so is the fourth. This is a case of material implication, for both propositions are absolute constants, not dependent for their meaning upon the assigning of a value to a variable. But each of them *states* a formal implication. The fourth states that if x and y be triangles fulfilling certain conditions, then x and y are triangles fulfilling certain other conditions, and that this implication holds for all values of x and y; and the fifth states that if x is an isosceles triangle, x has the angles at the base equal. The formal implication involved in each of these two propositions is quite a different thing from the material implication holding between the propositions as wholes; both notions are required in the propositional calculus, but it is the study of material implication which specially distinguishes this subject, for formal implication occurs throughout the whole of mathematics.

It has been customary, in treatises on logic, to confound the two kinds of implication, and often to be really considering the formal kind where the material kind only was apparently involved. For example, when it is said that "Socrates is a man, therefore Socrates is a mortal," Socrates is *felt* as a variable: he is a type of humanity, and one feels that any other man would have done as well. If, instead of *therefore*, which implies the truth of hypothesis and consequent, we put "Socrates is a man implies Socrates is a mortal," it appears at once that we may substitute not only another man, but any other entity whatever, in the place of Socrates. Thus although what is explicitly stated, in such a case, is a material implication, what is meant is a formal implication; and some effort is needed to confine our imagination to material implication.

16. A definition of implication is quite impossible. If p implies q, then if p is true q is true, *i.e.* p's truth implies q's truth; also if q is false p is false, *i.e.* q's falsehood implies p's falsehood*. Thus truth and falsehood give us merely new implications, not a definition of implication.

---

* The reader is recommended to observe that the main implications in these statements are formal, *i.e.* "p implies q" *formally* implies "p's truth implies q's truth," while the subordinate implications are material.

If p implies q, then both are false or both true, or p is false and q true ; it is impossible to have q false and p true, and it is necessary to have q true or p false*. In fact, the assertion that q is true or p false turns out to be strictly equivalent to "p implies q"; but as equivalence means mutual implication, this still leaves implication fundamental, and not definable in terms of disjunction. Disjunction, on the other hand, is definable in terms of implication, as we shall shortly see. It follows from the above equivalence that of any two propositions there must be one which implies the other, that false propositions imply all propositions, and true propositions are implied by all propositions. But these are results to be demonstrated; the premisses of our subject deal exclusively with rules of inference.

It may be observed that, although implication is indefinable, *proposition* can be defined. Every proposition implies itself, and whatever is not a proposition implies nothing. Hence to say "p is a proposition" is equivalent to saying "p implies p"; and this equivalence may be used to define propositions. As the mathematical sense of *definition* is widely different from that current among philosophers, it may be well to observe that, in the mathematical sense, a new propositional function is said to be defined when it is stated to be equivalent to (*i.e.* to imply and be implied by) a propositional function which has either been accepted as indefinable or has been defined in terms of indefinables. The definition of entities which are not propositional functions is derived from such as are in ways which will be explained in connection with classes and relations.

17. We require, then, in the propositional calculus, no indefinables except the two kinds of implication—remembering, however, that formal implication is a complex notion, whose analysis remains to be undertaken. As regards our two indefinables, we require certain indemonstrable propositions, which hitherto I have not succeeded in reducing to less than ten. Some indemonstrables there must be; and some propositions, such as the syllogism, must be of the number, since no demonstration is possible without them. But concerning others, it may be doubted whether they are indemonstrable or merely undemonstrated; and it should be observed that the method of supposing an axiom false, and deducing the consequences of this assumption, which has been found admirable in such cases as the axiom of parallels, is here not universally available. For all our axioms are principles of deduction; and if they are true, the consequences which appear to follow from the employment of an opposite principle will not really follow, so that arguments from the supposition of the falsity of an axiom are here subject to special fallacies. Thus the number of indemonstrable propositions may be capable of further reduction, and in regard to some of them I know of

---

* I may as well state once for all that the alternatives of a disjunction will never be considered as mutually exclusive unless expressly said to be so.

no grounds for regarding them as indemonstrable except that they have hitherto remained undemonstrated.

18. The ten axioms are the following. (1) If p implies q, then p implies q*; in other words, whatever p and q may be, "p implies q" is a proposition. (2) If p implies q, then p implies p; in other words, whatever implies anything is a proposition. (3) If p implies q, then q implies q; in other words, whatever is implied by anything is a proposition. (4) A true hypothesis in an implication may be dropped, and the consequent asserted. This is a principle incapable of formal symbolic statement, and illustrating the essential limitations of formalism—a point to which I shall return at a later stage. Before proceeding further, it is desirable to define the joint assertion of two propositions, or what is called their logical product. This definition is highly artificial, and illustrates the great distinction between mathematical and philosophical definitions. It is as follows: If p implies p, then, if q implies q, pq (the logical product of p and q) means that if p implies that q implies r, then r is true. In other words, if p and q are propositions, their joint assertion is equivalent to saying that every proposition is true which is such that the first implies that the second implies it. We cannot, with formal correctness, state our definition in this shorter form, for the hypothesis "p and q are propositions" is already the logical product of "p is a proposition" and "q is a proposition." We can now state the six main principles of inference, to each of which, owing to its importance, a name is to be given; of these all except the last will be found in Peano's accounts of the subject. (5) If p implies p and q implies q, then pq implies p. This is called *simplification*, and asserts merely that the joint assertion of two propositions implies the assertion of the first of the two. (6) If p implies q and q implies r, then p implies r. This will be called the *syllogism*. (7) If q implies q and r implies r, and if p implies that q implies r, then pq implies r. This is the principle of *importation*. In the hypothesis, we have a product of three propositions; but this can of course be defined by means of the product of two. The principle states that if p implies that q implies r, then r follows from the joint assertion of p and q. For example: "If I call on so-and-so, then if she is at home I shall be admitted" implies "If I call on so-and-so and she is at home, I shall be admitted." (8) If p implies p and q implies q, then, if pq implies r, then p implies that q implies r. This is the converse of the preceding principle, and is called *exportation*†. The previous illustration reversed will illustrate this principle. (9) If p implies q and p implies r, then p implies qr: in other words, a

---

* Note that the implications denoted by *if* and *then*, in these axioms, are formal, while those denoted by *implies* are material.

† (7) and (8) cannot (I think) be deduced from the definition of the logical product, because they are required for passing from "If p is a proposition, then 'q is a proposition' implies etc." to "If p and q are propositions, then etc."

proposition which implies each of two propositions implies them both. This is called the principle of *composition*. (10) If p implies p and q implies q, then "'p implies q' implies p" implies p. This is called the principle of *reduction*; it has less self-evidence than the previous principles, but is equivalent to many propositions that are self-evident. I prefer it to these, because it is explicitly concerned, like its predecessors, with implication, and has the same kind of logical character as they have. If we remember that "p implies q" is equivalent to "q or not-p," we can easily convince ourselves that the above principle is true; for "'p implies q' implies p" is equivalent to "p or the denial of 'q or not-p,'" *i.e.* to "p or 'p and not q,'" *i.e.* to p. But this way of persuading ourselves that the principle of reduction is true involves many logical principles which have not yet been demonstrated, and cannot be demonstrated except by reduction or some equivalent. The principle is especially useful in connection with negation. Without its help, by means of the first nine principles, we can prove the law of contradiction; we can prove, if p and q be propositions, that p implies not-not-p; that "p implies not-q" is equivalent to "q implies not-p" and to not-pq; that "p implies q" implies "not-q implies not-p"; that p implies that not-p implies p; that not-p is equivalent to "p implies not-p"; and that "p implies not-q" is equivalent to "not-not-p implies not-q." But we cannot prove without reduction or some equivalent (so far at least as I have been able to discover) that p or not-p must be true (the law of excluded middle); that every proposition is equivalent to the negation of some other proposition; that not-not-p implies p; that "not-q implies not-p" implies "p implies q"; that "not-p implies p" implies p, or that "p implies q" implies "q or not-p." Each of these assumptions is equivalent to the principle of reduction, and may, if we choose, be sub-stituted for it. Some of them—especially excluded middle and double negation—appear to have far more self-evidence. But when we have seen how to define disjunction and negation in terms of implication, we shall see that the supposed simplicity vanishes, and that, for formal purposes at any rate, reduction is simpler than any of the possible alternatives. For this reason I retain it among my premisses in preference to more usual and more superficially obvious propositions.

19. Disjunction or logical addition is defined as follows: "p or q" is equivalent to "'p implies q' implies q." It is easy to persuade ourselves of this equivalence, by remembering that a false proposition implies every other; for if p is false, p does imply q, and therefore, if "p implies q" implies q, it follows that q is true. But this argument again uses principles which have not yet been demonstrated, and is merely designed to elucidate the definition by anticipation. From this definition, by the help of reduction, we can prove that "p or q" is equivalent to "q or p." An alternative definition, deducible from the above, is: "Any proposition implied by p and implied by q is true," or, in other words, "'*p* implies *s*' and '*q* implies *s*' together imply *s*, whatever *s* may be." Hence we proceed to the definition of negation: not-*p* is equivalent to the assertion that *p* implies all propositions, *i.e.* that "*r* implies *r*" implies "*p* implies *r*" whatever *r* may be*. From this point we can prove the laws of contradiction and excluded middle and double negation, and establish all the formal properties of logical multiplication and addition—the associative, commutative and distributive laws. Thus the logic of propositions is now complete.

Philosophers will object to the above definitions of disjunction and negation on the ground that what we *mean* by these notions is something quite distinct from what the definitions assign as their meanings, and that the equivalences stated in the definitions are, as a matter of fact, significant propositions, not mere indications as to the way in which symbols are going to be used. Such an objection is, I think, well-founded, if the above account is advocated as giving the true philosophic analysis of the matter. But where a purely formal purpose is to be served, any equivalence in which a certain notion appears on one side but not on the other will do for a definition. And the advantage of having before our minds a strictly formal development is that it provides the data for philosophical analysis in a more definite shape than would be otherwise possible. Criticism of the procedure of formal logic, therefore, will be best postponed until the present brief account has been brought to an end.

B. *The Calculus of Classes.*

20. In this calculus there are very much fewer new primitive propositions—in fact, two seem sufficient—but there are much greater difficulties in the way of non-symbolic exposition of the ideas embedded in our symbolism. These difficulties, as far as possible, will be postponed to later chapters. For the present, I shall try to make an exposition which is to be as straightforward and simple as possible.

The calculus of classes may be developed by regarding as fundamental the notion of *class*, and also the relation of a member of a class to its class. This method is adopted by Professor Peano, and is perhaps more philosophically correct than a different method which, for formal purposes, I have found more convenient. In this method we still take as

* The principle that false propositions imply all propositions solves Lewis Carroll's logical paradox in *Mind*, N. S. No. 11 (1894). The assertion made in that paradox is that, if *p*, *q*, *r* be propositions, and *q* implies *r*, while *p* implies that *q* implies not-*r*, then *p* must be false, on the supposed ground that "*q* implies *r*" and "*q* implies not-*r*" are incompatible. But in virtue of our definition of negation, if *q* be false both these implications will hold : the two together, in fact, whatever proposition *r* may be, are equivalent to not-*q*. Thus the only inference warranted by Lewis Carroll's premisses is that if *p* be true, *q* must be false, *i.e.* that *p* implies not-*q*; and this is the conclusion, oddly enough, which common sense would have drawn in the particular case which he discusses.

fundamental the relation (which, following Peano, I shall denote by ϵ) of an individual to a class to which it belongs, *i.e.* the relation of Socrates to the human race which is expressed by saying that Socrates is a man. In addition to this, we take as indefinables the notion of a propositional function and the notion of *such that*. It is these three notions that characterize the class-calculus. Something must be said in explanation of each of them.

21. The insistence on the distinction between ϵ and the relation of whole and part between classes is due to Peano, and is of very great importance to the whole technical development and the whole of the applications to mathematics. In the scholastic doctrine of the syllogism, and in all previous symbolic logic, the two relations are confounded, except in the work of Frege*. The distinction is the same as that between the relation of individual to species and that of species to genus, between the relation of Socrates to the class of Greeks and the relation of Greeks to men. On the philosophical nature of this distinction I shall enlarge when I come to deal critically with the nature of classes; for the present it is enough to observe that the relation of whole and part is transitive, while ϵ is not so: we have Socrates is a a man, and men are a class, but not Socrates is a class. It is to be observed that the class must be distinguished from the class-concept or predicate by which it is to be defined: thus men are a class, while *man* is a class-concept. The relation ϵ must be regarded as holding between Socrates and men considered collectively, not between Socrates and *man*. I shall return to this point in Chapter vi. Peano holds that all propositional functions containing only a single variable are capable of expression in the form "x is an a," where a is a constant class; but this view we shall find reason to doubt.

22. The next fundamental notion is that of a propositional function. Although propositional functions occur in the calculus of propositions, they are there each defined as it occurs, so that the general notion is not required. But in the class-calculus it is necessary to introduce the general notion explicitly. Peano, does not require it, owing to his assumption that the form "x is an a" is general for one variable, and that extensions of the same form are available for any number of variables. But we must avoid this assumption, and must therefore introduce the notion of a propositional function. We may explain (but not define) this notion as follows: ϕx is a propositional function if, for every value of x, ϕx is a proposition, determinate when x is given. Thus "x is a man" is a propositional function. In any proposition, however complicated, which contains no real variables, we may imagine one of the terms, not a verb or adjective, to be replaced by other terms: instead of "Socrates is a man" we may put "Plato is a man," "the number 2

---

* See his *Begriffsschrift*, Halle, 1879, and *Grundgesetze der Arithmetik*, Jena, 1893, p. 2.

is a man," and so on*. Thus we get successive propositions all agreeing except as to the one variable term. Putting x for the variable term, " x is a man " expresses the type of all such propositions. A propositional function in general will be true for some values of the variable and false for others. The instances where it is true for *all* values of the variable, so far as they are known to me, all express implications, such as " x is a man implies x is a mortal"; but I know of no *à priori* reason for asserting that no other propositional functions are true for all values of the variable.

23. This brings me to the notion of *such that*. The values of x which render a propositional function ϕx true are like the roots of an equation—indeed the latter are a particular case of the former—and we may consider all the values of x which are *such that ϕx* is true. In general, these values form a *class*, and in fact a class may be defined as all the terms satisfying some propositional function. There is, however, some limitation required in this statement, though I have not been able to discover precisely what the limitation is. This results from a certain contradiction which I shall discuss at length at a later stage (Chap. x). The reasons for defining *class* in this way are, that we require to provide for the null-class, which prevents our defining a class as a term to which some other has the relation ϵ, and that we wish to be able to define classes by relations, *i.e.* all the terms which have to other terms the relation R are to form a class, and such cases require somewhat complicated propositional functions.

24. With regard to these three fundamental notions, we require two primitive propositions. The first asserts that if x belongs to the class of terms satisfying a propositional function ϕx, then ϕx is true. The second asserts that if ϕx and ψx are equivalent propositions for all values of x, then the class of x's such that ϕx is true is identical with the class of x's such that ψx is true. Identity, which occurs here, is defined as follows: x is identical with y if y belongs to every class to which x belongs, on other words, if " x is a u " implies "y is a u " for all values of u. With regard to the primitive proposition itself, it is to be observed that it decides in favour of an extensional view of classes. Two class concepts need not be identical when their extensions are so: *man* and *featherless biped* are by no means identical, and no more are *even prime and integer between* 1 *and* 3. These are class-*concepts*, and if our axiom is to hold, it must not be of these that we are to speak in dealing with classes. We must be concerned with the actual assemblage of terms, not with any concept denoting that assemblage. For mathematical purposes, this is quite essential. Consider, for example, the problem as to how many combinations can be formed of a given set

---

* Verbs and adjectives occurring as such are distinguished by the fact that, if they be taken as variable, the resulting function is only a proposition for *some* values of the variable, *i.e.* for such as are verbs or adjectives respectively. See Chap. IV.

of terms taken any number at a time, *i.e.* as to how many classes are
contained in a given class. If distinct classes may have the same extension, this problem becomes utterly indeterminate. And certainly
common usage would regard a class as determined when all its terms are
given. The extensional view of classes, in some form, is thus essential to
Symbolic Logic and to mathematics, and its necessity is expressed in the
above axiom. But the axiom itself is not employed until we come to
Arithmetic; at least it need not be employed, if we choose to distinguish
the equality of classes, which is defined as mutual inclusion, from the
identity of individuals. Formally, the two are totally distinct: identity
is defined as above, equality of a and b is defined by the equivalence of
" x is an a" and " x is a b" for all values of x.

25. Most of the propositions of the class-calculus are easily
deduced from those of the propositional calculus. The logical product
or common part of two classes a and b is the class of x's such that the
logical product of " x is an a" and "x is a b" is true. Similarly we define
the logical sum of two classes (a or b), and the negation of a class (not-a).
A new idea is introduced by the logical product and sum of a class of
classes. If k is a class of classes, its logical product is the class of terms belonging to each of the classes of k, *i.e.* the class of terms x such that " u
is a k" implies " x is a u" for all values of u. The logical sum is the class
which is contained in every class in which every class of the class k is
contained, *i.e.* the class of terms x such that, if " u is a k" implies "u is
contained in c" for all values of u, then, for all values of c, x is a c.
And we say that a class a is contained in a class b when " x is an a"
implies " x is a b" for all values of x. In like manner with the above
we may define the product and sum of a class of propositions. Another
very important notion is what is called the *existence* of a class—a word
which must not be supposed to mean what existence means in philosophy.
A class is said to exist when it has at least one term. A formal definition is as follows: a is an existent class when and only when any
proposition is true provided " x is an a" always implies it whatever value
we may give to x. It must be understood that the proposition implied
must be a genuine proposition, not a propositional function of x. A
class a exists when the logical sum of all propositions of the form " x is
an a" is true, *i.e.* when not all such propositions are false.

It is important to understand clearly the manner in which propositions in the class-calculus are obtained from those in the propositional calculus. Consider, for example, the syllogism. We have
" p implies q" and "q implies r" imply "p implies r." Now put " x is
an a," "x is a b," "x is a c" for p, q, r, where x must have some definite
value, but it is not necessary to decide what value. We then find that
if, for the value of x in question, x is an a implies x is a b, and x is a b
implies x is a c, then x is an a implies x is a c. Since the value of x is
irrelevant, we may vary x, and thus we find that if a is contained in b,

and b in c, then a is contained in c. This is the class-syllogism. But in applying this process it is necessary to employ the utmost caution, if fallacies are to be successfully avoided. In this connection it will be instructive to examine a point upon which a dispute has arisen between Schröder and Mr McColl*. Schröder asserts that if p, q, r are propositions, "pq implies r" is equivalent to the disjunction "p implies r or q implies r." Mr McColl admits that the disjunction implies the other, but denies the converse implication. The reason for the divergence is, that Schröder is thinking of propositions and material implication, while Mr McColl is thinking of propositional functions and formal implication. As regards propositions, the truth of the principle may be easily made plain by the following considerations. If pq implies r, then, if either p or q be false, the one of them which is false implies r, because false propositions imply all propositions. But if both be true, pq is true, and therefore r is true, and therefore p implies r and q implies r, because true propositions are implied by every proposition. Thus in any case, one at least of the propositions p and q must imply r. (This is not a proof, but an elucidation.) But Mr McColl objects: Suppose p and q to be mutually contradictory, and r to be the null proposition, then pq implies r but neither p nor q implies r. Here we are dealing with propositional functions and formal implication. A propositional function is said to be null when it is false for all values of x; and the class of x's satisfying the function is called the null-class, being in fact a class of no terms. Either the function or the class, following Peano, I shall denote by Λ. Now let our r be replaced by Λ, our p by ϕx, and our q by not-ϕx, where ϕx is any propositional function. Then pq is false for all values of x, and therefore implies Λ. But it is not in general the case that ϕx is always false, nor yet that not-ϕx is always false; hence neither always implies Λ. Thus the above formula can only be truly interpreted in the propositional calculus : in the class-calculus it is false. This may be easily rendered obvious by the following considerations: Let ϕx, ψx, χx be three propositional functions. Then "ϕx . ψx implies χx" implies, for all values of x, that either ϕx implies χx or ψx implies χx. But it does not imply that either ϕx implies χx for all values of x, or ψx implies χx for all values of x. The disjunction is what I shall call a *variable* disjunction, as opposed to a constant one: that is, in some cases one alternative is true, in others the other, whereas in a constant disjunction there is one of the alternatives (though it is not stated which) that is always true. Wherever disjunctions occur in regard to propositional functions, they will only be transformable into statements in the class-calculus in cases where the disjunction is constant. This is a point which is both important in itself and instructive in its bearings. Another way of stating the matter is this: In the proposition: If

* Schröder, *Algebra der Logik*, Vol. ii, pp. 258-9 ; McColl, "Calculus of Equivalent Statements," fifth paper, *Proc. Lond. Math. Soc.* Vol. xxviii, p. 182.

ϕx . ψx implies χx, then either ϕx implies χx or ψx implies χx, the implication indicated by *if* and *then* is formal, while the subordinate implications are material; hence the subordinate implications do not lead to the inclusion of one class in another, which results only from formal implication.

The formal laws of addition, multiplication, tautology and negation are the same as regards classes and propositions. The law of tautology states that no change is made when a class or proposition is added to or multiplied by itself. A new feature of the class-calculus is the null-class, or class having no terms. This may be defined as the class of terms that belong to every class, as the class which does not exist (in the sense defined above), as the class which is contained in every class, as the class Λ which is such that the propositional function "x is a Λ" is false for all values of x, or as the class of x's satisfying any propositional function ϕx which is false for all values of x. All these definitions are easily shown to be equivalent.

26. Some important points arise in connection with the theory of identity. We have already defined two terms as identical when the second belongs to every class to which the first belongs. It is easy to show that this definition is symmetrical, and that identity is transitive and reflexive (*i.e.* if x and y, y and z are identical, so are x and z; and whatever x may be, x is identical with x). Diversity is defined as the negation of identity. If x be any term, it is necessary to distinguish from x the class whose only member is x: this may be defined as the class of terms which are identical with x. The necessity for this distinction, which results primarily from purely formal considerations, was discovered by Peano; I shall return to it at a later stage. Thus the class of even primes is not to be identified with the number 2, and the class of numbers which are the sum of 1 and 2 is not to be identified with 3. In what, philosophically speaking, the difference consists, is a point to be considered in Chapter VI.

C. *The Calculus of Relations.*

27. The calculus of relations is a more modern subject than the calculus of classes. Although a few hints for it are to be found in De Morgan *, the subject was first developed by C. S. Peirce†. A careful analysis of mathematical reasoning shows (as we shall find in the course of the present work) that types of relations are the true subject-matter discussed, however a bad phraseology may disguise this fact; hence the logic of relations has a more immediate bearing on mathematics than

---

* *Camb. Phil. Trans.* Vol. x, "On the Syllogism, No. iv, and on the Logic of Relations." Cf. *ib.* Vol. ix, p. 104; also his *Formal Logic* (London, 1847), p. 50.

† See especially his articles on the Algebra of Logic, *American Journal of Mathematics*, Vols. iii and vii. The subject is treated at length by C. S. Peirce's methods in Schröder, *op. cit.*, Vol. iii.

that of classes or propositions, and aι.y theoretically correct and adequate expression of mathematical truths is only possible by its means. Peirce and Schröder have realized the great importance of the subject, but unfortunately their methods, being based, not on Peano, but on the older Symbolic Logic derived (with modifications) from Boole, are so cumbrous and difficult that most of the applications which ought to be made are practically not feasible. In addition to the defects of the old Symbolic Logic, their method suffers technically (whether philosophically or not I do not at present discuss) from the fact that they regard a relation essentially as a class of couples, thus requiring elaborate formulae of summation for dealing with single relations. This view is derived, I think, probably unconsciously, from a philosophical error : it has always been customary to suppose relational propositions less ultimate than class-propositions (or subject-predicate propositions, with which class-propositions are habitually confounded), and this has led to a desire to treat relations as a kind of classes. However this may be, it was certainly from the opposite philosophical belief, which I derived from my friend Mr G. E. Moore*, that I was led to a different formal treatment of relations. This treatment, whether more philo-sophically correct or not, is certainly far more convenient and far more powerful as an engine of discovery in actual mathematics†.

28. If R be a relation, we express by xRy the propositional function " x has the relation R to y." We require a primitive (*i.e.* indemonstrable) proposition to the effect that xRy is a proposition for all values of x and y. We then have to consider the following classes : The class of terms which have the relation R to some term or other, which I call the class of *referents* with respect to R; and the class of terms to which some term has the relation R, which I call the class of *relata*. Thus if R be paternity, the referents will be fathers and the relata will be children. We have also to consider the corresponding classes with respect to particular terms or classes of terms : so-and-so's children, or the children of Londoners, afford illustrations.

The intensional view of relations here advocated leads to the result that two relations may have the same extension without being identical. Two relations R, R' are said to be equal or equivalent, or to have the same extension, when xRy implies and is implied by $xR'y$ for all values of x and y. But there is no need here of a primitive proposition, as there was in the case of classes, in order to obtain a relation which is determinate when the extension is determinate. We may replace a relation R by the logical sum or product of the class of relations equivalent to R, *i.e.* by the assertion of some or of all such relations; and this is identical with the logical sum or product of the class of relations equivalent to R', if R' be equivalent to R. Here we use

---

* See his article "On the Nature of Judgment," *Mind*, N. S. No. 30.
† See my articles in *R. d. M.* Vol. vii, No. 2 and subsequent numbers.

the identity of two classes, which results from the primitive proposition as to identity of classes, to establish the identity of two relations— a procedure which could not have been applied to classes themselves without a vicious circle.

A primitive proposition in regard to relations is that every relation has a converse, *i.e.* that, if R be any relation, there is a relation R' such that xRy is equivalent to $yR'x$ for all values of x and y. Following Schröder, I shall denote the converse of R by $\breve{R}$. Greater and less, before and after, implying and implied by, are mutually converse relations. With some relations, such as identity, diversity, equality, inequality, the converse is the same as the original relation: such relations are called *symmetrical*. When the converse is incompatible with the original relation, as in such cases as greater and less, I call the relation *asymmetrical*; in intermediate cases, *not-symmetrical*.

The most important of the primitive propositions in this subject is that between any two terms there is a relation not holding between any two other terms. This is analogous to the principle that any term is the only member of some class; but whereas that could be proved, owing to the extensional view of classes, this principle, so far as I can discover, is incapable of proof. In this point, the extensional view of relations has an advantage; but the advantage appears to me to be outweighed by other considerations. When relations are considered intensionally, it may seem possible to doubt whether the above principle is true at all. It will, however, be generally admitted that, of any two terms, some propositional function is true which is not true of a certain given different pair of terms. If this be admitted, the above principle follows by considering the logical product of all the relations that hold between our first pair of terms. Thus the above principle may be replaced by the following, which is equivalent to it: If xRy implies $x'Ry'$, whatever R may be, so long as R is a relation, then x and x', y and y' are respectively identical. But this principle introduces a logical difficulty from which we have been hitherto exempt, namely a variable with a restricted field; for unless R is a relation, xRy is not a proposition at all, true or false, and thus R, it would seem, cannot take *all* values, but only such as are relations. I shall return to the discussion of this point at a later stage.

29. Other assumptions required are that the negation of a relation is a relation, and that the logical product of a class of relations (*i.e.* the assertion of all of them simultaneously) is a relation. Also the *relative product* of two relations must be a relation. The relative product of two relations R, S is the relation which holds between x and z whenever there is a term y to which x has the relation R and which has to z the relation S. Thus the relation of a maternal grandfather to his grandson is the relative product of father and mother; that of a paternal grandmother to her grandson is the relative product of mother and father;

that of grandparent to grandchild is the relative product of parent and parent. The relative product, as these instances show, is not in general commutative, and does not in general obey the law of tautology. The relative product is a notion of very great importance. Since it does not obey the law of tautology, it leads to powers of relations : the square of the relation of parent and child is the relation of grandparent and grandchild, and so on. Peirce and Schröder consider also what they call the relative sum of two relations R and S, which holds between x and z, when, if y be any other term whatever, either x has to y the relation R, or y has to z the relation S. This is a complicated notion, which I have found no occasion to employ, and which is introduced only in order to preserve the duality of addition and multiplication. This duality has a certain technical charm when the subject is considered as an independent branch of mathematics ; but when it is considered solely in relation to the principles of mathematics, the duality in question appears devoid of all philosophical importance.

30. Mathematics requires, so far as I know, only two other primitive propositions, the one that material implication is a relation, the other that ϵ (the relation of a term to a class to which it belongs) is a relation*. We can now develop the whole of mathematics without further assumptions or indefinables. Certain propositions in the logic of relations deserve to be mentioned, since they are important, and it might be doubted whether they were capable of formal proof. If u, v be any two classes, there is a relation R the assertion of which between any two terms x and y is equivalent to the assertion that x belongs to u and y to v. If u be any class which is not null, there is a relation which all its terms have to it, and which holds for no other pairs of terms. If R be any relation, and u any class contained in the class of referents with respect to R, there is a relation which has u for the class of its referents, and is equivalent to R throughout that class ; this relation is the same as R where it holds, but has a more restricted domain. (I use *domain* as synonymous with *class of referents*.) From this point onwards, the development of the subject is technical : special types of relations are considered, and special branches of mathematics result.

D. *Peano's Symbolic Logic.*

31. So much of the above brief outline of Symbolic Logic is inspired by Peano, that it seems desirable to discuss his work explicitly, justifying by criticism the points in which I have departed from him.

The question as to which of the notions of symbolic logic are to be taken as indefinable, and which of the propositions as indemonstrable, is, as Professor Peano has insisted†, to some extent arbitrary. But it is

---

* There is a difficulty in regard to this primitive proposition, discussed in §§ 53, 94 below.

† *E.g. F.* 1901, p. 6 ; *F.* 1897, Part I, pp. 62–3.

important to establish all the mutual relations of the simpler notions of logic, and to examine the consequence of taking various notions as indefinable. It is necessary to realize that definition, in mathematics, does not mean, as in philosophy, an analysis of the idea to be defined into constituent ideas. This notion, in any case, is only applicable to concepts, whereas in mathematics it is possible to define terms which are not concepts *. Thus also many notions are defined by symbolic logic which are not capable of philosophical definition, since they are simple and unanalyzable. Mathematical definition consists in pointing out a fixed relation to a fixed term, of which one term only is capable: this term is then defined by means of the fixed relation and the fixed term. The point in which this differs from philosophical definition may be elucidated by the remark that the mathematical definition does not point out the term in question, and that only what may be called philosophical insight reveals which it is among all the terms there are. This is due to the fact that the term is defined by a concept which *denotes* it unambiguously, not by actually mentioning the term denoted. What is meant by *denoting*, as well as the different ways of denoting, must be accepted as primitive ideas in any symbolic logic†: in this respect, the order adopted seems not in any degree arbitrary.

32. For the sake of definiteness, let us now examine some one of Professor Peano's expositions of the subject. In his later expositions‡ he has abandoned the attempt to distinguish clearly certain ideas and propositions as primitive, probably because of the realization that any such distinction is largely arbitrary. But the distinction appears useful, as introducing greater definiteness, and as showing that a certain set of primitive ideas and propositions are sufficient; so far from being abandoned, it ought rather to be made in every possible way. I shall, therefore, in what follows, expound one of his earlier expositions, that cf 1897§.

The primitive notions with which Peano starts are the following: Class, the relation of an individual to a class of which it is a member, the notion of a term, implication where both propositions contain the same variables, *i.e.* formal implication, the simultaneous affirmation of two propositions, the notion of definition, and the negation of a proposition. From these notions, together with the division of a complex proposition into parts, Peano professes to deduce all symbolic logic by means of certain primitive propositions. Let us examine the deduction in outline.

We may observe, to begin with, that the simultaneous affirmation of *two* propositions might seem, at first sight, not enough to take as a primitive idea. For although this can be extended, by successive steps, to the simultaneous affirmation of any finite number of propositions,

---

* See Chap. IV.	† See Chap. V.
‡ *F.* 1901 and *R. d. M.* Vol. VII, No. 1 (1900).	§ *F.* 1897, Part I.

yet this is not all that is wanted; we require to be able to affirm simultaneously all the propositions of any class, finite or infinite. But the simultaneous assertion of a class of propositions, oddly enough, is much easier to define than that of two propositions (see § 34, (3)). If k be a class of propositions, their simultaneous affirmation is the assertion that "p is a k" implies p. If this holds, all propositions of the class are true; if it fails, one at least must be false. We have seen that the logical product of two propositions can be defined in a highly artificial manner; but it might almost as well be taken as indefinable, since no further property can be proved by means of the definition. We may observe, also, that formal and material implication are combined by Peano into one primitive idea, whereas they ought to be kept separate.

33. Before giving any primitive propositions, Peano proceeds to some definitions. (1) If a is a class, "x and y are a's" is to mean "x is an a and y is an a." (2) If a and b are classes, "every a is a b" means "x is an a implies that x is a b." If we accept formal implication as a primitive notion, this definition seems unobjectionable; but it may well be held that the relation of inclusion between classes is simpler than formal implication, and should not be defined by its means. This is a difficult question, which I reserve for subsequent discussion. A formal implication appears to be the assertion of a whole class of material implications. The complication introduced at this point arises from the nature of the variable, a point which Peano, though he has done very much to show its importance, appears not to have himself sufficiently considered. The notion of one proposition containing a variable implying another such proposition, which he takes as primitive, is complex, and should therefore be separated into its constituents; from this separation arises the necessity of considering the simultaneous affirmation of a whole class of propositions before interpreting such a proposition as "x is an a implies that x is a b." (3) We come next to a perfectly worthless definition, which has been since abandoned*. This is the definition of *such that*. The x's such that x is an a, we are told, are to mean the class a. But this only gives the meaning of *such that* when placed before a proposition of the type "x is an a." Now it is often necessary to consider an x such that some proposition is true of it, where this proposition is not of the form "x is an a." Peano holds (though he does not lay it down as an axiom) that every proposition containing only one variable is reducible to the form "x is an a†." But we shall see (Chap. x) that at least one such proposition is not reducible to this form. And in any case, the only utility of *such that* is to effect the reduction, which cannot therefore be assumed to be already effected without it. The fact is that *such that* contains a primitive idea, but one which it is not easy clearly to disengage from other ideas.

* In consequence of the criticisms of Padoa, *R. d. M.* Vol. vi, p. 112.

† *R. d. M.* Vol. vii, No. 1, p. 25; *F.* 1901, p. 21, § 2, Prop. 4. 0, Note.

In order to grasp the meaning of *such that*, it is necessary to observe, first of all, that what Peano and mathematicians generally call *one* proposition containing a variable is really, if the variable is apparent, the conjunction of a certain class of propositions defined by some constancy of form; while if the variable is real, so that we have a propositional function, there is not a proposition at all, but merely a kind of schematic representation of *any* proposition of a certain type. "The sum of the angles of a triangle is two right angles," for example, when stated by means of a variable, becomes: Let x be a triangle; then the sum of the angles of x is two right angles. This expresses the conjunction of all the propositions in which it is said of particular definite entities that if they are triangles, the sum of their angles is two right angles. But a propositional function, where the variable is real, represents *any* proposition of a certain form, not *all* such propositions (see §§ 59–62). There is, for each propositional function, an indefinable relation between propositions and entities, which may be expressed by saying that all the propositions have the same form, but different entities enter into them. It is this that gives rise to propositional functions. Given, for example, a constant relation and a constant term, there is a one-one correspondence between the propositions asserting that various terms have the said relation to the said term, and the various terms which occur in these propositions. It is this notion which is requisite for the comprehension of *such that*. Let x be a variable whose values form the class a, and let $f(x)$ be a one-valued function of x which is a true proposition for all values of x within the class a, and which is false for all other values of x. Then the terms of a are the class of terms *such that* $f(x)$ is a true proposition. This gives an explanation of *such that*. But it must always be remembered that the appearance of having *one* proposition $f(x)$ satisfied by a number of values of x is fallacious: $f(x)$ is not a proposition at all, but a propositional function. What is fundamental is the relation of various propositions of given form to the various terms entering severally into them as arguments or values of the variable; this relation is equally required for interpreting the propositional function $f(x)$ and the notion *such that*, but is itself ultimate and inexplicable. (4) We come next to the definition of the logical product, or common part, of two classes. If a and b be two classes, their common part consists of the class of terms x such that x is an a and x is a b. Here already, as Padoa points out (*loc. cit.*), it is necessary to extend the meaning of *such that* beyond the case where our proposition asserts membership of a class, since it is only by means of the definition that the common part is shown to be a class.

34. The remainder of the definitions preceding the primitive propositions are less important, and may be passed over. Of the primitive propositions, some appear to be merely concerned with the symbolism, and not to express any real properties of what is symbolized; others, on the contrary, are of high logical importance.

(1) The first of Peano's axioms is "every class is contained in itself." This is equivalent to "every proposition implies itself." There seems no way of evading this axiom, which is equivalent to the law of identity, except the method adopted above, of using self-implication to define propositions. (2) Next we have the axiom that the product of two classes is a class. This ought to have been stated, as ought also the definition of the logical product, for a class of classes; for when stated for only two classes, it cannot be extended to the logical product of an infinite class of classes. If *class* is taken as indefinable, it is a genuine axiom, which is very necessary to reasoning. But it might perhaps be somewhat generalized by an axiom concerning the terms satisfying propositions of a given form: *e.g.* "the terms having one or more given relations to one or more given terms form a class." In Section B, above, the axiom was wholly evaded by using a generalized form of the axiom as the definition of *class*. (3) We have next two axioms which are really only one, and appear distinct only because Peano defines the common part of two classes instead of the common part of a class of classes. These two axioms state that, if *a*, *b* be classes, their logical product, *ab*, is contained in *a* and is contained in *b*. These appear as different axioms, because, as far as the symbolism shows, *ab* might be different from *ba*. It is one of the defects of most symbolisms that they give an order to terms which intrinsically have none, or at least none that is relevant. So in this case: if *K* be a class of classes, the logical product of *K* consists of all terms belonging to *every* class that belongs to *K*. With this definition, it becomes at once evident that no order of the terms of *K* is involved. Hence if *K* has only two terms, *a* and *b*, it is indifferent whether we represent the logical product of *K* by *ab* or by *ba*, since the order exists only in the symbols, not in what is symbolized. It is to be observed that the corresponding axiom with regard to propositions is, that the simultaneous assertion of a class of propositions implies any proposition of the class; and this is perhaps the best form of the axiom. Nevertheless, though an axiom is not required, it is necessary, here as elsewhere, to have a means of connecting the case where we start from a class of classes or of propositions or of relations with the case where the class results from enumeration of its terms. Thus although no order is involved in the product of a *class* of propositions, there is an order in the product of two definite propositions *p*, *q*, and it is significant to assert that the products *pq* and *qp* are equivalent. But this can be proved by means of the axioms with which we began the calculus of propositions (§ 18). It is to be observed that this proof is prior to the proof that the class whose terms are *p* and *q* is identical with the class whose terms are *q* and *p*. (4) We have next two forms of syllogism, both primitive propositions. The first asserts

that, if a, b, c be classes, and u is contained in b, and x is an a, then x is a b; the second asserts that if a, b, c be classes, and a is contained in b, b in c, then a is contained in c. It is one of the greatest of Peano's merits to have clearly distinguished the relation of the individual to its class from the relation of inclusion between classes. The difference is exceedingly fundamental : the former relation is the simplest and most essential of all relations, the latter a complicated relation derived from logical implication. It results from the distinction that the syllogism in Barbara has two forms, usually confounded : the one the time-honoured assertion that Socrates is a man, and therefore mortal, the other the assertion that Greeks are men, and therefore mortal. These two forms are stated by Peano's axioms. It is to be observed that, in virtue of the definition of what is meant by one class being contained in another, the first form results from the axiom that, if p, q, r be propositions, and p implies that q implies r, then the product of p and q implies r. This axiom is now substituted by Peano for the first form of the syllogism*: it is more general and cannot be deduced from the said form. The second form of the syllogism, when applied to propositions instead of classes, asserts that implication is transitive. This principle is, of course, the very life of all chains of reasoning. (5) We have next a principle of reasoning which Peano calls *composition* : this asserts that if a is contained in b and also in c, then it is contained in the common part of both. Stating this principle with regard to propositions, it asserts that if a proposition implies each of two others, then it implies their joint assertion or logical product ; and this is the principle which was called *composition* above.

35. From this point, we advance successfully until we require the idea of *negation*. This is taken, in the edition of the *Formulaire* we are considering, as a new primitive idea, and disjunction is defined by its means. By means of the negation of a proposition, it is of course easy to define the negation of a class : for " x is a not-a " is equivalent to " x is not an a." But we require an axiom to the effect that not-a is a class, and another to the effect that not-not-a is a. Peano gives also a third axiom, namely : If a, b, c be classes, and ab is contained in c, and x is an a but not a c, then x is not a b. This is simpler in the form : If p, q, r be propositions, and p, q together imply r, and q is true while r is false, then q is false. This would be still further improved by being put in the form : If q, r are propositions, and q implies r, then not-r implies not-q ; a form which Peano obtains as a deduction. By dealing with propositions before classes or propositional functions, it is possible, as we saw, to avoid treating negation as a primitive idea, and to replace all axioms respecting negation by the principle of reduction.

We come next to the definition of the disjunction or logical sum of two classes. On this subject Peano has many times changed his

* See *e.g.* F. 1901, Part I, § 1, Prop. 3. 3 (p. 10).

procedure. In the edition we are considering, " *a* or *b* " is defined as the negation of the logical product of not-*a* and not-*b*, *i.e.* as the class of terms which are not both not-*a* and not-*b*. In later editions (*e.g. F.* 1901, p. 19), we find a somewhat less artificial definition, namely : " *a* or *b* " consists of all terms which belong to any class which contains *a* and contains *b*. Either definition seems logically unobjectionable. It is to be observed that *a* and *b* are classes, and that it remains a question for philosophical logic whether there is not a quite different notion of the disjunction of individuals, as *e.g.* "Brown or Jones." I shall consider this question in Chapter v. It will be remembered that, when we begin by the calculus of propositions, disjunction is defined before negation ; with the above definition (that of 1897), it is plainly necessary to take negation first.

36. The connected notions of the null-class and the existence of a class are next dealt with. In the edition of 1897, a class is defined as null when it is contained in every class. When we remember the definition of one class *a* being contained in another *b* (" *x* is an *a* " implies " *x* is a *b* " for all values of *x*), we see that we are to regard the implication as holding for *all* values, and not only for those values for which *x* really is an *a*. This is a point upon which Peano is not explicit, and I doubt whether he has made up his mind on it. If the implication were only to hold when *x* really is an *a*, it would not give a definition of the null-class, for which this hypothesis is false for all values of *x*. I do not know whether it is for this reason or for some other that Peano has since abandoned the definition of the inclusion of classes by means of formal implication between propositional functions : the inclusion of classes appears to be now regarded as indefinable. Another definition which Peano has sometimes favoured (*e.g. F.* 1895, Errata, p. 116) is, that the null-class is the product of any class into its negation—a definition to which similar remarks apply. In *R. d. M.* vii, No. 1 (§ 3, Prop. 1. 0), the null-class is defined as the class of those terms that belong to every class, *i.e.* the class of terms *x* such that " *a* is a class " implies " *x* is an *a* " for all values of *a*. There are of course no such terms *x* ; and there is a grave logical difficulty in trying to interpret extensionally a class which has no extension. This point is one to which I shall return in Chapter vi.

From this point onward, Peano's logic proceeds by a smooth development. But in one respect it is still defective : it does not recognize as ultimate relational propositions not asserting membership of a class. For this reason, the definitions of a function* and of other essentially relational notions are defective. But this defect is easily remedied by applying, in the manner explained above, the principles of the *Formulaire* to the logic of relations†.

---

* *E.g. F.* 1901, Part I, § 10, Props. 1. 0. 01 (p. 33).
† See my article "Sur la logique des relations," *R. d. M.* Vol. vii, 2 (1901).

CHAPTER III.

IMPLICATION AND FORMAL IMPLICATION.

37. In the preceding chapter I endeavoured to present, briefly and uncritically, all the data, in the shape of formally fundamental ideas and propositions, that pure mathematics requires. In subsequent Parts I shall show that these are all the data by giving definitions of the various mathematical concepts—number, infinity, continuity, the various spaces of geometry, and motion. In the remainder of Part I, I shall give indications, as best I can, of the philosophical problems arising in the analysis of the data, and of the directions in which I imagine these problems to be probably soluble. Some logical notions will be elicited which, though they seem quite fundamental to logic, are not commonly discussed in works on the subject; and thus problems no longer clothed in mathematical symbolism will be presented for the consideration of philosophical logicians.

Two kinds of implication, the material and the formal, were found to be essential to every kind of deduction. In the present chapter I wish to examine and distinguish these two kinds, and to discuss some methods of attempting to analyze the second of them.

In the discussion of inference, it is common to permit the intrusion of a psychological element, and to consider our acquisition of new knowledge by its means. But it is plain that where we validly infer one proposition from another, we do so in virtue of a relation which holds between the two propositions whether we perceive it or not: the mind, in fact, is as purely receptive in inference as common sense supposes it to be in perception of sensible objects. The relation in virtue of which it is possible for us validly to infer is what I call material implication. We have already seen that it would be a vicious circle to define this relation as meaning that *if* one proposition is true, *then* another is true, for *if* and *then* already involve implication. The relation holds, in fact, when it does hold, without any reference to the truth or falsehood of the propositions involved.

But in developing the consequences of our assumptions as to implication, we were led to conclusions which do not by any means agree with what is commonly held concerning implication, for we found that any false proposition implies every proposition and any true proposition is implied by every proposition. Thus propositions are formally like a set of lengths each of which is one inch or two, and implication is like the relation "equal to or less than" among such lengths. It would certainly not be commonly maintained that "$2 + 2 = 4$" can be deduced from "Socrates is a man," or that both are implied by "Socrates is a triangle." But the reluctance to admit such implications is chiefly due, I think, to preoccupation with formal implication, which is a much more familiar notion, and is really before the mind, as a rule, even where material implication is what is explicitly mentioned. In inferences from "Socrates is a man," it is customary not to consider the philosopher who vexed the Athenians, but to regard Socrates merely as a symbol, capable of being replaced by any other man; and only a vulgar prejudice in favour of true propositions stands in the way of replacing Socrates by a number, a table, or a plum-pudding. Nevertheless, wherever, as in Euclid, one particular proposition is deduced from another, material implication is involved, though as a rule the material implication may be regarded as a particular instance of some formal implication, obtained by giving some constant value to the variable or variables involved in the said formal implication. And although, while relations are still regarded with the awe caused by unfamiliarity, it is natural to doubt whether any such relation as implication is to be found, yet, in virtue of the general principles laid down in Section C of the preceding chapter, there must be a relation holding between nothing except propositions, and holding between any two propositions of which either the first is false or the second true. Of the various equivalent relations satisfying these conditions, one is to be called *implication*, and if such a notion seems unfamiliar, that does not suffice to prove that it is illusory.

38. At this point, it is necessary to consider a very difficult logical problem, namely, the distinction between a proposition actually asserted, and a proposition considered merely as a complex concept. One of our indemonstrable principles was, it will be remembered, that if the hypothesis in an implication is true, it may be dropped, and the consequent asserted. This principle, it was observed, eludes formal statement, and points to a certain failure of formalism in general. The principle is employed whenever a proposition is said to be *proved*; for what happens is, in all such cases, that the proposition is shown to be implied by some true proposition. Another form in which the principle is constantly employed is the substitution of a constant, satisfying the hypothesis, in the consequent of a formal implication. If ϕx implies ψx for all values of x, and if a is a constant satisfying ϕx, we can assert ψa, dropping the true hypothesis ϕa. This occurs, for example, whenever any of those rules of inference which employ the hypothesis that the variables involved are propositions, are applied to particular propositions. The principle in question is, therefore, quite vital to any kind of demonstration.

The independence of this principle is brought out by a consideration of Lewis Carroll's puzzle, "What the Tortoise said to Achilles *." The principles of inference which we accepted lead to the proposition that, if p and q be propositions, then p together with "p implies q" implies q. At first sight, it might be thought that this would enable us to assert q provided p is true and implies q. But the puzzle in question shows that this is not the case, and that, until we have some new principle, we shall only be led into an endless regress of more and more complicated implications, without ever arriving at the assertion of q. We need, in fact, the notion of *therefore*, which is quite different from the notion of *implies*, and holds between different entities. In grammar, the distinction is that between a verb and a verbal noun, between, say, "A is greater than B" and "A's being greater than B." In the first of these, a proposition is actually asserted, whereas in the second it is merely considered. But these are psychological terms, whereas the difference which I desire to express is genuinely logical. It is plain that, if I may be allowed to use the word *assertion* in a non-psychological sense, the proposition "p implies q" *asserts* an implication, though it does not *assert* p or q. The p and the q which enter into this proposition are not strictly the same as the p or the q which are separate propositions, at least, if they are true. The question is: How does a proposition differ by being actually true from what it would be as an entity if it were not true? It is plain that true and false propositions alike are entities of a kind, but that true propositions have a quality not belonging to false ones, a quality which, in a non-psychological sense, may be called being *asserted*. Yet there are grave difficulties in forming a consistent theory on this point, for if assertion in any way changed a proposition, no proposition which can possibly in any context be unasserted could be true, since when asserted it would become a different proposition. But this is plainly false; for in "p implies q," p and q are not asserted, and yet they may be true. Leaving this puzzle to logic, however, we must insist that there is a difference of some kind between an asserted and an unasserted proposition†. When we say *therefore*, we state a relation which can only hold between asserted propositions, and which thus differs from implication. Wherever *therefore* occurs, the hypothesis may be dropped, and the conclusion asserted by itself. This seems to be the first step in answering Lewis Carroll's puzzle.

39. It is commonly said that an inference must have premisses and a conclusion, and it is held, apparently, that two or more premisses are necessary, if not to all inferences, yet to most. This view is borne out, at first sight, by obvious facts: every syllogism, for example, is held

---

* *Mind*, N. S. Vol. iv, p. 278.
† Frege (*loc. cit.*) has a special symbol to denote assertion.

to have two premises. Now such a theory greatly complicates the relation of implication, since it renders it a relation which may have any number of terms, and is symmetrical with respect to all but one of them, but not symmetrical with respect to that one (the conclusion). This complication is, however, unnecessary, first, because every simultaneous assertion of a number of propositions is itself a single proposition, and secondly, because, by the rule which we called *exportation*, it is always possible to exhibit an implication explicitly as holding between single propositions. To take the first point first: if k be a class of propositions, all the propositions of the class k are asserted by the single proposition "for all values of x, if x implies x, then 'x is a k' implies x," or, in more ordinary language, "every k is true." And as regards the second point, which assumes the number of premisses to be finite, "pq implies r" is equivalent, if q be a proposition, to "p implies that q implies r," in which latter form the implications hold explicitly between single propositions. Hence we may safely hold implication to be a relation between two propositions, not a relation of an arbitrary number of premisses to a single conclusion.

40. I come now to formal implication, which is a far more difficult notion than material implication. In order to avoid the general notion of propositional function, let us begin by the discussion of a particular instance, say "x is a man implies x is a mortal for all values of x." This proposition is equivalent to "all men are mortal" "every man is mortal" and "any man is mortal." But it seems highly doubtful whether it is the same proposition. It is also connected with a purely intensional proposition in which *man* is asserted to be a complex notion of which *mortal* is a constituent, but this proposition is quite distinct from the one we are discussing. Indeed, such intensional propositions are not always present where one class is included in another: in general, either class may be defined by various different predicates, and it is by no means necessary that every predicate of the smaller class should contain every predicate of the larger class as a factor. Indeed, it may very well happen that both predicates are philosophically simple: thus *colour* and *existent* appear to be both simple, yet the class of colours is part of the class of existents. The intensional view, derived from predicates, is in the main irrelevant to Symbolic Logic and to Mathematics, and I shall not consider it further at present.

41. It may be doubted, to begin with, whether "x is a man implies x is a mortal" is to be regarded as asserted strictly of all possible terms, or only of such terms as are men. Peano, though he is not explicit, appears to hold the latter view. But in this case, the hypothesis ceases to be significant, and becomes a mere definition of x: x is to mean any man. The hypothesis then becomes a mere assertion concerning the meaning of the symbol x, and the whole of what is asserted concerning the matter dealt with by our symbol is put into the conclusion. The premiss says: x is to mean any man. The conclusion says: x is mortal. But the implication is merely concerning the symbolism: since any man is mortal, if x denotes any man, x is mortal. Thus formal implication, on this view, has wholly disappeared, leaving us the proposition "any man is mortal" as expressing the whole of what is relevant in the proposition with a variable. It would now only remain to examine the proposition "any man is mortal," and if possible to explain this proposition without reintroducing the variable and formal implication. It must be confessed that some grave difficulties are avoided by this view. Consider, for example, the simultaneous assertion of all the propositions of some class k: this is not expressed by "'x is a k' implies x for all values of x." For as it stands, this proposition does not express what is meant, since, if x be not a proposition, "x is a k" cannot imply x; hence the range of variability of x must be confined to propositions, unless we prefix (as above, § 39) the hypothesis "x implies x." This remark applies generally, throughout the propositional calculus, to all cases where the conclusion is represented by a single letter: unless the letter does actually represent a proposition, the implication asserted will be false, since only propositions can be implied. The point is that, if x be our variable, x itself is a proposition for all values of x which are propositions, but not for other values. This makes it plain what the limitations are to which our variable is subject: it must vary only within the range of values for which the two sides of the principal implication are propositions, in other words, the two sides, when the variable is not replaced by a constant, must be genuine propositional functions. If this restriction is not observed, fallacies quickly begin to appear. It should be noticed that there may be any number of subordinate implications which do not require that their terms should be propositions: it is only of the principal implication that this is required. Take, for example, the first principle of inference: If p implies q, then p implies q. This holds equally whether p and q be propositions or not; for if either is not a proposition, "p implies q" becomes false, but does not cease to be a proposition. In fact, in virtue of the definition of a proposition, our principle states that "p implies q" is a propositional function, *i.e.* that it is a proposition for *all* values of p and q. But if we apply the principle of importation to this proposition, so as to obtain "'p implies q,' together with p, implies q," we have a formula which is only true when p and q are propositions: in order to make it true universally, we must preface it by the hypothesis "p implies p and q implies q." In this way, in many cases, if not in all, the restriction on the variability of the variable can be removed; thus, in the assertion of the logical product of a class of propositions, the formula "if x implies x, then 'x is a k' implies x" appears unobjectionable, and allows x to vary without restriction. Here the subordinate implications in the premiss and the conclusion are material: only the principal implication is formal.

Returning now to " x is a man implies x is a mortal," it is plain that no restriction is required in order to insure our having a genuine proposition. And it is plain that, although we *might* restrict the values of x to men, and although this seems to be done in the proposition "all men are mortal," yet there is no reason, so far as the truth of our proposition is concerned, why we should so restrict our x. Whether x be a man or not, " x is a man " is always, when a constant is substituted for x, a proposition implying, for that value of x, the proposition " x is a mortal." And unless we admit the hypothesis equally in the cases where it is false, we shall find it impossible to deal satisfactorily with the null-class or with null propositional functions. We must, therefore, allow our x, wherever the truth of our formal implication is thereby unimpaired, to take *all* values without exception; and where any restriction on variability is required, the implication is not to be regarded as formal until the said restriction has been removed by being prefixed as hypothesis. (If ψx be a proposition whenever x satisfies ϕx, where ϕx is a propositional function, and if ψx, whenever it is a proposition, implies χx, then " ψx implies χx" is not a formal implication, but " ϕx implies that ψx implies χx" is a formal implication.)

42. It is to be observed that " x is a man implies x is a mortal " is not a relation of two propositional functions, but is itself a single propositional function having the elegant property of being always true. For " x is a man" is, as it stands, not a proposition at all, and does not imply anything ; and we must not first vary our x in " x is a man," and then independently vary it in " x is a mortal," for this would lead to the proposition that "everything is a man " implies "everything is a mortal," which, though true, is not what was meant. This proposition would have to be expressed, if the language of variables were retained, by two variables, as " x is a man implies y is a mortal." But this formula too is unsatisfactory, for its natural meaning would be : "If anything is a man, then everything is a mortal." The point to be emphasized is, of course, that our x, though variable, must be the same on both sides of the implication, and this requires that we should not obtain our formal implication by first varying (say) Socrates in "Socrates is a man," and then in "Socrates is a mortal," but that we should start from the whole proposition "Socrates is a man implies Socrates is a mortal," and vary Socrates in this proposition as a whole. Thus our formal implication asserts a class of implications, not a single implication at all. We do not, in a word, have one implication containing a variable, but rather a variable implication. We have a class of implications, no one of which contains a variable, and we assert that every member of this class is true. This is a first step towards the analysis of the mathematical notion of the variable.

But, it may be asked, how comes it that Socrates may be varied in the proposition " Socrates is a man implies Socrates is mortal "? In virtue of the fact that true propositions are implied by all others, we have "Socrates is a man implies Socrates is a philosopher"; but in this proposition, alas, the variability of Socrates is sadly restricted. This seems to show that formal implication involves something over and above the relation of implication, and that some additional relation must hold where a term can be varied. In the case in question, it is natural to say that what is involved is the relation of inclusion between the classes *men* and *mortals*—the very relation which was to be defined and explained by our formal implication. But this view is too simple to meet all cases, and is therefore not required in any case. A larger number of cases, though still not all cases, can be dealt with by the notion of what I shall call *assertions*. This notion must now be briefly explained, leaving its critical discussion to Chapter VII.

43. It has always been customary to divide propositions into subject and predicate; but this division has the defect of omitting the verb. It is true that a graceful concession is sometimes made by loose talk about the copula, but the verb deserves far more respect than is thus paid to it. We may say, broadly, that every proposition may be divided, some in only one way, some in several ways, into a term (the subject) and something which is said about the subject, which something I shall call the *assertion*. Thus "Socrates is a man" may be divided into *Socrates* and *is a man*. The verb, which is the distinguishing mark of propositions, remains with the assertion; but the assertion itself, being robbed of its subject, is neither true nor false. In logical discussions, the notion of assertion often occurs, but as the word *proposition* is used for it, it does not obtain separate consideration. Consider, for example, the best statement of the identity of indiscernibles : " If x and y be any two diverse entities, some assertion holds of x which does not hold of y." But for the word *assertion*, which would ordinarily be replaced by *proposition*, this statement is one which would commonly pass unchallenged. Again, it might be said: "Socrates was a philosopher, and the same is true of Plato." Such statements require the analysis of a proposition into an assertion and a subject, in order that there may be something identical which can be said to be affirmed of two subjects.

44. We can now see how, where the analysis into subject and assertion is legitimate, to distinguish implications in which there is a term which can be varied from others in which this is not the case. Two ways of making the distinction may be suggested, and we shall have to decide between them. It may be said that there is a relation between the two assertions "is a man" and "is a mortal," in virtue of which, when the one holds, so does the other. Or again, we may analyze the whole proposition "Socrates is a man implies Socrates is a mortal" into Socrates and an assertion about him, and say that the assertion in question holds of all terms. Neither of these theories replaces the above analysis of "x is a man implies x is a mortal" into a class of material implications; but whichever of the two is true carries the analysis one step further. The first theory suffers from the difficulty that it is essential to the relation of assertions involved that both assertions should be made of the *same* subject, though it is otherwise irrelevant what subject we choose. The second theory appears objectionable on the ground that the suggested analysis of "Socrates is a man implies Socrates is a mortal" seems scarcely possible. The proposition in question consists of two terms and a relation, the terms being "Socrates is a man" and "Socrates is a mortal"; and it would seem that when a relational proposition is analyzed into a subject and an assertion, the subject must be one of the terms of the relation which is asserted. This objection seems graver than that against the former view; I shall therefore, at any rate for the present, adopt the former view, and regard formal implication as derived from a relation between assertions.

We remarked above that the relation of inclusion between classes is insufficient. This results from the irreducible nature of relational propositions. Take *e.g.* "Socrates is married implies Socrates had a father." Here it is affirmed that because Socrates has one relation, he must have another. Or better still, take "A is before B implies B is after A." This is a formal implication, in which the assertions are (superficially at least) concerning different subjects; the only way to avoid this is to say that both propositions have both A and B as subjects, which, by the way, is quite different from saying that they have the one subject "A and B." Such instances make it plain that the notion of a propositional function, and the notion of an assertion, are more fundamental than the notion of *class*, and that the latter is not adequate to explain all cases of formal implication. I shall not enlarge upon this point now, as it will be abundantly illustrated in subsequent portions of the present work.

It is important to realize that, according to the above analysis of formal implication, the notion of *every term* is indefinable and ultimate. A formal implication is one which holds of every term, and therefore *every* cannot be explained by means of formal implication. If a and b be classes, we can explain "every a is a b" by means of "x is an a implies x is a b"; but the *every* which occurs here is a derivative and subsequent notion, presupposing the notion of *every term*. It seems to be the very essence of what may be called a *formal* truth, and of formal reasoning generally, that some assertion is affirmed to hold of every term; and unless the notion of *every term* is admitted, formal truths are impossible.

45. The fundamental importance of formal implication is brought out by the consideration that it is involved in all the rules of inference. This shows that we cannot hope wholly to define it in terms of material implication, but that some further element or elements must be involved.

We may observe, however, that, in a particular inference, the rule according to which the inference proceeds is not required as a premiss. This point has been emphasized by Mr Bradley*; it is closely connected with the principle of dropping a true premiss, being again a respect in which formalism breaks down. In order to apply a rule of inference, it is formally necessary to have a premiss asserting that the present case is an instance of the rule; we shall then need to affirm the rule by which we can go from the rule to an instance, and also to affirm that here we have an instance of this rule, and so on into an endless process. The fact is, of course, that any implication warranted by a rule of inference does actually hold, and is not merely implied by the rule. This is simply an instance of the non-formal principle of dropping a true premiss: if our rule implies a certain implication, the rule may be dropped and the implication asserted. But it remains the case that the fact that our rule does imply the said implication, if introduced at all, must be simply perceived, and is not guaranteed by any formal deduction; and often it is just as easy, and consequently just as legitimate, to perceive immediately the implication in question as to perceive that it is implied by one or more of the rules of inference.

To sum up our discussion of formal implication : a formal implication, we said, is the affirmation of *every* material implication of a certain class; and the class of material implications involved is, in simple cases, the class of all propositions in which a given fixed assertion, made concerning a certain subject or subjects, is affirmed to imply another given fixed assertion concerning the same subject or subjects. Where a formal implication holds, we agreed to regard it, wherever possible, as due to some relation between the assertions concerned. This theory raises many formidable logical problems, and requires, for its defence, a thorough analysis of the constituents of propositions. To this task we must now address ourselves.

---

* *Logic*, Book II, Part I, Chap. II (p. 227).

CHAPTER IV.

PROPER NAMES, ADJECTIVES, AND VERBS.

46. In the present chapter, certain questions are to be discussed belonging to what may be called philosophical grammar. The study of grammar, in my opinion, is capable of throwing far more light on philosophical questions than is commonly supposed by philosophers. Although a grammatical distinction cannot be uncritically assumed to correspond to a genuine philosophical difference, yet the one is *primâ facie* evidence of the other, and may often be most usefully employed as a source of discovery. Moreover, it must be admitted, I think, that every word occurring in a sentence must have *some* meaning: a perfectly meaningless sound could not be employed in the more or less fixed way in which language employs words. The correctness of our philosophical analysis of a proposition may therefore be usefully checked by the exercise of assigning the meaning of each word in the sentence expressing the proposition. On the whole, grammar seems to me to bring us much nearer to a correct logic than the current opinions of philosophers; and in what follows, grammar, though not our master, will yet be taken as our guide*.

Of the parts of speech, three are specially important: substantives, adjectives, and verbs. Among substantives, some are derived from adjectives or verbs, as humanity from human, or sequence from *follows*. (I am not speaking of an etymological derivation, but of a logical one.) Others, such as proper names, or space, time, and matter, are not derivative, but appear primarily as substantives. What we wish to obtain is a classification, not of words, but of ideas; I shall therefore call adjectives or predicates all notions which are capable of being such, even in a form in which grammar would call them substantives. The fact is, as we shall see, that *human* and *humanity* denote precisely the same concept, these words being employed respectively according to the kind of relation in which this concept stands to the other constituents of a proposition in which it occurs. The distinction which we require

---

* The excellence of grammar as a guide is proportional to the paucity of inflexions, *i.e.* to the degree of analysis effected by the language considered.

is not identical with the grammatical distinction between substantive and adjective, since one single concept may, according to circumstances, be either substantive or adjective: it is the distinction between proper and general names that we require, or rather between the objects indicated by such names. In every proposition, as we saw in Chapter III, we may make an analysis into something asserted and something about which the assertion is made. A proper name, when it occurs in a proposition, is always, at least according to one of the possible ways of analysis (where there are several), the subject that the proposition or some subordinate constituent proposition is about, and not what is said about the subject. Adjectives and verbs, on the other hand, are capable of occurring in propositions in which they cannot be regarded as subject, but only as parts of the assertion. Adjectives are distinguished by capacity for *denoting*—a term which I intend to use in a technical sense to be discussed in Chapter v. Verbs are distinguished by a special kind of connection, exceedingly hard to define, with truth and falsehood, in virtue of which they distinguish an asserted proposition from an unasserted one, *e.g.* "Caesar died" from "the death of Caesar." These distinctions must now be amplified, and I shall begin with the distinction between general and proper names.

47. Philosophy is familiar with a certain set of distinctions, all more or less equivalent: I mean, the distinctions of subject and predicate, substance and attribute, substantive and adjective, *this* and *what**. I wish now to point out briefly what appears to me to be the truth concerning these cognate distinctions. The subject is important, since the issues between monism and monadism, between idealism and empiricism, and between those who maintain and those who deny that all truth is concerned with what exists, all depend, in whole or in part, upon the theory we adopt in regard to the present question. But the subject is treated here only because it is essential to any doctrine of number or of the nature of the variable. Its bearings on general philosophy, important as they are, will be left wholly out of account.

Whatever may be an object of thought, or may occur in any true or false proposition, or can be counted as *one*, I call a *term*. This, then, is the widest word in the philosophical vocabulary. I shall use as synonymous with it the words unit, individual, and entity. The first two emphasize the fact that every term is *one*, while the third is derived from the fact that every term has being, *i.e. is* in some sense. A man, a moment, a number, a class, a relation, a chimaera, or anything else that can be mentioned, is sure to be a term ; and to deny that such and such a thing is a term must always be false.

It might perhaps be thought that a word of such extreme generality could not be of any great use. Such a view, however, owing to certain

* This last pair of terms is due to Mr Bradley.

wide-spread philosophical doctrines, would be erroneous. A term is, in fact, possessed of all the properties commonly assigned to substances or substantives. Every term, to begin with, is a logical subject: it is, for example, the subject of the proposition that itself is one. Again every term is immutable and indestructible. What a term is, it is, and no change can be conceived in it which would not destroy its identity and make it another term *. Another mark which belongs to terms is numerical identity with themselves and numerical diversity from all other terms †. Numerical identity and diversity are the source of unity and plurality ; and thus the admission of many terms destroys monism. And it seems undeniable that every constituent of every proposition can be counted as one, and that no proposition contains less than two constituents. *Term* is, therefore, a useful word, since it marks dissent from various philosophies, as well as because, in many statements, we wish to speak of *any* term or *some* term.

48. Among terms, it is possible to distinguish two kinds, which I shall call respectively *things* and *concepts*. The former are the terms indicated by proper names, the latter those indicated by all other words. Here proper names are to be understood in a somewhat wider sense than is usual, and things also are to be understood as embracing all particular points and instants, and many other entities not commonly called things. Among concepts, again, two kinds at least must be distinguished, namely those indicated by adjectives and those indicated by verbs. The former kind will often be called predicates or class-concepts ; the latter are always or almost always relations. (In intransitive verbs, the notion expressed by the verb is complex, and usually asserts a definite relation to an indefinite relatum, as in "Smith breathes.")

In a large class of propositions, we agreed, it is possible, in one or more ways, to distinguish a subject and an assertion about the subject. The assertion must always contain a verb, but except in this respect, assertions appear to have no universal properties. In a relational proposition, say "*A* is greater than *B*," we may regard *A* as the subject, and "is greater than *B*" as the assertion, or *B* as the subject and "*A* is greater than" as the assertion. There are thus, in the case proposed, two ways of analyzing the proposition into subject and assertion. Where a relation has more than two terms, as in "*A* is here now‡," there will be more than two ways of making the analysis. But in some propositions, there is only a single way : these are the subject-

---

* The notion of a term here set forth is a modification of Mr G. E. Moore's notion of a *concept* in his article "On the Nature of Judgment," *Mind*, N. S. No. 30, from which notion, however, it differs in some important respects.

† On identity, see Mr G. E. Moore's article in the *Proceedings of the Aristotelian Society*, 1900–1901.

‡ This proposition means "*A* is in this place at this time." It will be shown in Part VII that the relation expressed is not reducible to a two-term relation.

predicate propositions, such as "Socrates is human." The proposition "humanity belongs to Socrates," which is equivalent to "Socrates is human," is an assertion about humanity; but it is a distinct proposition. In "Socrates is human," the notion expressed by *human* occurs in a different way from that in which it occurs when it is called *humanity*, the difference being that in the latter case, but not in the former, the proposition is *about* this notion. This indicates that humanity is a concept, not a thing. I shall speak of the *terms* of a proposition as those terms, however numerous, which occur in a proposition and may be regarded as subjects about which the proposition is. It is a characteristic of the terms of a proposition that any one of them may be replaced by any other entity without our ceasing to have a proposition. Thus we shall say that "Socrates is human" is a proposition having only one term; of the remaining components of the proposition, one is the verb, the other is a *predicate*. With the sense which *is* has in this proposition, we no longer have a proposition at all if we replace *human* by something other than a predicate. Predicates, then, are concepts, other than verbs, which occur in propositions having only one term or subject. Socrates is a thing, because Socrates can never occur otherwise than as term in a proposition : Socrates is not capable of that curious twofold use which is involved in *human* and *humanity*. Points, instants, bits of matter, particular states of mind, and particular existents generally, are things in the above sense, and so are many terms which do not exist, for example, the points in a non-Euclidean space and the pseudo-existents of a novel. All classes, it would seem, as numbers, men, spaces, etc., when taken as single terms, are things; but this is a point for Chapter vi.

Predicates are distinguished from other terms by a number of very interesting properties, chief among which is their connection with what I shall call *denoting*. One predicate always gives rise to a host of cognate notions : thus in addition to *human* and *humanity*, which only differ grammatically, we have *man, a man, some man, any man, every man, all men**, all of which appear to be genuinely distinct one from another. The study of these various notions is absolutely vital to any philosophy of mathematics; and it is on account of them that the theory of predicates is important.

49. It might be thought that a distinction ought to be made between a concept as such and a concept used as a term, between, *e.g.*, such pairs as *is* and *being, human* and *humanity, one* in such a proposition as "this is one" and ˙1 in "1 is a number." But inextricable difficulties will envelop us if we allow such a view. There is,

---

* I use *all men* as collective, *i.e.* as nearly synonymous with *the human race*, but differing therefrom by being many and not one. I shall always use *all* collectively, confining myself to *every* for the distributive sense. Thus I shall say " every man is mortal," not " all men are mortal."

of course, a grammatical difference, and this corresponds to a difference as regards relations. In the first case, the concept in question is used as a concept, that is, it is actually predicated of a term or asserted to relate two or more terms; while in the second case, the concept is itself said to have a predicate or a relation. There is, therefore, no difficulty in accounting for the grammatical difference. But what I wish to urge is, that the difference lies solely in external relations, and not in the intrinsic nature of the terms. For suppose that *one* as adjective differed from 1 as term. In this statement, *one* as adjective has been made into a term; hence either it has become 1, in which case the supposition is self-contradictory; or there is some other difference between *one* and 1 in addition to the fact that the first denotes a concept not a term while the second denotes a concept which is a term. But in this latter hypothesis, there must be propositions concerning *one* as term, and we shall still have to maintain propositions concerning *one* as adjective as opposed to *one* as term; yet all such propositions must be false, since a proposition about *one* as adjective makes *one* the subject, and is therefore really about *one* as term. In short, if there were any adjectives which could not be made into substantives without change of meaning, all propositions concerning such adjectives (since they would necessarily turn them into substantives) would be false, and so would the proposition that all such propositions are false, since this itself turns the adjectives into substantives. But this state of things is self-contradictory.

The above argument proves that we were right in saying that terms embrace everything that can occur in a proposition, with the possible exception of complexes of terms of the kind denoted by *any* and cognate words*. For if *A* occurs in a proposition, then, in this statement, *A* is the subject; and we have just seen that, if *A* is ever not the subject, it is exactly and numerically the same *A* which is not subject in one proposition and is subject in another. Thus the theory that there are adjectives or attributes or ideal things, or whatever they may be called, which are in some way less substantial, less self-subsistent, less self-identical, than true substantives, appears to be wholly erroneous, and to be easily reduced to a contradiction. Terms which are concepts differ from those which are not, not in respect of self-subsistence, but in virtue of the fact that, in certain true or false propositions, they occur in a manner which is different in an indefinable way from the manner in which subjects or terms of relations occur.

50. Two concepts have, in addition to the numerical diversity which belongs to them as terms, another special kind of diversity which may be called conceptual. This may be characterized by the fact that two propositions in which the concepts occur otherwise than as terms, even if, in all other respects, the two propositions are identical,

* See the next chapter.

yet differ in virtue of the fact that the concepts which occur in them are conceptually diverse. Conceptual diversity implies numerical diversity, but the converse implication does not hold, since not all terms are concepts. Numerical diversity, as its name implies, is the source of plurality, and conceptual diversity is less important to mathematics. But the whole possibility of making different assertions about a given term or set of terms depends upon conceptual diversity, which is therefore fundamental in general logic.

51. It is interesting and not unimportant to examine very briefly the connection of the above doctrine of adjectives with certain traditional views on the nature of propositions. It is customary to regard all propositions as having a subject and a predicate, *i.e.* as having an immediate *this*, and a general concept attached to it by way of description. This is, of course, an account of the theory in question which will strike its adherents as extremely crude; but it will serve for a general indication of the view to be discussed. This doctrine develops by internal logical necessity into the theory of Mr Bradley's Logic, that all words stand for ideas having what he calls *meaning*, and that in every judgment there is a something, the true subject of the judgment, which is not an idea and does not have meaning. To have meaning, it seems to me, is a notion confusedly compounded of logical and psychological elements. *Words* all have meaning, in the simple sense that they are symbols which stand for something other than themselves. But a proposition, unless it happens to be linguistic, does not itself contain words: it contains the entities indicated by words. Thus meaning, in the sense in which words have meaning, is irrelevant to logic. But such concepts as *a man* have meaning in another sense: they are, so to speak, symbolic in their own logical nature, because they have the property which I call *denoting*. That is to say, when *a man* occurs in a proposition (*e.g.* "I met a man in the street"), the proposition is not about the concept *a man*, but about something quite different, some actual biped denoted by the concept. Thus concepts of this kind have meaning in a non-psychological sense. And in this sense, when we say "this is a man," we are making a proposition in which a concept is in some sense attached to what is not a concept. But when meaning is thus understood, the entity indicated by *John* does not have meaning, as Mr Bradley contends*; and even among concepts, it is only those that denote that have meaning. The confusion is largely due, I believe, to the notion that *words* occur in propositions, which in turn is due to the notion that propositions are essentially mental and are to be identified with cognitions. But these topics of general philosophy must be pursued no further in this work.

52. It remains to discuss the verb, and to find marks by which it is distinguished from the adjective. In regard to verbs also, there is

---

* *Logic*, Book I, Chap. i, §§ 17, 18 (pp. 58–60).

a twofold grammatical form corresponding to a difference in merely external relations. There is the verb in the form which it has as verb (the various inflexions of this form may be left out of account), and there is the verbal noun, indicated by the infinitive or (in English) the present participle. The distinction is that between "Felton killed Buckingham" and "Killing no murder." By analyzing this difference, the nature and function of the verb will appear.

It is plain, to begin with, that the concept which occurs in the verbal noun is the very same as that which occurs as verb. This results from the previous argument, that every constituent of every proposition must, on pain of self-contradiction, be capable of being made a logical subject. If we say "*kills* does not mean the same as *to kill*," we have already made *kills* a subject, and we cannot say that the concept expressed by the word *kills* cannot be made a subject. Thus the very verb which occurs as verb can occur also as subject. The question is: What logical difference is expressed by the difference of grammatical form? And it is plain that the difference must be one in external relations. But in regard to verbs, there is a further point. By transforming the verb, as it occurs in a proposition, into a verbal noun, the whole proposition can be turned into a single logical subject, no longer asserted, and no longer containing in itself truth or falsehood. But here too, there seems to be no possibility of maintaining that the logical subject which results is a different entity from the proposition. "Caesar died" and "the death of Caesar" will illustrate this point. If we ask: What is asserted in the proposition "Caesar died"? the answer must be "the death of Caesar is asserted." In that case, it would seem, it is the death of Caesar which is true or false; and yet neither truth nor falsity belongs to a mere logical subject. The answer here seems to be that the death of Caesar has an external relation to truth or falsehood (as the case may be), whereas "Caesar died" in some way or other contains its own truth or falsehood as an element. But if this is the correct analysis, it is difficult to see how "Caesar died" differs from "the truth of Caesar's death" in the case where it is true, or "the falsehood of Caesar's death" in the other case. Yet it is quite plain that the latter, at any rate, is never equivalent to "Caesar died." There appears to be an ultimate notion of assertion, given by the verb, which is lost as soon as we substitute a verbal noun, and is lost when the proposition in question is made the subject of some other proposition. This does not depend upon grammatical form; for if I say "*Caesar died* is a proposition," I do not assert that Caesar did die, and an element which is present in "Caesar died" has disappeared. Thus the contradiction which was to have been avoided, of an entity which cannot be made a logical subject, appears to have here become inevitable. This difficulty, which seems to be inherent in the very nature of truth and falsehood, is one with which I do not know how to deal satisfactorily. The most obvious course would be to say that the difference between an asserted and an unasserted proposition is not logical, but psychological. In the sense in which false propositions may be asserted, this is doubtless true. But there is another sense of assertion, very difficult to bring clearly before the mind, and yet quite undeniable, in which only true propositions are asserted. True and false propositions alike are in some sense entities, and are in some sense capable of being logical subjects; but when a proposition happens to be true, it has a further quality, over and above that which it shares with false propositions, and it is this further quality which is what I mean by assertion in a logical as opposed to a psychological sense. The nature of truth, however, belongs no more to the principles of mathematics than to the principles of everything else. I therefore leave this question to the logicians with the above brief indication of a difficulty.

53. It may be asked whether everything that, in the logical sense we are concerned with, is a verb, expresses a relation or not. It seems plain that, if we were right in holding that "Socrates is human" is a proposition having only one term, the *is* in this proposition cannot express a relation in the ordinary sense. In fact, subject-predicate propositions are distinguished by just this non-relational character. Nevertheless, a relation between Socrates and humanity is certainly *implied*, and it is very difficult to conceive the proposition as expressing no relation at all. We may perhaps say that it is a relation, although it is distinguished from other relations in that it does not permit itself to be regarded as an assertion concerning either of its terms indifferently, but only as an assertion concerning the referent. A similar remark may apply to the proposition "*A* is," which holds of every term without exception. The *is* here is quite different from the *is* in "Socrates is human"; it may be regarded as complex, and as really predicating Being of *A*. In this way, the true logical verb in a proposition may be always regarded as asserting a relation. But it is so hard to know exactly what is meant by *relation* that the whole question is in danger of becoming purely verbal.

54. The twofold nature of the verb, as actual verb and as verbal noun, may be expressed, if all verbs are held to be relations, as the difference between a relation in itself and a relation actually relating. Consider, for example, the proposition "*A* differs from *B*." The constituents of this proposition, if we analyze it, appear to be only *A*, difference, *B*. Yet these constituents, thus placed side by side, do not reconstitute the proposition. The difference which occurs in the proposition actually relates *A* and *B*, whereas the difference after analysis is a notion which has no connection with *A* and *B*. It may be said that we ought, in the analysis, to mention the relations which difference has to *A* and *B*, relations which are expressed by *is* and *from* when we say "*A* is different from *B*." These relations consist in the fact that A is referent and B relatum with respect to difference. But "A, referent, difference, relatum, B" is still merely a list of terms, not a proposition. A proposition, in fact, is essentially a unity, and when analysis has destroyed the unity, no enumeration of constituents will restore the proposition. The verb, when used as a verb, embodies the unity of the proposition, and is thus distinguishable from the verb considered as a term, though I do not know how to give a clear account of the precise nature of the distinction.

55. It may be doubted whether the general concept *difference* occurs at all in the proposition "A differs from B," or whether there is not rather a specific difference of A and B, and another specific difference of C and D, which are respectively affirmed in "A differs from B" and "C differs from D." In this way, *difference* becomes a class-concept of which there are as many instances as there are pairs of different terms; and the instances may be said, in Platonic phrase, to partake of the nature of difference. As this point is quite vital in the theory of relations, it may be well to dwell upon it. And first of all, I must point out that in "A differs from B" I intend to consider the bare numerical difference in virtue of which they are two, not difference in this or that respect.

Let us first try the hypothesis that a difference is a complex notion, compounded of difference together with some special quality distinguishing a particular difference from every other particular difference. So far as the relation of difference itself is concerned, we are to suppose that no distinction can be made between different cases; but there are to be different associated qualities in different cases. But since cases are distinguished by their terms, the quality must be primarily associated with the terms, not with difference. If the quality be not a relation, it can have no special connection with the difference of A and B, which it was to render distinguishable from bare difference, and if it fails in this it becomes irrelevant. On the other hand, if it be a new relation between A and B, over and above difference, we shall have to hold that any two terms have two relations, difference and a specific difference, the latter not holding between any other pair of terms. This view is a combination of two others, of which the first holds that the abstract general relation of difference itself holds between A and B, while the second holds that when two terms differ they have, corresponding to this fact, a specific relation of difference, unique and unanalyzable and not shared by any other pair of terms. Either of these views may be held with either the denial or the affirmation of the other. Let us see what is to be said for and against them.

Against the notion of specific differences, it may be urged that, if differences differ, their differences from each other must also differ, and thus we are led into an endless process. Those who object to endless processes will see in this a proof that differences do not differ. But in the present work, it will be maintained that there are no contradictions peculiar to the notion of infinity, and that an endless process is not to be objected to unless it arises in the analysis of the actual meaning of a proposition. In the present case, the process is one of implications, not one of analysis; it must therefore be regarded as harmless.

Against the notion that the abstract relation of difference holds between A and B, we have the argument derived from the analysis of " A differs from B," which gave rise to the present discussion. It is to be observed that the hypothesis which combines the general and the specific difference must suppose that there are two distinct propositions, the one affirming the general, the other the specific difference. Thus if there cannot be a general difference between A and B, this mediating hypothesis is also impossible. And we saw that the attempt to avoid the failure of analysis by including in the *meaning* of " A differs from B" the relations of difference to A and B was vain. This attempt, in fact, leads to an endless process of the inadmissible kind; for we shall have to include the relations of the said relations to A and B and difference, and so on, and in this continually increasing complexity we are supposed to be only analyzing the *meaning* of our original proposition. This argument establishes a point of very great importance, namely, that when a relation holds between two terms, the relations of the relation to the terms, and of these relations to the relation and the terms, and so on *ad infinitum*, though all implied by the proposition affirming the original relation, form no part of the *meaning* of this proposition.

But the above argument does not suffice to prove that the relation of A to B cannot be abstract difference: it remains tenable that, as was suggested to begin with, the true solution lies in regarding every proposition as having a kind of unity which analysis cannot preserve, and which is lost even though it be mentioned by analysis as an element in the proposition. This view has doubtless its own difficulties, but the view that no two pairs of terms can have the same relation both contains difficulties of its own and fails to solve the difficulty for the sake of which it was invented. For, even if the difference of A and B be absolutely peculiar to A and B, still the three terms A, B, difference of A from B, do not reconstitute the proposition " A differs from B," any more than A and B and difference did. And it seems plain that, even if differences did differ, they would still have to have something in common. But the most general way in which two terms can have something in common is by both having a given relation to a given term. Hence if no two pairs of terms can have the same relation, it follows that no two terms can have anything in common, and hence different differences will not be in any definable sense *instances* of difference*. I conclude, then, that

* The above argument appears to prove that Mr Moore's theory of universals with numerically diverse instances in his paper on Identity (*Proceedings of the*

the relation affirmed between A and B in the proposition "A differs from B" is the general relation of difference, and is precisely and numerically the same as the relation affirmed between C and D in "C differs from D." And this doctrine must be held, for the same reasons, to be true of all other relations; relations do not have instances, but are strictly the same in all propositions in which they occur.

We may now sum up the main points elicited in our discussion of the verb. The verb, we saw, is a concept which, like the adjective, may occur in a proposition without being one of the terms of the proposition, though it may also be made into a logical subject. One verb, and one only, must occur as verb in every proposition; but every proposition, by turning its verb into a verbal noun, can be changed into a single logical subject, of a kind which I shall call in future a propositional concept. Every verb, in the logical sense of the word, may be regarded as a relation; when it occurs as verb, it actually relates, but when it occurs as verbal noun it is the bare relation considered independently of the terms which it relates. Verbs do not, like adjectives, have instances, but are identical in all the cases of their occurrence. Owing to the way in which the verb actually relates the terms of a proposition, every proposition has a unity which renders it distinct from the sum of its constituents. All these points lead to logical problems, which, in a treatise on logic, would deserve to be fully and thoroughly discussed.

Having now given a general sketch of the nature of verbs and adjectives, I shall proceed, in the next two chapters, to discussions arising out of the consideration of adjectives, and in Chapter VII to topics connected with verbs. Broadly speaking, classes are connected with adjectives, while propositional functions involve verbs. It is for this reason that it has been necessary to deal at such length with a subject which might seem, at first sight, to be somewhat remote from the principles of mathematics.

Aristotelian Society, 1900—1901) must not be applied to all concepts. The relation of an instance to its universal, at any rate, must be actually and numerically the same in all cases where it occurs.

CHAPTER V.

DENOTING.

56. THE notion of denoting, like most of the notions of logic, has been obscured hitherto by an undue admixture of psychology. There is a sense in which *we* denote, when we point or describe, or employ words as symbols for concepts; this, however, is not the sense that I wish to discuss. But the fact that description is possible—that we are able, by the employment of concepts, to designate a thing which is not a concept —is due to a logical relation between some concepts and some terms, in virtue of which such concepts inherently and logically *denote* such terms. It is this sense of denoting which is here in question. This notion lies at the bottom (I think) of all theories of substance, of the subject-predicate logic, and of the opposition between things and ideas, discursive thought and immediate perception. These various developments, in the main, appear to me mistaken, while the fundamental fact itself, out of which they have grown, is hardly ever discussed in its logical purity.

A concept *denotes* when, if it occurs in a proposition, the proposition is not *about* the concept, but about a term connected in a certain peculiar way with the concept. If I say " I met a man," the proposition is not about *a man*: this is a concept which does not walk the streets, but lives in the shadowy limbo of the logic-books. What I met was a thing, not a concept, an actual man with a tailor and a bank-account or a public-house and a drunken wife. Again, the proposition " any finite number is odd or even " is plainly true; yet the *concept* " any finite number " is neither odd nor even. It is only particular numbers that are odd or even; there is not, in addition to these, another entity, *any number*, which is either odd or even, and if there were, it is plain that it could not be odd and could not be even. Of the *concept* " any number," almost all the propositions that contain the *phrase* " any number " are false. If we wish to speak of the concept, we have to indicate the fact by italics or inverted commas. People often assert that man is mortal; but what is mortal will die, and yet we should be surprised to find in the " Times " such a notice as the following: " Died at his residence of

Camelot, Gladstone Road, Upper Tooting, on the 18th of June 19—, Man, eldest son of Death and Sin." *Man*, in fact, does not die; hence if "man is mortal" were, as it appears to be, a proposition about *man*, it would be simply false. The fact is, the proposition is about men; and here again, it is not about the concept *men*, but about what this concept denotes. The whole theory of definition, of identity, of classes, of symbolism, and of the variable is wrapped up in the theory of denoting. The notion is a fundamental notion of logic, and, in spite of its difficulties, it is quite essential to be as clear about it as possible.

57. The notion of denoting may be obtained by a kind of logical genesis from subject-predicate propositions, upon which it seems more or less dependent. The simplest of propositions are those in which one predicate occurs otherwise than as a term, and there is only one term of which the predicate in question is asserted. Such propositions may be called subject-predicate propositions. Instances are: *A* is, *A* is one, *A* is human. Concepts which are predicates might also be called class-concepts, because they give rise to classes, but we shall find it necessary to distinguish between the words *predicate* and *class-concept*. Propositions of the subject-predicate type always imply and are implied by other propositions of the type which asserts that an individual belongs to a class. Thus the above instances are equivalent to: *A* is an entity, *A* is a unit, *A* is a man. These new propositions are not identical with the previous ones, since they have an entirely different form. To begin with, *is* is now the only concept not used as a term. *A man*, we shall find, is neither a concept nor a term, but a certain kind of combination of certain terms, namely of those which are human. And the relation of Socrates to *a man* is quite different from his relation to humanity; indeed "Socrates is human" must be held, if the above view is correct, to be not, in the most usual sense, a judgment of relation between Socrates and humanity, since this view would make *human* occur as term in "Socrates is human." It is, of course, undeniable that a relation to humanity is implied by "Socrates is human," namely the relation expressed by "Socrates has humanity"; and this relation conversely implies the subject-predicate proposition. But the two propositions can be clearly distinguished, and it is important to the theory of classes that this should be done. Thus we have, in the case of every predicate, three types of propositions which imply one another, namely, "Socrates is human," "Socrates has humanity," and "Socrates is a man." The first contains a term and a predicate, the second two terms and a relation (the second term being identical with the predicate of the first proposition)*, while the third contains a term, a relation, and what I shall call a disjunction (a term which will be explained shortly)†. The class-concept differs little, if at

---

* Cf. § 49.

† There are two allied propositions expressed by the same words, namely "Socrates is a-man" and "Socrates is-a man." The above remarks apply to the

all, from the predicate, while the class, as opposed to the class-concept, is the sum or conjunction of all the terms which have the given predicate. The relation which occurs in the second type (Socrates has humanity) is characterized completely by the fact that it implies and is implied by a proposition with only one term, in which the other term of the relation has become a predicate. A class is a certain combination of terms, a class-concept is closely akin to a predicate, and the terms whose combination forms the class are determined by the class-concept. Predicates are, in a certain sense, the simplest type of concepts, since they occur in the simplest type of proposition.

58. There is, connected with every predicate, a great variety of closely allied concepts, which, in so far as they are distinct, it is important to distinguish. Starting, for example, with *human*, we have man, men, all men, every man, any man, the human race, of which all except the first are twofold, a denoting concept and an object denoted ; we have also, less closely analogous, the notions "a man" and "some man," which again denote objects* other than themselves. This vast apparatus connected with every predicate must be borne in mind, and an endeavour must be made to give an analysis of all the above notions. But for the present, it is the property of denoting, rather than the various denoting concepts, that we are concerned with.

The combination of concepts as such to form new concepts, of greater complexity than their constituents, is a subject upon which writers on logic have said many things. But the combination of terms as such, to form what by analogy may be called complex terms, is a subject upon which logicians, old and new, give us only the scantiest discussion. Nevertheless, the subject is of vital importance to the philosophy of mathematics, since the nature both of number and of the variable turns upon just this point. Six words, of constant occurrence in daily life, are also characteristic of mathematics : these are the words *all*, *every*, *any*, *a*, *some* and *the*. For correctness of reasoning, it is essential that these words should be sharply distinguished one from another; but the subject bristles with difficulties, and is almost wholly neglected by logicians†.

It is plain, to begin with, that a phrase containing one of the above

former ; but in future, unless the contrary is indicated by a hyphen or otherwise, the latter will always be in question. The former expresses the identity of Socrates with an ambiguous individual ; the latter expresses a relation of Socrates to the class-concept *man*.

* I shall use the word *object* in a wider sense than *term*, to cover both singular and plural, and also cases of ambiguity, such as "a man." The fact that a word can be framed with a wider meaning than *term* raises grave logical problems. Cf. § 47.

† On the indefinite article, some good remarks are made by Meinong, "Abstrahiren und Vergleichen," *Zeitschrift für Psychologie und Physiologie der Sinnesorgane*, Vol. xxiv, p. 63.

six words always denotes. It will be convenient, for the present discussion, to distinguish a class-concept from a predicate: I shall call *human* a predicate, and *man* a class-concept, though the distinction is perhaps only verbal. The characteristic of a class-concept, as distinguished from terms in general, is that "*x* is a *u*" is a propositional function when, and only when, *u* is a class-concept. It must be held that when *u* is not a class-concept, we do not have a false proposition, but simply no proposition at all, whatever value we may give to *x*. This enables us to distinguish a class-concept belonging to the null-class, for which all propositions of the above form are false, from a term which is not a class-concept at all, for which there are no propositions of the above form. Also it makes it plain that a class-concept is not a term in the proposition "*x* is a *u*," for *u* has a restricted variability if the formula is to remain a proposition. A denoting phrase, we may now say, consists always of a class-concept preceded by one of the above six words or some synonym of one of them.

59. The question which first meets us in regard to denoting is this: Is there one way of denoting six different kinds of objects, or are the ways of denoting different? And in the latter case, is the object denoted the same in all six cases, or does the object differ as well as the way of denoting it? In order to answer this question, it will be first necessary to explain the differences between the six words in question. Here it will be convenient to omit the word *the* to begin with, since this word is in a different position from the others, and is liable to limitations from which they are exempt.

In cases where the class defined by a class-concept has only a finite number of terms, it is possible to omit the class-concept wholly, and indicate the various objects denoted by enumerating the terms and connecting them by means of *and* or *or* as the case may be. It will help to isolate a part of our problem if we first consider this case, although the lack of subtlety in language renders it difficult to grasp the difference between objects indicated by the same form of words.

Let us begin by considering two terms only, say Brown and Jones. The objects denoted by *all, every, any, a* and *some** are respectively involved in the following five propositions. (1) Brown and Jones are two of Miss Smith's suitors; (2) Brown and Jones are paying court to Miss Smith; (3) if it was Brown or Jones you met, it was a very ardent lover; (4) if it was one of Miss Smith's suitors, it must have been Brown or Jones; (5) Miss Smith will marry Brown or Jones. Although only two forms of words, *Brown and Jones* and *Brown or Jones*, are involved in these propositions, I maintain that five different combinations are involved. The distinctions, some of which are rather subtle, may be

---

* I intend to distinguish between *a* and *some* in a way not warranted by language; the distinction of *all* and *every* is also a straining of usage. Both are necessary to avoid circumlocution.

brought out by the following considerations. In the first proposition, it is Brown *and* Jones who are two, and this is not true of either separately; nevertheless it is not the whole composed of Brown and Jones which is two, for this is only one. The two are a genuine combination of Brown with Jones, the kind of combination which, as we shall see in the next chapter, is characteristic of classes. In the second proposition, on the contrary, what is asserted is true of Brown and Jones severally; the proposition is equivalent to, though not (I think) identical with, "Brown is paying court to Miss Smith and Jones is paying court to Miss Smith." Thus the combination indicated by *and* is not the same here as in the first case: the first case concerned *all* of them collectively, while the second concerns *all* distributively, *i.e.* each or every one of them. For the sake of distinction, we may call the first a *numerical* conjunction, since it gives rise to number, the second a *propositional* conjunction, since the proposition in which it occurs is equivalent to a conjunction of propositions. (It should be observed that the conjunction of propositions in question is of a wholly different kind from any of the combinations we are considering, being in fact of the kind which is called the logical product. The propositions are combined *quâ* propositions, not *quâ* terms.)

The third proposition gives the kind of conjunction by which *any* is defined. There is some difficulty about this notion, which seems half-way between a conjunction and a disjunction. This notion may be further explained as follows. Let *a* and *b* be two different propositions, each of which implies a third proposition *c*. Then the disjunction "*a* or *b*" implies *c*. Now let *a* and *b* be propositions assigning the same predicate to two different subjects, then there is a combination of the two subjects to which the given predicate may be assigned so that the resulting proposition is equivalent to the disjunction "*a* or *b*." Thus suppose we have "if you met Brown, you met a very ardent lover," and "if you met Jones, you met a very ardent lover." Hence we infer "if you met Brown or if you met Jones, you met a very ardent lover," and we regard this as equivalent to "if you met Brown or Jones, etc." The combination of Brown and Jones here indicated is the same as that indicated by *either* of them. It differs from a disjunction by the fact that it implies and is implied by a statement concerning *both*; but in some more complicated instances, this mutual implication fails. The method of combination is, in fact, different from that indicated by *both*, and is also different from both forms of disjunction. I shall call it the *variable* conjunction. The first form of disjunction is given by (4): this is the form which I shall denote by *a* suitor. Here, although it must have been Brown or Jones, it is not true that it must have been Brown, nor yet that it must have been Jones. Thus the proposition is not equivalent to the disjunction of propositions "it must have been Brown or it must have been Jones." The proposition, in fact, is not capable of statement either as a disjunction or as a conjunction of propositions, except in the very roundabout form : "if it was not Brown, it was Jones, and if it was not Jones, it was Brown," a form which rapidly becomes intolerable when the number of terms is increased beyond two, and becomes theoretically inadmissible when the number of terms is infinite. Thus this form of disjunction denotes a variable term, that is, whichever of the two terms we fix upon, it does not denote this term, and yet it does denote one or other of them. This form accordingly I shall call the *variable* disjunction. Finally, the second form of disjunction is given by (5). This is what I shall call the *constant* disjunction, since here either Brown is denoted, or Jones is denoted, but the alternative is undecided. That is to say, our proposition is now equivalent to a disjunction of propositions, namely "Miss Smith will marry Brown, or she will marry Jones." She will marry *some* one of the two, and the disjunction denotes a particular one of them, though it may denote either particular one. Thus all the five combinations are distinct.

It is to be observed that these five combinations yield neither terms nor concepts, but strictly and only combinations of terms. The first yields many terms, while the others yield something absolutely peculiar, which is neither one nor many. The combinations are combinations of terms, effected without the use of relations. Corresponding to each combination there is, at least if the terms combined form a class, a perfectly definite concept, which *denotes* the various terms of the combination combined in the specified manner. To explain this, let us repeat our distinctions in a case where the terms to be combined are not enumerated, as above, but are defined as the terms of a certain class.

60.   When a class-concept *a* is given, it must be held that the various terms belonging to the class are also given. That is to say, any term being proposed, it can be decided whether or not it belongs to the class. In this way, a collection of terms can be given otherwise than by enumeration. Whether a collection can be given otherwise than by enumeration or by a class-concept, is a question which, for the present, I leave undetermined. But the possibility of giving a collection by a class-concept is highly important, since it enables us to deal with infinite collections, as we shall see in Part V. For the present, I wish to examine the meaning of such phrases as *all a's, every a, any a, an a*, and *some a*. *All a's*, to begin with, denotes a numerical conjunction; it is definite as soon as *a* is given. The concept *all a's* is a perfectly definite single concept, which denotes the terms of *a* taken all together. The terms so taken have a number, which may thus be regarded, if we choose, as a property of the class-concept, since it is determinate for any given class-concept. *Every a*, on the contrary, though it still denotes all the *a's*, denotes them in a different way, *i.e.* severally instead of collectively. *Any a* denotes only one *a*, but it is wholly irrelevant which it denotes, and what is said will be equally true whichever it may be. Moreover,

any a denotes a variable *a*, that is, whatever particular *a* we may fasten upon, it is certain that *any a* does not denote that one; and yet of that one any proposition is true which is true of any *a*. *An a* denotes a variable disjunction: that is to say, a proposition which holds of *an a* may be false concerning each particular *a*, so that it is not reducible to a disjunction of propositions. For example, a point lies between any point and any other point; but it would not be true of any one particular point that it lay between any point and any other point. since there would be many pairs of points between which it did not lie. This brings us finally to *some a*, the constant disjunction. This denotes just one term of the class *a*, but the term it denotes may be any term of the class. Thus "some moment does not follow any moment" would mean that there was a first moment in time, while "a moment precedes any moment" means the exact opposite, namely, that every moment has predecessors.

61. In the case of a class *a* which has a finite number of terms—say $a_1, a_2, a_3, \ldots a_n$, we can illustrate these various notions as follows:

(1) *All a*'s denotes a_1 and a_2 and ... and a_n.

(2) *Every a* denotes a_1 and denotes a_2 and ... and denotes a_n.

(3) *Any a* denotes a_1 or a_2 or ... or a_n, where *or* has the meaning that it is irrelevant which we take.

(4) *An a* denotes a_1 or a_2 or ... or a_n, where *or* has the meaning that no one in particular must be taken, just as in *all a*'s we must not take any one in particular.

(5) *Some a* denotes a_1 or denotes a_2 or ... or denotes a_n, where it is not irrelevant which is taken, but on the contrary some one particular *a* must be taken.

As the nature and properties of the various ways of combining terms are of vital importance to the principles of mathematics, it may be well to illustrate their properties by the following important examples.

(*a*) Let *a* be a class, and *b* a class of classes. We then obtain in all six possible relations of *a* to *b* from various combinations of *any*, *a* and *some*. *All* and *every* do not, in this case, introduce anything new. The six cases are as follows.

(1) Any *a* belongs to any class belonging to *b*, in other words, the class *a* is wholly contained in the common part or logical product of the various classes belonging to *b*.

(2) Any *a* belongs to a *b*, *i.e.* the class *a* is contained in any class which contains all the *b*'s, or, is contained in the logical sum of all the *b*'s.

(3) Any *a* belongs to some *b*, *i.e.* there is a class belonging to *b*, in which the class *a* is contained. The difference between this case and the second arises from the fact that here there is one *b* to which every *a* belongs, whereas before it was only decided that every *a* belonged to a *b*, and different *a*'s might belong to different *b*'s.

(4) An *a* belongs to any *b*, *i.e.* whatever *b* we take, it has a part in common with *a*.

(5) An *a* belongs to a *b*, *i.e.* there is a *b* which has a part in common with *a*. This is equivalent to " some (or an) *a* belongs to some *b*."

(6) Some *a* belongs to any *b*, *i.e.* there is an *a* which belongs to the common part of all the *b*'s, or *a* and all the *b*'s have a common part. These are all the cases that arise here.

(β) It is instructive, as showing the generality of the type of relations here considered, to compare the above case with the following. Let *a*, *b* be two series of real numbers; then six precisely analogous cases arise.

(1) Any *a* is less than any *b*, or, the series *a* is contained among numbers less than every *b*.

(2) Any *a* is less than a *b*, or, whatever *a* we take, there is a *b* which is greater, or, the series *a* is contained among numbers less than a (variable) term of the series *b*. It does not follow that some term of the series *b* is greater than all the *a*'s.

(3) Any *a* is less than some *b*, or, there is a term of *b* which is greater than all the *a*'s. This case is not to be confounded with (2).

(4) An *a* is less than any *b*, *i.e.* whatever *b* we take, there is an *a* which is less than it.

(5) An *a* is less than a *b*, *i.e.* it is possible to find an *a* and a *b* such that the *a* is less than the *b*. This merely denies that any *a* is greater than any *b*.

(6) Some *a* is less than any *b*, *i.e.* there is an *a* which is less than all the *b*'s. This was not implied in (4), where the *a* was variable, whereas here it is constant.

In this case, actual mathematics have compelled the distinction between the variable and the constant disjunction. But in other cases, where mathematics have not obtained sway, the distinction has been neglected; and the mathematicians, as was natural, have not investigated the logical nature of the disjunctive notions which they employed.

(γ) I shall give one other instance, as it brings in the difference between *any* and *every*, which has not been relevant in the previous cases. Let *a* and *b* be two classes of classes; then twenty different relations between them arise from different combinations of the terms of their terms. The following technical terms will be useful. If *a* be a class of classes, its logical sum consists of all terms belonging to any *a*, *i.e.* all terms such that there is an *a* to which they belong, while its logical product consists of all terms belonging to every *a*, *i.e.* to the common part of all the *a*'s. We have then the following cases.

(1) Any term of any *a* belongs to every *b*, *i.e.* the logical sum of *a* is contained in the logical product of *b*.

(2) Any term of any *a* belongs to a *b*, *i.e.* the logical sum of *a* is contained in the logical sum of *b*.

(3) Any term of any *a* belongs to some *b, i.e.* there is a *b* which contains the logical sum of *a*.

(4) Any term of some (or an) *a* belongs to every *b, i.e.* there is an *a* which is contained in the product of *b*.

(5) Any term of some (or an) *a* belongs to a *b, i.e.* there is an *a* which is contained in the sum of *b*.

(6) Any term of some (or an) *a* belongs to some *b, i.e.* there is a *b* which contains one class belonging to *a*.

(7) A term of any *a* belongs to any *b, i.e.* any class of *a* and any class of *b* have a common part.

(8) A term of any *a* belongs to a *b, i.e.* any class of *a* has a part in common with the logical sum of *b*.

(9) A term of any *a* belongs to some *b, i.e.* there is a *b* with which any *a* has a part in common.

(10) A term of an *a* belongs to every *b, i.e.* the logical sum of *a* and the logical product of *b* have a common part.

(11) A term of an *a* belongs to any *b, i.e.* given any *b*, an *a* can be found with which it has a common part.

(12) A term of an *a* belongs to a *b, i.e.* the logical sums of *a* and of *b* have a common part.

(13) Any term of every *a* belongs to every *b, i.e.* the logical product of *a* is contained in the logical product of *b*.

(14) Any term of every *a* belongs to a *b, i.e.* the logical product of *a* is contained in the logical sum of *b*.

(15) Any term of every *a* belongs to some *b, i.e.* there is a term of *b* in which the logical product of *a* is contained.

(16) A (or some) term of every *a* belongs to every *b, i.e.* the logical products of *a* and of *b* have a common part.

(17) A (or some) term of every *a* belongs to a *b, i.e.* the logical product of *a* and the logical sum of *b* have a common part.

(18) Some term of any *a* belongs to every *b, i.e.* any *a* has a part in common with the logical product of *b*.

(19) A term of some *a* belongs to any *b, i.e.* there is some term of *a* with which any *b* has a common part.

(20) A term of every *a* belongs to any *b, i.e.* any *b* has a part in common with the logical product of *a*.

The above examples show that, although it may often happen that there is a mutual implication (which has not always been stated) of corresponding propositions concerning *some* and *a*, or concerning *any* and *every*, yet in other cases there is no such mutual implication. Thus the five notions discussed in the present chapter are genuinely distinct, and to confound them may lead to perfectly definite fallacies.

62. It appears from the above discussion that, whether there are different ways of denoting or not, the objects denoted by *all men, every man*, etc. are certainly distinct. It seems therefore legitimate to say that the whole difference lies in the objects, and that denoting itself is the same in all cases. There are, however, many difficult problems connected with the subject, especially as regards the nature of the objects denoted. *All men*, which I shall identify with the class of men, seems to be an unambiguous object, although grammatically it is plural. But in the other cases the question is not so simple: we may doubt whether an ambiguous object is unambiguously denoted, or a definite object ambiguously denoted. Consider again the proposition "I met a man." It is quite certain, and is implied by this proposition, that what I met was an unambiguous perfectly definite man: in the technical language which is here adopted, the proposition is expressed by "I met some man." But the actual man whom I met forms no part of the proposition in question, and is not specially denoted by *some man*. Thus the concrete event which happened is not asserted in the proposition. What is asserted is merely that some one of a class of concrete events took place. The whole human race is involved in my assertion: if any man who ever existed or will exist had not existed or been going to exist, the purport of my proposition would have been different. Or, to put the same point in more intensional language, if I substitute for *man* any of the other class-concepts applicable to the individual whom I had the honour to meet, my proposition is changed, although the individual in question is just as much denoted as before. What this proves is, that *some man* must not be regarded as actually denoting Smith and actually denoting Brown, and so on: the whole procession of human beings throughout the ages is always relevant to every proposition in which *some man* occurs, and what is denoted is essentially not each separate man, but a kind of combination of all men. This is more evident in the case of *every*, *any*, and *a*. There is, then, a definite something, different in each of the five cases, which must, in a sense, be an object, but is characterized as a set of terms combined in a certain way, which something is denoted by *all men*, *every man*, *any man*, *a man* or *some man*; and it is with this very paradoxical object that propositions are concerned in which the corresponding concept is used as denoting.

63. It remains to discuss the notion of *the*. This notion has been symbolically emphasized by Peano, with very great advantage to his calculus; but here it is to be discussed philosophically. The use of identity and the theory of definition are dependent upon this notion, which has thus the very highest philosophical importance.

The word *the*, in the singular, is correctly employed only in relation to a class-concept of which there is only one instance. We speak of *the* King, *the* Prime Minister, and so on (understanding *at the present time*); and in such cases there is a method of denoting one single definite term by means of a concept, which is not given us by any of our other five words. It is owing to this notion that mathematics can give definitions of terms which are not concepts—a possibility which illustrates the difference between mathematical and philosophical definition. Every term is the only instance of *some* class-concept, and thus every term, theoretically, is capable of definition, provided we have not adopted a system in which the said term is one of our indefinables. It is a curious paradox, puzzling to the symbolic mind, that definitions, theoretically, are nothing but statements of symbolic abbreviations, irrelevant to the reasoning and inserted only for practical convenience, while yet, in the development of a subject, they always require a very large amount of thought, and often embody some of the greatest achievements of analysis. This fact seems to be explained by the theory of denoting. An object may be present to the mind, without our knowing any concept of which the said object is *the* instance; and the discovery of such a concept is not a mere improvement in notation. The reason why this appears to be the case is that, as soon as the definition is found, it becomes wholly unnecessary to the reasoning to remember the actual object defined, since only concepts are relevant to our deductions. In the moment of discovery, the definition is seen to be *true*, because the object to be defined was already in our thoughts; but as part of our reasoning it is not true, but merely symbolic, since what the reasoning requires is not that it should deal with *that* object, but merely that it should deal with the object denoted by the definition.

In most actual definitions of mathematics, what is defined is a *class* of entities, and the notion of *the* does not then explicitly appear. But even in this case, what is really defined is *the* class satisfying certain conditions; for a class, as we shall see in the next chapter, is always a term or conjunction of terms and never a concept. Thus the notion of *the* is always relevant in definitions; and we may observe generally that the adequacy of concepts to deal with things is wholly dependent upon the unambiguous denoting of a single term which this notion gives.

64. The connection of denoting with the nature of identity is important, and helps, I think, to solve some rather serious problems. The question whether identity is or is not a relation, and even whether there is such a concept at all, is not easy to answer. For, it may be said, identity cannot be a relation, since, where it is truly asserted, we have only one term, whereas two terms are required for a relation. And indeed identity, an objector may urge, cannot be anything at all: two terms plainly are not identical, and one term cannot be, for what is it identical with? Nevertheless identity must be something. We might attempt to remove identity from terms to relations, and say that two terms are identical in some respect when they have a given relation to a given term. But then we shall have to hold either that there is strict identity between the two cases of the given relation, or that the two cases have identity in the sense of having a given relation to a given term; but the latter view leads to an endless process of the illegitimate kind. Thus identity must be admitted, and the difficulty as to the two terms of a relation must be met by a sheer denial that two different terms are necessary. There must always be a referent and a relatum, but these need not be distinct; and where identity is affirmed, they are not so*.

But the question arises: Why is it ever worth while to affirm identity? This question is answered by the theory of denoting. If we say " Edward VII is the King," we assert an identity; the reason why this assertion is worth making is, that in the one case the actual term occurs, while in the other a denoting concept takes its place. (For purposes of discussion, I ignore the fact that Edwards form a class, and that seventh Edwards form a class having only one term. Edward VII is practically, though not formally, a proper name.) Often two denoting concepts occur, and the term itself is not mentioned, as in the proposition " the present Pope is the last survivor of his generation." When a term is given, the assertion of its identity with itself, though true, is perfectly futile, and is never made outside the logic-books ; but where denoting concepts are introduced, identity is at once seen to be significant. In this case, of course, there is involved, though not asserted, a relation of the denoting concept to the term, or of the two denoting concepts to each other. But the *is* which occurs in such propositions does not itself state this further relation, but states pure identity†.

65. To sum up. When a class-concept, preceded by one of the six words *all, every, any, a, some, the,* occurs in a proposition, the proposition is, as a rule, not *about* the concept formed of the two words together, but about an object quite different from this, in general not a concept at all, but a term or complex of terms. This may be seen by the fact that propositions in which such concepts occur are in general false concerning the concepts themselves. At the same time, it is possible to consider and make propositions about the concepts themselves, but these are not the natural propositions to make in employing the concepts. " Any number is odd or even " is a perfectly natural proposition, whereas "*Any number* is a variable conjunction" is a proposition only to be made in a logical discussion. In such cases, we say that the concept in question *denotes*. We decided that denoting is a perfectly

---

* On relations of terms to themselves, *v. inf.* Chap. ix, § 95.

† The word *is* is terribly ambiguous, and great care is necessary in order not to confound its various meanings. We have (1) the sense in which it asserts Being, as in "*A* is"; (2) the sense of identity; (3) the sense of predication, in "*A* is human"; (4) the sense of "*A* is a-man" (cf. p. 54, note), which is very like identity. In addition to these there are less common uses, as "to be good is to be happy," where a relation of assertions is meant, that relation, in fact, which, where it exists, gives rise to formal implication. Doubtless there are further meanings which have not occurred to me. On the meanings of *is*, cf. De Morgan, *Formal Logic*, pp. 49, 50.

definite relation, the same in all six cases, and that it is the nature of the denoted object and the denoting concept which distinguishes the cases. We discussed at some length the nature and the differences of the denoted objects in the five cases in which these objects are combinations of terms. In a full discussion, it would be necessary also to discuss the denoting concepts: the actual meanings of these concepts, as opposed to the nature of the objects they denote, have not been discussed above. But I do not know that there would be anything further to say on this topic. Finally, we discussed *the*, and showed that this notion is essential to what mathematics calls definition, as well as to the possibility of uniquely determining a term by means of concepts; the actual use of identity, though not its meaning, was also found to depend upon this way of denoting a single term. From this point we can advance to the discussion of classes, thereby continuing the development of the topics connected with adjectives.

CHAPTER VI.

CLASSES.

66. To bring clearly before the mind what is meant by *class*, and to distinguish this notion from all the notions to which it is allied, is one of the most difficult and important problems of mathematical philosophy. Apart from the fact that *class* is a very fundamental concept, the utmost care and nicety is required in this subject on account of the contradiction to be discussed in Chapter x. I must ask the reader, therefore, not to regard as idle pedantry the apparatus of somewhat subtle discriminations to be found in what follows.

It has been customary, in works on logic, to distinguish two standpoints, that of extension and that of intension. Philosophers have usually regarded the latter as more fundamental, while Mathematics has been held to deal specially with the former. M. Couturat, in his admirable work on Leibniz, states roundly that Symbolic Logic can only be built up from the standpoint of extension*; and if there really were only these two points of view, his statement would be justified. But as a matter of fact, there are positions intermediate between pure intension and pure extension, and it is in these intermediate regions that Symbolic Logic has its lair. It is essential that the classes with which we are concerned should be composed of terms, and should not be predicates or concepts, for a class must be definite when its terms are given, but in general there will be many predicates which attach to the given terms and to no others. We cannot of course attempt an intensional definition of a class as the class of predicates attaching to the terms in question and to no others, for this would involve a vicious circle; hence the point of view of extension is to some extent unavoidable. On the other hand, if we take extension pure, our class is defined by enumeration of its terms, and this method will not allow us to deal, as Symbolic Logic does, with infinite classes. Thus our classes must in general be regarded as objects denoted by concepts, and to this extent the point of view of intension is essential. It is owing to this con-

* *La Logique de Leibniz*, Paris, 1901, p. 387.

sideration that the theory of denoting is of such great importance. In the present chapter we have to specify the precise degree in which extension and intension respectively enter into the definition and employment of classes; and throughout the discussion, I must ask the reader to remember that whatever is said has to be applicable to infinite as well as to finite classes.

67. When an object is unambiguously denoted by a concept, I shall speak of the concept as a concept (or sometimes, loosely, as *the* concept) of the object in question. Thus it will be necessary to distinguish the concept of a class from a class-concept. We agreed to call *man* a class-concept, but *man* does not, in its usual employment, denote anything. On the other hand, *men* and *all men* (which I shall regard as synonyms) do denote, and I shall contend that what they denote is the class composed of all men. Thus *man* is the class-concept, *men* (the concept) is the concept of the class, and men (the object denoted by the concept *men*) are the class. It is no doubt confusing, at first, to use *class-concept* and *concept of a class* in different senses; but so many distinctions are required that some straining of language seems unavoidable. In the phraseology of the preceding chapter, we may say that a class is a numerical conjunction of terms. This is the thesis which is to be established.

68. In Chapter II we regarded classes as derived from assertions, *i.e.* as all the entities satisfying some assertion, whose form was left wholly vague. I shall discuss this view critically in the next chapter; for the present, we may confine ourselves to classes as they are derived from predicates, leaving open the question whether every assertion is equivalent to a predication. We may, then, imagine a kind of genesis of classes, through the successive stages indicated by the typical propositions "Socrates is human," "Socrates has humanity," "Socrates is a man," "Socrates is one among men." Of these propositions, the last only, we should say, explicitly contains the class as a constituent; but every subject-predicate proposition gives rise to the other three equivalent propositions, and thus every predicate (provided it can be sometimes truly predicated) gives rise to a class. This is the genesis of classes from the intensional standpoint.

On the other hand, when mathematicians deal with what they call a manifold, aggregate, *Menge, ensemble,* or some equivalent name, it is common, especially where the number of terms involved is finite, to regard the object in question (which is in fact a class) as defined by the enumeration of its terms, and as consisting possibly of a single term, which in that case *is* the class. Here it is not predicates and denoting that are relevant, but terms connected by the word *and,* in the sense in which this word stands for a *numerical* conjunction. Thus Brown and Jones are a class, and Brown singly is a class. This is the extensional genesis of classes.

69. The best formal treatment of classes in existence is that of Peano*. But in this treatment a number of distinctions of great philosophical importance are overlooked. Peano, not I think quite consciously, identifies the class with the class-concept; thus the relation of an individual to its class is, for him, expressed by *is a*. For him, "2 is a number" is a proposition in which a term is said to belong to the class *number*. Nevertheless, he identifies the equality of classes, which consists in their having the same terms, with identity—a proceeding which is quite illegitimate when the class is regarded as the class-concept. In order to perceive that *man* and *featherless biped* are not identical, it is quite unnecessary to take a hen and deprive the poor bird of its feathers. Or, to take a less complex instance, it is plain that *even prime* is not identical with *integer next after* 1. Thus when we identify the class with the class-concept, we must admit that two classes may be equal without being identical. Nevertheless, it is plain that when two class-concepts are equal, some identity is involved, for we say that they have the *same* terms. Thus there is some object which is positively identical when two class-concepts are equal; and this object, it would seem, is more properly called the class. Neglecting the plucked hen, the class of featherless bipeds, every one would say, is the *same* as the class of men; the class of even primes is the *same* as the class of integers next after 1. Thus we must not identify the class with the class-concept, or regard "Socrates is a man" as expressing the relation of an individual to a class of which it is a member. This has two consequences (to be established presently) which prevent the philosophical acceptance of certain points in Peano's formalism. The first consequence is, that there is no such thing as the null-class, though there are null class-concepts. The second is, that a class having only one term is to be identified, contrary to Peano's usage, with that one term. I should not propose, however, to alter his practice or his notation in consequence of either of these points; rather I should regard them as proofs that Symbolic Logic ought to concern itself, as far as notation goes, with class-concepts rather than with classes.

70. A class, we have seen, is neither a predicate nor a class-concept, for different predicates and different class-concepts may correspond to the same class. A class also, in one sense at least, is distinct from the whole composed of its terms, for the latter is only and essentially one, while the former, where it has many terms, is, as we shall see later, the very kind of object of which *many* is to be asserted. The distinction of a class as many from a class as a whole is often made by language: space and points, time and instants, the army and the soldiers, the navy and the sailors, the Cabinet and the Cabinet Ministers, all illustrate the distinction. The notion of a whole, in the sense of a pure aggregate

---

* Neglecting Frege, who is discussed in the Appendix.

which is here relevant, is, we shall find, not always applicable where the notion of the class as many applies (see Chapter x). In such cases, though terms may be said to belong to the class, the class must not be treated as itself a single logical subject*. But this case never arises where a class can be generated by a predicate. Thus we may for the present dismiss this complication from our minds. In a class as many, the component terms, though they have some kind of unity, have less than is required for a whole. They have, in fact, just so much unity as is required to make them many, and not enough to prevent them from remaining many. A further reason for distinguishing wholes from classes as many is that a class as one may be one of the terms of itself as many, as in " classes are one among classes " (the extensional equivalent of " class is a class-concept "), whereas a complex whole can never be one of its own constituents.

71. *Class* may be defined either extensionally or intensionally. That is to say, we may define the kind of object which is a class, or the kind of concept which denotes a class: this is the precise meaning of the opposition of extension and intension in this connection. But although the general notion can be defined in this two-fold manner, particular classes, except when they happen to be finite, can only be defined intensionally, *i.e.* as the objects denoted by such and such concepts. I believe this distinction to be purely psychological: logically, the extensional definition appears to be equally applicable to infinite classes, but practically, if we were to attempt it, Death would cut short our laudable endeavour before it had attained its goal. Logically, therefore, extension and intension seem to be on a par. I will begin with the extensional view.

When a class is regarded as defined by the enumeration of its terms, it is more naturally called a *collection*. I shall for the moment adopt this name, as it will not prejudge the question whether the objects denoted by it are truly classes or not. By a collection I mean what is conveyed by "*A* and *B*" or "*A* and *B* and *C*," or any other enumeration of definite terms. The collection is defined by the actual mention of the terms, and the terms are connected by *and*. It would seem that *and* represents a fundamental way of combining terms, and that just this way of combination is essential if anything is to result of which a number other than 1 can be asserted. Collections do not presuppose numbers, since they result simply from the terms together with *and*: they could only presuppose numbers in the particular case where the terms of the collection themselves presupposed numbers. There is a grammatical difficulty which, since no method exists of avoiding it, must be pointed out and allowed for. A collection, grammatically, is

---

* A plurality of terms is not the logical subject when a number is asserted of it : such propositions have not one subject, but many subjects. See end of § 74.

singular, whereas A and B, A and B and C, etc. are essentially plural. This grammatical difficulty arises from the logical fact (to be discussed presently) that whatever is many in general forms a whole which is one; it is, therefore, not removable by a better choice of technical terms.

The notion of *and* was brought into prominence by Bolzano*. In order to understand what infinity is, he says, " we must go back to one of the simplest conceptions of our understanding, in order to reach an agreement concerning the word that we are to use to denote it. This is the conception which underlies the conjunction *and*, which, however, if it is to stand out as clearly as is required, in many cases, both by the purposes of mathematics and by those of philosophy, I believe to be best expressed by the words : ' A system (*Inbegriff*) of certain things,' or ' a whole consisting of certain parts.' But we must add that every arbitrary object A can be combined in a system with any others B, C, D, ..., or (speaking still more correctly) already forms a system by itself †, of which some more or less important truth can be enunciated, provided only that each of the presentations A, B, C, D, ... in fact represents a *different* object, or in so far as none of the propositions ' A is the same as B,' ' A is the same as C,' ' A is the same as D,' etc., is true. For if *e.g.* A is the same as B, then it is certainly unreasonable to speak of a system of the things A and B."

The above passage, good as it is, neglects several distinctions which we have found necessary. First and foremost, it does not distinguish the many from the whole which they form. Secondly, it does not appear to observe that the method of enumeration is not practically applicable to infinite systems. Thirdly, and this is connected with the second point, it does not make any mention of intensional definition nor of the notion of a class. What we have to consider is the difference, if any, of a class from a collection on the one hand, and from the whole formed of the collection on the other. But let us first examine further the notion of *and*.

Anything of which a finite number other than 0 or 1 can be asserted would be commonly said to be many, and many, it might be said, are always of the form " A and B and C and ...." Here A, B, C, ... are each one and are all different. To say that A is one seems to amount to much the same as to say that A is not of the form " A_1 and A_2 and A_3 and ...." To say that A, B, C, ... are all different seems to amount only to a condition as regards the symbols: it should be held that " A and A " is meaningless, so that diversity is implied by *and*, and need not be specially stated.

A term A which is one may be regarded as a particular case of a

---

* *Paradoxien des Unendlichen*, Leipzig, 1854 (2nd ed., Berlin, 1889), § 3.
† *i.e.* the combination of A with B, C, D, ... already forms a system.

collection, namely as a collection of one term. Thus every collection which is many presupposes many collections which are each one : *A and B* presupposes *A* and presupposes *B*. Conversely some collections of one term presuppose many, namely those which are complex : thus "*A* differs from *B*" is one, but presupposes *A and difference and B*. But there is not symmetry in this respect, for the ultimate presuppositions of anything are always simple terms.

Every pair of terms, without exception, can be combined in the manner indicated by *A and B*, and if neither *A* nor *B* be many, then *A* and *B* are two. *A* and *B* may be any conceivable entities, any possible objects of thought, they may be points or numbers or true or false propositions or events or people, in short anything that can be counted. A teaspoon and the number 3, or a chimaera and a four-dimensional space, are certainly two. Thus no restriction whatever is to be placed on *A* and *B*, except that neither is to be many. It should be observed that *A* and *B* need not exist, but must, like anything that can be mentioned, have Being. The distinction of Being and existence is important, and is well illustrated by the process of counting. What can be counted must be something, and must certainly *be*, though it need by no means be possessed of the further privilege of existence. Thus what we demand of the terms of our collection is merely that each should be an entity.

The question may now be asked : What is meant by *A and B* ? Does this mean anything more than the juxtaposition of *A* with *B* ? That is, does it contain any element over and above that of *A* and that of *B* ? Is *and* a separate concept, which occurs besides *A*, *B* ? To either answer there are objections. In the first place, *and*, we might suppose, cannot be a new concept, for if it were, it would have to be some kind of relation between *A* and *B* ; *A and B* would then be a proposition, or at least a propositional concept, and would be one, not two. Moreover, if there are two concepts, there *are* two, and no third mediating concept seems necessary to make them two. Thus *and* would seem meaningless. But it is difficult to maintain this theory. To begin with, it seems rash to hold that any word is meaningless. When we use the word *and*, we do not seem to be uttering mere idle breath, but some idea seems to correspond to the word. Again some kind of combination seems to be implied by the fact that *A and B* are two, which is not true of either separately. When we say "*A* and *B* are yellow," we can replace the proposition by "*A* is yellow" and "*B* is yellow"; but this cannot be done for "*A* and *B* are two"; on the contrary, *A* is *one* and *B* is *one*. Thus it seems best to regard *and* as expressing a definite unique kind of combination, not a relation, and not combining *A* and *B* into a whole, which would be one. This unique kind of combination will in future be called *addition of individuals*. It is important to observe that it applies to terms, and only applies to numbers in consequence of their being terms. Thus for the present, 1 and 2 are two, and 1 and 1 is meaningless.

As regards what is meant by the combination indicated by *and*, it is indistinguishable from what we before called a numerical conjunction. That is, *A and B* is what is denoted by the concept of a class of which *A* and *B* are the only members. If *u* be a class-concept of which the propositions " *A* is a *u* " " *B* is a *u* " are true, but of which all other propositions of the same form are false, then " all *u*'s " is the concept of a class whose only terms are *A* and *B* ; this concept *denotes* the terms *A*, *B* combined in a certain way, and " *A* and *B* " *are* those terms combined in just that way. Thus " *A* and *B* " are the class, but are distinct from the class-concept and from the concept of the class.

The notion of *and*, however, does not enter into the *meaning* of a class, for a single term is a class, although it is not a numerical conjunction. If *u* be a class-concept, and only one proposition of the form " *x* is a *u* " be true, then " all *u*'s " is a concept denoting a single term, and this term is the class of which " all *u*'s " is a concept. Thus what seems essential to a class is not the notion of *and*, but the being denoted by some concept of a class. This brings us to the intensional view of classes.

72. We agreed in the preceding chapter that there are not different ways of denoting, but only different kinds of denoting concepts and correspondingly different kinds of denoted objects. We have discussed the kind of denoted object which constitutes a class ; we have now to consider the kind of denoting concept.

The consideration of classes which results from denoting concepts is more general than the extensional consideration, and that in two respects. In the first place it allows, what the other *practically* excludes, the admission of infinite classes ; in the second place it introduces the null concept of a class. But, before discussing these matters, there is a purely logical point of some importance to be examined.

If *u* be a class-concept, is the concept " all *u*'s " analyzable into two constituents, *all* and *u*, or is it a new concept, defined by a certain relation to *u*, and no more complex than *u* itself? We may observe, to begin with, that " all *u*'s " is synonymous with " *u*'s," at least according to a very common use of the plural. Our question is, then, as to the meaning of the plural. The word *all* has certainly some definite meaning, but it seems highly doubtful whether it means more than the indication of a relation. " All men " and " all numbers " have in common the fact that they both have a certain relation to a class-concept, namely to *man* and *number* respectively. But it is very difficult to isolate any further element of *all-ness* which both share, unless we take as this element the mere fact that both are concepts of classes. It would seem, then, that " all *u*'s " is not validly analyzable into *all* and *u*, and that language, in this case as in some others, is a misleading guide. The same remark will apply to *every, any, some, a,* and *the.*

It might perhaps be thought that a class ought to be considered, not merely as a numerical conjunction of terms, but as a numerical conjunction denoted by the concept of a class. This complication, however, would serve no useful purpose, except to preserve Peano's distinction between a single term and the class whose only term it is— a distinction which is easy to grasp when the class is identified with the class-concept, but which is inadmissible in our view of classes. It is evident that a numerical conjunction considered as denoted is either the same entity as when not so considered, or else is a complex of denoting together with the object denoted; and the object denoted is plainly what we mean by a class.

With regard to infinite classes, say the class of numbers, it is to be observed that the concept *all numbers,* though not itself infinitely complex, yet denotes an infinitely complex object. This is the inmost secret of our power to deal with infinity. An infinitely complex concept, though there may be such, can certainly not be manipulated by the human intelligence; but infinite collections, owing to the notion of denoting, can be manipulated without introducing any concepts of infinite complexity. Throughout the discussions of infinity in later Parts of the present work, this remark should be borne in mind: if it is forgotten, there is an air of magic which causes the results obtained to seem doubtful.

73. Great difficulties are associated with the null-class, and generally with the idea of *nothing.* It is plain that there is such a concept as *nothing,* and that in some sense nothing is something. In fact, the proposition "nothing is not nothing" is undoubtedly capable of an interpretation which makes it true—a point which gives rise to the contradictions discussed in Plato's *Sophist.* In Symbolic Logic the null-class is the class which has no terms at all; and symbolically it is quite necessary to introduce some such notion. We have to consider whether the contradictions which naturally arise can be avoided.

It is necessary to realize, in the first place, that a concept may denote although it does not denote anything. This occurs when there are propositions in which the said concept occurs, and which are not about the said concept, but all such propositions are false. Or rather, the above is a first step towards the explanation of a denoting concept which denotes nothing. It is not, however, an adequate explanation. Consider, for example, the proposition "chimaeras are animals" or "even primes other than 2 are numbers." These propositions appear to be true, and it would seem that they are not concerned with the denoting concepts, but with what these concepts denote; yet that is impossible, for the concepts in question do not denote anything.

Symbolic Logic says that these concepts denote the null-class, and that the propositions in question assert that the null-class is contained in certain other classes. But with the strictly extensional view of classes propounded above, a class which has no terms fails to be anything at all : what is merely and solely a collection of terms cannot subsist when all the terms are removed. Thus we must either find a different interpretation of classes, or else find a method of dispensing with the null-class.

The above imperfect definition of a concept which denotes, but does not denote anything, may be amended as follows. All denoting concepts, as we saw, are derived from class-concepts; and a is a class-concept when "x is an a" is a propositional function. The denoting concepts associated with a will not denote anything when and only when "x is an a" is false for all values of x. This is a complete definition of a denoting concept which does not denote anything; and in this case we shall say that a is a null class-concept, and that "all a's" is a null concept of a class. Thus for a system such as Peano's, in which what are called classes are really class-concepts, technical difficulties need not arise ; but for us a genuine logical problem remains.

The proposition " chimaeras are animals " may be easily interpreted by means of formal implication, as meaning "x is a chimaera implies x is an animal for all values of x." But in dealing with classes we have been assuming that propositions containing *all* or *any* or *every*, though equivalent to formal implications, were yet distinct from them, and involved ideas requiring independent treatment. Now in the case of chimaeras, it is easy to substitute the pure intensional view, according to which what is really stated is a relation of predicates : in the case in question the adjective *animal* is part of the definition of the adjective *chimerical* (if we allow ourselves to use this word, contrary to usage, to denote the defining predicate of chimaeras). But here again it is fairly plain that we are dealing with a proposition which implies that chimaeras are animals, but is not the same proposition—indeed, in the present case, the implication is not even reciprocal. By a negation we can give a kind of extensional interpretation : nothing is denoted by *a chimaera* which is not denoted by *an animal*. But this is a very roundabout interpretation. On the whole, it seems most correct to reject the proposition altogether, while retaining the various other propositions that would be equivalent to it if there were chimaeras. By symbolic logicians, who have experienced the utility of the null-class, this will be felt as a reactionary view. But I am not at present discussing what should be done in the logical calculus, where the established practice appears to me the best, but what is the philosophical truth concerning the null-class. We shall say, then, that, of the bundle of normally equivalent interpretations of logical symbolic formulae, the class of interpretations considered in the present chapter, which are dependent upon actual classes, fail where we are concerned with null class-concepts, on the ground that there is no actual null-class.

We may now reconsider the proposition " nothing is not nothing "— a proposition plainly true, and yet, unless carefully handled, a source of apparently hopeless antinomies. *Nothing* is a denoting concept, which denotes nothing. The concept which denotes is of course not nothing, *i.e.* it is not denoted by itself. The proposition which looks so para-doxical means no more than this : *Nothing*, the denoting concept, is not nothing, *i.e.* is not what itself denotes. But it by no means follows from this that there is an actual null-class : only the null class-concept and the null concept of a class are to be admitted.

But now a new difficulty has to be met. The equality of class-concepts, like all relations which are reflexive, symmetrical, and transitive, indicates an underlying identity, *i.e.* it indicates that every class-concept has to some term a relation which all equal class-concepts also have to that term—the term in question being different for different sets of equal class-concepts, but the same for the various members of a single set of equal class-concepts. Now for all class-concepts which are not null, this term is found in the corresponding class ; but where are we to find it for null class-concepts ? To this question several answers may be given, any of which may be adopted. For we now know what a class is, and we may therefore adopt as our term the class of all null class-concepts or of all null propositional functions These are not null-classes, but genuine classes, and to either of them all null class-concepts have the same relation. If we then wish to have an entity analogous to what is elsewhere to be called a class, but corresponding to null class-concepts, we shall be forced, wherever it is necessary (as in counting classes) to introduce a term which is identical for equal class-concepts, to substitute everywhere the class of class-concepts equal to a given class-concept for the class corresponding to that class-concept. The class corresponding to the class-concept remains logically fundamental, but need not be actually employed in our symbolism. The null-class, in fact, is in some ways analogous to an irrational in Arithmetic : it cannot be interpreted on the same principles as other classes, and if we wish to give an analogous interpretation elsewhere, we must substitute for classes other more complicated entities—in the present case, certain correlated classes. The object of such a procedure will be mainly technical ; but failure to understand the procedure will lead to in-extricable difficulties in the interpretation of the symbolism. A very closely analogous procedure occurs constantly in Mathematics, for example with every generalization of number ; and so far as I know, no single case in which it occurs has been rightly interpreted either by philosophers or by mathematicians. So many instances will meet us in the course of the present work that it is unnecessary to linger longer over the point at present. Only one possible misunderstanding must be guarded against. No vicious circle is involved in the above account of the null-class; for the general notion of *class* is first laid down, is found to involve what is called existence, is then symbolically, not philosophically, replaced by the notion of a class of equal class-concepts, and is found, in this new form, to be applicable to what corresponds to null class-concepts, since what corresponds is now a class which is not null. Between classes *simpliciter* and classes of equal class-concepts there is a one-one correlation, which breaks down in the sole case of the class of null class-concepts, to which no null-class corresponds; and this fact is the reason for the whole complication.

74. A question which is very fundamental in the philosophy of Arithmetic must now be discussed in a more or less preliminary fashion. Is a class which has many terms to be regarded as itself one or many? Taking the class as equivalent simply to the numerical conjunction "*A* and *B* and *C* and etc.," it seems plain that it is many; yet it is quite necessary that we should be able to count classes as one each, and we do habitually speak of *a* class. Thus classes would seem to be one in one sense and many in another.

There is a certain temptation to identify the class as many and the class as one, *e.g.*, *all men* and *the human race*. Nevertheless, wherever a class consists of more than one term, it can be proved that no such identification is permissible. A concept of a class, if it denotes a class as one, is not the same as any concept of the class which it denotes. That is to say, *classes of all rational animals*, which denotes the human race as one term, is different from *men*, which denotes men, *i.e.* the human race as many. But if the human race were identical with men, it would follow that whatever denotes the one must denote the other, and the above difference would be impossible. We might be tempted to infer that Peano's distinction, between a term and a class of which the said term is the only member, must be maintained, at least when the term in question is a class*. But it is more correct, I think, to infer an ultimate distinction between a class as many and a class as one, to hold that the many are only many, and are not also one. The class as one may be identified with the whole composed of the terms of the class, *i.e.*, in the case of men, the class as one will be the human race.

But can we now avoid the contradiction always to be feared, where there is something that cannot be made a logical subject? I do not myself see any way of eliciting a precise contradiction in this case. In the case of concepts, we were dealing with what was plainly one entity; in the present case, we are dealing with a complex essentially capable of analysis into units. In such a proposition as "*A* and *B* are two," there is no logical subject: the assertion is not about *A*, nor

---

* This conclusion is actually drawn by Frege from an analogous argument: *Archiv für syst. Phil.* 1, p. 444. See Appendix.

about *B*, nor about the whole composed of both, but strictly and only about *A* and *B*. Thus it would seem that assertions are not necessarily *about* single subjects, but may be about many subjects; and this removes the contradiction which arose, in the case of concepts, from the impossibility of making assertions about them unless .they were turned into subjects. This impossibility being here absent, the contradiction which was to be feared does not arise.

75. We may ask, as suggested by the above discussion, what is to be said of the objects denoted by *a man, every man, some man*, and *any man*. Are these objects one or many or neither? Grammar treats them all as one. But to this view, the natural objection is, which one? Certainly not Socrates, nor Plato, nor any other particular person. Can we conclude that no one is denoted? As well might we conclude that every one is denoted, which in fact is true of the concept *every man*. I think one is denoted in every case, but in an impartial distributive manner. *Any number* is neither 1 nor 2 nor any other particular number, whence it is easy to conclude that *any number* is not any one number, a proposition at first sight contradictory, but really resulting from an ambiguity in *any*, and more correctly expressed by " *any number* is not *some* one number." There are, however, puzzles in this subject which I do not yet know how to solve.

A logical difficulty remains in regard to the nature of the whole composed of all the terms of a class. Two propositions appear self-evident: (1) Two wholes composed of different terms must be different; (2) A whole composed of one term only is that one term. It follows that the whole composed of a class considered as one term is that class considered as one term, and is therefore identical with the whole composed of the terms of the class; but this result contradicts the first of our supposed self-evident principles. The answer in this case, however, is not difficult. The first of our principles is only universally true when all the terms composing our two wholes are simple. A given whole is capable, if it has more than two parts, of being analyzed in a plurality of ways; and the resulting constituents, so long as analysis is not pushed as far as possible, will be different for different ways of analyzing. This proves that different sets of constituents may constitute the same whole, and thus disposes of our difficulty.

76. Something must be said as to the relation of a term to a class of which it is a member, and as to the various allied relations. One of the allied relations is to be called ϵ, and is to be fundamental in Symbolic Logic. But it is to some extent optional which of them we take as symbolically fundamental.

Logically, the fundamental relation is that of subject and predicate, expressed in "Socrates is human"—a relation which, as we saw in Chapter IV, is peculiar in that the relatum cannot be regarded as a term in the proposition. The first relation that grows out of this is the one expressed by "Socrates has humanity," which is distinguished by the fact that here the relation is a term. Next comes "Socrates is a man." This proposition, considered as a relation between Socrates and the concept *man*, is the one which Peano regards as fundamental ; and his ϵ expresses the relation *is a* between Socrates and *man*. So long as we use class-concepts for classes in our symbolism, this practice is unobjectionable ; but if we give ϵ this meaning, we must not assume that two symbols representing equal class-concepts both represent one and the same entity. We may go on to the relation between Socrates and the human race, *i.e.* between a term and its class considered as a whole ; this is expressed by "Socrates belongs to the human race." This relation might equally well be represented by ϵ. It is plain that, since a class, except when it has one term, is essentially many, it cannot be *as such* represented by a single letter : hence in any possible Symbolic Logic the letters which do duty for classes cannot represent the classes *as many*, but must represent either class-concepts, or the wholes composed of classes, or some other allied single entities. And thus ϵ cannot represent the relation of a term to its class as many ; for this would be a relation of one term to many terms, not a two-term relation such as we want. This relation might be expressed by "Socrates is one among men" ; but this, in any case, cannot be taken to be the meaning of ϵ.

77. A relation which, before Peano, was almost universally confounded with ϵ, is the relation of inclusion between classes, as *e.g.* between men and mortals. This is a time-honoured relation, since it occurs in the traditional form of the syllogism : it has been a battleground between intension and extension, and has been so much discussed that it is astonishing how much remains to be said about it. Empiricists hold that such propositions mean an actual enumeration of the terms of the contained class, with the assertion, in each case, of membership of the containing class. They must, it is to be inferred, regard it as doubtful whether all primes are integers, since they will scarcely have the face to say that they have examined all primes one by one. Their opponents have usually held, on the contrary, that what is meant is a relation of whole and part between the defining predicates, but turned in the opposite sense from the relation between the classes : *i.e.* the defining predicate of the larger class is part of that of the smaller. This view seems far more defensible than the other ; and wherever such a relation does hold between the defining predicates, the relation of inclusion follows. But two objections may be made, first, that in some cases of inclusion there is no such relation between the defining predicates, and secondly, that in any case what is *meant* is a relation between the classes, not a relation of their defining predicates. The first point may be easily established by instances.

The concept *even prime* does not contain as a constituent the concept *integer between* 1 *and* 10; the concept " English King whose head was cut off" does not contain the concept "people who died in 1649"; and so on through innumerable obvious cases. This might be met by saying that, though the relation of the defining predicates is not one of whole and part, it is one more or less analogous to implication, and is always what is really meant by propositions of inclusion. Such a view represents, I think, what is said by the better advocates of intension, and I am not concerned to deny that a relation of the kind in question does always subsist between defining predicates of classes one of which is contained in the other. But the second of the above points remains valid as against any intensional interpretation. When we say that men are mortals, it is evident that we are saying something about men, not about the concept *man* or the predicate *human*. The question is, then, what exactly are we saying?

Peano held, in earlier editions of his *Formulaire*, that what is asserted is the formal implication " x is a man implies x is a mortal." This is certainly implied, but I cannot persuade myself that it is the same proposition. For in this proposition, as we saw in Chapter III, it is essential that x should take *all* values, and not only such as are men. But when we say " all men are mortals," it seems plain that we are only speaking of men, and not of all other imaginable terms. We may, if we wish for a genuine relation of classes, regard the assertion as one of whole and part between the two classes each considered as a single term. Or we may give a still more purely extensional form to our proposition, by making it mean : Every (or any) man is a mortal. This proposition raises very interesting questions in the theory of denoting : for it appears to assert an identity, yet it is plain that what is denoted by *every man* is different from what is denoted by *a mortal*. These questions, however, interesting as they are, cannot be pursued here. It is only necessary to realize clearly what are the various equivalent propositions involved where one class is included in another. The form most relevant to Mathematics is certainly the one with formal implication, which will receive a fresh discussion in the following chapter.

Finally, we must remember that classes are to be derived, by means of the notion of *such that*, from other sources than subject-predicate propositions and their equivalents. Any propositional function in which a fixed assertion is made of a variable term is to be regarded, as was explained in Chapter II, as giving rise to a class of values satisfying it. This topic requires a discussion of assertions; but one strange contradiction, which necessitates the care in discrimination aimed at in the present chapter, may be mentioned at once.

78. Among predicates, most of the ordinary instances cannot be predicated of themselves, though, by introducing negative predicates, it will be found that there are just as many instances of predicates which are predicable of themselves. One at least of these, namely predicability, or the property of being a predicate, is not negative: predicability, as is evident, is predicable, *i.e.* it is a predicate of itself. But the most common instances are negative: thus non-humanity is non-human, and so on. The predicates which are not predicable of themselves are, therefore, only a selection from among predicates, and it is natural to suppose that they form a class having a defining predicate. But if so, let us examine whether this defining predicate belongs to the class or not. If it belongs to the class, it is not predicable of itself, for that is the characteristic property of the class. But if it is not predicable of itself, then it does not belong to the class whose defining predicate it is, which is contrary to the hypothesis. On the other hand, if it does not belong to the class whose defining predicate it is, then it is not predicable of itself, *i.e.* it *is* one of those predicates that are not predicable of themselves, and therefore it does belong to the class whose defining predicate it is—again contrary to the hypothesis. Hence from either hypothesis we can deduce its contradictory. I shall return to this contradiction in Chapter x; for the present, I have introduced it merely as showing that no subtlety in distinguishing is likely to be excessive.

79. To sum up the above somewhat lengthy discussion. A class, we agreed, is essentially to be interpreted in extension; it is either a single term, or that kind of combination of terms which is indicated when terms are connected by the word *and*. But practically, though not theoretically, this purely extensional method can only be applied to finite classes. All classes, whether finite or infinite, can be obtained as the objects denoted by the plurals of class-concepts—men, numbers, points, etc. Starting with predicates, we distinguished two kinds of proposition, typified by "Socrates is human" and "Socrates has humanity," of which the first uses *human* as predicate, the second as a term of a relation. These two classes of propositions, though very important logically, are not so relevant to Mathematics as their derivatives. Starting from *human*, we distinguished (1) the class-concept *man*, which differs slightly, if at all, from *human*; (2) the various denoting concepts *all men, every man, any man, a man* and *some man*; (3) the objects denoted by these concepts, of which the one denoted by *all men* was called the *class as many*, so that *all men* (the concept) was called the *concept of the class*; (4) the class as one, *i.e.* the human race. We had also a classification of propositions about Socrates, dependent upon the above distinctions, and approximately parallel with them: (1) "Socrates is-a man" is nearly, if not quite, identical with "Socrates has humanity"; (2) "Socrates is a-man" expresses identity between

Socrates and one of the terms denoted by *a man*; (3) "Socrates is one among men," a proposition which raises difficulties owing to the plurality of men; (4) "Socrates belongs to the human race," which alone expresses a relation of an individual to its class, and, as the possibility of relation requires, takes the class as one, not as many. We agreed that the null-class, which has no terms, is a fiction, though there are null class-concepts. It appeared throughout that, although any symbolic treatment must work largely with class-concepts and intension, classes and extension are logically more fundamental for the principles of Mathematics; and this may be regarded as our main general conclusion in the present chapter.

CHAPTER VII.

PROPOSITIONAL FUNCTIONS.

80. In the preceding chapter an endeavour was made to indicate the kind of object that is to be called a class, and for purposes of discussion classes were considered as derived from subject-predicate propositions. This did not affect our view as to the notion of *class* itself; but if adhered to, it would greatly restrict the extension of the notion. It is often necessary to recognize as a class an object not defined by means of a subject-predicate proposition. The explanation of this necessity is to be sought in the theory of assertions and *such that*.

The general notion of an assertion has been already explained in connection with formal implication. In the present chapter its scope and legitimacy are to be critically examined, and its connection with classes and *such that* is to be investigated. The subject is full of difficulties, and the doctrines which I intend to advocate are put forward with a very limited confidence in their truth.

The notion of *such that* might be thought, at first sight, to be capable of definition; Peano used, in fact, to define the notion by the proposition " the x's such that x is an a are the class a." Apart from further objections, to be noticed immediately, it is to be observed that the class as obtained from *such that* is the genuine class, taken in extension and as many, whereas the a in " x is an a" is not the class, but the class-concept. Thus it is formally necessary, if Peano's procedure is to be permissible, that we should substitute for " x's such that so-and-so " the genuine class-concept " x such that so-and-so," which may be regarded as obtained from the predicate " such that so-and-so " or rather, " being an x such that so-and-so," the latter form being necessary because so-and-so is a propositional function containing x. But when this purely formal emendation has been made the point remains that *such that* must often be put before such propositions as xRa, where R is a given relation and a a given term. We cannot reduce this proposition to the form " x is an a' " without using *such that*; for if we ask what a' must be, the answer is: a' must be such that each of its terms, and no other terms, have the relation R to a. To take examples from daily life: the children of Israel are a class defined by a certain relation to Israel, and the class can only be defined as the terms such that they have this relation. *Such that* is roughly equivalent to *who* or *which*, and represents the general notion of satisfying a propositional function. But we may go further: given a class a, we cannot define, in terms of a, the class of propositions " x is an a " for different values of x. It is plain that there is a relation which each of these propositions has to the x which occurs in it, and that the relation in question is determinate when a is given. Let us call the relation R. Then any entity which is a referent with respect to R is a proposition of the type " x is an a." But here the notion of *such that* is already employed. And the relation R itself can only be defined as the relation which holds between " x is an a " and x for all values of x, and does not hold between any other pairs of terms. Here *such that* again appears. The point which is chiefly important in these remarks is the indefinability of propositional functions. When these have been admitted, the general notion of one-valued functions is easily defined. Every relation which is many-one, *i.e.* every relation for which a given referent has only one relatum, defines a function: the relatum is that function of the referent which is defined by the relation in question. But where the function is a proposition, the notion involved is presupposed in the symbolism, and cannot be defined by means of it without a vicious circle: for in the above general definition of a function propositional functions already occur. In the case of propositions of the type " x is an a," if we ask *what* propositions are of this type, we can only answer " all propositions in which a term is said to be a "; and here the notion to be defined reappears.

81. Can the indefinable element involved in propositional functions be identified with assertion together with the notion of *every* proposition containing a given assertion, or an assertion made concerning *every* term? The only alternative, so far as I can see, is to accept the general notion of a propositional function itself as indefinable, and for formal purposes this course is certainly the best; but philosophically, the notion appears at first sight capable of analysis, and we have to examine whether or not this appearance is deceptive.

We saw in discussing verbs, in Chapter IV, that when a proposition is completely analyzed into its simple constituents, these constituents taken together do not reconstitute it. A less complete analysis of propositions into subject and assertion has also been considered; and this analysis does much less to destroy the proposition. A subject and an assertion, if simply juxtaposed, do not, it is true, constitute a proposition; but as soon as the assertion is actually asserted of the subject, the proposition reappears. The assertion is everything that remains of the proposition when the subject is omitted: the verb remains an asserted verb, and is not turned into a verbal noun; or at any rate the verb retains that curious indefinable intricate relation to the other terms of the proposition which distinguishes a relating relation from the same relation abstractly considered. It is the scope and legitimacy of this notion of assertion which is now to be examined. Can every proposition be regarded as an assertion concerning any term occurring in it, or are limitations necessary as to the form of the proposition and the way in which the term enters into it?

In some simple cases, it is obvious that the analysis into subject and assertion is legitimate. In "Socrates is a man," we can plainly distinguish Socrates and something that is asserted about him; we should admit unhesitatingly that the *same* thing may be said about Plato or Aristotle. Thus we can consider a class of propositions containing this assertion, and this will be the class of which a typical number is represented by "*x* is a man." It is to be observed that the assertion must appear *as* assertion, not as term : thus "to be a man is to suffer" contains the same assertion, but used as term, and this proposition does not belong to the class considered. In the case of propositions asserting a fixed relation to a fixed term, the analysis seems equally undeniable. To be more than a yard long, for example, is a perfectly definite assertion, and we may consider the class of propositions in which this assertion is made, which will be represented by the propositional function "*x* is more than a yard long." In such phrases as "snakes which are more than a yard long," the assertion appears very plainly; for it is here explicitly referred to a variable subject, not asserted of any one definite subject. Thus if *R* be a fixed relation and *a* a fixed term, ... *Ra* is a perfectly definite assertion. (I place dots before the *R*, to indicate the place where the subject must be inserted in order to make a proposition.) It may be doubted whether a relational proposition can be regarded as an assertion concerning the relatum. For my part, I hold that this can be done except in the case of subject-predicate propositions; but this question is better postponed until we have discussed relations *.

82. More difficult questions must now be considered. Is such a proposition as "Socrates is a man implies Socrates is a mortal," or "Socrates has a wife implies Socrates has a father," an assertion concerning Socrates or not? It is quite certain that, if we replace Socrates by a variable, we obtain a propositional function; in fact, the truth of this function for all values of the variable is what is asserted in the corresponding formal implication, which does not, as might be thought at first sight, assert a relation between two propositional functions. Now it was our intention, if possible, to explain propositional functions by means of assertions; hence, if our intention can be carried out, the

* See § 96.

above propositions must be assertions concerning Socrates. There is, however, a very great difficulty in so regarding them. An assertion was to be obtained from a proposition by simply omitting one of the terms occurring in the proposition. But when we omit Socrates, we obtain "... is a man implies ... is a mortal." In this formula it is essential that, in restoring the proposition, the *same* term should be substituted in the two places where dots indicate the necessity of a term. It does not matter what term we choose, but it must be identical in both places. Of this requisite, however, no trace whatever appears in the would-be assertion, and no trace can appear, since all mention of the term to be inserted is necessarily omitted. When an x is inserted to stand for the variable, the identity of the term to be inserted is indicated by the repetition of the letter x ; but in the assertional form no such method is available. And yet, at first sight, it seems very hard to deny that the proposition in question tells us a fact *about* Socrates, and that the *same* fact is true about Plato or a plum-pudding or the number 2. It is certainly undeniable that "Plato is a man implies Plato is a mortal" is, in some sense or other, the *same* function of Plato as our previous proposition is of Socrates. The natural interpretation of this statement would be that the one proposition has to Plato the same relation as the other has to Socrates. But this requires that we should regard the propositional function in question as definable by means of its relation to the variable. Such a view, however, requires a propositional function more complicated than the one we are considering. If we represent "x is a man implies x is a mortal" by ϕx, the view in question maintains that ϕx is the term having to x the relation R, where R is some definite relation. The formal statement of this view is as follows : For all values of x and y, "y is identical with ϕx" is equivalent to "y has the relation R to x." It is evident that this will not do as an explanation, since it has far greater complexity than what it was to explain. It would seem to follow that propositions may have a certain constancy of form, expressed in the fact that they are instances of a given propositional function, without its being possible to analyze the propositions into a constant and a variable factor. Such a view is curious and difficult : constancy of form, in all other cases, is reducible to constancy of relations, but the constancy involved here is presupposed in the notion of constancy of relation, and cannot therefore be explained in the usual way.

The same conclusion, I think, will result from the case of two variables. The simplest instance of this case is xRy, where R is a constant relation, while x and y are independently variable. It seems evident that this is a propositional function of two independent variables : there is no difficulty in the notion of the class of all propositions of the form xRy. This class is involved—or at least all those members of the class that are true are involved—in the notion of the classes of referents and relata with respect to R, and these classes are unhesitatingly admitted in such words as parents and children, masters and servants, husbands and wives, and innumerable other instances from daily life, as also in logical notions such as premisses and conclusions, causes and effects, and so on. All such notions depend upon the class of propositions typified by xRy, where R is constant while x and y are variable. Yet it is very difficult to regard xRy as analyzable into the assertion R concerning x and y, for the very sufficient reason that this view destroys the *sense* of the relation, *i.e.* its direction from x to y, leaving us with some assertion which is symmetrical with respect to x and y, such as "the relation R holds between x and y." Given a relation and its terms, in fact, two distinct propositions are possible. Thus if we take R itself to be an assertion, it becomes an ambiguous assertion : in supplying the terms, if we are to avoid ambiguity, we must decide which is referent and which relatum. We may quite legitimately regard ...Ry as an assertion, as was explained before ; but here y has become constant. We may then go on to vary y, considering the class of assertions ...Ry for different values of y ; but this process does not seem to be identical with that which is indicated by the independent variability of x and y in the propositional function xRy. Moreover, the suggested process requires the variation of an element in an assertion, namely of y in ...Ry, and this is in itself a new and difficult notion.

A curious point arises, in this connection, from the consideration, often essential in actual Mathematics, of a relation of a term to itself. Consider the propositional function xRx, where R is a constant relation. Such functions are required in considering, *e.g.*, the class of suicides or of self-made men ; or again, in considering the values of the variable for which it is equal to a certain function of itself, which may often be necessary in ordinary Mathematics. It seems exceedingly evident, in this case, that the proposition contains an element which is lost when it is analyzed into a term x and an assertion R. Thus here again, the propositional function must be admitted as fundamental.

83. A difficult point arises as to the variation of the concept in a proposition. Consider, for example, all propositions of the type aRb, where a and b are fixed terms, and R is a variable relation. There seems no reason to doubt that the class-concept " relation between a and b" is legitimate, and that there is a corresponding class ; but this requires the admission of such propositional functions as aRb, which, moreover, are frequently required in actual Mathematics, as, for example, in counting the number of many-one relations whose referents and relata are given classes. But if our variable is to have, as we normally require, an unrestricted field, it is necessary to substitute the propositional function " R is a relation implies aRb." In this proposition the implication involved is material, not formal. If the implication were formal, the proposition would not be a function of R, but would be equivalent to the (necessarily false) proposition: "All relations hold between a and b." Generally we have some such proposition as "aRb implies $\phi(R)$ provided R is a relation," and we wish to turn this into a formal implication. If $\phi(R)$ is a proposition for all values of R, our object is effected by substituting "If 'R is a relation' implies 'aRb,' then $\phi(R)$." Here R can take all values*, and the *if* and *then* is a formal implication, while the *implies* is a material implication. If $\phi(R)$ is not a propositional function, but is a proposition only when R satisfies $\psi(R)$, where $\psi(R)$ is a propositional function implied by "R is a relation" for all values of R, then our formal implication can be put in the form "If 'R is a relation' implies aRb, then, for all values of R, $\psi(R)$ implies $\phi(R)$," where both the subordinate implications are material. As regards the material implication "'R is a relation' implies aRb," this is always a proposition, whereas aRb is only a proposition when R is a relation. The new propositional function will only be true when R is a relation which does hold between a and b: when R is not a relation, the antecedent is false and the consequent is not a proposition, so that the implication is false; when R is a relation which does not hold between a and b, the antecedent is true and the consequent false, so that again the implication is false; only when both are true is the implication true. Thus in defining the class of relations holding between a and b, the formally correct course is to define them as the values satisfying "R is a relation implies aRb"—an implication which, though it contains a variable, is not formal, but material, being satisfied by some only of the possible values of R. The variable R in it is, in Peano's language, real and not apparent.

The general principle involved is: If ϕx is only a proposition for some values of x, then "'ϕx implies ϕx' implies ϕx" is a proposition for *all* values of x, and is true when and only when ϕx is true. (The implications involved are both material.) In some cases, "ϕx implies ϕx" will be equivalent to some simpler propositional function ψx (such as "R is a relation" in the above instance), which may then be substituted for it†.

Such a propositional function as "R is a relation implies aRb" appears even less capable than previous instances of analysis into R and an assertion about R, since we should have to assign a meaning to "$a...b$," where the blank space may be filled by anything, not necessarily by a relation. There is here, however, a suggestion of an entity which has not yet been considered, namely the couple with sense. It may be doubted whether there is any such entity, and yet such phrases as

---

* It is necessary to assign some meaning (other than a proposition) to aRb when R is not a relation.

† A propositional function, though for every value of the variable it is true or false, is not itself true or false, being what is denoted by "any proposition of the type in question," which is not itself a proposition.

"*R* is a relation holding from *a* to *b*" seem to show that its rejection would lead to paradoxes. This point, however, belongs to the theory of relations, and will be resumed in Chapter IX (§ 98).

From what has been said, it appears that propositional functions must be accepted as ultimate data. It follows that formal implication and the inclusion of classes cannot be generally explained by means of a relation between assertions, although, where a propositional function asserts a fixed relation to a fixed term, the analysis into subject and assertion is legitimate and not unimportant.

84. It only remains to say a few words concerning the derivation of classes from propositional functions. When we consider the *x*'s *such that* ϕx, where ϕx is a propositional function, we are introducing a notion of which, in the calculus of propositions, only a very shadowy use is made—I mean the notion of *truth*. We are considering, among all the propositions of the type ϕx, those that are true: the corresponding values of *x* give the class defined by the function ϕx. It must be held, I think, that every propositional function which is not null defines a class, which is denoted by "*x*'s such that ϕx." There is thus always a concept of the class, and the class-concept corresponding will be the singular, " *x* such that ϕx." But it may be doubted—indeed the contradiction with which I ended the preceding chapter gives reason for doubting—whether there is always a defining predicate of such classes. Apart from the contradiction in question, this point might appear to be merely verbal: " being an *x* such that ϕx," it might be said, may always be taken to be a predicate. But in view of our contradiction, all remarks on this subject must be viewed with caution. This subject, however, will be resumed in Chapter X.

85. It is to be observed that, according to the theory of propositional functions here advocated, the ϕ in ϕx is not a separate and distinguishable entity : it lives in the propositions of the form ϕx, and cannot survive analysis. I am highly doubtful whether such a view does not lead to a contradiction, but it appears to be forced upon us, and it has the merit of enabling us to avoid a contradiction arising from the opposite view. If ϕ were a distinguishable entity, there would be a proposition asserting ϕ of itself, which we may denote by $\phi (\phi)$; there would also be a proposition not-$\phi (\phi)$, denying $\phi (\phi)$. In this proposition we may regard ϕ as variable; we thus obtain a propositional function. The question arises: Can the assertion in this propositional function be asserted of itself? The assertion is non-assertibility of self, hence if it can be asserted of itself, it cannot, and if it cannot, it can. This contradiction is avoided by the recognition that the functional part of a propositional function is not an independent entity. As the contradiction in question is closely analogous to the other, concerning predicates not predicable of themselves, we may hope that a similar solution will apply there also.

CHAPTER VIII.

THE VARIABLE.

86. THE discussions of the preceding chapter elicited the fundamental nature of the variable; no apparatus of assertions enables us to dispense with the consideration of the varying of one or more elements in a proposition while the other elements remain unchanged. The variable is perhaps the most distinctively mathematical of all notions; it is certainly also one of the most difficult to understand. The attempt, if not the deed, belongs to the present chapter.

The theory as to the nature of the variable, which results from our previous discussions, is in outline the following. When a given term occurs as term in a proposition, that term may be replaced by any other while the remaining terms are unchanged. The class of propositions so obtained have what may be called constancy of form, and this constancy of form must be taken as a primitive idea. The notion of a class of propositions of constant form is more fundamental than the general notion of *class*, for the latter can be defined in terms of the former, but not the former in terms of the latter. Taking *any* term, a certain member of any class of propositions of constant form will contain that term. Thus x, the variable, is what is denoted by *any term*, and ϕx, the propositional function, is what is denoted by *the* proposition of the form ϕ in which x occurs. We may say that x is *the x* is *any* ϕx, where ϕx denotes the class of propositions resulting from different values of x. Thus in addition to propositional functions, the notions of *any* and of denoting are presupposed in the notion of the variable. This theory, which, I admit, is full of difficulties, is the least objectionable that I have been able to imagine. I shall now set it forth more in detail.

87. Let us observe, to begin with, that the explicit mention of *any, some*, etc., need not occur in Mathematics: formal implication will express all that is required. Let us recur to an instance already discussed in connection with denoting, where a is a class and b a class of classes. We have

"Any a belongs to any b" is equivalent to "'x is an a' implies that 'u is a b' implies 'x is a u'";

" Any *a* belongs to a *b*" is equivalent to " '*x* is an *a*' implies 'there
is a *b*, say *u*, such that *x* is a *u*' "*;

" Any *a* belongs to some *b*" is equivalent to "there is a *b*, say *u*, such
that '*x* is an *a*' implies '*x* is a *u*' ";

and so on for the remaining relations considered in Chapter v. The
question arises: How far do these equivalences constitute definitions of
any, *a*, *some*, and how far are these notions involved in the symbolism
itself ?

The variable is, from the formal standpoint, *the* characteristic notion
of Mathematics. Moreover it is *the* method of stating general theorems,
which always *mean* something different from the intensional propositions
to which such logicians as Mr Bradley endeavour to reduce them. That
the meaning of an assertion about all men or any man is different from
the meaning of an equivalent assertion about the concept *man*, appears
to me, I must confess, to be a self-evident truth—as evident as the fact
that propositions about John are not about the *name* John. This point,
therefore, I shall not argue further. That the variable characterizes
Mathematics will be generally admitted, though it is not generally
perceived to be present in elementary Arithmetic. Elementary Arithmetic, as taught to children, is characterized by the fact that the *numbers*
occurring in it are constants; the answer to any schoolboy's sum is
obtainable without propositions concerning *any* number. But the fact
that this is the case can only be proved by the help of propositions
about *any* number, and thus we are led from schoolboy's Arithmetic to
the Arithmetic which uses letters for numbers and proves general
theorems. How very different this subject is from childhood's enemy may
be seen at once in such works as those of Dedekind† and Stolz‡. Now
the difference consists simply in this, that our numbers have now become
variables instead of being constants. We now prove theorems concerning *n*, not concerning 3 or 4 or any other particular number. Thus it is
absolutely essential to any theory of Mathematics to understand the
nature of the variable.

Originally, no doubt, the variable was conceived dynamically, as
something which changed with the lapse of time, or, as is said, as something which successively assumed all values of a certain class. This
view cannot be too soon dismissed. If a theorem· is proved concerning
n, it must not be supposed that *n* is a kind of arithmetical Proteus,
which is 1 on Sundays and 2 on Mondays, and so on. Nor must it be
supposed that *n* simultaneously assumes all its values. If *n* stands for
any integer, we cannot say that *n* is 1, nor yet that it is 2, nor yet that

---

* Here "there is a *c*," where *c* is any class, is defined as equivalent to "If *p*
implies *p*, and '*x* is a *c*' implies *p* for all values of *x*, then *p* is true."

† *Was sind und was sollen die Zahlen?* Brunswick, 1893.

‡ *Allgemeine Arithmetik*, Leipzig, 1885.

it is any other particular number. In fact, n just denotes *any* number, and this is something quite distinct from each and all of the numbers. It is not true that 1 is any number, though it is true that whatever holds of any number holds of 1. The variable, in short, requires the indefinable notion of *any* which was explained in Chapter v.

88. We may distinguish what may be called the true or formal variable from the restricted variable. *Any term* is a concept denoting the true variable; if u be a class not containing all terms, *any u* denotes a restricted variable. The terms included in the object denoted by the defining concept of a variable are called the *values* of the variable: thus every value of a variable is a constant. There is a certain difficulty about such propositions as "any number is a number." Interpreted by formal implication, they offer no difficulty, for they assert merely that the propositional function "x is a number implies x is a number" holds for all values of x. But if "any number" be taken to be a definite object, it is plain that it is not identical with 1 or 2 or 3 or any number that may be mentioned. Yet these are all the numbers there are, so that "any number" cannot be a number at all. The fact is that the concept "any number" does denote one number, but not a particular one. This is just the distinctive point about *any*, that it denotes a term of a class, but in an impartial distributive manner, with no preference for one term over another. Thus although x is a number, and no one number is x, yet there is here no contradiction, so soon as it is recognized that x is not one definite term.

The notion of the restricted variable can be avoided, except in regard to propositional functions, by the introduction of a suitable hypothesis, namely the hypothesis expressing the restriction itself. But in respect of propositional functions this is not possible. The x in ϕx, where ϕx is a propositional function, is an unrestricted variable; but the ϕx itself is restricted to the class which we may call ϕ. (It is to be remembered that the *class* is here fundamental, for we found it impossible, without a vicious circle, to discover any common characteristic by which the class could be defined, since the statement of any common characteristic is itself a propositional function.) By making our x always an unrestricted variable, we can speak of *the* variable, which is conceptually identical in Logic, Arithmetic, Geometry, and all other formal subjects. The *terms* dealt with are always *all* terms; only the complex concepts that occur distinguish the various branches of Mathematics.

89. We may now return to the apparent definability of *any*, *some*, and *a*, in terms of formal implication. Let a and b be class-concepts, and consider the proposition "any a is a b." This is to be interpreted as meaning "x is an a implies x is a b." It is plain that, to begin with, the two propositions do not *mean* the same thing: for *any a* is a concept denoting only a's, whereas in the formal implication x need not be an a. But we might, in Mathematics, dispense altogether with "any a is a b," and content ourselves with the formal implication: this is, in fact, symbolically the best course. The question to be examined, therefore, is: How far, if at all, do *any* and *some* and *a* enter into the formal implication? (The fact that the indefinite article appears in "*x* is an *a*" and "*x* is a *b*" is irrelevant, for these are merely taken as typical propositional functions.) We have, to begin with, a class of true propositions, each asserting of some constant term that if it is an *a* it is a *b*. We then consider the restricted variable, "any proposition of this class." We assert the truth of any term included among the values of this restricted variable. But in order to obtain the suggested formula, it is necessary to transfer the variability from the proposition as a whole to its variable term. In this way we obtain "*x* is an *a* implies *x* is *b*." But the genesis remains essential, for we are not here expressing a relation of two propositional functions "*x* is an *a*" and "*x* is a *b*." If this were expressed, we should not require the *same x* both times. Only one propositional function is involved, namely the whole formula. Each proposition of the class expresses a relation of one term of the propositional function "*x* is an *a*" to one of "*x* is a *b*"; and we may say, if we choose, that the whole formula expresses a relation of *any* term of "*x* is an *a*" to *some* term of "*x* is a *b*." We do not so much have an implication containing a variable as a variable implication. Or again, we may say that the first *x* is *any* term, but the second is *some* term, namely the first *x*. We have a class of implications not containing variables, and we consider *any* member of this class. If *any* member is true, the fact is indicated by introducing a typical implication containing a variable. This typical implication is what is called a *formal* implication: it is *any* member of a class of material implications. Thus it would seem that *any* is presupposed in mathematical formalism, but that *some* and *a* may be legitimately replaced by their equivalents in terms of formal implications.

90. Although *some may* be replaced by its equivalent in terms of *any*, it is plain that this does not give the meaning of *some*. There is, in fact, a kind of duality of *any* and *some*: given a certain propositional function, if *all* terms belonging to the propositional function are asserted, we have *any*, while if one at least is asserted (which gives what is called an existence-theorem), we get *some*. The proposition ϕx asserted without comment, as in "*x* is a man implies *x* is a mortal," is to be taken to mean that ϕx is true for *all* values of *x* (or for *any* value), but it might equally well have been taken to mean that ϕx is true for *some* value of *x*. In this way we might construct a calculus with two kinds of variable, the conjunctive and the disjunctive, in which the latter would occur wherever an existence-theorem was to be stated. But this method does not appear to possess any practical advantages.

91. It is to be observed that what is fundamental is not particular propositional functions, but the class-concept *propositional function*. A

propositional function is the class of all propositions which arise from
the variation of a single term, but this is not to be considered as a
definition, for reasons explained in the preceding chapter.

92. From propositional functions all other classes can be derived
by definition, with the help of the notion of *such that*. Given a propositional function ϕx, the terms such that, when x is identified with
any one of them, ϕx is true, are the class defined by ϕx. This is the
class as many, the class in extension. It is not to be assumed that every
class so obtained has a defining predicate : this subject will be discussed
afresh in Chapter x. But it must be assumed, I think, that a class in
extension is defined by any propositional function, and in particular
that *all* terms form a class, since many propositional functions (*e.g.*
all formal implications) are true of *all* terms. Here, as with formal
implications, it is necessary that the whole propositional function whose
truth defines the class should be kept intact; and not, even where this
is possible for every value of x, divided into separate propositional
functions. For example, if a and b be two classes, defined by ϕx and ψx
respectively, their common part is defined by the product $\phi x \cdot \psi x$, where
the product has to be made for every value of x, and then x varied
afterwards. If this is not done, we do not necessarily have the *same*
x in ϕx and ψx. Thus we do not multiply propositional functions, but
propositions : the new propositional function is the class of products
of corresponding propositions belonging to the previous functions, and
is by no means the product of ϕx and ψx. It is only in virtue of
a definition that the logical product of the classes defined by ϕx and ψx
is the class defined by $\phi x \cdot \psi x$. And wherever a proposition containing
an apparent variable is asserted, what is asserted is the truth, for all
values of the variable or variables, of the propositional function corresponding to the whole proposition, and is never a relation of propositional
functions.

93. It appears from the above discussion that the variable is a
very complicated logical entity, by no means easy to analyze correctly.
The following appears to be as nearly correct as any analysis I can make.
Given any proposition (not a propositional function), let a be one of
its terms, and let us call the proposition $\phi(a)$. Then in virtue of the
primitive idea of a propositional function, if x be any term, we can
consider the proposition $\phi(x)$, which arises from the substitution of x
in place of a. We thus arrive at the class of all propositions $\phi(x)$.
If all are true, $\phi(x)$ is asserted simply : $\phi(x)$ may then be called a
formal truth. In a formal implication, $\phi(x)$, *for every value of x*, states
an implication, and the assertion of $\phi(x)$ is the assertion of a *class* of
implications, not of a single implication. If $\phi(x)$ is sometimes true,
the values of x which make it true form a class, which is the class defined
by $\phi(x)$: the class is said to *exist* in this case. If $\phi(x)$ is false for all
values of x, the class defined by $\phi(x)$ is said not to exist, and as a

matter of fact, as we saw in Chapter VI, there is no such class, if classes are taken in extension. Thus x is, in some sense, the object denoted by *any term*; yet this can hardly be strictly maintained, for different variables may occur in a proposition, yet the object denoted by *any term*, one would suppose, is unique. This, however, elicits a new point in the theory of denoting, namely that *any term* does not denote, properly speaking, an assemblage of terms, but denotes one term, only not one particular definite term. Thus *any term* may denote different terms in different places. We may say : any term has some relation to any term ; and this is quite a different proposition from : any term has some relation to itself. Thus variables have a kind of individuality. This arises, as I have tried to show, from propositional functions. When a propositional function has two variables, it must be regarded as obtained by successive steps. If the propositional function $\phi(x, y)$ is to be asserted for all values of x and y, we must consider the assertion, for all values of y, of the propositional function $\phi(a, y)$, where a is a constant. This does not involve y, and may be represented by $\psi(a)$. We then vary a, and assert $\psi(x)$ for all values of x. The process is analogous to double integration ; and it is necessary to prove formally that the order in which the variations are made makes no difference to the result. The individuality of variables appears to be thus explained. A variable is not *any term* simply, but any term as entering into a propositional function. We may say, if ϕx be a propositional function, that x is *the* term in *any* proposition of the class of propositions whose type is ϕx. It thus appears that, as regards propositional functions, the notions of class, of denoting, and of *any*, are fundamental, being presupposed in the symbolism employed. With this conclusion, the analysis of formal implication, which has been one of the principal problems of Part I, is carried as far as I am able to carry it. May some reader succeed in rendering it more complete, and in answering the many questions which I have had to leave unanswered.

CHAPTER IX.

RELATIONS.

94. NEXT after subject-predicate propositions come two types of propositions which appear equally simple. These are the propositions in which a relation is asserted between two terms, and those in which two terms are said to be two. The latter class of propositions will be considered hereafter; the former must be considered at once. It has often been held that every proposition can be reduced to one of the subject-predicate type, but this view we shall, throughout the present work, find abundant reason for rejecting. It might be held, however, that all propositions not of the subject-predicate type, and not asserting numbers, could be reduced to propositions containing two terms and a relation. This opinion would be more difficult to refute, but this too, we shall find, has no good grounds in its favour*. We may therefore allow that there are relations having more than two terms; but as these are more complex, it will be well to consider first such as have two terms only.

A relation between two terms is a concept which occurs in a proposition in which there are two terms not occurring as concepts†, and in which the interchange of the two terms gives a different proposition. This last mark is required to distinguish a relational proposition from one of the type "*a* and *b* are two," which is identical with "*b* and *a* are two." A relational proposition may be symbolized by *aRb*, where *R* is the relation and *a* and *b* are the terms; and *aRb* will then always, provided *a* and *b* are not identical, denote a different proposition from *bRa*. That is to say, it is characteristic of a relation of two terms that it proceeds, so to speak, *from* one *to* the other. This is what may be called the *sense* of the relation, and is, as we shall find, the source of order and series. It must be held as an axiom that *aRb* implies and is implied by a relational proposition *bR'a*, in which the

---

* See *inf.*, Part IV, Chap. xxv, § 200.

† This description, as we saw above (§ 48), excludes the pseudo-relation of subject to predicate.

relation R' proceeds from b to a, and may or may not be the same
relation as R. But even when aRb implies and is implied by bRa,
it must be strictly maintained that these are different propositions.
We may distinguish the term *from* which the relation proceeds as the
referent, and the term *to* which it proceeds as the *relatum*. The sense
of a relation is a fundamental notion, which is not capable of definition.
The relation which holds between b and a whenever R holds between
a and b will be called the *converse* of R, and will be denoted (following
Schröder) by $\breve{R}$. The relation of R to $\breve{R}$ is the relation of oppositeness,
or difference of sense; and this must not be defined (as would seem at
first sight legitimate) by the above mutual implication in any single
case, but only by the fact of its holding for all cases in which the given
relation occurs. The grounds for this view are derived from certain
propositions in which terms are related to themselves not-symmetrically,
i.e. by a relation whose converse is not identical with itself. These
propositions must now be examined.

95. There is a certain temptation to affirm that no term can be
related to itself; and there is a still stronger temptation to affirm that,
if a term can be related to itself, the relation must be symmetrical,
i.e. identical with its converse. But both these temptations must be
resisted. In the first place, if no term were related to itself, we should
never be able to assert self-identity, since this is plainly a relation.
But since there is such a notion as identity, and since it seems undeniable
that every term is identical with itself, we must allow that a term may
be related to itself. Identity, however, is still a symmetrical relation,
and may be admitted without any great qualms. The matter becomes
far worse when we have to admit not-symmetrical relations of terms
to themselves. Nevertheless the following propositions seem undeniable;
Being is, or has being; 1 is one, or has unity; concept is conceptual:
term is a term; class-concept is a class-concept. All these are of one
of the three equivalent types which we distinguished at the beginning of
Chapter v, which may be called respectively subject-predicate propositions, propositions asserting the relation of predication, and propositions
asserting membership of a class. What we have to consider is, then,
the fact that a predicate may be predicable of itself. It is necessary, for
our present purpose, to take our propositions in the second form (Socrates
has humanity), since the subject-predicate form is not in the above sense
relational. We may take, as the type of such propositions, " unity has
unity." Now it is certainly undeniable that the relation of predication
is asymmetrical, since subjects cannot in general be predicated of their
predicates. Thus " unity has unity " asserts one relation of unity to
itself, and implies another, namely the converse relation: unity has
to itself both the relation of subject to predicate, and the relation of
predicate to subject. Now if the referent and the relatum are identical,
it is plain that the relatum has to the referent the same relation as the

referent has to the relatum. Hence if the converse of a relation in a particular case were defined by mutual implication in that particular case, it would appear that, in the present case, our relation has two converses, since two different relations of relatum to referent are implied by "unity has unity." We must therefore define the converse of a relation by the fact that aRb implies and is implied by $b\breve{R}a$ *whatever* a and b may be, and whether or not the relation R holds between them. That is to say, a and b are here essentially variables, and if we give them any constant value, we may find that aRb implies and is implied by $bR'a$, where R' is some relation other than $\breve{R}$.

Thus three points must be noted with regard to relations of two terms: (1) they all have sense, so that, provided a and b are not identical, we can distinguish aRb from bRa; (2) they all have a converse, *i.e.* a relation $\breve{R}$ such that aRb implies and is implied by $b\breve{R}a$, whatever a and b may be; (3) some relations hold between a term and itself, and such relations are not necessarily symmetrical, *i.e.* there may be two different relations, which are each other's converses, and which both hold between a term and itself.

96. For the general theory of relations, especially in its mathematical developments, certain axioms relating classes and relations are of great importance. It is to be held that to have a given relation to a given term is a predicate, so that all terms having this relation to this term form a class. It is to be held further that to have a given relation at all is a predicate, so that all referents with respect to a given relation form a class. It follows, by considering the converse relation, that all relata also form a class. These two classes I shall call respectively the *domain* and the *converse domain* of the relation; the logical sum of the two I shall call the *field* of the relation.

The axiom that all referents with respect to a given relation form a class seems, however, to require some limitation, and that on account of the contradiction mentioned at the end of Chapter VI. This contradiction may be stated as follows. We saw that some predicates can be predicated of themselves. Consider now those of which this is not the case. These are the referents (and also the relata) in what seems like a complex relation, namely the combination of non-predicability with identity. But there is no predicate which attaches to all of them and to no other terms. For this predicate will either be predicable or not predicable of itself. If it is predicable of itself, it is one of those referents by relation to which it was defined, and therefore, in virtue of their definition, it is not predicable of itself. Conversely, if it is not predicable of itself, then again it is one of the said referents, of all of which (by hypothesis) it is predicable, and therefore again it is predicable of itself. This is a contradiction, which shows that all the referents considered have no exclusive common predicate, and therefore, if defining predicates are essential to classes, do not form a class.

The matter may be put otherwise. In defining the would-be class of predicates, all those not predicable of themselves have been used up. The common predicate of all these predicates cannot be one of them, since for each of them there is at least one predicate (namely itself) of which it is not predicable. But again, the supposed common predicate cannot be any other predicate, for if it were, it would be predicable of itself, *i.e.* it would be a member of the supposed class of predicates, since these were defined as those of which it is predicable. Thus no predicate is left over which could attach to all the predicates considered.

It follows from the above that not every definable collection of terms forms a class defined by a common predicate. This fact must be borne in mind, and we must endeavour to discover what properties a collection must have in order to form such a class. The exact point established by the above contradiction may be stated as follows: A proposition apparently containing only one variable may not be equivalent to any proposition asserting that the variable in question has a certain predicate. It remains an open question whether every class must have a defining predicate.

That all terms having a given relation to a given term form a class defined by an exclusive common predicate results from the doctrine of Chapter VII, that the proposition aRb can be analyzed into the subject a and the assertion Rb. To be a term of which Rb can be asserted appears to be plainly a predicate. But it does not follow, I think, that to be a term of which, for some value of y, Ry can be asserted, is a predicate. The doctrine of propositional functions requires, however, that all terms having the latter property should form a class. This class I shall call the *domain* of the relation R as well as the class of referents. The domain of the converse relation will be also called the converse domain, as well as the class of relata. The two domains together will be called the *field* of the relation—a notion chiefly important as regards series. Thus if paternity be the relation, fathers form its domain, children its converse domain, and fathers and children together its field.

It may be doubted whether a proposition aRb can be regarded as asserting aR of b, or whether only $\breve{R}a$ can be asserted of b. In other words, is a relational proposition only an assertion concerning the referent, or also an assertion concerning the relatum? If we take the latter view, we shall have, connected with (say) "a is greater than b," four assertions, namely "is greater than b," "a is greater than," "is less than a" and "b is less than." I am inclined myself to adopt this view, but I know of no argument on either side.

97. We can form the logical sum and product of two relations or of a class of relations exactly as in the case of classes, except that here we have to deal with double variability. In addition to these ways of combination, we have also the relative product, which is in general non-commutative, and therefore requires that the number of factors should be finite. If R, S be two relations, to say that their relative product RS holds between two terms x, z is to say that there is a term y to which x has the relation R, and which itself has the relation S to z. Thus brother-in-law is the relative product of wife and brother or of sister and husband: father-in-law is the relative product of wife and father, whereas the relative product of father and wife is mother or step-mother.

98. There is a temptation to regard a relation as definable in extension as a class of couples. This has the formal advantage that it avoids the necessity for the primitive proposition asserting that every couple has a relation holding between no other pair of terms. But it is necessary to give sense to the couple, to distinguish the referent from the relatum: thus a couple becomes essentially distinct from a class of two terms, and must itself be introduced as a primitive idea. It would seem, viewing the matter philosophically, that sense can only be derived from some relational proposition, and that the assertion that a is referent and b relatum already involves a purely relational proposition in which a and b are terms, though the relation asserted is only the general one of referent to relatum. There are, in fact, concepts such as *greater*, which occur otherwise than as terms in propositions having two terms (§§ 48, 54); and no doctrine of couples can evade such propositions. It seems therefore more correct to take an intensional view of relations, and to identify them rather with class-concepts than with classes. This procedure is formally more convenient, and seems also nearer to the logical facts. Throughout Mathematics there is the same rather curious relation of intensional and extensional points of view: the symbols other than variable terms (*i.e.* the variable class-concepts and relations) stand for intensions, while the actual objects dealt with are always extensions. Thus in the calculus of relations, it is classes of couples that are relevant, but the symbolism deals with them by means of relations. This is precisely similar to the state of things explained in relation to classes, and it seems unnecessary to repeat the explanations at length.

99. Mr Bradley, in *Appearance and Reality*, Chapter III, has based an argument against the reality of relations upon the endless regress arising from the fact that a relation which relates two terms must be related to each of them. The endless regress is undeniable, if relational propositions are taken to be ultimate, but it is very doubtful whether it forms any logical difficulty. We have already had occasion (§ 55) to distinguish two kinds of regress, the one proceeding merely to perpetually new implied propositions, the other in the meaning of a proposition itself; of these two kinds, we agreed that the former, since the solution of the problem of infinity, has ceased to be objectionable, while the latter remains inadmissible. We have to inquire which kind of regress occurs in the present instance. It may be urged that it is part of the very meaning of a relational proposition that the relation

involved should have to the terms the relation expressed in saying that
it relates them, and that this is what makes the distinction, which we
formerly (§ 54) left unexplained, between a relating relation and a relation
in itself. It may be urged, however, against this view, that the assertion
of a relation between the relation and the terms, though implied, is no
part of the original proposition, and that a relating relation is distinguished from a relation in itself by the indefinable element of assertion
which distinguishes a proposition from a concept. Against this it
might be retorted that, in the concept " difference of a and b," difference
relates a and b just as much as in the proposition " a and b differ"; but
to this it may be rejoined that we found the difference of a and b, except
in so far as some specific point of difference may be in question, to be
indistinguishable from bare difference. Thus it seems impossible to
prove that the endless regress involved is of the objectionable kind.
We may distinguish, I think, between " a exceeds b " and " a is greater
than b," though it would be absurd to deny that people usually mean
the same thing by these two propositions. On the principle, from which
I can see no escape, that every genuine word must have some meaning,
the *is* and *than* must form part of " a is greater than b," which thus
contains more than two terms and a relation. The *is* seems to state
that a has to *greater* the relation of referent, while the *than* states
similarly that b has to *greater* the relation of relatum. But " a exceeds
b" may be held to express solely the relation of a to b, without including any of the implications of further relations. Hence we shall
have to conclude that a relational proposition aRb does not include
in its *meaning* any relation of a or b to R, and that the endless regress,
though undeniable, is logically quite harmless. With these remarks,
we may leave the further theory of relations to later Parts of the present
work.

CHAPTER X.

THE CONTRADICTION.

100. Before taking leave of fundamental questions, it is necessary to examine more in detail the singular contradiction, already mentioned, with regard to predicates not predicable of themselves. Before attempting to solve this puzzle, it will be well to make some deductions connected with it, and to state it in various different forms. I may mention that I was led to it in the endeavour to reconcile Cantor's proof that there can be no greatest cardinal number with the very plausible supposition that the class of all terms (which we have seen to be essential to all formal propositions) has necessarily the greatest possible number of members[*].

Let w be a class-concept which can be asserted of itself, *i.e.* such that " w is a w." Instances are *class-concept*, and the negations of ordinary class-concepts, *e.g.* not-man. Then (a) if w be contained in another class v, since w is a w, w is a v; consequently there is a term of v which is a class-concept that can be asserted of itself. Hence by contraposition, (β) if u be a class-concept none of whose members are class-concepts that can be asserted of themselves, no class-concept contained in u can be asserted of itself. Hence further, (γ) if u be any class-concept whatever, and u' the class-concept of those members of u which are not predicable of themselves, this class-concept is contained in itself, and none of its members are predicable of themselves; hence by (β) u' is not predicable of itself. Thus u' is not a u', and is therefore not a u; for the terms of u that are not terms of u' are all predicable of themselves, which u' is not. Thus (δ) if u be any class-concept whatever, there is a class-concept contained in u which is not a member of u, and is also one of those class-concepts that are not predicable of themselves. So far, our deductions seem scarcely open to question. But if we now take the last of them, and admit the class of those class-concepts that cannot be asserted of themselves, we find that this class must contain a class-concept not a member of itself and yet not belonging to the class in question.

We may observe also that, in virtue of what we have proved in (β), the class of class-concepts which cannot be asserted of themselves, which we

---

[*] See Part V, Chap. xliii, § 344 ff.

will call w, contains as members of itself all its sub-classes, although it is easy to prove that every class has more sub-classes than terms. Again, if y be any term of w, and w' be the whole of w except y, then w', being a sub-class of w, is not a w' but is a w, and therefore is y. Hence each class-concept which is a term of w has all other terms of w as its extension. It follows that the concept *bicycle* is a teaspoon, and *teaspoon* is a bicycle. This is plainly absurd, and any number of similar absurdities can be proved.

101. Let us leave these paradoxical consequences, and attempt the exact statement of the contradiction itself. We have first the statement in terms of predicates, which has been given already. If x be a predicate, x may or may not be predicable of itself. Let us assume that "not-predicable of oneself" is a predicate. Then to suppose either that this predicate is, or that it is not, predicable of itself, is self-contradictory. The conclusion, in this case, seems obvious: "not-predicable of oneself" is not a predicate.

Let us now state the same contradiction in terms of class-concepts. A class-concept may or may not be a term of its own extension. "Class-concept which is not a term of its own extension" appears to be a class-concept. But if it is a term of its own extension, it is a class-concept which is not a term of its own extension, and *vice versâ*. Thus we must conclude, against appearances, that "class-concept which is not a term of its own extension" is not a class-concept.

In terms of classes the contradiction appears even more extraordinary. A class as one may be a term of itself as many. Thus the class of all classes is a class; the class of all the terms that are not men is not a man, and so on. Do all the classes that have this property form a class? If so, is it as one a member of itself as many or not? If it is, then it is one of the classes which, as ones, are not members of themselves as many, and *vice versâ*. Thus we must conclude again that the classes which as ones are not members of themselves as many do not form a class—or rather, that they do not form a class as one, for the argument cannot show that they do not form a class as many.

102. A similar result, which, however, does not lead to a contradiction, may be proved concerning any relation. Let R be a relation, and consider the class w of terms which do not have the relation R to themselves. Then it is impossible that there should be any term a to which all of them and no other terms have the relation R. For, if there were such a term, the propositional function "x does not have the relation R to x" would be equivalent to "x has the relation R to a." Substituting a for x throughout, which is legitimate since the equivalence is formal, we find a contradiction. When in place of R we put ϵ, the relation of a term to a class-concept which can be asserted of it, we get the above contradiction. The reason that a contradiction emerges here is that we have taken it as an axiom that any propositional function containing only one variable is equivalent to asserting membership of a class defined by the propositional function. Either this axiom, or the principle that every class can be taken as one term, is plainly false, and there is no fundamental objection to dropping either. But having dropped the former, the question arises : Which propositional functions define classes which are single terms as well as many, and which do not? And with this question our real difficulties begin.

Any method by which we attempt to establish a one-one or many-one correlation of all terms and all propositional functions must omit at least one propositional function. Such a method would exist if all propositional functions could be expressed in the form ...ϵu, since this form correlates u with ...ϵu. But the impossibility of any such correlation is proved as follows. Let ϕ_x be a propositional function correlated with x; then, if the correlation covers all terms, the denial of $\phi_x(x)$ will be a propositional function, since it is a proposition for all values of x. But it cannot be included in the correlation ; for if it were correlated with a, $\phi_a(x)$ would be equivalent, for all values of x, to the denial of $\phi_x(x)$; but this equivalence is impossible for the value a, since it makes $\phi_a(a)$ equivalent to its own denial. It follows that there are more propositional functions than terms—a result which seems plainly impossible, although the proof is as convincing as any in Mathematics. We shall shortly see how the impossibility is removed by the doctrine of logical types.

103. The first method which suggests itself is to seek an ambiguity in the notion of ϵ. But in Chapter vi we distinguished the various meanings as far as any distinction seemed possible, and we have just seen that with each meaning the same contradiction emerges. Let us, however, attempt to state the contradiction throughout in terms of propositional functions. Every propositional function which is not null, we supposed, defines a class, and every class can certainly be defined by a propositional function. Thus to say that a class as one is not a member of itself as many is to say that the class as one does not satisfy the function by which itself as many is defined. Since all propositional functions except such as are null define classes, all will be used up, in considering all classes having the above property, except such as do not have the above property. If any propositional function were satisfied by every class having the above property, it would therefore necessarily be one satisfied also by the class w of all such classes considered as a single term. Hence the class w does not itself belong to the class w, and therefore there must be some propositional function satisfied by the terms of w but not by w itself. Thus the contradiction re-emerges, and we must suppose, either that there is no such entity as w, or that there is no propositional function satisfied by its terms and by no others.

It might be thought that a solution could be found by denying the legitimacy of variable propositional functions. If we denote by k_ϕ, for the moment, the class of values satisfying ϕ, our propositional function is the denial of $\phi(k_\phi)$, where ϕ is the variable. The doctrine of Chapter VII, that ϕ is not a separable entity, might make such a variable seem illegitimate; but this objection can be overcome by substituting for ϕ the class of propositions ϕx, or the relation of ϕx to x. Moreover it is impossible to exclude variable propositional functions altogether. Wherever a variable class or a variable relation occurs, we have admitted a variable propositional function, which is thus essential to assertions about every class or about every relation. The definition of the domain of a relation, for example, and all the general propositions which constitute the calculus of relations, would be swept away by the refusal to allow this type of variation. Thus we require some further characteristic by which to distinguish two kinds of variation. This characteristic is to be found, I think, in the independent variability of the function and the argument. In general, ϕx is itself a function of two variables, ϕ and x; of these, either may be given a constant value, and either may be varied without reference to the other. But in the type of propositional functions we are considering in this Chapter, the argument is itself a function of the propositional function: instead of ϕx, we have $\phi\{f(\phi)\}$, where $f(\phi)$ is defined as a function of ϕ. Thus when ϕ is varied, the argument of which ϕ is asserted is varied too. Thus "x is an x" is equivalent to: "ϕ can be asserted of the class of terms satisfying ϕ," this class of terms being x. If here ϕ is varied, the argument is varied at the same time in a manner dependent upon the variation of ϕ. For this reason, $\phi\{f(\phi)\}$, though it is a definite proposition when x is assigned, is not a propositional function, in the ordinary sense, when x is variable. Propositional functions of this doubtful type may be called *quadratic forms*, because the variable enters into them in a way somewhat analogous to that in which, in Algebra, a variable appears in an expression of the second degree.

104. Perhaps the best way to state the suggested solution is to say that, if a collection of terms can only be defined by a variable propositional function, then, though a class as many may be admitted, a class as one must be denied. When so stated, it appears that propositional functions may be varied, provided the resulting collection is never itself made into the subject in the original propositional function. In such cases there is only a class as many, not a class as one. We took it as axiomatic that the class as one is to be found wherever there is a class as many; but this axiom need not be universally admitted, and appears to have been the source of the contradiction. By denying it, therefore, the whole difficulty will be overcome.

A class as one, we shall say, is an object of the same *type* as its terms; *i.e.* any propositional function $\phi(x)$ which is significant when one of the terms is substituted for x is also significant when the class as one is substituted. But the class as one does not always exist, and the class as many is of a different type from the terms of the class, even when the class has only one term, *i.e.* there are propositional functions $\phi(u)$ in which u may be the class as many, which are meaningless if, for u, we substitute one of the terms of the class. And so "x is one among x's" is not a proposition at all if the relation involved is that of a term to its class as many; and this is the only relation of whose presence a propositional function always assures us. In this view, a class as many may be a logical subject, but in propositions of a different kind from those in which its terms are subjects; of any object other than a single term, the question whether it is one or many will have different answers according to the proposition in which it occurs. Thus we have "Socrates is one among men," in which men are plural; but "men are one among species of animals," in which men are singular. It is the distinction of logical types that is the key to the whole mystery*.

105. Other ways of evading the contradiction, which might be suggested, appear undesirable, on the ground that they destroy too many quite necessary kinds of propositions. It might be suggested that identity is introduced in "x is not an x" in a way which is not permissible. But it has been already shown that relations of terms to themselves are unavoidable, and · it may be observed that suicides or self-made men or the heroes of Smiles's *Self-Help* are all defined by relations to themselves. And generally, identity enters in a very similar way into formal implication, so that it is quite impossible to reject it.

A natural suggestion for escaping from the contradiction would be to demur to the notion of *all* terms or of *all* classes. It might be urged that no such sum-total is conceivable; and if *all* indicates a whole, our escape from the contradiction requires us to admit this. But we have already abundantly seen that if this view were maintained against *any* term, all formal truth would be impossible, and Mathematics, whose characteristic is the statement of truths concerning *any* term, would be abolished at one stroke. Thus the correct statement of formal truths requires the notion of *any* term or *every* term, but not the collective notion of *all* terms.

It should be observed, finally, that no peculiar philosophy is involved in the above contradiction, which springs directly from common sense, and can only be solved by abandoning some common-sense assumption. Only the Hegelian philosophy, which nourishes itself on contradictions, can remain indifferent, because it finds similar problems everywhere. In any other doctrine, so direct a challenge demands an answer, on pain of a confession of impotence. Fortunately, no other similar difficulty, so far as I know, occurs in any other portion of the Principles of Mathematics.

* On this subject, see Appendix.

106. We may now briefly review the conclusions arrived at in Part I. Pure Mathematics was defined as the class of propositions asserting formal implications and containing no constants except logical constants. And logical constants are: Implication, the relation of a term to a class of which it is a member, the notion of *such that*, the notion of relation, and such further notions as are involved in formal implication, which we found (§ 93) to be the following: propositional function, class *, denoting, and *any* or *every term*. This definition brought Mathematics into very close relation to Logic, and made it practically identical with Symbolic Logic. An examination of Symbolic Logic justified the above enumeration of mathematical indefinables. In Chapter III we distinguished implication and formal implication. The former holds between any two propositions provided the first be false or the second true. The latter is not a relation, but the assertion, for every value of the variable or variables, of a propositional function which, for every value of the variable or variables, asserts an implication. Chapter IV distinguished what may be called *things* from predicates and relations (including the *is* of predications among relations for this purpose). It was shown that this distinction is connected with the doctrine of substance and attributes, but does not lead to the traditional results. Chapters V and VI developed the theory of predicates. In the former of these chapters it was shown that certain concepts, derived from predicates, occur in propositions not *about* themselves, but about combinations of terms, such as are indicated by *all, every, any, a, some,* and *the*. Concepts of this kind, we found, are fundamental in Mathematics, and enable us to deal with infinite classes by means of propositions of finite complexity. In Chapter VI we distinguished predicates, class-concepts, concepts of classes, classes as many, and classes as one. We agreed that single terms, or such combinations as result from *and,* are classes, the latter being classes as many; and that classes as many are the objects denoted by concepts of classes, which are the plurals of class-concepts. But in the present chapter we decided that it is necessary to distinguish a single term from the class whose only member it is, and that consequently the null-class may be admitted.

In Chapter VII we resumed the study of the verb. Subject-predicate propositions, and such as express a fixed relation to a fixed term, could be analyzed, we found, into a subject and an assertion; but this analysis becomes impossible when a given term enters into a proposition in a more complicated manner than as referent of a relation. Hence it became necessary to take *propositional function* as a primitive notion. A propositional function of one variable is any proposition of a set defined by the variation of a single term, while the other terms remain

---

* The notion of *class* in general, we decided, could be replaced, as an indefinable, by that of the class of propositions defined by a propositional function.

constant. But in general it is impossible to define or isolate the constant element in a propositional function, since what remains, when a certain term, wherever it occurs, is left out of a proposition, is in general no discoverable kind of entity. Thus the term in question must be not simply omitted, but replaced by a *variable*.

The notion of the variable, we found, is exceedingly complicated. The x is not simply *any* term, but any term with a certain individuality; for if not, any two variables would be indistinguishable. We agreed that a variable is any term *quâ* term in a certain propositional function, and that variables are distinguished by the propositional functions in which they occur, or, in the case of several variables, by the place they occupy in a given multiply variable propositional function. A variable, we said, is *the* term in *any* proposition of the set denoted by a given propositional function.

Chapter IX pointed out that relational propositions are ultimate, and that they all have *sense*: *i.e.* the relation being the concept as such in a proposition with two terms, there is another proposition containing the same terms and the same concept as such, as in "*A* is greater than *B*" and "*B* is greater than *A*." These two propositions, though different, contain precisely the same constituents. This is a characteristic of relations, and an instance of the loss resulting from analysis. Relations, we agreed, are to be taken intensionally, not as classes of couples *.

Finally, in the present chapter, we examined the contradiction resulting from the apparent fact that, if w be the class of all classes which as single terms are not members of themselves as many, then w as one can be proved both to be and not to be a member of itself as many. The solution suggested was that it is necessary to distinguish various types of objects, namely terms, classes of terms, classes of classes, classes of couples of terms, and so on; and that a propositional function ϕx in general requires, if it is to have any meaning, that x should belong to some one type. Thus $x \epsilon x$ was held to be meaningless, because ϵ requires that the relatum should be a class composed of objects which are of the type of the referent. The class as one, where it exists, is, we said, of the same type as its constituents; but a quadratic propositional function in general appears to define only a class as many, and the contradiction proves that the class as one, if it ever exists, is certainly sometimes absent.

* On this point, however, see Appendix.

PART II.

NUMBER.

CHAPTER XI.

DEFINITION OF CARDINAL NUMBERS.

107. WE have now briefly reviewed the apparatus of general logical notions with which Mathematics operates. In the present Part, it is to be shown how this apparatus suffices, without new indefinables or new postulates, to establish the whole theory of cardinal integers as a special branch of Logic*. No mathematical subject has made, in recent years, greater advances than the theory of Arithmetic. The movement in favour of correctness in deduction, inaugurated by Weierstrass, has been brilliantly continued by Dedekind, Cantor, Frege, and Peano, and attains what seems its final goal by means of the logic of relations. As the modern mathematical theory is but imperfectly known even by most mathematicians, I shall begin this Part by four chapters setting forth its outlines in a non-symbolic form. I shall then examine the process of deduction from a philosophical standpoint, in order to discover, if possible, whether any unperceived assumptions have covertly intruded themselves in the course of the argument.

108. It is often held that both number and particular numbers are indefinable. Now definability is a word which, in Mathematics, has a precise sense, though one which is relative to some given set of notions†. Given any set of notions, a term is definable by means of these notions when, and only when, it is the only term having to certain of these notions a certain relation which itself is one of the said notions. But philosophically, the word *definition* has not, as a rule, been employed in this sense; it has, in fact, been restricted to the analysis of an idea into its constituents. This usage is inconvenient and, I think, useless; moreover it seems to overlook the fact that wholes are *not*, as a

---

* Cantor has shown that it is necessary to separate the study of Cardinal and Ordinal numbers, which are distinct entities, of which the former are simpler, but of which both are essential to ordinary Mathematics. On Ordinal numbers, cf. Chaps. XXIX, XXXVIII, *infra*.

† See Peano, *F.* 1901, p. 6 ff. and Padoa, "Théorie Algébrique des Nombres Entiers," *Congrès*, Vol. III, p. 314 ff.

rule, determinate when their constituents are given, but are themselves new entities (which may be in some sense simple), defined, in the mathematical sense, by certain relations to their constituents. I shall, therefore, in future, ignore the philosophical sense, and speak only of mathematical definability. I shall, however, restrict this notion more than is done by Professor Peano and his disciples. They hold that the various branches of Mathematics have various indefinables, by means of which the remaining ideas of the said subjects are defined. I hold— and it is an important part of my purpose to prove—that all Pure Mathematics (including Geometry and even rational Dynamics) contains only one set of indefinables, namely the fundamental logical concepts discussed in Part I. When the various logical constants have been enumerated, it is somewhat arbitrary which of them we regard as indefinable, though there are apparently some which must be indefinable in any theory. But my contention is, that the indefinables of Pure Mathematics are all of this kind, and that the presence of any other indefinables indicates that our subject belongs to Applied Mathematics. Moreover, of the three kinds of definition admitted by Peano—the nominal definition, the definition by postulates, and the definition by abstraction*—I recognize only the nominal: the others, it would seem, are only necessitated by Peano's refusal to regard relations as part of the fundamental apparatus of logic, and by his somewhat undue haste in regarding as an individual what is really a class. These remarks will be best explained by considering their application to the definition of cardinal numbers.

109. It has been common in the past, among those who regarded numbers as definable, to make an exception as regards the number 1, and to define the remainder by its means. Thus 2 was $1 + 1$, 3 was $2 + 1$, and so on. This method was only applicable to finite numbers, and made a tiresome difference between 1 and other numbers; moreover the meaning of + was commonly not explained. We are able now-a-days to improve greatly upon this method. In the first place, since Cantor has shown how to deal with the infinite, it has become both desirable and possible to deal with the fundamental properties of numbers in a way which is equally applicable to finite and infinite numbers. In the second place, the logical calculus has enabled us to give an exact definition of arithmetical addition; and in the third place, it has become as easy to define 0 and 1 as to define any other number. In order to explain how this is done, I shall first set forth the definition of numbers by abstraction; I shall then point out formal defects in this definition, and replace it by a nominal definition.

Numbers are, it will be admitted, applicable essentially to classes. It is true that, where the number is finite, individuals may be enumerated

* Cf. Burali-Forti, "Sur les différentes définitions du nombre réel," *Congrès*, III, p. 294 ff.

to make up the given number, and may be counted one by one without any mention of a class-concept. But all finite collections of individuals form classes, so that what results is after all the number of a class. And where the number is infinite, the individuals cannot be enumerated, but must be defined by intension, *i.e.* by some common property in virtue of which they form a class. Thus when any class-concept is given, there is a certain number of individuals to which this class-concept is applicable, and the number may therefore be regarded as a property of the class. It is this view of numbers which has rendered possible the whole theory of infinity, since it relieves us of the necessity of enumerating the individuals whose number is to be considered. This view depends fundamentally upon the notion of *all*, the numerical conjunction as we agreed to call it (§ 59). *All men*, for example, denotes men conjoined in a certain way; and it is as thus denoted that they have a number. Similarly *all numbers* or *all points* denotes numbers or points conjoined in a certain way, and as thus conjoined numbers or points have a number. Numbers, then, are to be regarded as properties of classes.

The next question is: Under what circumstances do two classes have the same number? The answer is, that they have the same number when their terms can be correlated one to one, so that any one term of either corresponds to one and only one term of the other. This requires that there should be some one-one relation whose domain is the one class and whose converse domain is the other class. Thus, for example, if in a community all the men and all the women are married, and polygamy and polyandry are forbidden, the number of men must be the same as the number of women. It might be thought that a one-one relation could not be defined except by reference to the number 1. But this is not the case. A relation is one-one when, if x and x' have the relation in question to y, then x and x' are identical; while if x has the relation in question to y and y', then y and y' are identical. Thus it is possible, without the notion of unity, to define what is meant by a one-one relation. But in order to provide for the case of two classes which have no terms, it is necessary to modify slightly the above account of what is meant by saying that two classes have the same number. For if there are no terms, the terms cannot be correlated one to one. We must say: Two classes have the same number when, and only when, there is a one-one relation whose domain includes the one class, and which is such that the class of correlates of the terms of the one class is identical with the other class. From this it appears that two classes having no terms have always the same number of terms; for if we take any one-one relation whatever, its domain includes the null-class, and the class of correlates of the null-class is again the null-class. When two classes have the same number, they are said to be *similar*.

Some readers may suppose that a definition of what is meant by saying that two classes have the same number is wholly unnecessary.

The way to find out, they may say, is to count both classes. It is such notions as this which have, until very recently, prevented the exhibition of Arithmetic as a branch of Pure Logic. For the question immediately arises: What is meant by counting? To this question we usually get only some irrelevant psychological answer, as, that counting consists in successive acts of attention. In order to count 10, I suppose that ten acts of attention are required: certainly a most useful definition of the number 10! Counting has, in fact, a good meaning, which is not psychological. But this meaning is highly complex; it is only applicable to classes which can be well-ordered, which are not known to be all classes; and it only gives the number of the class when this number is finite—a rare and exceptional case. We must not, therefore, bring in counting where the definition of numbers is in question.

The relation of similarity between classes has the three properties of being reflexive, symmetrical, and transitive; that is to say, if u, v, w be classes, u is similar to itself; if u be similar to v, v is similar to u; and if u be similar to v, and v to w, then u is similar to w. These properties all follow easily from the definition. Now these three properties of a relation are held by Peano and common sense to indicate that when the relation holds between two terms, those two terms have a certain common property, and *vice versâ*. This common property we call their number* This is the definition of numbers by abstraction.

110. Now this definition by abstraction, and generally the process employed in such definitions, suffers from an absolutely fatal formal defect: it does not show that only one object satisfies the definition†. Thus instead of obtaining *one* common property of similar classes, which is *the* number of the classes in question, we obtain a *class* of such properties, with no means of deciding how many terms this class contains. In order to make this point clear, let us examine what is meant, in the present instance, by a common property. What is meant is, that any class has to a certain entity, its number, a relation which it has to nothing else, but which all similar classes (and no other entities) have to the said number. That is, there is a many-one relation which every class has to its number and to nothing else. Thus, so far as the definition by abstraction can show, any set of entities to each of which some class has a certain many-one relation, and to one and only one of which any given class has this relation, and which are such that all classes similar to a given class have this relation to one and the same entity of the set, appear as the set of numbers, and any entity of this set is *the* number of some class. If, then, there are many such sets of entities—and it is easy

---

* Cf. Peano, *F.* 1901, § 32, ·0, Note.
† On the necessity of this condition, cf. Padoa, *loc. cit.*, p. 324. Padoa appears not to perceive, however, that *all* definitions define the single individual of a class: when what is defined is a class, this must be the only term of some class of classes.

to prove that there are an infinite number of them—every class will have many numbers, and the definition wholly fails to define *the* number of a class. This argument is perfectly general, and shows that definition by abstraction is never a logically valid process.

111. There are two ways in which we may attempt to remedy this defect. One of these consists in defining as *the* number of a class the whole class of entities, chosen one from each of the above sets of entities, to which all classes similar to the given class (and no others) have some many-one relation or other. But this method is practically useless, since all entities, without exception, belong to every such class, so that every class will have as its number the class of all entities of every sort and description. The other remedy is more practicable, and applies to all the cases in which Peano employs definition by abstraction. This method is, to define as the number of a class the class of all classes similar to the given class. Membership of this class of classes (considered as a predicate) is a common property of all the similar classes and of no others; moreover every class of the set of similar classes has to the set a relation which it has to nothing else, and which every class has to its own set. Thus the conditions are completely fulfilled by this class of classes, and it has the merit of being determinate when a class is given, and of being different for two classes which are not similar. This, then, is an irreproachable definition of the number of a class in purely logical terms.

To regard a number as a class of classes must appear, at first sight, a wholly indefensible paradox. Thus Peano (*F.* 1901, § 32) remarks that " we cannot identify the number of [a class] *a* with the class of classes in question [*i.e.* the class of classes similar to *a*], for these objects have different properties." He does not tell us what these properties are, and for my part I am unable to discover them. Probably it appeared to him immediately evident that a number is not a class of classes. But something may be said to mitigate the appearance of paradox in this view. In the first place, such a word as *couple* or *trio* obviously does denote a class of classes. Thus what we have to say is, for example, that " two men" means "logical product of class of men and couple," and "there are two men" means "there is a class of men which is also a couple." In the second place, when we remember that a class-concept is not itself a collection, but a property by which a collection is defined, we see that, if we define the number as the class-concept, not the class, a number is really defined as a common property of a set of similar classes and of nothing else. This view removes the appearance of paradox to a great degree. There is, however, a philosophical difficulty in this view, and generally in the connection of classes and predicates. It may be that there are many predicates common to a certain collection of objects and to no others. In this case, these predicates are all regarded by Symbolic Logic as equivalent, and any one of them is said to be equal to any other. Thus if the predicate were defined by the collection of objects, we should not obtain, in general, a single predicate, but a class of predicates; for this class of predicates we should require a new class-concept, and so on. The only available class-concept would be "predicability of the given collection of terms and of no others." But in the present case, where the collection is defined by a certain relation to one of its terms, there is some danger of a logical error. Let u be a class; then the number of u, we said, is the class of classes similar to u. But "similar to u" cannot be the actual concept which constitutes the number of u; for, if v be similar to u, "similar to v" defines the same class, although it is a different concept. Thus we require, as the defining predicate of the class of similar classes, some concept which does not have any special relation to one or more of the constituent classes. In regard to every particular number that may be mentioned, whether finite or infinite, such a predicate is, as a matter of fact, discoverable; but when all we are told about a number is that it is the number of some class u, it is natural that a special reference to u should appear in the definition. This, however, is not the point at issue. The real point is, that what is defined is the same whether we use the predicate "similar to u" or "similar to v," provided u is similar to v. This shows that it is not the class-concept or defining predicate that is defined, but the class itself whose terms are the various classes which are similar to u or to v. It is such classes, therefore, and not predicates such as "similar to u," that must be taken to constitute numbers.

Thus, to sum up: Mathematically, a number is nothing but a class of similar classes: this definition allows the deduction of all the usual properties of numbers, whether finite or infinite, and is the only one (so far as I know) which is possible in terms of the fundamental concepts of general logic. But philosophically we may admit that every collection of similar classes has some common predicate applicable to no entities except the classes in question, and if we can find, by inspection, that there is a certain class of such common predicates, of which one and only one applies to each collection of similar classes, then we may, if we see fit, call this particular class of predicates the class of numbers. For my part, I do not know whether there is any such class of predicates, and I do know that, if there be such a class, it is wholly irrelevant to Mathematics. Wherever Mathematics derives a common property from a reflexive, symmetrical, and transitive relation, all mathematical purposes of the supposed common property are completely served when it is replaced by the class of terms having the given relation to a given term; and this is precisely the case presented by cardinal numbers. For the future, therefore, I shall adhere to the above definition, since it is at once precise and adequate to all mathematical uses.

CHAPTER XII.

ADDITION AND MULTIPLICATION.

112. In most mathematical accounts of arithmetical operations we find the error of endeavouring to give at once a definition which shall be applicable to rationals, or even to real numbers, without dwelling at sufficient length upon the theory of integers. For the present, integers alone will occupy us. The definition of integers, given in the preceding chapter, obviously does not admit of extension to fractions; and in fact the absolute difference between integers and fractions, even between integers and fractions whose denominator is unity, cannot possibly be too strongly emphasized. What rational fractions are, and what real numbers are, I shall endeavour to explain at a later stage; positive and negative numbers also are at present excluded. The integers with which we are now concerned are not positive, but signless. And so the addition and multiplication to be defined in this chapter are only applicable to integers; but they have the merit of being equally applicable to finite and infinite integers. Indeed, for the present, I shall rigidly exclude all propositions which involve either the finitude or the infinity of the numbers considered.

113. There is only one fundamental kind of addition, namely the logical kind. All other kinds can be defined in terms of this and logical multiplication. In the present chapter the addition of integers is to be defined by its means. Logical addition, as was explained in Part I, is the same as disjunction; if p and q are propositions, their logical sum is the proposition "p or q," and if u and v are classes, their logical sum is the class "u or v," i.e. the class to which belongs every term which either belongs to u or belongs to v. The logical sum of two classes u and v may be defined in terms of the logical product of two propositions, as the class of terms belonging to every' class in which both u and v are contained*. This definition is not essentially confined to two classes, but may be extended to a class of classes, whether finite or infinite. Thus if k be a class of classes, the logical sum of the classes composing k (called for short the sum of k) is

* F. 1901, § 2, Prop. 1·0.

the class of terms belonging to every class which contains every class which is a term of k. It is this notion which underlies arithmetical addition. If k be a class of classes no two of which have any common terms (called for short an exclusive class of classes), then the arithmetical sum of the numbers of the various classes of k is the number of terms in the logical sum of k. This definition is absolutely general, and applies equally whether k or any of its constituent classes be finite or infinite. In order to assure ourselves that the resulting number depends only upon the *numbers* of the various classes belonging to k, and not upon the particular class k that happens to be chosen, it is necessary to prove (as is easily done) that if k' be another exclusive class of classes, similar to k, and every member of k is similar to its correlate in k', and *vice versâ*, then the number of terms in the sum of k is the same as the number in the sum of k'. Thus, for example, suppose k has only two terms, u and v, and suppose u and v have no common part. Then the number of terms in the logical sum of u and v is the sum of the number of terms in u and in v; and if u' be similar to u, and v' to v, and u', v' have no common part, then the sum of u' and v' is similar to the sum of u and v.

114. With regard to this definition of a sum of numbers, it is to be observed that it cannot be freed from reference to classes which have the numbers in question. The number obtained by summation is essentially the number of the logical sum of a certain class of classes or of some similar class of similar classes. The necessity of this reference to classes emerges when one number occurs twice or oftener in the summation. It is to be observed that the numbers concerned have no *order* of summation, so that we have no such proposition as the commutative law: this proposition, as introduced in Arithmetic, results only from a defective symbolism, which causes an order among the symbols which has no correlative order in what is symbolized. But owing to the absence of order, if one number occurs twice in a summation, we cannot distinguish a first and a second occurrence of the said number. If we exclude a reference to classes which have the said number, there is no sense in the supposition of its occurring twice: the summation of a class of numbers can be defined, but in that case, no number can be repeated. In the above definition of a sum, the numbers concerned are defined as the numbers of certain classes, and therefore it is not necessary to decide whether any number is repeated or not. But in order to define, without reference to particular classes, a sum of numbers of which some are repeated, it is necessary first to define multiplication.

This point may be made clearer by considering a special case, such as $1 + 1$. It is plain that we cannot take the number 1 itself twice over, for there is one number 1, and there are not two instances of it. And if the logical addition of 1 to itself were in question, we should find that 1 and 1 is 1, according to the general principle of Symbolic Logic. Nor can we define 1 + 1 as the arithmetical sum of a certain class of numbers. This method can be employed as regards 1 + 2, or any sum in which no number is repeated; but as regards 1 + 1, the only class of numbers involved is the class whose only member is 1, and since this class has one member, not two, we cannot define 1 + 1 by its means. Thus the full definition of 1 + 1 is as follows: 1 + 1 is the number of a class w which is the logical sum of two classes u and v which have no common term and have each only one term. The chief point to be observed is, that logical addition of classes is the fundamental notion, while the arithmetical addition of numbers is wholly subsequent.

115. The general definition of multiplication is due to Mr A. N. Whitehead*. It is as follows. Let k be a class of classes, no two of which have any term in common. Form what is called the multiplicative class of k, *i.e.* the class each of whose terms is a class formed by choosing one and only one term from each of the classes belonging to k. Then the number of terms in the multiplicative class of k is the product of all the numbers of the various classes composing k. This definition, like that of addition given above, has two merits, which make it preferable to any other hitherto suggested. In the first place, it introduces no order among the numbers multiplied, so that there is no need of the commutative law, which, here as in the case of addition, is concerned rather with the symbols than with what is symbolized. In the second place, the above definition does not require us to decide, concerning any of the numbers involved, whether they are finite or infinite. Cantor has given† definitions of the sum and product of *two* numbers, which do not require a decision as to whether these numbers are finite or infinite. These definitions can be extended to the sum and product of any *finite* number of finite or infinite numbers; but they do not, as they stand, allow the definition of the sum or product of an infinite number of numbers. This grave defect is remedied in the above definitions, which enable us to pursue Arithmetic, as it ought to be pursued, without introducing the distinction of finite and infinite until we wish to study it. Cantor's definitions have also the formal defect of introducing an order among the numbers summed or multiplied: but this is, in his case, a mere defect in the symbols chosen, not in the ideas which he symbolizes. Moreover it is not practically desirable, in the case of the sum or product of *two* numbers, to avoid this formal defect, since the resulting cumbrousness becomes intolerable.

116. It is easy to deduce from the above definitions the usual connection of addition and multiplication, which may be thus stated. If k be a class of b mutually exclusive classes, each of which contains a terms, then the logical sum of k contains $a \times b$ terms‡. It is also

---

* *American Journal of Mathematics,* Oct. 1902.
† *Math. Annalen,* Vol. XLVI, § 3.   ‡ See Whitehead, *loc. cit.*

easy to obtain the definition of a^b, and to prove the associative and distributive laws, and the formal laws for powers, such as $a^b a^c = a^{b+c}$. But it is to be observed that exponentiation is not to be regarded as a new independent operation, since it is merely an application of multiplication. It is true that exponentiation can be independently defined, as is done by Cantor*, but there is no advantage in so doing. Moreover exponentiation unavoidably introduces ordinal notions, since a^b is not in general equal to b^a. For this reason we cannot define the result of an infinite number of exponentiations. Powers, therefore, are to be regarded simply as abbreviations for products in which all the numbers multiplied together are equal.

From the data which we now possess, all those propositions which hold equally of finite and infinite numbers can be deduced. The next step, therefore, is to consider the distinction between the finite and the infinite.

* *Loc. cit.*, § 4.

CHAPTER XIII.

FINITE AND INFINITE.

117. THE purpose of the present chapter is not to discuss the philosophical difficulties concerning the infinite, which are postponed to Part V. For the present I wish merely to set forth briefly the mathematical theory of finite and infinite as it appears in the theory of cardinal numbers. This is its most fundamental form, and must be understood before the ordinal infinite can be adequately explained*.

Let u be any class, and let u' be a class formed by taking away one term x from u. Then it may or may not happen that u is similar to u'. For example, if u be the class of all finite numbers, and u' the class of all finite numbers except 0, the terms of u' are obtained by adding 1 to each of the terms of u, and this correlates one term of u with one of u' and *vice versâ*, no term of either being omitted or taken twice over. Thus u' is similar to u. But if u consists of all finite numbers up to n, where n is some finite number, and u' consists of all these except 0, then u' is not similar to u. If there is one term x which can be taken away from u to leave a similar class u', it is easily proved that if any other term y is taken away instead of x we also get a class similar to u. When it is possible to take away one term from u and leave a class u' similar to u, we say that u is an *infinite* class. When this is not possible, we say that u is a *finite* class. From these definitions it follows that the null-class is finite, since no term can be taken from it. It is also easy to prove that if u be a finite class, the class formed by adding one term to u is finite ; and conversely if this class is finite, so is u. It follows from the definition that the numbers of finite classes other than the null-class are altered by subtracting 1, while those of infinite classes are unaltered by this operation. It is easy to prove that the same holds of the addition of 1.

118. Among finite classes, if one is a proper part of another, the one has a smaller number of terms than the other. (A proper part is a part not the whole.) But among infinite classes, this no longer holds.

---

* On the present topic cf. Cantor, *Math. Annalen*, Vol. XLVI, §§ 5, 6, where most of what follows will be found.

This distinction is, in fact, an essential part of the above definitions of the finite and the infinite. Of two infinite classes, one may have a greater or a smaller number of terms than the other. A class u is said to be greater than a class v, or to have a number greater than that of v, when the two are not similar, but v is similar to a proper part of u. It is known that if u is similar to a proper part of v, and v to a proper part of u (a case which can only arise when u and v are infinite), then u is similar to v; hence "u is greater than v" is inconsistent with "v is greater than u." It is not at present known whether, of two different infinite numbers, one must be greater and the other less. But it is known that there is a least infinite number, *i.e.* a number which is less than any different infinite number. This is the number of finite integers, which will be denoted, in the present work, by α_0*. This number is capable of several definitions in which no mention is made of the finite numbers. In the first place it may be defined (as is implicitly done by Cantor†) by means of the principle of mathematical induction. This definition is as follows: α_0 is the number of any class u which is the domain of a one-one relation R, whose converse domain is contained in but not coextensive with u, and which is such that, calling the term to which x has the relation R the *successor* of x, if s be any class to which belongs a term of u which is not a successor of any other term of u, and to which belongs the successor of every term of u which belongs to s, then every term of u belongs to s. Or again, we may define α_0 as follows. Let P be a transitive and asymmetrical relation, and let any two different terms of the field of P have the relation P or its converse. Further let any class u contained in the field of P and having successors (*i.e.* terms to which every term of u has the relation P) have an immediate successor, *i.e.* a term whose predecessors either belong to u or precede some term of u; let there be one term of the field of P which has no predecessors, but let every term which has predecessors have successors and also have an immediate predecessor; then the number of terms in the field of P is α_0. Other definitions may be suggested, but as all are equivalent it is not necessary to multiply them. The following characteristic is important: Every class whose number is α_0 can be arranged in a series having consecutive terms, a beginning but no end, and such that the number of predecessors of any term of the series is finite; and any series having these characteristics has the number α_0.

It is very easy to show that every infinite class contains classes whose number is α_0. For let u be such a class, and let x_0 be a term of u. Then u is similar to the class obtained by taking away x_0, which we will call the class u_1. Thus u_1 is an infinite class. From this we can take

---

* Cantor employs for this number the Hebrew Aleph with the suffix 0, but this notation is inconvenient.

† *Math. Annalen,* Vol. XLVI, § 6.

away a term x_1, leaving an infinite class u_2, and so on. The series of terms x_1, x_2, ... is contained in u, and is of the type which has the number α_0. From this point we can advance to an alternative definition of the finite and the infinite by means of mathematical induction, which must now be explained.

119. If n be any finite number, the number obtained by adding 1 to n is also finite, and is different from n. Thus beginning with 0 we can form a series of numbers by successive additions of 1. We may define finite numbers, if we choose, as those numbers that can be obtained from 0 by such steps, and that obey mathematical induction. That is, the class of finite numbers is the class of numbers which is contained in every class s to which belongs 0 and the successor of every number belonging to s, where the successor of a number is the number obtained by adding 1 to the given number. Now α_0 is not such a number, since, in virtue of propositions already proved, no such number is similar to a part of itself. Hence also no number greater than α_0 is finite according to the new definition. But it is easy to prove that every number less than α_0 is finite with the new definition as with the old. Hence the two definitions are equivalent. Thus we may define finite numbers either as those that can be reached by mathematical induction, starting from 0 and increasing by 1 at each step, or as those of classes which are not similar to the parts of themselves obtained by taking away single terms. These two definitions are both frequently employed, and it is important to realize that either is a consequence of the other. Both will occupy us much hereafter; for the present it is only intended, without controversy, to set forth the bare outlines of the mathematical theory of finite and infinite, leaving the details to be filled in during the course of the work.

CHAPTER XIV.

THEORY OF FINITE NUMBERS.

120. HAVING now clearly distinguished the finite from the infinite, we can devote ourselves to the consideration of finite numbers. It is customary, in the best treatises on the elements of Arithmetic*, not to define number or particular finite numbers, but to begin with certain axioms or primitive propositions, from which all the ordinary results are shown to follow. This method makes Arithmetic into an independent study, instead of regarding it, as is done in the present work, as merely a development, without new axioms or indefinables, of a certain branch of general Logic. For this reason, the method in question seems to indicate a less degree of analysis than that adopted here. I shall nevertheless begin by an exposition of the more usual method, and then proceed to definitions and proofs of what are usually taken as indefinables and indemonstrables. For this purpose, I shall take Peano's exposition in the *Formulaire*†, which is, so far as I know, the best from the point of view of accuracy and rigour. This exposition has the inestimable merit of showing that all Arithmetic can be developed from three fundamental notions (in addition to those of general Logic) and five fundamental propositions concerning these notions. It proves also that, if the three notions be regarded as determined by the five propositions, these five propositions are mutually independent. This is shown by finding, for each set of four out of the five propositions, an interpretation which renders the remaining proposition false. It therefore only remains, in order to connect Peano's theory with that here adopted, to give a definition of the three fundamental notions and a demonstration of the five fundamental propositions. When once this has been accomplished, we know with certainty that everything in the theory of finite integers follows.

* Except Frege's *Grundgesetze der Arithmetik* (Jena, 1893).
† *F.* 1901, Part II and *F.* 1899, § 20 ff. *F.* 1901 differs from earlier editions in making "number is a class" a primitive proposition. I regard this as unnecessary, since it is implied by "0 is a number." I therefore follow the earlier editions.

Peano's three indefinables are 0, *finite integer**, and *successor of.*
It is assumed, as part of the idea of succession (though it would,
I think, be better to state it as a separate axiom), that every number
has one and only one successor. (By *successor* is meant, of course,
immediate successor.) Peano's primitive propositions are then the
following. (1) 0 is a number. (2) If *a* is a number, the successor of
a is a number. (3) If two numbers have the same successor, the two
numbers are identical. (4) 0 is not the successor of any number.
(5) If *s* be a class to which belongs 0 and also the successor of every
number belonging to *s*, then every number belongs to *s*. The last of
these propositions is the principle of mathematical induction.

121. The mutual independence of these five propositions has been
demonstrated by Peano and Padoa as follows†. (1) Giving the usual
meanings to 0 and *successor*, but denoting by *number* finite integers
other than 0, all the above propositions except the first are true.
(2) Giving the usual meanings to 0 and *successor*, but denoting by
number only finite integers less than 10, or less than any other specified
finite integer, all the above propositions are true except the second.
(3) A series which begins by an antiperiod and then becomes periodic
(for example, the digits in a decimal which becomes recurring after a
certain number of places) will satisfy all the above propositions except
the third. (4) A periodic series (such as the hours on the clock)
satisfies all except the fourth of the primitive propositions. (5) Giving
to *successor* the meaning *greater by* 2, so that the successor of 0 is 2,
and of 2 is 4, and so on, all the primitive propositions are satisfied
except the fifth, which is not satisfied if *s* be the class of even numbers
including 0. Thus no one of the five primitive propositions can be
deduced from the other four.

122. Peano points out (*loc. cit.*) that other classes besides that of
the finite integers satisfy the above five propositions. What he says
is as follows: " There is an infinity of systems satisfying all the primitive
propositions. They are all verified, *e.g.*, by replacing *number* and 0 by
number other than 0 and 1. All the systems which satisfy the primitive
propositions have a one-one correspondence with the numbers. Number
is what is obtained from all these systems by abstraction; in other
words, number is the system which has all the properties enunciated
in the primitive propositions, and those only." This observation appears
to me lacking in logical correctness. In the first place, the question
arises: How are the various systems distinguished, which agree in satisfying the primitive propositions? How, for example, is the system
beginning with 1 distinguished from that beginning with 0? To this

---

* Throughout the rest of this chapter, I shall use *number* as synonymous with
finite integer.
† *F.* 1899, p. 30.

question two different answers may be given. We may say that 0 and 1 are both primitive ideas, or at least that 0 is so, and that therefore 0 and 1 can be intrinsically distinguished, as yellow and blue are distinguished. But if we take this view—which, by the way, will have to be extended to the other primitive ideas, number and succession—we shall have to say that these three notions are what I call constants, and that there is no need of any such process of abstraction as Peano speaks of in the definition of number. In this method, 0, number, and succession appear, like other indefinables, as ideas which must be simply recognized. Their recognition yields what mathematicians call the existence-theorem, *i.e.* it assures us that there really are numbers. But this process leaves it doubtful whether numbers are *logical* constants or not, and therefore makes Arithmetic, according to the definition in Part I, Chapter I, *primâ facie* a branch of Applied Mathematics. Moreover it is evidently not the process which Peano has in mind. The other answer to the question consists in regarding 0, number, and succession as a class of three ideas belonging to a certain class of trios defined by the five primitive propositions. It is very easy so to state the matter that the five primitive propositions become transformed into the nominal definition of a certain class of trios. There are then no longer any indefinables or indemonstrables in our theory, which has become a pure piece of Logic. But 0, number and succession become variables, since they are only determined as one of the class of trios: moreover the existence-theorem now becomes doubtful, since we cannot know, except by the discovery of at least one actual trio of this class, that there are any such trios at all. One *actual* trio, however, would be a constant, and thus we require some method of giving constant values to 0, number, and succession. What we can show is that, if there is one such trio, there are an infinite number of them. For by striking out the first term from any class satisfying the conditions laid down concerning number, we always obtain a class which again satisfies the conditions in question. But even this statement, since the meaning of number is still in question, must be differently worded if circularity is to be avoided. Moreover we must ask ourselves: Is any process of abstraction from all systems satisfying the five axioms, such as Peano contemplates, logically possible? Every term of a class is the term it is, and satisfies some proposition which becomes false when another term of the class is substituted. There is therefore no term of a class which has merely the properties defining the class and no others. What Peano's process of abstraction really amounts to is the consideration of the class and variable members of it, to the exclusion of constant members. For only a variable member of the class will have only the properties by which the class is defined. Thus Peano does not succeed in indicating any constant meaning for 0, number, and succession, nor in showing that any constant meaning is possible, since the existence-theorem is not proved. His only method, therefore, is to say that at least one such constant meaning can be immediately perceived, but is not definable. This method is not logically unsound, but it is wholly different from the impossible abstraction which he suggests. And the proof of the mutual independence of his five primitive propositions is only necessary in order to show that the definition of the class of trios determined by them is not redundant. Redundancy is not a logical error, but merely a defect of what may be called style. My object, in the above account of cardinal numbers, has been to prove, from general Logic, that there is one constant meaning which satisfies the above five propositions, and that this constant meaning should be called number, or rather finite cardinal number. And in this way, new indefinables and indemonstrables are wholly avoided; for when we have shown that the class of trios in question has at least one member, and when this member has been used to define number, we easily show that the class of trios has an infinite number of members, and we define the class by means of the five properties enumerated in Peano's primitive propositions. For the comprehension of the connection between Mathematics and Logic, this point is of very great importance, and similar points will occur constantly throughout the present work.

123. In order to bring out more clearly the difference between Peano's procedure and mine, I shall here repeat the definition of the class satisfying his five primitive propositions, the definition of *finite number*, and the proof, in the case of finite numbers, of his five primitive propositions.

The class of classes satisfying his axioms is the same as the class of classes whose cardinal number is α_0, *i.e.* the class of classes, according to my theory, which *is* α_0. It is most simply defined as follows : α_0 is the class of classes u each of which is the domain of some one-one relation R (the relation of a term to its successor) which is such that there is at least one term which succeeds no other term, every term which succeeds has a successor, and u is contained in any class s which contains a term of u having no predecessors, and also contains the successor of every term of u which belongs to s. This definition includes Peano's five primitive propositions and no more. Thus of every such class all the usual propositions in the arithmetic of finite numbers can be proved : addition, multiplication, fractions, etc. can be defined, and the whole of analysis can be developed, in so far as complex numbers are not involved. But in this whole development, the meaning of the entities and relations which occur is to a certain degree indeterminate, since the entities and the relation with which we start are variable members of a certain class. Moreover, in this whole development, nothing shows that there are such classes as the definition speaks of.

In the logical theory of cardinals, we start from the opposite end. We first define a certain class of entities, and then show that this class of entities belongs to the class α_0 above defined. This is done as follows. (1) 0 is the class of classes whose only member is the null-class. (2) A number is the class of all classes similar to any one of themselves. (3) 1 is the class of all classes which are not null and are such that, if x belongs to the class, the class without x is the null-class ; or such that, if x and y belong to the class, then x and y are identical. (4) Having shown that if two classes be similar, and a class of one term be added to each, the sums are similar, we define that, if n be a number, $n + 1$ is the number resulting from adding a unit to a class of n terms. (5) Finite numbers are those belonging to every class s to which belongs 0, and to which $n + 1$ belongs if n belongs. This completes the definition of finite numbers. We then have, as regards the five propositions which Peano assumes : (1) 0 is a number. (2) Meaning $n + 1$ by the successor of n, if n be a number, then $n + 1$ is a number. (3) If $n + 1 = m + 1$, then $n = m$. (4) If n be any number, $n + 1$ is different from 0. (5) If s be a class, and 0 belongs to this class, and if when n belongs to it, $n + 1$ belongs to it, then all finite numbers belong to it. Thus all the five essential properties are satisfied by the class of finite numbers as above defined. Hence the class of classes α_0 has members, and the class *finite number* is one definite member of α_0. There is, therefore, from the mathematical standpoint, no need whatever of new indefinables or indemonstrables in the whole of Arithmetic and Analysis.

CHAPTER XV.

ADDITION OF TERMS AND ADDITION OF CLASSES.

124. Having now briefly set forth the mathematical theory of cardinal numbers, it is time to turn our attention to the philosophical questions raised by this theory. I shall begin by a few preliminary remarks as to the distinction between philosophy and mathematics, and as to the function of philosophy in such a subject as the foundations of mathematics. The following observations are not necessarily to be regarded as applicable to other branches of philosophy, since they are derived specially from the consideration of the problems of logic.

The distinction of philosophy and mathematics is broadly one of point of view: mathematics is constructive and deductive, philosophy is critical, and in a certain impersonal sense controversial. Wherever we have deductive reasoning, we have mathematics; but the principles of deduction, the recognition of indefinable entities, and the distinguishing between such entities, are the business of philosophy. Philosophy is, in fact, mainly a question of insight and perception. Entities which are perceived by the so-called senses, such as colours and sounds, are, for some reason, not commonly regarded as coming within the scope of philosophy, except as regards the more abstract of their relations; but it seems highly doubtful whether any such exclusion can be maintained. In any case, however, since the present work is essentially unconcerned with sensible objects, we may confine our remarks to entities which are not regarded as existing in space and time. Such entities, if we are to know anything about them, must be also in some sense perceived, and must be distinguished one from another; their relations also must be in part immediately apprehended. A certain body of indefinable entities and indemonstrable propositions must form the starting-point for any mathematical reasoning; and it is this starting-point that concerns the philosopher. When the philosopher's work has been perfectly accomplished, its results can be wholly embodied in premises from which deduction may proceed. Now it follows from the very nature of such inquiries that results may be disproved, but can never be proved. The disproof will consist in pointing out contradictions and inconsistencies;

but the absence of these can never amount to proof. All depends, in
the end, upon immediate perception; and philosophical argument,
strictly speaking, consists mainly of an endeavour to cause the reader to
perceive what has been perceived by the author. The argument, in
short, is not of the nature of proof, but of exhortation. Thus the
question of the present chapter: Is there any indefinable set of entities
commonly called numbers, and different from the set of entities above
defined? is an essentially philosophical question, to be settled by inspection rather than by accurate chains of reasoning.

125. In the present chapter, we shall examine the question whether
the above definition of cardinal numbers in any way presupposes some
more fundamental sense of number. There are several ways in which
this may be supposed to be the case. In the first place, the individuals
which compose classes seem to be each in some sense *one*, and it might
be thought that a one-one relation could not be defined without introducing the number 1. In the second place, it may very well be
questioned whether a class which has only one term can be distinguished
from that one term. And in the third place, it may be held that the
notion of *class* presupposes number in a sense different from that above
defined: it may be maintained that classes arise from the addition of
individuals, as indicated by the word *and*, and that the logical addition
of classes is subsequent to this addition of individuals. These questions
demand a new inquiry into the meaning of *one* and of *class*, and here,
I hope, we shall find ourselves aided by the theories set forth in Part I.

As regards the fact that any individual or term is in some sense *one*,
this is of course undeniable. But it does not follow that the notion of
one is presupposed when individuals are spoken of: it may be, on the
contrary, that the notion of term or individual is the fundamental one,
from which that of *one* is derived. This view was adopted in Part I,
and there seems no reason to reject it. And as for one-one relations,
they are defined by means of identity, without any mention of *one*, as
follows: R is a one-one relation if, when x and x' have the relation R to
y, and x has the relation R to y and y', then x and x' are identical, and
so are y and y'. It is true that here x, y, x', y' are each *one* term, but
this is not (it would seem) in any way presupposed in the definition.
This disposes (pending a new inquiry into the nature of classes) of the
first of the above objections.

The next question is as to the distinction between a class containing
only one member, and the one member which it contains. If we could
identify a class with its defining predicate or class-concept, no difficulty
would arise on this point. When a certain predicate attaches to one
and only one term, it is plain that that term is not identical with the
predicate in question. But if two predicates attach to precisely the
same terms, we should say that, although the predicates are different,
the classes which they define are identical, *i.e.* there is only one class

which both define. If, for example, all featherless bipeds are men, and all men are featherless bipeds, the classes *men* and *featherless bipeds* are identical, though *man* differs from *featherless biped*. This shows that a class cannot be identified with its class-concept or defining predicate. There might seem to be nothing left except the actual terms, so that when there is only one term, that term would have to be identical with the class. Yet for many formal reasons this view cannot give the meaning of the symbols which stand for classes in symbolic logic. For example, consider the class of numbers which, added to 3, give 5. This is a class containing no terms except the number 2. But we can say that 2 is a member of this class, *i.e.* it has to the class that peculiar indefinable relation which terms have to the classes they belong to. This seems to indicate that the class is different from the one term. The point is a prominent one in Peano's Symbolic Logic, and is connected with his distinction between the relation of an individual to its class and the relation of a class to another in which it is contained. Thus the class of numbers which, added to 3, give 5, is contained in the class of numbers, but is not a number; whereas 2 is a number, but is not a class contained in the class of numbers. To identify the two relations which Peano distinguishes is to cause havoc in the theory of infinity, and to destroy the formal precision of many arguments and definitions. It seems, in fact, indubitable that Peano's distinction is just, and that some way must be found of discriminating a term from a class containing that term only.

126. In order to decide this point, it is necessary to pass to our third difficulty, and reconsider the notion of *class* itself. This notion appears to be connected with the notion of *denoting*, explained in Part I, Chapter v. We there pointed out five ways of denoting, one of which we called the *numerical conjunction*. This was the kind indicated by *all*. This kind of conjunction appears to be that which is relevant in the case of classes. For example, *man* being the class-concept, *all men* will be the class. But it will not be *all men* quâ concept which will be the class, but what this concept denotes, *i.e.* certain terms combined in the particular way indicated by *all*. The way of combination is essential, since *any man* or *some man* is plainly not the class, though either denotes combinations of precisely the same terms. It might seem as though, if we identify a class with the numerical conjunction of its terms, we must deny the distinction of a term from a class whose only member is that term. But we found in Chapter x that a class must be always an object of a different logical type from its members, and that, in order to avoid the proposition $x \epsilon x$, this doctrine must be extended even to classes which have only one member. How far this forbids us to identify classes with numerical conjunctions, I do not profess to decide; in any case, the distinction between a term and the class whose only member it is must be made, and yet classes must be taken extensionally to the degree involved in their being determinate when their members are given. Such classes are called by Frege *Werthverläufe*; and cardinal numbers are to be regarded as classes in this sense.

127. There is still, however, a certain difficulty, which is this: a class *seems* to be not many terms, but to be itself a single term, even when many terms are members of the class. This difficulty would seem to indicate that the class cannot be identified with all its members, but is rather to be regarded as the whole which they compose. In order, however, to state the difficulty in an unobjectionable manner, we must exclude unity and plurality from the statement of it, since these notions were to be defined by means of the notion of *class*. And here it may be well to clear up a point which is likely to occur to the reader. Is the notion of *one* presupposed every time we speak of *a* term? A term, it may be said, means *one* term, and thus no statement can be made concerning a term without presupposing *one*. In some sense of *one*, this proposition seems indubitable. Whatever is, is one: being and one, as Leibniz remarks, are convertible terms*. It is difficult to be sure how far such statements are merely grammatical. For although whatever is, is one, yet it is equally true that whatever are, are many. But the truth seems to be that the kind of object which is a class, *i.e.* the kind of object denoted by *all men*, or by any concept of a class, is not *one* except where the class has only one term, and must not be made a single logical subject. There is, as we said in Part I, Chapter vi, in simple cases an associated single term which is the class as a whole; but this is sometimes absent, and is in any case not identical with the class as many. But in this view there is not a contradiction, as in the theory that verbs and adjectives cannot be made subjects; for assertions can be made about classes as many, but the subject of such assertions is many, not one only as in other assertions. "Brown and Jones are two of Miss Smith's suitors" is an assertion about the class "Brown and Jones," but not about this class considered as a single term. Thus one-ness belongs, in this view, to a certain type of logical subjects, but classes which are not one may yet have assertions made about them. Hence we conclude that one-ness is implied, but not presupposed, in statements about a term, and "a term" is to be regarded as an indefinable.

128. It seems necessary, however, to make a distinction as regards the use of *one*. The sense in which every object is *one*, which is apparently involved in speaking of *an* object, is, as Frege urges†, a very shadowy sense, since it is applicable to everything alike. But the sense in which a class may be said to have one member is quite precise. A class u has one member when u is not null, and "x and y are u's" implies "x is identical with y." Here the one-ness is a property of the

---

* Ed. Gerhardt, ii, p. 300.
† *Grundlagen der Arithmetik*, Breslau, 1884, p. 40.

class, which may therefore be called a unit-class. The x which is its only member may be itself a class of many terms, and this shows that the sense of *one* involved in *one term* or *a term* is not relevant to Arithmetic, for many terms as such may be a single member of a class of classes. *One*, therefore, is not to be asserted of terms, but of classes having one member in the above-defined sense; *i.e.* "u is one," or better "u is a unit" means "u is not null, and 'x and y are u's' implies 'x and y are identical'." The member of u, in this case, will itself be none or one or many if u is a class of classes; but if u is a class of terms, the member of u will be neither none nor one nor many, but simply a term.

129. The commonly received view, as regards finite numbers, is that they result from counting, or, as some philosophers would prefer to say, from synthesizing. Unfortunately, those who hold this view have not analyzed the notion of counting: if they had done so, they would have seen that it is very complex, and presupposes the very numbers which it is supposed to generate.

The process of counting has, of course, a psychological aspect, but this is quite irrelevant to the theory of Arithmetic. What I wish now to point out is the logical process involved in the act of counting, which is as follows. When we say one, two, three, etc., we are necessarily considering some one-one relation which holds between the numbers used in counting and the objects counted. What is meant by the "one, two, three" is that the objects indicated by these numbers are their correlates with respect to the relation which we have in mind. (This relation, by the way, is usually extremely complex, and is apt to involve a reference to our state of mind at the moment.) Thus we correlate a class of objects with a class of numbers; and the class of numbers consists of all the numbers from 1 up to some number n. The only immediate inference to be drawn from this correlation is, that the number of objects is the same as the number of numbers from 1 up to n. A further process is required to show that this number of numbers is n, which is only true, as a matter of fact, when n is finite, or, in a certain wider sense, when n is α_0 (the smallest of infinite numbers). Moreover the process of counting gives us no indication as to what the numbers are, as to why they form a series, or as to how it is to be proved (in the cases where it is true) that there are n numbers from 1 up to n. Hence counting is irrelevant in the foundations of Arithmetic; and with this conclusion, it may be dismissed until we come to order and ordinal numbers.

130. Let us return to the notion of the numerical conjunction. It is plain that it is of such objects as "A and B," "A and B and C," that numbers other than one are to be asserted. We examined such objects, in Part I, in relation to classes, with which we found them to be identical. Now we must investigate their relation to numbers and plurality.

The notion to be now examined is the notion of a numerical conjunction or, more shortly, a *collection*. This is not to be identified, to begin with, with the notion of a *class*, but is to receive a new and independent treatment. By a collection I mean what is conveyed by "*A* and *B*" or "*A* and *B* and *C*," or any other enumeration of definite terms. The collection is defined by the actual mention of the terms, and the terms are connected by *and*. It would seem that *and* represents a fundamental way of combining terms, and it might be urged that just this way of combination is essential if anything is to result of which a number other than 1 is to be asserted. Collections do not presuppose numbers, since they result simply from the terms together with *and*: they could only *presuppose* numbers in the particular case where the terms of the collection themselves presupposed numbers. There is a grammatical difficulty which, since no method exists of avoiding it, must be pointed out and allowed for. A collection, grammatically, is one, whereas *A* and *B*, or *A* and *B* and *C*, are essentially many. The strict meaning of *collection* is the whole composed of many, but since a word is needed to denote the many themselves, I choose to use the word *collection* in this sense, so that a collection, according to the usage here adopted, is many and not one.

As regards what is meant by the combination indicated by *and*, it gives what we called before the numerical conjunction. That is *A* and *B* is what is denoted by the concept of a class of which *A* and *B* are the only terms, and is precisely *A* and *B* denoted in the way which is indicated by *all*. We may say, if *u* be the class-concept corresponding to a class of which *A* and *B* are the only terms, that "all *u*'s" is a concept which denotes the terms *A*, *B* combined in a certain way, and *A* and *B* are those terms combined in precisely that way. Thus *A* and *B* appears indistinguishable from the class, though distinguishable from the class-concept and from the concept of the class. Hence if *u* be a class of more than one term, it seems necessary to hold that *u* is not one, but many, since *u* is distinguished both from the class-concept and from the whole composed of the terms of *u**. Thus we are brought back to the dependence of numbers upon classes; and where it is not said that the classes in question are finite, it is practically necessary to begin with class-concepts and the theory of denoting, not with the theory of *and* which has just been given. The theory of *and* applies practically only to finite numbers, and gives to finite numbers a position which is different, at least psychologically, from that of infinite numbers. There

---

* A conclusive reason against identifying a class with the whole composed of its terms is, that one of these terms may be the class itself, as in the case "class is a class," or rather "classes are one among classes." The logical type of the class *class* is of an infinite order, and therefore the usual objection to "*x ε x*" does not apply in this case.

are, in short, two ways of defining particular finite classes, but there is only one practicable way of defining. particular infinite classes, namely by intension. It is largely the habit of considering classes primarily from the side of extension which has hitherto stood in the way of a correct logical theory of infinity.

131. Addition, it should be carefully observed, is not primarily a method of forming numbers, but of forming classes or collections. If we add B to A, we do not obtain the number 2, but we obtain A and B, which is a collection of two terms, or a couple. And a couple is defined as follows: u is a couple if u has terms, and if, if x be a term of u, there is a term of u different from x, but if x, y be different terms of u, and z differs from x and from y, then every class to which z belongs differs from u. In this definition, only diversity occurs, together with the notion of a class having terms. It might no doubt be objected that we have to take just two terms x, y in the above definition: but as a matter of fact any finite number can be defined by induction without introducing more than one term. For, if n has been defined, a class u has $n + 1$ terms when, if x be a term of u, the number of terms of u which differ from x is n. And the notion of the arithmetical sum $n + 1$ is obtained from that of the logical sum of a class of n terms and a class of one term. When we say $1 + 1 = 2$, it is not possible that we should mean 1 and 1, since there is only one 1: if we take 1 as an individual, 1 and 1 is nonsense, while if we take it as a class, the rule of Symbolic Logic applies, according to which 1 and 1 is 1. Thus in the corresponding logical proposition, we have on the left-hand side terms of which 1 can be asserted, and on the right-hand side we have a couple. That is, $1 + 1 = 2$ means "one term and one term are two terms," or, stating the proposition in terms of variables, "if u has one term and v has one term, and u differs from v, their logical sum has two terms." It is to be observed that on the left-hand side we have a numerical conjunction of propositions, while on the right-hand side we have a proposition concerning a numerical conjunction of terms. But the true premiss, in the above proposition, is not the conjunction of the three propositions, but their logical product. This point, however, has little importance in the present connection.

132. Thus the only point which remains is this: Does the notion of a term presuppose the notion of 1? For we have seen that all numbers except 0 involve in their definitions the notion of a term, and if this in turn involves 1, the definition of 1 becomes circular, and 1 will have to be allowed to be indefinable. This objection to our procedure is answered by the doctrine of § 128, that a term is not *one* in the sense which is relevant to Arithmetic, or in the sense which is opposed to *many*. The notion of *any term* is a logical indefinable, presupposed in formal truth and in the whole theory of the variable; but this notion is that of the variable conjunction of terms, which in no way involves the number 1. There is therefore nothing circular in defining the number 1 by means of the notion of *a term* or of *any term*.

To sum up: Numbers are classes of classes, namely of all classes similar to a given class. Here classes have to be understood in the sense of numerical conjunctions in the case of classes having many terms; but a class may have no terms, and a class of one term is distinct from that term, so that a class is not simply the sum of its terms. Only classes have numbers; of what is commonly called one object, it is not true, at least in the sense required, to say that it is one, as appears from the fact that the object may be a class of many terms. "One object" seems to mean merely "a logical subject in some proposition." Finite numbers are not to be regarded as generated by counting, which on the contrary presupposes them; and addition is primarily logical addition, first of propositions, then of classes, from which latter arithmetical addition is derivative. The assertion of numbers depends upon the fact that a class of many terms can be a logical subject without being arithmetically one. Thus it appeared that no philosophical argument could overthrow the mathematical theory of cardinal numbers set forth in Chapters xi to xiv.

CHAPTER XVI.

WHOLE AND PART.

133. For the comprehension of analysis, it is necessary to investigate the notion of whole and part, a notion which has been wrapped in obscurity—though not without certain more or less valid logical reasons—by the writers who may be roughly called Hegelian. In the present chapter I shall do my best to set forth a straightforward and non-mystical theory of the subject, leaving controversy as far as possible on one side. It may be well to point out, to begin with, that I shall use the word *whole* as strictly correlative to *part*, so that nothing will be called a whole unless it has parts. Simple terms, such as points, instants, colours, or the fundamental concepts of logic, will not be called wholes.

Terms which are not classes may be, as we saw in the preceding chapter, of two kinds. The first kind are simple: these may be characterized, though not defined, by the fact that the propositions asserting the being of such terms have no presuppositions. The second kind of terms that are not classes, on the other hand, are complex, and in their case, their being presupposes the being of certain other terms. Whatever is not a class is called a *unit*, and thus units are either simple or complex. A complex unit is a *whole*; its parts are other units, whether simple or complex, which are presupposed in it. This suggests the possibility of defining whole and part by means of logical priority, a suggestion which, though it must be ultimately rejected, it will be necessary to examine at length.

134. Wherever we have a one-sided formal implication, it may be urged, if the two propositional functions involved are obtainable one from the other by the variation of a single constituent, then what is implied is simpler than what implies it. Thus "Socrates is a man" implies "Socrates is a mortal," but the latter proposition does not imply the former: also the latter proposition is simpler than the former, since *man* is a concept of which *mortal* forms part. Again, if we take a proposition asserting a relation of two entities A and B, this proposition implies the being of A and the being of B, and the being of the relation, none of which implies the proposition, and each of which is simpler than the proposition. There will only be equal complexity—according to the theory that intension and extension vary inversely as one another—in cases of mutual implication, such as "*A* is greater than *B*" and "*B* is less than *A*." Thus we might be tempted to set up the following definition: *A* is said to be part of *B* when *B is* implies *A is*, but *A is* does not imply *B is*. If this definition could be maintained, whole and part would not be a new indefinable, but would be derivative from logical priority. There are, however, reasons why such an opinion is untenable.

The first objection is, that logical priority is not a simple relation: implication is simple, but logical priority of *A* to *B* requires not only "*B* implies *A*," but also "*A* does not imply *B*." (For convenience, I shall say that *A* implies *B* when *A is* implies *B is*.) This state of things, it is true, is realized when *A* is part of *B*; but it seems necessary to regard the relation of whole to part as something simple, which must be different from any possible relation of one whole to another which is not part of it. This would not result from the above definition. For example, "*A* is greater and better than *B*" implies "*B* is less than *A*," but the converse implication does not hold: yet the latter proposition is not part of the former*.

Another objection is derived from such cases as redness and colour. These two concepts appear to be equally simple: there is no specification, other and simpler than redness itself, which can be added to colour to produce redness, in the way in which specifications will turn *mortal* into *man*. Hence *A is red* is no more complex than *A is coloured*, although there is here a one-sided implication. Redness, in fact, appears to be (when taken to mean one particular shade) a simple concept, which, although it implies colour, does not contain colour as a constituent. The inverse relation of extension and intension, therefore, does not hold in all cases. For these reasons, we must reject, in spite of their very close connection, the attempt to define whole and part by means of implication.

135. Having failed to define wholes by logical priority, we shall not, I think, find it possible to define them at all. The relation of whole and part is, it would seem, an indefinable and ultimate relation, or rather, it is several relations, often confounded, of which one at least is indefinable. The relation of a part to a whole must be differently discussed according to the nature both of the whole and of the parts. Let us begin with the simplest case, and proceed gradually to those that are more elaborate.

(1) Whenever we have any collection of many terms, in the sense explained in the preceding chapter, there the terms, provided there is

---

* See Part IV, Chap. xxvii.

some non-quadratic propositional function which they all satisfy, together form a whole. In the preceding chapter we regarded the class as formed by all the terms, but usage seems to show no reason why the class should not equally be regarded as the whole composed of all the terms in those cases where there is such a whole. The first is the class as many, the second the class as one. Each of the terms then has to the whole a certain indefinable relation*, which is one meaning of the relation of whole and part. The whole is, in this case, a whole of a particular kind, which I shall call an *aggregate*: it differs from wholes of other kinds by the fact that it is definite as soon as its constituents are known.

(2) But the above relation holds only between the aggregate and the single terms of the collection composing the aggregate: the relation to our aggregate of aggregates containing some but not all the terms of our aggregate, is a different relation, though also one which would be commonly called a relation of part to whole. For example, the relation of the Greek nation to the human race is different from that of Socrates to the human race; and the relation of the whole of the primes to the whole of the numbers is different from that of 2 to the whole of the numbers. This most vital distinction is due to Peano†. The relation of a subordinate aggregate to one in which it is contained can be defined, as was explained in Part I, by means of implication and the first kind of relation of part to whole. If u, v be two aggregates, and for every value of x "x is a u" implies "x is a v," then, provided the converse implication does not hold, u is a proper part (in the second sense) of v. This sense of whole and part, therefore, is derivative and definable.

(3) But there is another kind of whole, which may be called a *unity*. Such a whole is always a proposition, though it need not be an *asserted* proposition. For example, "A differs from B," or "A's difference from B," is a complex of which the parts are A and B and difference; but this sense of whole and part is different from the previous senses, since "A differs from B" is not an aggregate, and has no parts at all in the first two senses of parts. It is parts in this third sense that are chiefly considered by philosophers, while the first two senses are those usually relevant in symbolic logic and mathematics. This third sense of *part* is the sense which corresponds to analysis: it appears to be indefinable, like the first sense—*i.e.*, I know no way of defining it. It must be held that the three senses are always to be kept distinct: *i.e.*, if A is part of B in one sense, while B is part of C in another, it must not be inferred (in general) that A is part of C in any of the three senses. But we may make a fourth general sense, in which anything which is part in

---

* Which may, if we choose, be taken as Peano's ϵ. The objection to this meaning for ϵ is that not every propositional function defines a whole of the kind required. The whole differs from the class as many by being of the same *type* as its terms.

† Cf. *e.g.* F. 1901, § 1, Prop. 4. 4, note (p. 12).

any sense, or part in one sense of part in another, is to be called a part. This sense, however, has seldom, if ever, any utility in actual discussion.

136. The difference between the kinds of wholes is important, and illustrates a fundamental point in Logic. I shall therefore repeat it in other words. Any collection whatever, if defined by a non-quadratic propositional function, though as such it is many, yet composes a whole, whose parts are the terms of the collection or any whole composed of some of the terms of the collection. It is highly important to realize the difference between a whole and all its parts, even in this case where the difference is a minimum. The word *collection*, being singular, applies more strictly to the whole than to all the parts; but convenience of expression has led me to neglect grammar, and speak of all the terms as the collection. The whole formed of the terms of the collection I call an *aggregate*. Such a whole is completely specified when all its simple constituents are specified; its parts have no direct connection *inter se*, but only the indirect connection involved in being parts of one and the same whole. But other wholes occur, which contain relations or what may be called predicates, not occurring simply as terms in a collection, but as relating or qualifying. Such wholes are always propositions. These are not completely specified when their parts are all known. Take, as a simple instance, the proposition "A differs from B," where A and B are simple terms. The simple parts of this whole are A and B and difference; but the enumeration of these three does not specify the whole, since there are two other wholes composed of the same parts, namely the aggregate formed of A and B and difference, and the proposition "B differs from A." In the former case, although the whole was different from all its parts, yet it was completely specified by specifying its parts; but in the present case, not only is the whole different, but it is not even specified by specifying its parts. We cannot explain this fact by saying that the parts stand in certain relations which are omitted in the analysis; for in the above case of "A differs from B," the relation was included in the analysis. The fact seems to be that a relation is one thing when it relates, and another when it is merely enumerated as a term in a collection. There are certain fundamental difficulties in this view, which however I leave aside as irrelevant to our present purpose*.

Similar remarks apply to A *is*, which is a whole composed of A and *Being*, but is different from the whole formed of the collection A and *Being*. *A is one* raises the same point, and so does A and B *are two*. Indeed all propositions raise this point, and we may distinguish them among complex terms by the fact that they raise it.

Thus we see that there are two very different classes of wholes, of which the first will be called *aggregates*, while the second will be called *unities*. (*Unit* is a word having a quite different application, since what-

---

* See Part I, Chap. IV, esp. § 54.

ever is a class which is not null, and is such that, if x and y be members
of it, x and y are identical, is a unit.)   Each class of wholes consists of
terms not simply equivalent to all their parts; but in the case of unities,
the whole is not even specified by its parts.   For example, the parts A,
greater than, B, may compose simply an aggregate, or either of the
propositions "A is greater than B," "B is greater than A."   Unities
thus involve problems from which aggregates are free.   As aggregates
are more specially relevant to mathematics than unities, I shall in
future generally confine myself to the former.

137.   It is important to realize that a whole is a new single term,
distinct from each of its parts and from all of them: it is one, not many *,
and is related to the parts, but has a being distinct from theirs.   The
reader may perhaps be inclined to doubt whether there is any need of
wholes other than unities; but the following reasons seem to make
aggregates logically unavoidable.   (1) We speak of one collection, one
manifold, etc., and it would seem that in all these cases there really is
something that is a single term.   (2) The theory of fractions, as we shall
shortly see, appears to depend partly upon aggregates.   (3) We shall find
it necessary, in the theory of extensive quantity, to assume that aggregates,
even when they are infinite, have what may be called magnitude of
divisibility, and that two infinite aggregates may have the same number
of terms without having the same magnitude of divisibility: this theory,
we shall find, is indispensable in metrical geometry.   For these reasons,
it would seem, the aggregate must be admitted as an entity distinct
from all its constituents, and having to each of them a certain ultimate
and indefinable relation.

138.   I have already touched on a very important logical doctrine,
which the theory of whole and part brings into prominence—I mean the
doctrine that analysis is falsification.   Whatever can be analyzed is a
whole, and we have already seen that analysis of wholes is in some
measure falsification.   But it is important to realize the very narrow
limits of this doctrine.   We cannot conclude that the parts of a whole
are not really its parts, nor that the parts are not presupposed in the
whole in a sense in which the whole is not presupposed in the parts, nor
yet that the logically prior is not usually simpler than the logically
subsequent.   In short, though analysis gives us the truth, and nothing
but the truth, yet it can never give us the whole truth.   This is the
only sense in which the doctrine is to be accepted.   In any wider sense,
it becomes merely a cloak for laziness, by giving an excuse to those who
dislike the labour of analysis.

139.   It is to be observed that what we called classes as one may
always, except where they contain one term or none, or are defined by
quadratic propositional functions, be interpreted as aggregates.   The

---

* *I.e.* it is of the same logical type as its simple parts.

logical product of two classes as one will be the common part (in the second of our three senses) of the two aggregates, and their sum will be the aggregate which is identical with or part of (again in the second sense) any aggregate of which the two given aggregates are parts, but is neither identical with nor part of any other aggregate*. The relation of whole and part, in the second of our three senses, is transitive and asymmetrical, but is distinguished from other such relations by the fact of allowing logical addition and multiplication. It is this peculiarity which forms the basis of the Logical Calculus as developed by writers previous to Peano and Frege (including Schröder)†. But wherever infinite wholes are concerned it is necessary, and in many other cases it is practically unavoidable, to begin with a class-concept or predicate or propositional function, and obtain the aggregate from this. Thus the theory of whole and part is less fundamental logically than that of predicates or class-concepts or propositional functions; and it is for this reason that the consideration of it has been postponed to so late a stage.

* Cf. Peano, *F.* 1901, § 2, Prop. 1·0 (p. 19).
† See *e.g.* his *Algebra der Logik*, Vol. ı (Leipzig, 1890).

CHAPTER XVII.

INFINITE WHOLES.

140. In the present chapter the special difficulties of infinity are not to be considered: all these are postponed to Part V. My object now is to consider two questions: (1) Are there any infinite wholes? (2) If so, must an infinite whole which contains parts in the second of our three senses be an aggregate of parts in the first sense? In order to avoid the reference to the first, second and third senses, I propose henceforward to use the following phraseology: A part in the first sense is to be called a *term* of the whole*; a part in the second sense is to be called a *part* simply; and a part in the third sense will be called a *constituent* of the whole. Thus terms and parts belong to aggregates, while constituents belong to unities. The consideration of aggregates and unities, where infinity is concerned, must be separately conducted. I shall begin with aggregates.

An infinite aggregate is an aggregate corresponding to an infinite class, *i.e.* an aggregate which has an infinite number of terms. Such aggregates are defined by the fact that they contain parts which have as many terms as themselves. Our first question is: Are there any such aggregates?

Infinite aggregates are often denied. Even Leibniz, favourable as he was to the actual infinite, maintained that, where infinite classes are concerned, it is possible to make valid statements about *any* term of the class, but not about *all* the terms, nor yet about the whole which (as he would say) they do *not* compose†. Kant, again, has been much criticised for maintaining that space is an infinite given whole. Many maintain that every aggregate must have a finite number of terms, and that where this condition is not fulfilled there is no true whole. But I do not believe that this view can be successfully defended. Among those who deny that space is a given whole, not a few would admit that what they are pleased to call a finite space may be a given whole, for instance,

---

* A part in this sense will also be sometimes called a *simple* or *indivisible* part.

† Cf. *Phil. Werke*, ed. Gerhardt, ii, p. 315; also i, p. 338, v, pp. 144–5.

the space in a room, a box, a bag, or a book-case. But such a space is
only finite in a psychological sense, *i.e.* in the sense that we can take it
in at a glance: it is not finite in the sense that it is an aggregate of a
finite number of terms, nor yet a unity of a finite number of constituents.
Thus to admit that such a space can be a whole is to admit that there
are wholes which are not finite. (This does not follow, it should be
observed, from the admission of material objects apparently occupying
finite spaces, for it is always possible to hold that such objects, though
apparently continuous, consist really of a large but finite number of
material points.) With respect to time, the same argument holds: to
say, for example, that a certain length of time elapses between sunrise
and sunset, is to admit an infinite whole, or at least a whole which is not
finite. It is customary with philosophers to deny the reality of space
and time, and to deny also that, if they were real, they would be
aggregates. I shall endeavour to show, in Part VI, that these denials
are supported by a faulty logic, and by the now resolved difficulties of
infinity. Since science and common sense join in the opposite view, it
will therefore be accepted; and thus, since no argument *à priori* can
now be adduced against infinite aggregates, we derive from space and
time an argument in their favour.

Again, the natural numbers, or the fractions between 0 and 1, or the
sum-total of all colours, are infinite, and seem to be true aggregates:
the position that, although true propositions can be made about *any*
number, yet there are no true propositions about *all* numbers, could be
supported formerly, as Leibniz supported it, by the supposed contradictions of infinity, but has become, since Cantor's solution of these
contradictions, a wholly unnecessary paradox. And where a collection
can be defined by a non-quadratic propositional function, this must be
held, I think, to imply that there is a genuine aggregate composed
of the terms of the collection. It may be observed also that, if there
were no infinite wholes, the word *Universe* would be wholly destitute of
meaning.

141. We must, then, admit infinite aggregates. It remains to ask
a more difficult question, namely: Are we to admit infinite unities?
This question may also be stated in the form: Are there any
infinitely complex propositions? This question is one of great logical
importance, and we shall require much care both in stating and in
discussing it.

The first point is to be clear as to the meaning of an infinite unity.
A unity will be infinite when the aggregate of all its constituents is
infinite, but this scarcely constitutes the meaning of an infinite unity.
In order to obtain the meaning, we must introduce the notion of a
simple constituent. We may observe, to begin with, that a constituent
of a constituent is a constituent of the unity, *i.e.* this form of the
relation of part to whole, like the second, but unlike the first form, is

transitive. A simple constituent may now be defined as a constituent which itself has no constituents. We may assume, in order to eliminate the question concerning aggregates, that no constituent of our unity is to be an aggregate, or, if there be a constituent which is an aggregate, then this constituent is to be taken as simple. (This view of an aggregate is rendered legitimate by the fact that an aggregate is a single term, and does not have that kind of complexity which belongs to propositions.) With this the definition of a simple constituent is completed.

We may now define an infinite unity as follows: A unity is finite when, and only when, the aggregate of its simple constituents is finite. In all other cases a unity is said to be infinite. We have to inquire whether there are any such unities *.

If a unity is infinite, it is possible to find a constituent unity, which again contains a constituent unity, and so on without end. If there be any unities of this nature, two cases are *primâ facie* possible. (1) There may be simple constituents of our unity, but these must be infinite in number. (2) There may be no simple constituents at all, but all constituents, without exception, may be complex; or, to take a slightly more complicated case, it may happen that, although there are some simple constituents, yet these and the unities composed of them do not constitute all the constituents of the original unity. A unity of either of these two kinds will be called infinite. The two kinds, though distinct, may be considered together.

An infinite unity will be an infinitely complex proposition: it will not be analyzable in any way into a finite number of constituents. It thus differs radically from assertions about infinite aggregates. For example, the proposition "any number has a successor" is composed of a finite number of constituents: the number of concepts entering into it can be enumerated, and in addition to these there is an infinite aggregate of terms denoted in the way indicated by *any*, which counts as one constituent. Indeed it may be said that the logical purpose which is served by the theory of denoting is, to enable propositions of finite complexity to deal with infinite classes of terms: this object is effected by *all*, *any*, and *every*, and if it were not effected, every general proposition about an infinite class would have to be infinitely complex. Now, for my part, I see no possible way of deciding whether propositions of infinite complexity are possible or not; but this at least is clear, that all the propositions known to us (and, it would seem, all propositions that we *can* know) are of finite complexity. It is only by obtaining such propositions about infinite classes that we are enabled to deal with infinity; and it is a remarkable and fortunate fact that this method is successful. Thus the question whether or not there are infinite unities must be left unresolved; the only thing we can say, on this subject, is

---

* In Leibniz's philosophy, all contingent things are infinite unities.

that no such unities occur in any department of human knowledge, and therefore none such are relevant to the foundations of mathematics.

142. I come now to our second question: Must an infinite whole which contains parts be an aggregate of terms? It is often held, for example, that spaces have parts, and can be divided *ad lib.*, but that they have no *simple* parts, *i.e.* they are not aggregates of points. The same view is put forward as regards periods of time. Now it is plain that, if our definition of a part by means of terms (*i.e.* of the second sense of part by means of the first) was correct, the present problem can never arise, since parts only belong to aggregates. But it may be urged that the notion of *part* ought to be taken as an indefinable, and that therefore it may apply to other wholes than aggregates. This will require that we should add to aggregates and unities a new kind of whole, corresponding to the second sense of *part*. This will be a whole which has parts in the second sense, but is not an aggregate or a unity. Such a whole seems to be what many philosophers are fond of calling a continuum, and space and time are often held to afford instances of such a whole.

Now it may be admitted that, among infinite wholes, we find a distinction which *seems* relevant, but which, I believe, is in reality merely psychological. In some cases, we feel no doubt as to the terms, but great doubt as to the whole, while in others, the whole seems obvious, but the terms seem a precarious inference. The ratios between 0 and 1, for instance, are certainly indivisible entities; but the whole aggregate of ratios between 0 and 1 seems to be of the nature of a construction or inference. On the other hand, sensible spaces and times seem to be obvious wholes; but the inference to indivisible points and instants is so obscure as to be often regarded as illegitimate. This distinction seems, however, to have no logical basis, but to be wholly dependent on the nature of our senses. A slight familiarity with co-ordinate geometry suffices to make a finite length seem strictly analogous to the stretch of fractions between 0 and 1. It must be admitted, nevertheless, that in cases where, as with the fractions, the indivisible parts are evident on inspection, the problem with which we are concerned does not arise. But to infer that all infinite wholes have indivisible parts merely because this is known to be the case with some of them, would certainly be rash. The general problem remains, therefore, namely: Given an infinite whole, is there a universal reason for supposing that it contains indivisible parts?

143. In the first place, the definition of an infinite whole must not be held to deny that it has an assignable number of simple parts which do not reconstitute it. For example, the stretch of fractions from 0 to 1 has three simple parts, $\frac{1}{3}$, $\frac{1}{2}$, $\frac{2}{3}$. But these do not reconstitute the whole, that is, the whole has other parts which are not parts of the assigned parts or of the sum of the assigned parts. Again, if we form a whole out of the number 1 and a line an inch long, this whole certainly has one simple part, namely 1. Such a case as this may be excluded by asking whether every part of our whole either is simple or contains simple parts. In this case, if our whole be formed by adding n simple terms to an infinite whole, the n simple terms can be taken away, and the question can be asked concerning the infinite whole which is left. But again, the meaning of our question seems hardly to be: Is our infinite whole an actual aggregate of innumerable simple parts? This is doubtless an important question, but it is subsequent to the question we are asking, which is: Are there always simple parts at all? We may observe that, if a finite number of simple parts be found, and taken away from the whole, the remainder is always infinite. For if not, it would have a finite number; and since the term of two finite numbers is finite, the original whole would then be finite. Hence if it can be shown that every infinite whole contains one simple part, it follows that it contains an infinite number of them. For, taking away the one simple part, the remainder is an infinite whole, and therefore has a new simple part, and so on. It follows that every part of the whole either is simple, or contains simple parts, provided that every infinite whole has at least one simple part. But it seems as hard to prove this as to prove that every infinite whole is an aggregate.

If an infinite whole be divided into a finite number of parts, one at least of these parts must be infinite. If this be again divided, one of its parts must be infinite, and so on. Thus no finite number of divisions will reduce all the parts to finitude. Successive divisions give an endless series of parts, and in such endless series there is (as we shall see in Parts IV and V) no manner of contradiction. Thus there is no method of proving by actual division that every infinite whole must be an aggregate. So far as this method can show, there is no more reason for simple constituents of infinite wholes than for a first moment in time or a last finite number.

But perhaps a contradiction may emerge in the present case from the connection of whole and part with logical priority. It certainly seems a greater paradox to maintain that infinite wholes do not have indivisible parts than to maintain that there is no first moment in time or furthest limit to space. This might be explained by the fact that we know many simple terms, and some infinite wholes undoubtedly composed of simple terms, whereas we know of nothing suggesting a beginning of time or space. But it may perhaps have a more solid basis in logical priority. For the simpler is always implied in the more complex, and therefore there can be no truth about the more complex unless there is truth about the simpler. Thus in the analysis of our infinite whole, we are always dealing with entities which would not be at all unless their constituents were. This makes a real difference from the time-series, for example: a moment does not logically presuppose a previous moment, and if it did it would perhaps be self-contradictory to deny a first moment, as it has been held (for the same reason) self-contradictory to deny a First Cause. It seems to follow that infinite wholes would not have Being at all, unless there were innumerable simple Beings whose Being is presupposed in that of the infinite wholes. For where the presupposition is false, the consequence is false also. Thus there seems a special reason for completing the infinite regress in the case of infinite wholes, which does not exist where other asymmetrical transitive relations are concerned. This is another instance of the peculiarity of the relation of whole and part: a relation so important and fundamental that almost all our philosophy depends upon the theory we adopt in regard to it.

The same argument may be otherwise stated by asking how our infinite wholes are to be defined. The definition must not be infinitely complex, since this would require an infinite unity. Now if there is any definition which is of finite complexity, this cannot be obtained from the parts, since these are either infinitely numerous (in the case of an aggregate), or themselves as complex as the whole (in the case of a whole which is not an aggregate). But any definition which is of finite complexity will necessarily be intensional, *i.e.* it will give some characteristic of a collection of terms. There seems to be no other known method of defining an infinite whole, or of obtaining such a whole in a way not involving any infinite unity.

The above argument, it must be admitted, is less conclusive than could be wished, considering the great importance of the point at issue. It may, however, be urged in support of it that all the arguments on the other side depend upon the supposed difficulties of infinity, and are therefore wholly fallacious; also that the procedure of Geometry and Dynamics (as will be shown in Parts VI and VII) imperatively demands points and instants. In all applications, in short, the results of the doctrine here advocated are far simpler, less paradoxical, and more logically satisfactory, than those of the opposite view. I shall therefore assume, throughout the remainder of this work, that all the infinite wholes with which we shall have to deal are aggregates of terms.

CHAPTER XVIII.

RATIOS AND FRACTIONS.

144. THE present chapter, in so far as it deals with relations of integers, is essentially confined to *finite* integers: those that are infinite have no relations strictly analogous to what are usually called ratios. But I shall distinguish ratios, as relations between integers, from fractions, which are relations between aggregates, or rather between their magnitudes of divisibility; and fractions, we shall find, may express relations which hold where both aggregates are infinite. It will be necessary to begin with the mathematical definition of ratio, before proceeding to more general considerations.

Ratio is commonly associated with multiplication and division, and in this way becomes indistinguishable from fractions. But multiplication and division are equally applicable to finite and infinite numbers, though in the case of infinite numbers they do not have the properties which connect them with ratio in the finite case. Hence it becomes desirable to develop a theory of ratio which shall be independent of multiplication and division.

Two finite numbers are said to be consecutive when, if u be a class having one of the numbers, and one term be added to u, the resulting class has the other number. To be consecutive is thus a relation which is one-one and asymmetrical. If now a number a has to a number b the nth power of this relation of consecutiveness (powers of relations being defined by relative multiplication), then we have $a + n = b$. This equation expresses, between a and b, a one-one relation which is determinate when n is given. If now the mth power of this relation holds between a' and b', we shall have $a' + mn = b'$. Also we may define mn as $0 + mn$. If now we have three numbers a, b, c such that $ab = c$, this equation expresses between a and c a one-one relation which is determinate when b is given. Let us call this relation B. Suppose we have also $a'b' = c$. Then a has to a' a relation which is the relative product of B and the converse of B', where B' is derived from b' as B was derived from b. This relation we define as the ratio of a' to a. This theory has the advantage that it applies not only to finite integers, but to all other series of the same type, *i.e.* all series of type the which I call progressions.

145. The only point which it is important, for our present purpose, to observe as regards the above definition of ratios is, that they are one-one relations between finite integers, which are with one exception asymmetrical, which are such that one and only one holds between any specified pair of finite integers, which are definable in terms of consecutiveness, and which themselves form a series having no first or last term and having a term, and therefore an infinite number of terms, between any two specified terms. From the fact that ratios are relations it results that no ratios are to be identified with integers: the ratio of 2 to 1, for example, is a wholly different entity from 2. When, therefore, we speak of the series of ratios as containing integers, the integers said to be contained are not cardinal numbers, but relations which have a certain one-one correspondence with cardinal numbers. The same remark applies to positive and negative numbers. The nth power of the relation of consecutiveness is the positive number $+ n$, which is plainly a wholly different concept from the cardinal number n. The confusion of entities with others to which they have some important one-one relation is an error to which mathematicians are very liable, and one which has produced the greatest havoc in the philosophy of mathematics. We shall find hereafter innumerable other instances of the same error, and it is well to realize, as early as possible, that any failure in subtlety of distinctions is sure, in this subject at least, to cause the most disastrous consequences.

There is no difficulty in connecting the above theory of ratio with the usual theory derived from multiplication and division. But the usual theory does not show, as the present theory does, why the infinite integers do not have ratios strictly analogous to those of finite integers. The fact is, that ratio depends upon consecutiveness, and consecutiveness as above defined does not exist among infinite integers, since these are unchanged by the addition of 1.

It should be observed that what is called addition of ratios demands a new set of relations among ratios, relations which may be called positive and negative ratios, just as certain relations among integers are positive and negative integers. This subject, however, need not be further developed.

146. The above theory of ratio has, it must be confessed, a highly artificial appearance, and one which makes it seem extraordinary that ratios should occur in daily life. The fact is, it is not ratios, but fractions, that occur, and fractions are not purely arithmetical, but are really concerned with relations of whole and part.

Propositions asserting fractions show an important difference from those asserting integers. We can say A is one, A and B are two, and so on; but we cannot say A is one-third, or A and B are two-thirds.

There is always need of some second entity, to which our first has some fractional relation. We say A is one-third of C, A and B together are two-thirds of C, and so on. Fractions, in short, are either relations of a simple part to a whole, or of two wholes to one another. But it is not necessary that the one whole, or the simple part, should be part of the other whole. In the case of finite wholes, the matter seems simple: the fraction expresses the ratio of the number of parts in the one to the number in the other. But the consideration of infinite wholes will show us that this simple theory is inadequate to the facts.

147. There is no doubt that the notion of half a league, or half a day, is a legitimate notion. It is therefore necessary to find some sense for fractions in which they do not essentially depend upon number. For, if a given period of twenty-four hours is to be divided into two continuous portions, each of which is to be half of the whole period, there is only one way of doing this: but Cantor has shown that every possible way of dividing the period into two continuous portions divides it into two portions having the same *number* of terms. There must be, therefore, some other respect in which two periods of twelve hours are equal, while a period of one hour and another of twenty-three hours are unequal. I shall have more to say upon this subject in Part III; for the present I will point out that what we want is of the nature of a magnitude, and that it must be essentially a property of ordered wholes. I shall call this property *magnitude of divisibility*. To say now that A is one-half of B means: B is a whole, and if B be divided into two similar parts which have both the same magnitude of divisibility as each other, then A has the same magnitude of divisibility as each of these parts. We may interpret the fraction $\frac{1}{2}$ somewhat more simply, by regarding it as a relation (analogous to ratio so long as finite wholes are concerned) between two magnitudes of divisibility. Thus finite integral fractions (such as $n/1$) will measure the relation of the divisibility of an aggregate of n terms to the divisibility of a single term; the converse relation will be $1/n$. Thus here again we have a new class of entities which is in danger of being confused with finite cardinal integers, though in reality quite distinct. Fractions, as now interpreted, have the advantage (upon which all metrical geometry depends) that they introduce a discrimination of greater and smaller among infinite aggregates having the same number of terms. We shall see more and more, as the logical inadequacy of the usual accounts of measurement is brought to light, how absolutely essential the notion of magnitude of divisibility really is. Fractions, then, in the sense in which they may express relations of infinite aggregates—and this is the sense which they usually have in daily life— are really of the nature of relations between magnitudes of divisibility; and magnitudes of divisibility are only measured by number of parts where the aggregates concerned are finite. It may also be observed (though this remark is anticipatory) that, whereas ratios, as above defined, are essentially rational, fractions, in the sense here given to them, are also capable of irrational values. But the development of this topic must be left for Part V.

148. We may now sum up the results obtained in Part II. In the first four chapters, the modern mathematical theory of cardinal integers, as it results from the joint labours of arithmeticians and symbolic logicians, was briefly set forth. Chapter xi explained the notion of similar classes, and showed that the usual formal properties of integers result from defining them as classes of similar classes. In Chapter xii, we showed how arithmetical addition and multiplication both depend upon logical addition, and how both may be defined in a way which applies equally to finite and infinite numbers, and to finite and infinite sums and products, and which moreover introduces nowhere any idea of order. In Chapter xiii, we gave the strict definition of an infinite class, as one which is similar to a class resulting from taking away one of its terms ; and we showed in outline how to connect this definition with the definition of finite numbers by mathematical induction. The special theory of finite integers was discussed in Chapter xiv, and it was shown how the primitive propositions, which Peano proves to be sufficient in this subject, can all be deduced from our definition of finite cardinal integers. This confirmed us in the opinion that Arithmetic contains no indefinables or indemonstrables beyond those of general logic.

We then advanced, in Chapter xv, to the consideration of philosophical questions, with a view of testing critically the above mathematical deductions. We decided to regard both *term* and *a term* as indefinable, and to define the number 1, as well as all other numbers, by means of these indefinables (together with certain others). We also found it necessary to distinguish a class from its class-concept, since one class may have several different class-concepts. We decided that a class consists of all the terms denoted by the class-concept, denoted in a certain indefinable manner; but it appeared that both common usage and the majority of mathematical purposes would allow us to identify a class with the whole formed of the terms denoted by the class-concept. The only reasons against this view were, the necessity of distinguishing a class containing only one term from that one term, and the fact that some classes are members of themselves. We found also a distinction between finite and infinite classes, that the former can, while the latter cannot, be defined extensionally, *i.e.* by actual enumeration of their terms. We then proceeded to discuss what may be called the addition of individuals, *i.e.* the notion involved in "*A* and *B*"; and we found that a more or less independent theory of *finite* integers can be based upon this notion. But it appeared finally, in virtue of our analysis of the notion of *class*, that this theory was really indistinguishable from the theory previously expounded, the only difference being that it adopted an extensional definition of classes.

Chapter xvi dealt with the relation of whole and part. We found that there are two indefinable senses of this relation, and one definable sense, and that there are two correspondingly different sorts of wholes, which we called unities and aggregates respectively. We saw also that, by extending the notion of aggregates to single terms and to the null-class, we could regard the whole of the traditional calculus of Symbolic Logic as an algebra specially applicable to the relations of wholes and parts in the definable sense. We considered next, in Chapter xvii, the notion of an infinite whole. It appeared that infinite unities, even if they be logically possible, at any rate never appear in anything accessible to human knowledge. But infinite aggregates, we found, must be admitted; and it seemed that all infinite wholes which are not unities must be aggregates of terms, though it is by no means necessary that the terms should be simple. (They must, however, owing to the exclusion of infinite unities, be assumed to be of *finite* complexity.)

In Chapter xviii, finally, we considered ratios and fractions : the former were found to be somewhat complicated relations of finite integers, while the latter were relations between the divisibilities of aggregates. These divisibilities being magnitudes, their further discussion belongs to Part III, in which the general nature of quantity is to be considered.

PART III.

QUANTITY.

CHAPTER XIX.

THE MEANING OF MAGNITUDE.

149. Among the traditional problems of mathematical philosophy, few are more important than the relation of quantity to number. Opinion as to this relation has undergone many revolutions. Euclid, as is evident from his definitions of ratio and proportion, and indeed from his whole procedure, was not persuaded of the applicability of numbers to spatial magnitudes. When Des Cartes and Vieta, by the introduction of co-ordinate Geometry, made this applicability a fundamental postulate of their systems, a new method was founded, which, however fruitful of results, involved, like most mathematical advances of the seventeenth century, a diminution of logical precision and a loss in subtlety of distinction. What was meant by measurement, and whether *all* spatial magnitudes were susceptible of a numerical measure, were questions for whose decision, until very lately, the necessary mathematical instrument was lacking; and even now much remains to be done before a complete answer can be given. The view prevailed that number and quantity were *the* objects of mathematical investigation, and that the two were so similar as not to require careful separation. Thus number was applied to quantities without any hesitation, and conversely, where existing numbers were found inadequate to measurement, new ones were created on the sole ground that every quantity must have a numerical measure.

All this is now happily changed. Two different lines of argument, conducted in the main by different men, have laid the foundations both for large generalizations, and for thorough accuracy in detail. On the one hand, Weierstrass, Dedekind, Cantor, and their followers, have pointed out that, if irrational numbers are to be significantly employed as measures of quantitative fractions, they must be defined without reference to quantity; and the same men who showed the necessity of such a definition have supplied the want which they had created. In this way, during the last thirty or forty years, a new subject, which has added quite immeasurably to theoretical correctness, has been created, which may legitimately be called Arithmetic; for, starting with integers, it succeeds in defining whatever else it requires—rationals, limits, ir-rationals, continuity, and so on.   It results that, for all Algebra and Analysis, it is unnecessary to assume any material beyond the integers, which, as we have seen, can themselves be defined in logical terms. It is this science, far more than non-Euclidean Geometry, that is really fatal to the Kantian theory of *à priori* intuitions as the basis of mathematics.   Continuity and irrationals were formerly the strongholds of the school who may be called intuitionists, but these strongholds are theirs no longer.   Arithmetic has grown so as to include all that can strictly be called pure in the traditional mathematics.

150.   But, concurrently with this purist's reform, an opposite advance has been effected.   New branches of mathematics, which deal neither with number nor with quantity, have been invented; such are the Logical Calculus, Projective Geometry, and—in its essence—the Theory of Groups.   Moreover it has appeared that measurement—if this means the correlation, with numbers, of entities which are not numbers or aggregates—is not a prerogative of quantities: some quantities cannot be measured, and some things which are not quantities (for example anharmonic ratios projectively defined) can be measured.   Measurement, in fact, as we shall see, is applicable to all series of a certain kind—a kind which excludes some quantities and includes some things which are not quantities.   The separation between number and quantity is thus complete: each is wholly independent of the other.   Quantity, moreover, has lost the mathematical importance which it used to possess, owing to the fact that most theorems concerning it can be generalized so as to become theorems concerning order.   It would therefore be natural to discuss order before quantity.   As all propositions concerning order can, however, be established independently for particular instances of order, and as quantity will afford an illustration, requiring slightly less effort of abstraction, of the principles to be applied to series in general; as, further, the theory of distance, which forms a part of the theory of order, presupposes somewhat controversial opinions as to the nature of quantity, I shall follow the more traditional course, and consider quantity first.   My aim will be to give, in the present chapter, a theory of quantity which does not depend upon number, and then to show the peculiar relation to number which is possessed by two special classes of quantities, upon which depends the measurement of quantities wherever this is possible.   The whole of this Part, however—and it is important to realize this—is a concession to tradition; for quantity, we shall find, is not definable in terms of logical constants, and is not properly a notion belonging to pure mathematics at all.   I shall discuss quantity because it is traditionally supposed to occur in mathematics, and because a thorough discussion is required for disproving this supposition; but if the supposition did not exist, I should avoid all mention of any such notion as quantity.

151. In fixing the meaning of such a term as *quantity* or *magnitude*, one is faced with the difficulty that, however one may define the word, one must appear to depart from usage. This difficulty arises wherever two characteristics have been commonly supposed inseparable which, upon closer examination, are discovered to be capable of existing apart. In the case of magnitude, the usual meaning appears to imply (1) a capacity for the relations of *greater* and *less*, (2) divisibility. Of these characteristics, the first is supposed to imply the second. But as I propose to deny the implication, I must either admit that some things which are indivisible are magnitudes, or that some things which are greater or less than others are not magnitudes. As one of these departures from usage is unavoidable, I shall choose the former, which I believe to be the less serious. A magnitude, then, is to be defined as anything which is greater or less than something else.

It might be thought that *equality* should be mentioned, along with greater and less, in the definition of magnitude. We shall see reason to think, however—paradoxical as such a view may appear—that what can be greater or less than some term, can never be equal to any term whatever, and *vice versâ*. This will require a distinction, whose necessity will become more and more evident as we proceed, between the kind of terms that can be equal, and the kind that can be greater or less. The former I shall call *quantities*, the latter *magnitudes*. An actual foot-rule is a quantity: its length is a magnitude. Magnitudes are more abstract than quantities: when two quantities are equal, they have the *same* magnitude. The necessity of this abstraction is the first point to be established.

152. Setting aside magnitudes for the moment, let us consider quantities. A quantity is anything which is capable of quantitative equality to something else. Quantitative equality is to be distinguished from other kinds, such as arithmetical or logical equality. All kinds of equality have in common the three properties of being reflexive, symmetrical, and transitive, *i.e.* a term which has this relation at all has this relation to itself; if A has the relation to B, B has it to A; if A has it to B, and B to C, A has it to C*. What it is that distinguishes quantitative equality from other kinds, and whether this kind of equality is analyzable, is a further and more difficult question, to which we must now proceed.

There are, so far as I know, three main views of quantitative equality. There is (1) the traditional view, which denies quantity as

---

* On the independence of these three properties, see Peano, *Revue de Mathématique*, vii, p. 22. The reflexive property is not strictly necessary ; what is properly necessary and what is alone (at first sight at any rate) true of quantitative equality, is, that there exists at least one pair of terms having the relation in question. It follows then from the other two properties that each of these terms has to itself the relation in question.

an independent idea, and asserts that two terms are equal when, and only when, they have the same number of parts. (2) There is what may be called the relative view of quantity, according to which equal, greater and less are all direct relations between quantities. In this view we have no need of magnitude, since sameness of magnitude is replaced by the symmetrical and transitive relation of equality. (3) There is the absolute theory of quantity, in which equality is not a direct relation, but is to be analyzed into possession of a common magnitude, *i.e.* into sameness of relation to a third term. In this case there will be a special kind of relation of a term to its magnitude; between two magnitudes of the same kind there will be the relation of greater and less; while equal, greater and less will apply to quantities only in virtue of their relation to magnitudes. The difference between the second and third theories is exactly typical of a difference which arises in the case of many other series, and notably in regard to space and time. The decision is, therefore, a matter of very considerable importance.

153. (1) The kind of equality which consists in having the same number of parts has been already discussed in Part II. If this be indeed the meaning of quantitative equality, then quantity introduces no new idea. But it may be shown, I think, that greater and less have a wider field than whole and part, and an independent meaning. The arguments may be enumerated as follows: (α) We must admit indivisible quantities; (β) where the number of simple parts is infinite, there is no generalization of number which will give the recognized results as to inequality; (γ) some relations must be allowed to be quantitative, and relations are not even conceivably divisible; (δ) even where there is divisibility, the axiom that the whole is greater than the part must be allowed to be significant, and not a result of definition.

(α) Some quantities are indivisible. For it is generally admitted that some psychical existents, such as pleasure and pain, are quantitative. If now equality means sameness in the number of indivisible parts, we shall have to regard a pleasure or a pain as consisting of a collection of units, all perfectly simple, and not, in any significant sense, equal *inter se*; for the equality of compound pleasures results on this hypothesis, solely from the number of simple ones entering into their composition, so that equality is formally inapplicable to indivisible pleasures. If, on the other hand, we allow pleasures to be infinitely divisible, so that no unit we can take is indivisible, then the number of units in any given pleasure is wholly arbitrary, and if there is to be any equality of pleasures, we shall have to admit that any two units may be significantly called equal or unequal*. Hence we shall require for equality some meaning other than sameness as to the number of parts. This latter

---

* I shall never use the word *unequal* to mean merely *not equal*, but always to mean *greater or less*, *i.e.* not equal, though of the same kind of quantities.

theory, however, seems unavoidable. For there is not only no reason to regard pleasures as consisting of definite sums of indivisible units, but further—as a candid consideration will, I think, convince anyone— two pleasures can *always* be significantly judged equal or unequal. However small two pleasures may be, it must always be significant to say that they are equal. But on the theory I am combating, the judgment in question would suddenly cease to be significant when both pleasures were indivisible units. Such a view seems wholly unwarrantable, and I cannot believe that it has been consciously held by those* who have advocated the premisses from which it follows.

(β) Some quantities are infinitely divisible, and in these, whatever definition we take of infinite number, equality is not coextensive with sameness in the number of parts. In the first place, equality or inequality must always be definite: concerning two quantities of the same kind, one answer must be right and the other wrong, though it is often not in our power to decide the alternative. From this it follows that, where quantities consist of an infinite number of parts, if equality or inequality is to be reduced to number of parts at all, it must be reduced to number of *simple* parts; for the number of complex parts that may be taken to make up the whole is wholly arbitrary. But equality, for example in Geometry, is far narrower than sameness in the number of parts. The cardinal number of parts in any two continuous portions of space is the same, as we know from Cantor; even the ordinal number or type is the same for any two lengths whatever. Hence if there is to be any spatial inequality of the kind to which Geometry and common-sense have accustomed us, we must seek some other meaning for equality than that obtained from the number of parts. At this point I shall be told that the meaning is very obvious: it is obtained from superposition. Without trenching too far on discussions which belong to a later part, I may observe (*a*) that superposition applies to matter, not to space, (*b*) that as a criterion of equality, it presupposes that the matter superposed is rigid, (*c*) that rigidity means constancy as regards metrical properties. This shows that we cannot, without a vicious circle, define spatial equality by superposition. Spatial magnitude is, in fact, as indefinable as every other kind; and number of parts, in this case as in all others where the number is infinite, is wholly inadequate even as a criterion.

(γ) Some relations are quantities. This is suggested by the above discussion of spatial magnitudes, where it is very natural to base equality upon distances. Although this view, as we shall see hereafter, is not wholly adequate, it is yet partly true. There appear to be in certain spaces, and there certainly are in some series (for instance that of the

---

* *E.g.* Mr Bradley, "What do we mean by the Intensity of Psychical States?" *Mind*, N. S. Vol. IV; see esp. p. 5.

rational numbers), quantitative relations of distance among the various terms. Also similarity and difference appear to be quantities. Consider for example two shades of colour. It seems undeniable that two shades of red are more similar to each other than either is to a shade of blue; yet there is no common property in the one case which is not found in the other also. *Red* is a mere collective name for a certain series of shades, and the only reason for giving a collective name to this series lies in the close resemblance between its terms. Hence *red* must not be regarded as a common property in virtue of which two shades of red resemble each other. And since relations are not even conceivably divisible, greater and less among relations cannot depend upon number of parts.

(δ) Finally, it is well to consider directly the meanings of greater and less on the one hand, and of whole and part on the other. Euclid's axiom, that the whole is greater than the part, seems undeniably significant; but on the traditional view of quantity, this axiom would be a mere tautology. This point is again connected with the question whether superposition is to be taken as the meaning of equality, or as a mere criterion. On the latter view, the axiom must be significant, and we cannot identify magnitude with number of parts*.

154. (2) There is therefore in quantity something over and above the ideas which we have hitherto discussed. It remains to decide between the relative and absolute theories of magnitude.

The relative theory regards equal quantities as not possessed of any common property over and above that of unequal quantities, but as distinguished merely by the mutual relation of equality. There is no such thing as a magnitude, shared by equal quantities. We must not say: This and that are both a yard long; we must say: This and that are equal, or are both equal to the standard yard in the Exchequer. Inequality is also a direct relation between quantities, not between magnitudes. There is nothing by which a set of equal quantities are distinguished from one which is not equal to them, except the relation of equality itself. The course of definition is, therefore, as follows: We have first a quality or relation, say pleasure, of which there are various instances, specialized, in the case of a quality, by temporal or spatio-temporal position, and in the case of a relation, by the terms between which it holds. Let us, to fix ideas, consider quantities of pleasure. Quantities of pleasure consist merely of the complexes *pleasure at such a time*, and *pleasure at such another time* (to which *place* may be added, if it be thought that pleasures have position in space). In the analysis of a particular pleasure, there is, according to the relational theory, no other element to be found. But on comparing these particular pleasures,

---

* Compare, with the above discussion, Meinong, *Ueber die Bedeutung des Weber'-schen Gesetzes*, Hamburg and Leipzig, 1896; especially Chap. I, § 3.

we find that any two have one and only one of three relations, equal, greater, and less. Why some have one relation, some another, is a question to which it is theoretically and strictly impossible to give an answer ; for there is, *ex hypothesi*, no point of difference except temporal or spatio-temporal position, which is obviously irrelevant. Equal quantities of pleasure do not agree in any respect in which unequal ones differ : it merely happens that some have one relation and some another.

This state of things, it must be admitted, is curious, and it becomes still more so when we examine the indemonstrable axioms which the relational theory obliges us to assume. They are the following (*A, B, C* being all quantities of one kind) :

(*a*)   *A* = *B*, or *A* is greater than *B*, or *A* is less than *B*.

(*b*)   *A* being given, there is always a *B*, which may be identical with *A*, such that *A* = *B*.

(*c*)   If *A* = *B*, then *B* = *A*.

(*d*)   If *A* = *B* and *B* = *C*, then *A* = *C*.

(*e*)   If *A* is greater than *B*, then *B* is less than *A*.

(*f*)   If *A* is greater than *B*, and *B* is greater than *C*, then *A* is greater than *C*.

(*g*)   If *A* is greater than *B*, and *B* = *C*, then *A* is greater than *C*.

(*h*)   If *A* = *B*, and *B* is greater than *C*, then *A* is greater than *C*.

From (*b*), (*c*), and (*d*) it follows that *A* = *A**. From (*e*) and (*f*) it follows that, if *A* is less than *B*, and *B* is less than *C*, then *A* is less than *C* ; from (*c*), (*e*), and (*h*) it follows that, if *A* is less than *B*, and *B* = *C*, then *A* is less than *C* ; from (*c*), (*e*), and (*g*) it follows that, if *A* = *B*, and *B* is less than *C*, then *A* is less than *C*. (In the place of (*b*) we may put the axiom : If *A* be a quantity, then *A* = *A*.) These axioms, it will be observed, lead to the conclusion that, in any proposition asserting equality, excess, or defect, an equal quantity may be substituted anywhere without affecting the truth or falsehood of the proposition. Further, the proposition *A* = *A* is an essential part of the theory. Now the first of these facts strongly suggests that what is relevant in quantitative propositions is not the actual quantity, but some property which it shares with other equal quantities. And this suggestion is almost demonstrated by the second fact, *A* = *A*. For it may be laid down that the only unanalyzable symmetrical and transitive relation which a term can have to itself is identity, if this be indeed a relation. Hence the relation of equality should be analyzable. Now to say that a relation is analyzable is to say either that it consists of two or more relations between its terms, which is plainly not the case here, or that, when it is said to hold between two terms, there is some third term to which both are related in ways which, when compounded, give the original relation.

---

* This does not follow from (*c*) and (*d*) alone, since they do not assert that *A* is ever equal to *B*. See Peano, *loc. cit.*

Thus to assert that A is B's grandparent is to assert that there is some third person C, who is A's son or daughter and B's father or mother. Hence if equality be analyzable, two equal terms must both be related to some third term; and since a term may be equal to itself, any two equal terms must have the *same* relation to the third term in question. But to admit this is to admit the absolute theory of magnitude.

A direct inspection of what we mean when we say that two terms are equal or unequal will reinforce the objections to the relational theory. It seems preposterous to maintain that equal quantities have absolutely nothing in common beyond what is shared by unequal quantities. Moreover unequal quantities are not merely different: they are different in the specific manner expressed by saying that one is greater, the other less. Such a difference seems quite unintelligible unless there is some point of difference, where unequal quantities are concerned, which is absent where quantities are equal. Thus the relational theory, though apparently not absolutely self-contradictory, is complicated and paradoxical. Both the complication and the paradox, we shall find, are entirely absent in the absolute theory.

155. (3) In the absolute theory, there is, belonging to a set of equal quantities, one definite concept, namely a certain magnitude. Magnitudes are distinguished among concepts by the fact that they have the relations of greater and less (or at least one of them) to other terms, which are therefore also magnitudes. Two magnitudes cannot be equal, for equality belongs to quantities, and is defined as possession of the *same* magnitude. Every magnitude is a simple and indefinable concept. Not any two magnitudes are one greater and the other less; on the contrary, given any magnitude, those which are greater or less than that magnitude form a certain definite class, within which any two are one greater and the other less. Such a class is called a *kind* of magnitude. A kind of magnitude may, however, be also defined in another way, which has to be connected with the above by an axiom. Every magnitude is a magnitude *of* something—pleasure, distance, area, etc.—and has thus a certain specific relation to the something of which it is a magnitude. This relation is very peculiar, and appears to be incapable of further definition. All magnitudes which have this relation to one and the same something (*e.g.* pleasure) are magnitudes of one kind; and with this definition, it becomes an axiom to say that, of two magnitudes of the same kind, one is greater and the other less.

156. An objection to the above theory may be based on the relation of a magnitude to that whose magnitude it is. To fix our ideas, let us consider pleasure. A magnitude of pleasure is so much pleasure, such and such an intensity of pleasure. It seems difficult to regard this, as the absolute theory demands, as a simple idea: there seem to be two constituents, pleasure and intensity. Intensity need not be intensity of pleasure, and intensity of pleasure is distinct from abstract pleasure. But what we require for the constitution of a certain magnitude of pleasure is, not intensity in general, but a certain specific intensity; and a *specific* intensity cannot be indifferently of pleasure or of something else. We cannot first settle how much we will have, and then decide whether it is to be pleasure or mass. A specific intensity must be of a specific kind. Thus intensity and pleasure are not independent and coordinate elements in the definition of a given amount of pleasure. There are different kinds of intensity, and different magnitudes in each kind; but magnitudes in different kinds must be different. Thus it seems that the common element, indicated by the term *intensity* or *magnitude*, is not any thing intrinsic, that can be discovered by analysis of a single term, but is merely the fact of being one term in a relation of inequality. Magnitudes are defined by the fact that they have this relation, and they do not, so far as the definition shows, agree in anything else. The class to which they all belong, like the married portion of a community, is defined by mutual relations among its terms, not by a common relation to some outside term—unless, indeed, inequality itself were taken as such a term, which would be merely an unnecessary complication. It is necessary to consider what may be called the extension or field of a relation, as well as that of a class-concept: and magnitude is the class which forms the extension of inequality. Thus *magnitude of pleasure* is complex, because it combines magnitude and pleasure; but a particular magnitude of pleasure is not complex, for magnitude does not enter into its concept at all. It is only a magnitude because it is greater or less than certain other terms; it is only a magnitude of *pleasure* because of a certain relation which it has to pleasure. This is more easily understood where the particular magnitude has a special name. A yard, for instance, is a magnitude, because it is greater than a foot; it is a magnitude of length, because it is what is called *a* length. Thus all magnitudes are simple concepts, and are classified into kinds by their relation to some quality or relation. The quantities which are instances of a magnitude are particularized by spatio-temporal position or (in the case of relations which are quantities) by the terms between which the relation holds. Quantities are not properly greater or less, for the relations of greater and less hold between their magnitudes, which are distinct from the quantities.

When this theory is applied in the enumeration of the necessary axioms, we find a very notable simplification. The axioms in which equality appears have all become demonstrable, and we require only the following (*L, M, N* being magnitudes of one kind):

(*a*)   No magnitude is greater or less than itself.
(*b*)   *L* is greater than *M* or *L* is less than *M*.
(*c*)   If *L* is greater than *M*, then *M* is less than *L*.
(*d*)   If *L* is greater than *M* and *M* is greater than *N*, then *L* is greater than *N*.

The difficult axiom which we formerly called (*b*) is avoided, as are the other axioms concerning equality; and those that remain are simpler than our former set.

157. The decision between the absolute and relative theories can be made at once by appealing to a certain general principle, of very wide application, which I propose to call the principle of Abstraction. This principle asserts that, whenever a relation, of which there are instances, has the two properties of being symmetrical and transitive, then the relation in question is not primitive, but is analyzable into sameness of relation to some other term ; and that this common relation is such that there is only one term at most to which a given term can be so related, though many terms may be so related to a given term. (That is, the relation is like that of son to father : a man may have many sons, but can have only one father.)

This principle, which we have already met with in connection with cardinals, may seem somewhat elaborate. It is, however, capable of proof, and is merely a careful statement of a very common assumption. It is generally held that all relations are analyzable into identity or diversity of content. Though I entirely reject this view, I retain, so far as symmetrical transitive relations are concerned, what is really a somewhat modified statement of the traditional doctrine. Such relations, to adopt more usual phraseology, are always constituted by possession of a common property. But a common property is not a very precise conception, and will not, in most of its ordinary significations, formally fulfil the function of analyzing the relations in question. A common quality of two terms is usually regarded as a predicate of those terms. But the whole doctrine of subject and predicate, as the only form of which propositions are capable, and the whole denial of the ultimate reality of relations, are rejected by the logic advocated in the present work. Abandoning the word *predicate*, we may say that the most general sense which can be given to a common property is this : A common property of two terms is any third term to which both have one and the same relation. In this general sense, the possession of a common property is symmetrical, but not necessarily transitive. In order that it may be transitive, the relation to the common property must be such that only one term at most can be the property of any given term*. Such is the relation of a quantity to its magnitude, or of an event to the time at which it occurs : given one term of the relation, namely the referent, the other is determinate, but given the other, the one is by no means determinate. Thus it is capable of demonstration that the possession of a common property of the type in question always

---

* The proof of these assertions is mathematical, and depends upon the Logic of Relations ; it will be found in my article "Sur la Logique des Relations," *R. d. M.* VII, No. 2, § 1, Props. 6. 1, and 6. 2.

leads to a symmetrical transitive relation. What the principle of abstraction asserts is the converse, that such relations only spring from common properties of the above type*. It should be observed that the relation of the terms to what I have called their common property can never be that which is usually indicated by the relation of subject to predicate, or of the individual to its class. For no subject (in the received view) can have only one predicate, and no individual can belong to only one class. The relation of the terms to their common property is, in general, different in different cases. In the present case, the quantity is a complex of which the magnitude forms an element: the relation of the quantity to the magnitude is further defined by the fact that the magnitude has to belong to a certain class, namely that of magnitudes. It must then be taken as an axiom (as in the case of colours) that two magnitudes of the same kind cannot coexist in one spatio-temporal place, or subsist as relations between the same pair of terms; and this supplies the required uniqueness of the magnitude. It is such synthetic judgments of incompatibility that lead to negative judgments; but this is a purely logical topic, upon which it is not necessary to enlarge in this connection.

158. We may now sum up the above discussion in a brief statement of results. There are a certain pair of indefinable relations, called *greater* and *less*; these relations are asymmetrical and transitive, and are inconsistent the one with the other. Each is the converse of the other, in the sense that, whenever the one holds between A and B, the other holds between B and A. The terms which are capable of these relations are *magnitudes*. Every magnitude has a certain peculiar relation to some concept, expressed by saying that it is a magnitude *of* that concept. Two magnitudes which have this relation to the same concept are said to be of the same kind; to be of the same kind is the necessary and sufficient condition for the relations of greater and less. When a magnitude can be particularized by temporal, spatial, or spatio-temporal position, or when, being a relation, it can be particularized by taking into a consideration a pair of terms between which it holds, then the magnitude so particularized is called a *quantity*. Two magnitudes of the same kind can never be particularized by exactly the same specifications. Two quantities which result from particularizing the same magnitude are said to be *equal*.

Thus our indefinables are (1) greater and less, (2) every particular magnitude. Our indemonstrable propositions are:

---

* The principle is proved by showing that, if R be a symmetrical transitive relation, and a a term of the field of R, a has, to the class of terms to which it has the relation R taken as a whole, a many-one relation which, relationally multiplied by its converse, is equal to R. Thus a magnitude may, so far as formal arguments are concerned, be identified with a class of equal quantities.

(1) Every magnitude has to some term the relation which makes it of a certain kind.

(2) Any two magnitudes of the same kind are one greater and the other less.

(3) Two magnitudes of the same kind, if capable of occupying space or time, cannot both have the same spatio-temporal position; if relations, can never be both relations between the same pair of terms.

(4) No magnitude is greater than itself.

(5) If A is greater than B, B is less than A, and *vice versâ*.

(6) If A is greater than B and B is greater than C, then A is greater than C*.

Further axioms characterize various species of magnitudes, but the above seem alone necessary to magnitude in general. None of them depend in any way upon number or measurement; hence we may be undismayed in the presence of magnitudes which cannot be divided or measured, of which, in the next chapter, we shall find an abundance of instances.

Note to Chapter XIX. The work of Herr Meinong on Weber's Law, already alluded to, is one from which I have learnt so much, and with which I so largely agree, that it seems desirable to justify myself on the points in which I depart from it. This work begins (§ 1) by a characterization of magnitude as that which is limited towards zero. Zero is understood as the negation of magnitude, and after a discussion, the following statement is adopted (p. 8):

" That is or has magnitude, which allows the interpolation of terms between itself and its contradictory opposite."

Whether this constitutes a definition, or a mere criterion, is left doubtful (*ib.*), but in either case, it appears to me to be undesirable as a fundamental characterization of magnitude. It derives support, as Herr Meinong points out (p. 6 *n.*), from its similarity to Kant's "Anticipations of Perception†." But it is, if I am not mistaken, liable to several grave objections. In the first place, the whole theory of zero is most difficult, and seems subsequent, rather than prior, to the theory of other magnitudes. And to regard zero as the contradictory opposite of other magnitudes seems erroneous. The phrase should denote the class obtained by negation of the class " magnitudes of such and such a kind "; but this obviously would not yield the zero of that kind of magnitude. Whatever interpretation we give to the phrase, it would seem to imply that we must regard zero as not a magnitude of the kind whose zero it is. But in that case it is not less than the magnitudes of the kind in question, and there seems no particular meaning in saying

---

* It is not necessary in (5) and (6) to add "*A, B, C* being magnitudes," for the above relations of greater and less are what define magnitudes, and the addition would therefore be tautological.

† *Reine Vernunft*, ed. Hartenstein (1867), p. 158.

that a lesser magnitude is *between* zero and a greater magnitude. And in any case, the notion of *between*, as we shall see in Part IV, demands asymmetrical relations among the terms concerned. These relations, it would seem, are, in the case of magnitude, none other than *greater* and *less*, which are therefore prior to the betweenness of magnitudes, and more suitable to definition. I shall endeavour at a later stage to give what I conceive to be the true theory of zero; and it will then appear how difficult this subject is. It can hardly be wise, therefore, to introduce zero in the first account of magnitude. Other objections might be urged, as, for instance, that it is doubtful whether all kinds of magnitude have a zero; that in discrete kinds of magnitude, zero is unimportant; and that among distances, where the zero is simply identity, there is hardly the same relation of zero to negation or non-existence as in the case of qualities such as pleasure. But the main reason must be the logical inversion involved in the introduction of *between* before any asymmetrical relations have been specified from which it could arise. This subject will be resumed in Chapter XXII.

CHAPTER XX.

THE RANGE OF QUANTITY.

159. THE questions to be discussed in the present chapter are these : What kinds of terms are there which, by their common relation to a number of magnitudes, constitute a class of quantities of one kind ? Have all such terms anything else in common ? Is there any mark which will ensure that a term is thus related to a set of magnitudes ? What sorts of terms are capable of degree, or intensity, or greater and less ?

The traditional view regards divisibility as a common mark of all terms having magnitude. We have already seen that there is no *à priori* ground for this view. We are now to examine the question inductively, to find as many undoubted instances of quantities as possible, and to inquire whether they all have divisibility or any other common mark.

Any term of which a greater or less degree is possible contains under it a collection of magnitudes of one kind. Hence the comparative form in grammar is *primâ facie* evidence of quantity. If this evidence were conclusive, we should have to admit that all, or almost all, qualities are susceptible of magnitude. The praises and reproaches addressed by poets to their mistresses would afford comparatives and superlatives of most known adjectives. But some circumspection is required in using evidence of this grammatical nature. There is always, I think, *some* quantitative comparison wherever a comparative or superlative occurs, but it is often not a comparison as regards the quality indicated by grammar.

> "O ruddier than the cherry,
> O sweeter than the berry,
> O nymph more bright
> Than moonshine light,"

are lines containing three comparatives. As regards sweetness and brightness, we have, I think, a genuine quantitative comparison ; but as regards ruddiness, this may be doubted. The comparative here—and generally where colours are concerned—indicates, I think, not more of a given colour, but more likeness to a standard colour. Various shades of colour are supposed to be arranged in a series, such that the difference of quality is greater or less according as the distance in the series is greater or less. One of these shades is the ideal " ruddiness," and others are called more or less ruddy according as they are nearer to or further from this shade in the series. The same explanation applies, I think, to such terms as *whiter, blacker, redder*. The true quantity involved seems to be, in all these cases, a relation, namely the relation of similarity. The difference between two shades of colour is certainly a difference of quality, not merely of magnitude; and when we say that one thing is redder than another, we do not imply that the two are of the same shade. If there were no difference of shade, we should probably say one was *brighter* than the other, which is quite a different kind of comparison. But though the difference of two shades is a difference of quality, yet, as the possibility of serial arrangement shows, this difference of quality is itself susceptible of degrees. Each shade of colour seems to be simple and unanalyzable; but neighbouring colours in the spectrum are certainly more similar than remote colours. It is this that gives continuity to colours. Between two shades of colour, *A* and *B*, we should say, there is always a third colour *C*; and this means that *C* resembles *A* or *B* more than *B* or *A* does. But for such relations of immediate resemblance, we should not be able to arrange colours in series. The resemblance must be immediate, since all shades of colour are unanalyzable, as appears from any attempt at description or definition*. Thus we have an indubitable case of relations which have magnitude. The difference or resemblance of two colours is a relation, and is a magnitude; for it is greater or less than other differences or resemblances.

160. I have dwelt upon this case of colours, since it is one instance of a very important class. When any number of terms can be arranged in a series, it frequently happens that any two of them have a relation which may, in a generalized sense, be called a *distance*. This relation suffices to generate a serial arrangement, and is always necessarily a magnitude. In all such cases, if the terms of the series have names, and if these names have comparatives, the comparatives indicate, not more of the term in question, but more likeness to that term. Thus, if we suppose the time-series to be one in which there is distance, when an event is said to be more recent than another, what is meant is that its distance from the present was less than that of the other. Thus recentness is not itself a quality of the time or of the event. What are quantitatively

---

* On the subject of the resemblances of colours, see Meinong, "Abstrahiren und Vergleichen," *Zeitschrift f. Psych. u. Phys. d. Sinnesorgane*, Vol. xxiv, p. 72 ff. I am not sure that I agree with the whole of Meinong's argument, but his general conclusion, "dass die Umfangscollective des Aehnlichen Allgemeinheiten darstellen, an denen die Abstraction wenigstens unmittelbar keinen Antheil hat" (p. 78), appears to me to be a correct and important logical principle.

compared in such cases are relations, not qualities. The case of colours is convenient for illustration, because colours have names, and the difference of two colours is generally admitted to be qualitative. But the principle is of very wide application. The importance of this class of magnitudes, and the absolute necessity of clear notions as to their nature, will appear more and more as we proceed. The whole philosophy of space and time, and the doctrine of so-called extensive magnitudes, depend throughout upon a clear understanding of series and distance.

Distance must be distinguished from mere difference or unlikeness. It holds only between terms in a series. It is intimately connected with order, and implies that the terms between which it holds have an ultimate and simple difference, not one capable of analysis into constituents. It implies also that there is a more or less continuous passage, through other terms belonging to the same series, from one of the distant terms to the other. Mere difference *per se* appears to be the bare *minimum* of a relation, being in fact a precondition of almost all relations. It is always absolute, and is incapable of degrees. Moreover it holds between any two terms whatever, and is hardly to be distinguished from the assertion that they are two. But distance holds only between the members of certain series, and its existence is then the source of the series. It is a specific relation, and it has *sense*; we can distinguish the distance of A from B from that of B from A. This last mark alone suffices to distinguish distance from bare difference.

It might perhaps be supposed that, in a series in which there is distance, although the distance AB must be greater than or less than AC, yet the distance BD need not be either greater or less than AC. For example, there is obviously more difference between the pleasure derivable from £5 and that derivable from £100 than between that from £5 and that from £20. But need there be either equality or inequality between the difference for £1 and £20 and that for £5 and £100? This question must be answered affirmatively. For AC is greater or less than BC, and BC is greater or less than BD; hence AC, BC and also BC, BD are magnitudes of the same kind. Hence AC, BD are magnitudes of the same kind, and if not identical, one must be the greater and the other the less. Hence, when there is distance in a series, any two distances are quantitatively comparable.

It should be observed that all the magnitudes of one kind form a series, and that their distances, therefore, if they have distances, are again magnitudes. But it must not be supposed that these can, in general, be obtained by subtraction, or are of the same kind as the magnitudes whose differences they express. Subtraction depends, as a rule, upon divisibility, and is therefore in general inapplicable to indivisible quantities. The point is important, and will be treated in detail in the following chapter.

Thus nearness and distance are relations which have magnitude.

Are there any other relations having magnitude? This may, I think, be doubted*. At least I am unaware of any other such relation, though I know no way of disproving their existence.

161. There is a difficult class of terms, usually regarded as magnitudes, apparently implying relations, though certainly not always relational. These are differential coefficients, such as velocity and acceleration. They must be borne in mind in all attempts to generalize about magnitude, but owing to their complexity they require a special discussion. This will be given in Part V; and we shall then find that differential coefficients are never magnitudes, but only real numbers, or segments in some series.

162. All the magnitudes dealt with hitherto have been, strictly speaking, indivisible. Thus the question arises: Are there any divisible magnitudes? Here I think a distinction must be made. A magnitude is essentially one, not many. Thus no magnitude is correctly expressed as a number of terms. But may not the quantity which has magnitude be a sum of parts, and the magnitude a magnitude of divisibility? If so, every whole consisting of parts will be a single term possessed of the property of divisibility. The more parts it consists of, the greater is its divisibility. On this supposition, divisibility is a magnitude, of which we may have a greater or less degree; and the degree of divisibility corresponds exactly, in finite wholes, to the number of parts. But though the whole which has divisibility is of course divisible, yet its divisibility, which alone is strictly a magnitude, is not properly speaking divisible. The divisibility does not itself consist of parts, but only of the property of having parts. It is necessary, in order to obtain divisibility, to take the whole strictly as *one*, and to regard divisibility as its adjective. Thus although, in this case, we have numerical measurement, and all the mathematical consequences of division, yet, philosophically speaking, our magnitude is still indivisible.

There are difficulties, however, in the way of admitting divisibility as a kind of magnitude. It seems to be not a property of the whole, but merely a relation to the parts. It is difficult to decide this point, but a good deal may be said, I think, in support of divisibility as a simple quality. The whole has a certain relation, which for convenience we may call that of inclusion, to all its parts. This relation is the same whether there be many parts or few; what distinguishes a whole of many parts is that it has many such relations of inclusion. But it seems reasonable to suppose that a whole of many parts differs from a whole of few parts in some intrinsic respect. In fact, wholes may be arranged in a series according as they have more or fewer parts, and the serial arrangement implies, as we have already seen, some series of properties differing more or less from each other, and agreeing when two wholes have the same

---

* Cf. Meinong, *Ueber die Bedeutung des Weber'schen Gesetzes*, Hamburg and Leipzig, 1896, p. 23.

finite number of parts, but distinct from number of parts in finite wholes. These properties can be none other than greater or less degrees of divisibility. Thus magnitude of divisibility would *appear* to be a simple property of a whole, distinct from the number of parts included in the whole, but correlated with it, provided this number be finite. If this view can be maintained, divisibility may be allowed to remain as a numerically measurable, but not divisible, class of magnitudes. In this class we should have to place lengths, areas and volumes, but not distances. At a later stage, however, we shall find that the divisibility of infinite wholes, in the sense in which this is not measured by cardinal numbers, must be derived through relations in a way analogous to that in which distance is derived, and must be really a property of relations*.

Thus it would appear, in any case, that all magnitudes are indivisible. This is one common mark which they all possess, and so far as I know, it is the only one to be added to those enumerated in Chapter xix. Concerning the range of quantity, there seems to be no further general proposition. Very many simple non-relational terms have magnitude, the principal exceptions being colours, points, instants and numbers.

163. Finally, it is important to remember that, on the theory adopted in Chapter xix, a given magnitude of a given kind is a simple concept, having to the kind a relation analogous to that of inclusion in a class. When the kind is a kind of existents, such as pleasure, what actually exists is never the kind, but various particular magnitudes of the kind. Pleasure, abstractly taken, does not exist, but various amounts of it exist. This degree of abstraction is essential to the theory of quantity: there must be entities which differ from each other in nothing except magnitude. The grounds for the theory adopted may perhaps appear more clearly from a further examination of this case.

Let us start with Bentham's famous proposition: "Quantity of pleasure being equal, pushpin is as good as poetry." Here the qualitative difference of the pleasures is the very point of the judgment; but in order to be able to say that the quantities of pleasure are equal, we must be able to abstract the qualitative differences, and leave a certain magnitude of pleasure. If this abstraction is legitimate, the qualitative difference must be not truly a difference of quality, but only a difference of relation to other terms, as, in the present case, a difference in the causal relation. For it is not the whole pleasurable states that are compared, but only—as the form of the judgment aptly illustrates—their quality of pleasure. If we suppose the magnitude of pleasure to be not a separate entity, a difficulty will arise. For the mere element of pleasure must be identical in the two cases, whereas we require a possible difference of magnitude. Hence we can neither hold that only the whole concrete state exists, and any part of it is an abstraction, nor that

---

* See Chap. xlvii.

what exists is abstract pleasure, not magnitude of pleasure. Nor can we say : We abstract, from the whole states, the two elements magnitude and pleasure. For then we should not get a quantitative comparison of the pleasures. The two states would agree in being pleasures, and in being magnitudes. But this would not give us a magnitude of pleasure ; and it would give a magnitude to the states as a whole, which is not admissible. Hence we cannot abstract magnitude in general from the states, since as wholes they have no magnitude. And we have seen that we must not abstract bare pleasure, if we are to have any possibility of different magnitudes. Thus what we have to abstract is a magnitude of pleasure as a whole. This must not be analyzed into magnitude and pleasure, but must be abstracted as a whole. And the magnitude of pleasure must exist as a part of the whole pleasurable states, for it is only where there is no difference save at most one of magnitude that quantitative comparison is possible. Thus the discussion of this particular case fully confirms the theory that every magnitude is unanalyzable, and has only the relation analogous to inclusion in a class to that abstract quality or relation of which it is a magnitude.

Having seen that all magnitudes are indivisible, we have next to consider the extent to which numbers can be used to express magnitudes, and the nature and limits of measurement.

CHAPTER XXI.

NUMBERS AS EXPRESSING MAGNITUDES:
MEASUREMENT.

164. It is one of the assumptions of educated common-sense that two magnitudes of the same kind must be numerically comparable. People are apt to say that they are thirty per cent. healthier or happier than they were, without any suspicion that such phrases are destitute of meaning. The purpose of the present chapter is to explain what is meant by measurement, what are the classes of magnitudes to which it applies, and how it is applied to those classes.

Measurement of magnitudes is, in its most general sense, any method by which a unique and reciprocal correspondence is established between all or some of the magnitudes of a kind and all or some of the numbers, integral, rational, or real, as the case may be. (It might be thought that complex numbers ought to be included; but what can *only* be measured by complex numbers is in fact always an aggregate of magnitudes of different kinds, not a single magnitude.) In this general sense, measurement demands some one-one relation between the numbers and magnitudes in question—a relation which may be direct or indirect, important or trivial, according to circumstances. Measurement in this sense can be applied to very many classes of magnitudes; to two great classes, distances and divisibilities, it applies, as we shall see, in a more important and intimate sense.

Concerning measurement in the most general sense, there is very little to be said. Since the numbers form a series, and since every kind of magnitude also forms a series, it will be desirable that the order of the magnitudes measured should correspond to that of the numbers, *i.e.* that all relations of *between* should be the same for magnitudes and their measures. Wherever there is a zero, it is well that this should be measured by the number zero. These and other conditions, which a measure should fulfil if possible, may be laid down; but they are of practical rather than theoretical importance.

165. There are two general metaphysical opinions, either of which, if accepted, shows that *all* magnitudes are theoretically capable of measurement in the above sense. The first of these is the theory that all events either are, or are correlated with, events in the dynamical causal series. In regard to the so-called secondary qualities, this view has been so far acted upon by physical science that it has provided most of the so-called intensive quantities that appear in space with spatial, and thence numerical, measures. And with regard to mental quantities the theory in question is that of psychophysical parallelism. Here the motion which is correlated with any psychical quantity always theoretically affords a means of measuring that quantity. The other metaphysical opinion, which leads to universal measurability, is one suggested by Kant's "Anticipations of Perception *," namely that, among intensive magnitudes, an increase is always accompanied by an increase of reality. Reality, in this connection, seems synonymous with existence; hence the doctrine may be stated thus: Existence is a kind of intensive magnitude, of which, where a greater magnitude exists, there is always more than where a less magnitude exists. (That this is exactly Kant's doctrine seems improbable; but it is at least a tenable view.) In this case, since two instances of the same magnitude (*i.e.* two equal quantities) must have more existence than one, it follows that, if a single magnitude of the same kind can be found having the same amount of existence as the two equal quantities together, then that magnitude may be called double that of each of the equal quantities. In this way all intensive magnitudes become theoretically capable of measurement. That this method has any practical importance it would be absurd to maintain; but it may contribute to the appearance of meaning belonging to *twice as happy*. It gives a sense, for example, in which we may say that a child derives as much pleasure from one chocolate as from two acid drops; and on the basis of such judgments the hedonistic Calculus could theoretically be built.

There is one other general observation of some importance. If it be maintained that all series of magnitudes are either continuous in Cantor's sense, or are similar to series which can be chosen out of continuous series, then it is theoretically possible to correlate any kind of magnitudes with all or some of the real numbers, so that the zeros correspond, and the greater magnitudes correspond to the greater numbers. But if any series of magnitudes, without being continuous, contains continuous series, then such a series of magnitudes will be strictly and theoretically incapable of measurement by the real numbers †.

166. Leaving now these somewhat vague generalities, let us examine the more usual and concrete sense of measurement. What we require is some sense in which we may say that one magnitude is double of another.

---

* *Reine Vernunft*, ed. Hart. (1867), p. 160. The wording of the first edition illustrates better than that of the second the doctrine to which I allude. See *e.g.* Erdmann's edition, p. 161.

† See Part V, Chap. xxxiii ff.

In the above instances this sense was derived by correlation with spatio-temporal magnitudes, or with existence. This presupposed that in these cases a meaning had been found for the phrase. Hence measurement demands that, in some cases, there should be an intrinsic meaning to the proposition "this magnitude is double of that." (In what sense the meaning is intrinsic will appear as we proceed.) Now so long as quantities are regarded as inherently divisible, there is a perfectly obvious meaning to such a proposition: a magnitude *A* is double of *B* when it is the magnitude of two quantities together, each of these having the magnitude *B*. (It should be observed that to divide a *magnitude* into two equal parts must always be impossible, since there are no such things as equal magnitudes.) Such an interpretation will still apply to magnitudes of divisibility; but since we have admitted other magnitudes, a different interpretation (if any) must be found for these. Let us first examine the case of divisibilities, and then proceed to the other cases where measurement is intrinsically possible.

167. The divisibility of a finite whole is immediately and inherently correlated with the number of simple parts in the whole. In this case, although the magnitudes are even now incapable of addition of the sort required, the quantities can be added in the manner explained in Part II. The addition of two magnitudes of divisibility yields merely two magnitudes, not a new magnitude. But the addition of two quantities of divisibility, *i.e.* two wholes, does yield a new single whole, provided the addition is of the kind which results from logical addition by regarding classes as the wholes formed by their terms. Thus there is a good meaning in saying that one magnitude of divisibility is double of another, when it applies to a whole containing twice as many parts. But in the case of infinite wholes, the matter is by no means so simple. Here the number of simple parts (in the only senses of infinite number hitherto discovered) may be equal without equality in the magnitude of divisibility. We require here a method which does not go back to simple parts. In actual space, we have immediate judgments of equality as regards two infinite wholes. When we have such judgments, we can regard the sum of *n* equal wholes as *n* times each of them; for addition of wholes does not demand their finitude. In this way numerical comparison of some pairs of wholes becomes possible. By the usual well-known methods, by continual subdivision and the method of limits, this is extended to all pairs of wholes which are such that immediate comparisons are possible. Without these immediate comparisons, which are necessary both logically and psychologically*, nothing can be accomplished: we are always reduced in the last resort to the immediate judgment that our foot-rule has not greatly changed its size during measurement, and this judgment is prior to the results of physical

---

* Cf. Meinong, *op. cit.*, pp. 63–4.

science as to the extent to which bodies do actually change their sizes. But where immediate comparison is psychologically impossible, we may theoretically substitute a logical variety of measurement, which, however, gives a property not of the divisible whole, but of some relation or class of relations more or less analogous to those that hold between points in space.

That divisibility, in the sense required for areas and volumes, is not a property of a whole, results from the fact (which will be established in Part VI) that between the points of a space there are always relations which generate a different space. Thus two sets of points which, with regard to one set of relations, form equal areas, form unequal areas with respect to another set, or even form one an area and the other a line or a volume. If divisibility in the relevant sense were an intrinsic property of wholes, this would be impossible. But this subject cannot be fully discussed until we come to Metrical Geometry.

Where our magnitudes are divisibilities, not only do numbers measure them, but the difference of two measuring numbers, with certain limitations, measures the magnitude of the difference (in the sense of dissimilarity) between the divisibilities. If one of the magnitudes be fixed, its difference from the other increases as the difference of the measuring numbers increases; for this difference depends upon the difference in the number of parts. But I do not think it can be shown generally that, if A, B, C, D be the numbers measuring four magnitudes, and $A - B = C - D$, then the differences of the magnitudes are equal. It would seem, for instance, that the difference between one inch and two inches is greater than that between 1001 inches and 1002 inches. This remark has no importance in the present case, since differences of divisibility are never required; but in the case of distances it has a curious connection with non-Euclidean Geometry. But it is theoretically important to observe that, if divisibility be indeed a magnitude—as the equality of areas and volumes seems to require—then there is strictly no ground for saying that the divisibility of a sum of two units is twice as great as that of one unit. Indeed this proposition cannot be strictly taken, for no magnitude *is* a sum of parts, and no magnitude therefore is double of another. We can only mean that the sum of two units contains twice as many parts, which is an arithmetical, not a quantitative, judgment, and is adequate only in the case where the number of parts is finite, since in other cases the double of a number is in general equal to it. Thus even the measurement of divisibility by numbers contains an element of convention; and this element, we shall find, is still more prominent in the case of distances.

168. In the above case we still had addition in one of its two fundamental senses, *i.e.* the combination of wholes to form a new whole. But in other cases of magnitude we do not have any such addition. The sum of two pleasures is not a new pleasure, but is merely two pleasures. The sum of two distances is also not properly one distance. But in this case we have an extension of the idea of addition. Some such extension must always be possible where measurement is to be effected in the more natural and restricted sense which we are now discussing. I shall first explain this generalized addition in abstract terms, and then illustrate its application to distances.

It sometimes happens that two quantities, which are not capable of addition proper, have a relation, which has itself a one-one relation to a quantity of the same kind as those between which it holds. Supposing a, b, c to be such quantities, we have, in the case supposed, some proposition aBc, where B is a relation which uniquely determines and is uniquely determined by some quantity b of the same kind as that to which a and c belong. Thus for example two ratios have a relation, which we may call their difference, which is itself wholly determined by another ratio, namely the difference, in the arithmetical sense, of the two given ratios. If α, β, γ be terms in a series in which there is distance, the distances $\alpha\beta$, $\alpha\gamma$ have a relation which is measured by (though not identical with) the distance $\beta\gamma$. In all such cases, by an extension of addition, we may put $a + b = c$ in place of aBc. Wherever a set of quantities have relations of this kind, if further aBc implies bAc, so that $a + b = b + a$, we shall be able to proceed as if we had ordinary addition, and shall be able in consequence to introduce numerical measurement.

The conception of distance will be discussed fully in Part IV, in connection with order: for the present I am concerned only to show how distances come to be measurable. The word will be used to cover a far more general conception than that of distance in space. I shall mean by a kind of distance a set of quantitative asymmetrical relations of which one and only one holds between any pair of terms of a given class; which are such that, if there is a relation of the kind between a and b, and also between b and c, then there is one of the kind between a and c, the relation between a and c being the relative product of those between a and b, b and c; this product is to be commutative, *i. e.* independent of the order of its factors; and finally, if the distance ab be greater than the distance ac, then, d being any other member of the class, db is greater than dc. Although distances are thus relations, and therefore indivisible and incapable of addition proper, there is a simple and natural convention by which such distances become numerically measurable.

The convention is this. Let it be agreed that, when the distances a_0a_1, a_1a_2 ... $a_{n-1}a_n$ are all equal and in the same sense, then a_0a_n is said to be n times each of the distances a_0a_1, etc., *i. e.* is to be measured by a number n times as great. This has generally been regarded as not a convention, but an obvious truth; owing, however, to the fact that distances are indivisible, no distance is really a sum of other distances, and numerical measurement must be in part conventional. With this convention, the numbers corresponding to distances, where there are such numbers, become definite, except as to a common factor dependent upon the choice of a unit. Numbers are also assigned by this method to the members of the class between which the distances hold; these numbers have, in addition to the arbitrary factor, an arbitrary additive constant, depending upon the choice of origin. This method, which is capable of still further generalization, will be more fully explained in Part IV. In order to show that *all* the distances of our kind, and *all* the terms of our set, can have numbers assigned to them, we require two further axioms, the axiom of Archimedes, and what may be called the axiom of linearity*.

169. The importance of the numerical measurement of distance, at least as applied to space and time, depends partly upon a further fact, by which it is brought into relation with the numerical measurement of divisibility. In all series there are terms intermediate between any two whose distance is not the minimum. These terms are determinate when the two distant terms are specified. The intermediate terms may be called the *stretch* from a_0 to a_n†. The whole composed of these terms is a quantity, and has a divisibility measured by the number of terms, provided their number is finite. If the series is such that the distances of consecutive terms are all equal, then, if there are $n-1$ terms between a_0 and a_n, the measure of the distance is proportional to n. Thus, if we include in the stretch one of the end terms, but not the other, the measures of the stretch and the distance are proportional, and equal stretches correspond to equal distances. Thus the number of terms in the stretch measures both the distance of the end terms and the amount of divisibility of the whole stretch. When the stretch contains an infinite number of terms, we estimate equal stretches as explained above. It then becomes an axiom, which may or may not hold in a given case, that equal stretches correspond to equal distances. In this case, co-ordinates measure two entirely distinct magnitudes, which, owing to their common measure, are perpetually confounded.

170. The above analysis explains a curious problem which must have troubled most people who have endeavoured to philosophize about Geometry. Starting from one-dimensional magnitudes connected with the straight line, most theories may be divided into two classes, those appropriate to areas and volumes, and those appropriate to angles

---

* See Part IV, Chap. XXXI. This axiom asserts that a magnitude can be divided into n equal parts, and forms part of Du Bois Reymond's definition of linear magnitude. See his *Allgemeine Functionentheorie* (Tübingen, 1882), Chap. I, § 16; also Bettazzi, *Teoria delle Grandezze* (Pisa, 1890), p. 44. The axiom of Archimedes asserts that, given any two magnitudes of a kind, some finite multiple of the lesser exceeds the greater.

† Called *Strecke* by Meinong, *op. cit.*, *e.g.* p. 22.

between lines or planes. Areas and volumes are radically different
from angles, and are generally neglected in philosophies which hold
to relational views of space or start from projective Geometry. The
reason of this is plain enough. On the straight line, if, as is usually
supposed, there is such a relation as distance, we have two philosophically distinct but practically conjoined magnitudes, namely the distance,
and the divisibility of the stretch. The former is similar to angles; the
latter, to areas and volumes. Angles may also be regarded as distances
between terms in a series, namely between lines through a point or
planes through a line. Areas and volumes, on the contrary, are sums,
or magnitudes of divisibility. Owing to the confusion of the two kinds
of magnitude connected with the line, either angles, or else areas and
volumes, are usually incompatible with the philosophy invented to
suit the line. By the above analysis, this incompatibility is at once
explained and overcome*.

171. We thus see how two great classes of magnitudes—divisibilities
and distances—are rendered amenable to measure. These two classes
practically cover what are usually called extensive magnitudes, and it
will be convenient to continue to allow the name to them. I shall
extend this name to cover all distances and divisibilities, whether they
have any relation to space and time or not. But the word *extensive*
must not be supposed to indicate, as it usually does, that the magnitudes
so designated are divisible. We have already seen that no magnitude is
divisible. *Quantities* are only divisible into other quantities in the one
case of wholes which are quantities of divisibility. Quantities which are
distances, though I shall call them extensive, are not divisible into
smaller distances; but they allow the important kind of addition explained above, which I shall call in future *relational* addition†.

All other magnitudes and quantities may be properly called *intensive*.
Concerning these, unless by some causal relation, or by means of some
more or less roundabout relation such as those explained at the beginning
of the present chapter, numerical measurement is impossible. Those
mathematicians who are accustomed to an exclusive emphasis on numbers,
will think that not much can be said with definiteness concerning magnitudes incapable of measurement. This, however, is by no means the
case. The immediate judgments of equality, upon which (as we saw)
all measurements depend, are still possible where measurement fails, as
are also the immediate judgments of greater and less. Doubt only
arises where the difference is small; and all that measurement does,

---

* In Part VI, we shall find reason to deny distance in most spaces. But there
is still a distinction between stretches, consisting of the terms of some series, and
such quantities as areas and volumes, where the terms do not, in any simple sense,
form a one-dimensional series.

† Not to be confounded with the *relative* addition of the Algebra of Relatives.
It is connected rather with relative multiplication.

in this respect, is to make the margin of doubt smaller—an achievement which is purely psychological, and of no philosophical importance. Quantities not susceptible of numerical measurement can thus be arranged in a scale of greater and smaller magnitudes, and this is the only strictly quantitative achievement of even numerical measurement. We can know that one magnitude is greater than another, and that a third is intermediate between them ; also, since the differences of magnitudes are always magnitudes, there is always (theoretically, at least) an answer to the question whether the difference of one pair of magnitudes is greater than, less than, or the same as the difference of another pair of the same kind. And such propositions, though to the mathematician they may appear approximate, are just as precise and definite as the propositions of Arithmetic. Without numerical measurement, therefore, the quantitative relations of magnitudes have all the definiteness of which they are capable—nothing is added, from the theoretical standpoint, by the assignment of correlated numbers. The whole subject of the measurement of quantities is, in fact, one of more practical than theoretical importance. What is theoretically important in it is merged in the wider question of the correlation of series, which will occupy us much hereafter. The chief reason why I have treated the subject thus at length is derived from its traditional importance, but for which it might have been far more summarily treated.

CHAPTER XXII.

ZERO.

172. The present chapter is concerned, not with any form of the numerical zero, nor yet with the infinitesimal, but with the pure zero of magnitude. This is the zero which Kant has in mind, in his refutation of Mendelssohn's proof of the immortality of the soul*. Kant points out that an intensive magnitude, while remaining of the same kind, can become zero; and that, though zero is a definite magnitude, no quantity whose magnitude is zero can exist. This kind of zero, we shall find, is a fundamental quantitative notion, and is one of the points in which the theory of quantity presents features peculiar to itself. The quantitative zero has a certain connection both with the number 0 and with the null-class in Logic, but it is not (I think) definable in terms of either. What is less universally realized is its complete independence of the infinitesimal. The latter notion will not be discussed until the following chapter.

The meaning of zero, in any kind of quantity, is a question of much difficulty, upon which the greatest care must be bestowed, if contradictions are to be avoided. Zero seems to be definable by some general characteristic, without reference to any special peculiarity of the kind of quantity to which it belongs. To find such a definition, however, is far from easy. Zero *seems* to be a radically distinct conception according as the magnitudes concerned are discrete or continuous. To prove that this is not the case, let us examine various suggested definitions.

173. (1) Herr Meinong (*op. cit.*, p. 8) regards zero as the contradictory opposite of each magnitude of its kind. The phrase "contradictory opposite" is one which is not free from ambiguity. The opposite of a class, in symbolic logic, is the class containing all individuals not belonging to the first class; and hence the opposite of an individual should be all other individuals. But this meaning is evidently inappropriate: zero is not everything except one magnitude of its kind, nor yet everything except the class of magnitudes of its kind. It can hardly be regarded as true to say that a pain is a zero

---

* *Kritik der Reinen Vernunft*, ed. Hartenstein, p. 281 ff.

pleasure. On the other hand, a zero pleasure is said to be *no pleasure*, and this is evidently what Herr Meinong means. But although we shall find this view to be correct, the meaning of the phrase is very difficult to seize. It does not mean something other than pleasure, as when our friends assure us that it is no pleasure to tell us our faults. It seems to mean what is neither pleasure, nor yet anything else. But this would be merely a cumbrous way of saying *nothing*, and the reference to pleasure might be wholly dropped. This gives a zero which is the same for all kinds of magnitude, and if this be the true meaning of zero, then zero is not one among the magnitudes of a kind, nor yet a term in the series formed by magnitudes of a kind. For though it is often true that there is nothing smaller than all the magnitudes of a kind, yet it is always false that nothing itself is smaller than all of them. This zero, therefore, has no special reference to any particular kind of magnitude, and is incapable of fulfilling the functions which Herr Meinong demands of it *. The phrase, however, as we shall see, is capable of an interpretation which avoids this difficulty. But let us first examine some other suggested meanings of the word.

174. (2) Zero may be defined as the least magnitude of its kind. Where a kind of magnitude is discrete, and generally when it has what Professor Bettazzi calls a *limiting* magnitude of the kind†, such a definition is insufficient. For in such a case, the limiting magnitude seems to be really the least of its kind. And in any case, the definition gives rather a characteristic than a true definition, which must be sought in some more purely logical notion, for zero cannot fail to be in some sense a denial of all other magnitudes of the kind. The phrase that zero is the smallest of magnitudes is like the phrase which De Morgan commends for its rhetoric: "Achilles was the strongest of all his enemies." Thus it would be obviously false to say that 0 is the least of the positive integers, or that the interval between A and A is the least interval between any two letters of the alphabet. On the other hand, where a kind of magnitude is continuous, and has no limiting magnitude, although we have apparently a gradual and unlimited approach to zero, yet now a new objection arises. Magnitudes of this kind are essentially such as have no minimum. Hence we cannot without express contradiction take zero as their minimum. We may, however, avoid this contradiction by saying that there is always a magnitude less than any other, but not zero, unless that other be zero. This emendation avoids any formal contradiction, and is only inadequate because it gives rather a mark of zero than its true meaning. Whatever else is a magnitude of the kind in question might have been diminished; and we wish to know what it is that makes zero obviously incapable of any further diminution. This the suggested definition does not tell us, and therefore, though it gives a

---

* See note to Chap. xix, *supra*.
† *Teoria delle Grandezze*, Pisa, 1890, p. 24.

characteristic which often belongs to no other magnitude of the kind, it cannot be considered philosophically sufficient. Moreover, where there are negative magnitudes, it precludes us from regarding these as less than zero.

175. (3) Where our magnitudes are differences or distances, zero has, at first sight, an obvious meaning, namely identity. But here again, the zero so defined seems to have no relation to one kind of distances rather than another: a zero distance in time would seem to be the same as a zero distance in space. This can, however, be avoided, by substituting, for identity simply, identity with some member of the class of terms between which the distances in question hold. By this device, the zero of any class of relations which are magnitudes is made perfectly definite and free from contradiction; moreover we have both zero quantities and zero magnitudes, for if A and B be terms of the class which has distances, identity with A and identity with B are distinct zero quantities*. This case, therefore, is thoroughly clear. And yet the definition must be rejected: for it is plain that zero has some general logical meaning, if only this could be clearly stated, which is the same for all classes of quantities; and that a zero distance is not actually the same concept as identity.

176. (4) In any class of magnitudes which is continuous, in the sense of having a term between any two, and which also has no limiting magnitude, we can introduce zero in the manner in which real numbers are obtained from rationals. Any collection of magnitudes defines a class of magnitudes less than all of them. This class of magnitudes can be made as small as we please, and can actually be made to be the null-class, *i.e.* to contain no members at all. (This is effected, for instance, if our collection consists of all magnitudes of the kind.) The classes so defined form a series, closely related to the series of original magnitudes, and in this new series the null-class is definitely the first term. Thus taking the classes as quantities, the null-class is a zero quantity. There is no class containing a finite number of members, so that there is not, as in Arithmetic, a discrete approach to the null-class; on the contrary, the approach is (in several senses of the word) continuous. This method of defining zero, which is identical with that by which the real number zero is introduced, is important, and will be discussed in Part V. But for the present we may observe that it again makes zero the same for all kinds of magnitude, and makes it not one among the magnitudes whose zero it is.

177. (5) We are compelled, in this question, to face the problem as to the nature of negation. "No pleasure" is obviously a different concept from "no pain," even when these terms are taken strictly as mere denials of pleasure and pain respectively. It would seem that "no

---

* On this point, however, see § 55 above.

pleasure " has the same relation to *pleasure* as the various magnitudes of pleasure have, though it has also, of course, the special relation of negation. If this be allowed, we see that, if a kind of magnitudes be defined by that of which they are magnitudes, then *no pleasure* is one among the various magnitudes of pleasure. If, then, we are to hold to our axiom, that all pairs of magnitudes of one kind have relations of inequality, we shall be compelled to admit that zero is less than all other magnitudes of its kind. It seems, indeed, to be rendered evident that this must be admitted, by the fact that zero is obviously *not greater* than all other magnitudes of its kind. This shows that zero has a connection with *less* which it does not have with *greater*. And if we adopt this theory, we shall no longer accept the clear and simple account of zero distances given above, but we shall hold that a zero distance is strictly and merely *no distance*, and is only *correlated* with identity.

Thus it would seem that Herr Meinong's theory, with which we began, is substantially correct; it requires emendation, on the above view, only in this, that a zero magnitude is the denial of the defining concept of a kind of magnitudes, not the denial of any one particular magnitude, or of all of them. We shall have to hold that any concept which defines a kind of magnitudes defines also, by its negation, a particular magnitude of the kind, which is called the zero of that kind, and is less than all other members of the kind. And we now reap the benefit of the absolute distinction which we made between the defining concept of a kind of magnitude, and the various magnitudes of the kind. The relation which we allowed between a particular magnitude and that of which it is a magnitude was not identified with the class-relation, but was held to be *sui generis*; there is thus no contradiction, as there would be in most theories, in supposing this relation to hold between *no pleasure* and *pleasure*, or between *no distance* and *distance*.

178. But finally, it must be observed that *no pleasure*, the zero magnitude, is not obtained by the logical denial of pleasure, and is not the same as the logical notion of *not pleasure*. On the contrary, *no pleasure* is essentially a quantitative concept, having a curious and intimate relation to logical denial, just as 0 has a very intimate relation to the null-class. The relation is this, that there is no *quantity* whose *magnitude* is zero, so that the class of zero quantities is the null-class*. The zero of any kind of magnitude is incapable of that relation to existence or to particulars, of which the other magnitudes are capable. But this is a synthetic proposition, to be accepted only on account of its self-evidence. The zero magnitude of any kind, like the other magnitudes, is properly speaking indefinable, but is capable of specification by means of its peculiar relation to the logical zero.

* This must be applied in correction of what was formerly said about zero distances.

CHAPTER XXIII.

INFINITY, THE INFINITESIMAL, AND CONTINUITY.

179. ALMOST all mathematical ideas present one great difficulty: the difficulty of infinity. This is usually regarded by philosophers as an antinomy, and as showing that the propositions of mathematics are not metaphysically true. From this received opinion I am compelled to dissent. Although all apparent antinomies, except such as are quite easily disposed of, and such as belong to the fundamentals of logic, are, in my opinion, reducible to the one difficulty of infinite number, yet this difficulty itself appears to be soluble by a correct philosophy of *any*, and to have been generated very largely by confusions due to the ambiguity in the meaning of finite integers. The problem in general will be discussed in Part V; the purpose of the present chapter is merely to show that quantity, which has been regarded as the true home of infinity, the infinitesimal, and continuity, must give place, in this respect, to order; while the statement of the difficulties which arise in regard to quantity can be made in a form which is at once ordinal and arithmetical, but involves no reference to the special peculiarities of quantity.

180. The three problems of infinity, the infinitesimal, and continuity, as they occur in connection with quantity, are closely related. None of them can be fully discussed at this stage, since all depend essentially upon order, while the infinitesimal depends also upon number. The question of infinite quantity, though traditionally considered more formidable than that of zero, is in reality far less so, and might be briefly disposed of, but for the great devotion commonly shown by philosophers to a proposition which I shall call the axiom of finitude. Of some kinds of magnitude (for example ratios, or distances in space and time), it appears to be true that there is a magnitude greater than any given magnitude. That is, any magnitude being mentioned, another can be found which is greater than it. The deduction of infinity from this fact is, when correctly performed, a mere fiction to facilitate compression in the statement of results obtained by the method of limits.

Any class u of magnitudes of our kind being defined, three cases may arise: (1) There may be a class of terms greater than any of our class u, and this new class of terms may have a smallest member; (2) there may be such a class, but it may have no smallest member; (3) there may be no magnitudes which are greater than *any* term of our class u. Supposing our kind of magnitudes to be one in which there is no greatest magnitude, case (2) will always arise where the class u contains a finite number of terms. On the other hand, if our series be what is called *condensed in itself*, case (2) will never arise when u is an infinite class, and has no greatest term; and if our series is not condensed in itself, but does have a term between any two, another which has this property can always be obtained from it*. Thus all infinite series which have no greatest term will have limits, except in case (3). To avoid circumlocution, case (3) is defined as that in which the limit is infinite. But this is a mere device, and it is generally admitted by mathematicians to be such. Apart from special circumstances, there is no reason, merely because a kind of magnitudes has no maximum, to admit that there is an infinite magnitude of the kind, or that there are many such. When magnitudes of a kind having no maximum are capable of numerical measurement, they very often obey the axiom of Archimedes, in virtue of which the ratio of any two magnitudes of the kind is finite. Thus, so far, there might appear to be no problem connected with infinity.

But at this point the philosopher is apt to step in, and to declare that, by all true philosophic principles, every well-defined series of terms must have a last term. If he insists upon creating this last term, and calling it infinity, he easily deduces intolerable contradictions, from which he infers the inadequacy of mathematics to obtain absolute truth. For my part, however, I see no reason for the philosopher's axiom. To show, if possible, that it is not a necessary philosophic principle, let us undertake its analysis, and see what it really involves.

The problem of infinity, as it has now emerged, is not properly a quantitative problem, but rather one concerning order. It is only because our magnitudes form a series having no last term that the problem arises: the fact that the series is composed of magnitudes is wholly irrelevant. With this remark I might leave the subject to a later stage. But it will be worth while now to elicit, if not to examine, the philosopher's axiom of finitude.

181. It will be well, in the first place, to show how the problem concerning infinity is the same as that concerning continuity and the infinitesimal. For this purpose, we shall find it convenient to ignore the absolute zero, and to mean, when we speak of any kind of magnitudes, all the magnitudes of the kind except zero. This is a mere change of

---

* This will be further explained in Part V, Chap. xxxvi.

diction, without which intolerable repetitions would be necessary. Now there certainly are some kinds of magnitude where the three following axioms hold :

(1) If A and B be any two magnitudes of the kind, and A is greater than B, there is always a third magnitude C such that A is greater than C and C greater than B. (This I shall call, for the present, the axiom of continuity.)

(2) There is always a magnitude less than any given magnitude B.

(3) There is always a magnitude greater than any given magnitude A.

From these it follows :—

(1) That no two magnitudes of the kind are consecutive.

(2) That there is no least magnitude.

(3) That there is no greatest magnitude.

The above propositions are certainly true of *some* kinds of magnitude ; whether they are true of *all* kinds remains to be examined. The following three propositions, which directly contradict the previous three, must be always true, if the philosopher's axiom of finitude is to be accepted :

(*a*) There are consecutive magnitudes, *i.e.* magnitudes such that no other magnitude of the same kind is greater than the less and less than the greater of the two given magnitudes.

(*b*) There is a magnitude smaller than any other of the same kind.

(*c*) There is a magnitude greater than any other of the same kind *

As these three propositions directly contradict the previous three, it would seem that both sets cannot be true. We have to examine the grounds for both, and let one set of alternatives fall.

182. Let us begin with the propositions (*a*), (*b*), (*c*), and examine the nature of their grounds.

(*a*) A definite magnitude A being given, all the magnitudes greater than A form a series, whose differences from A are magnitudes of a new kind. If there be a magnitude B consecutive to A, its difference from A will be the least magnitude of its kind, provided equal stretches correspond to equal distances in the series. And conversely, if there be a smallest difference between two magnitudes, A, B, then these two magnitudes must always be consecutive ; for if not, any intermediate

---

* Those Hegelians who search for a chance of an antinomy may proceed to the definition of zero and infinity by means of the above propositions. When (2) and (*b*) both hold, they may say, the magnitude satisfying (*b*) is called zero ; when (3) and (*c*) both hold, the magnitude satisfying (*c*) is called infinity. We have seen, however, that zero is to be otherwise defined, and has to be excluded before (2) becomes true ; while infinity is not a magnitude of the kind in question at all, but merely a piece of mathematical shorthand. (Not infinity in general, that is, but infinite magnitude in the cases we are discussing.)

magnitude would have a smaller difference from A than B has. Thus if (*b*) is universally true, (*a*) must also be true; and conversely, if (*a*) is true, and if the series of magnitudes be such that equal stretches correspond to equal distances, then (*b*) is true of the distances between the magnitudes considered. We might rest content with the reduction of (*a*) to (*b*), and proceed to the proof of (*b*); but it seems worth while to offer a direct proof, such as presumably the finitist philosopher has in his mind.

Between A and B there is a certain number of magnitudes, unless A and B are consecutive. The intermediate magnitudes all have order, so that in passing from A to B all the intermediate magnitudes would be met with. In such an enumeration, there must be *some* magnitude which comes next after any magnitude C; or, to put the matter otherwise, since the enumeration has to begin, it must begin somewhere, and the term with which it begins must be the magnitude next to A. If this were not the case, there would be no definite series; for if all the terms have an order, some of them must be consecutive.

In the above argument, what is important is its dependence upon number. The whole argument turns upon the principle by which infinite number is shown to be self-contradictory, namely: *A given collection of many terms must contain some finite number of terms.* We say: All the magnitudes between A and B form a given collection. If there are no such magnitudes, A and B are consecutive, and the question is decided. If there are such magnitudes, there must be a finite number of them, say n. Since they form a series, there is a definite way of assigning to them the ordinal numbers from 1 to n. The mth and $(m+1)$th are then consecutive.

If the axiom in italics be denied, the whole argument collapses; and this, we shall find, is also the case as regards (*b*) and (*c*).

(*b*) The proof here is precisely similar to the proof of (*a*). If there are no magnitudes less than A, then A is the least of its kind, and the question is decided. If there are any, they form a definite collection, and therefore (by our axiom) have a finite number, say n. Since they form a series, ordinal numbers may be assigned to them growing higher as the magnitudes become more distant from A. Thus the nth magnitude is the smallest of its kind.

(*c*) The proof here is obtained as in (*b*), by considering the collection of magnitudes greater than A. Thus everything depends upon our axiom, without which no case can be made out against continuity, or against the absence of a greatest and least magnitude.

As regards the axiom itself, it will be seen that it has no particular reference to quantity, and at first sight it might seem to have no reference to order. But the word *finite*, which occurs in it, requires definition; and this definition, in the form suited to the present discussion, has, we shall find, an essential reference to order.

183. Of all the philosophers who have inveighed against infinite number, I doubt whether there is one who has known the difference between finite and infinite numbers. The difference is simply this. Finite numbers obey the law of mathematical induction; infinite numbers do not. That is to say, given any number n, if n belongs to every class s to which 0 belongs, and to which belongs also the number next after any number which is an s, then n is finite; if not, not. It is in this *alone*, and in its consequences, that finite and infinite numbers differ*.

The principle may be otherwise stated thus: If every proposition which holds concerning 0, and also holds concerning the immediate successor of every number of which it holds, holds concerning the number n, then n is finite; if not, not. This is the precise sense of what may be popularly expressed by saying that every finite number can be reached from 0 by successive steps, or by successive additions of 1. This is the principle which the philosopher must be held to lay down as obviously applicable to all numbers, though he will have to admit that the more precisely his principle is stated, the less obvious it becomes.

184. It may be worth while to show exactly how mathematical induction enters into the above proofs. Let us take the proof of (*a*), and suppose there are n magnitudes between A and B. Then to begin with, we supposed these magnitudes capable of enumeration, *i.e.* of an order in which there are consecutive terms and a first term, and a term immediately preceding any term except the first. This property presupposes mathematical induction, and was in fact the very property in dispute. Hence we must not presuppose the possibility of enumeration, which would be a *petitio principii*. But to come to the kernel of the argument: we supposed that, in any series, there must be a definite way of assigning ordinal numbers to the terms. This property belongs to a series of one term, and belongs to every series having $m + 1$ terms, if it belongs to every series having m terms. Hence, by mathematical induction, it belongs to all series having a finite number of terms. But if it be allowed that the number of terms may not be finite, the whole argument collapses.

As regards (*b*) and (*c*), the argument is similar. Every series having a finite number of terms can be shown by mathematical induction to have a first and last term; but no way exists of proving this concerning other series, or of proving that all series are finite. Mathematical induction, in short, like the axiom of parallels, is useful and convenient in its proper place; but to suppose it always true is to yield to the

---

* It must, however, be mentioned that one of these consequences gives a logical difference between finite and infinite numbers, which may be taken as an independent definition. This has been already explained in Part II, Chap. xiii, and will be further discussed in Part V.

tyranny of mere prejudice. The philosopher's finitist arguments, therefore, rest on a principle of which he is ignorant, which there is no reason to affirm, and every reason to deny. With this conclusion, the apparent antinomies may be considered solved.

185. It remains to consider what kinds of magnitude satisfy the propositions (1), (2), (3). There is no general principle from which these can be proved or disproved, but there are certainly cases where they are true, and others where they are false. It is generally held by philosophers that numbers are essentially discrete, while magnitudes are essentially continuous. This we shall find to be not the case. Real numbers possess the most complete continuity known, while many kinds of magnitude possess no continuity at all. The word *continuity* has many meanings, but in mathematics it has only two—one old, the other new. For present purposes the old meaning will suffice. I therefore set up, for the present, the following definition :

Continuity applies to series (and only to series) whenever these are such that there is a term between any two given terms*. Whatever is not a series, or a compound of series, or whatever is a series not fulfilling the above condition, is discontinuous.

Thus the series of rational numbers is continuous, for the arithmetic mean of two of them is always a third rational number between the two. The letters of the alphabet are not continuous.

We have seen that any two terms in a series have a distance, or a stretch which has magnitude. Since there are certainly discrete series (*e.g.* the alphabet), there are certainly discrete magnitudes, namely, the distances or the stretches of terms in discrete series. The distance between the letters A and C is greater than that between the letters A and B, but there is no magnitude which is greater than one of these and less than the other. In this case, there is also a greatest possible and a least possible distance, so that all three propositions (1), (2), (3) fail. It must not be supposed, however, that the three propositions have any necessary connection. In the case of the integers, for example, there are consecutive distances, and there is a least possible distance, namely, that between consecutive integers, but there is no greatest possible distance. Thus (3) is true, while (1) and (2) are false. In the case of the series of notes, or of colours of the rainbow, the series has a beginning and end, so that there is a greatest distance ; but there is no least distance, and there is a term between any two. Thus (1) and (2) are true, while (3) is false. Or again, if we take the series composed of zero and the fractions having one for numerator, there is a

---

* The objection to this definition (as we shall see in Part V) is, that it does not give the usual properties of the existence of limits to convergent series which are commonly associated with continuity. Series of the above kind will be called *compact*, except in the present discussion.

greatest distance, but no least distance, though the series is discrete. Thus (2) is true, while (1) and (3) are false. And other combinations might be obtained from other series.

Thus the three propositions (1), (2), (3), have no necessary connection, and all of them, or any selection, may be false as applied to any given kind of magnitude. We cannot hope, therefore, to prove their truth from the nature of magnitude. If they are ever to be true, this must be proved independently, or discovered by mere inspection in each particular case. That they are sometimes true, appears from a consideration of the distances between terms of the number-continuum or of the rational numbers. Either of these series is continuous in the above sense, and has no first or last term (when zero is excluded). Hence its distances or stretches fulfil all three conditions. The same might be inferred from space and time, but I do not wish to anticipate what is to be said of these. Quantities of divisibility do not fulfil these conditions when the wholes which are divisible consist of a finite number of indivisible parts. But where the number of parts is infinite in a whole class of differing magnitudes, all three conditions are satisfied, as appears from the properties of the number-continuum.

We thus see that the problems of infinity and continuity have no essential connection with quantity, but are due, where magnitudes present them at all, to characteristics depending upon number and order. Hence the discussion of these problems can only be undertaken after the pure theory of order has been set forth*. To do this will be the aim of the following Part.

186. We may now sum up the results obtained in Part III. In Chapter XIX we determined to define a magnitude as whatever is either greater or less than something else. We found that magnitude has no necessary connection with divisibility, and that greater and less are indefinable. Every magnitude, we saw, has a certain relation—analogous to, but not identical with, that of inclusion in a class—to a certain quality or relation; and this fact is expressed by saying that the magnitude in question is a magnitude *of* that quality or relation. We defined a *quantity* as a particular contained under a magnitude, *i.e.* as the complex consisting of a magnitude with a certain spatio-temporal position, or with a pair of terms between which it is a relation. We decided, by means of a general principle concerning transitive symmetrical relations, that it is impossible to content ourselves with quantities, and deny the further abstraction involved in magnitudes; that equality is not a direct relation between quantities, but consists in being particularizations of the same magnitude. Thus equal quantities are instances of the same

---

* Cf. Couturat, "*Sur la Définition du Continu,*" *Revue de Métaphysique et de Morale,* 1900.

magnitude. Similarly greater and less are not direct relations between quantities, but between magnitudes: quantities are only greater and less in virtue of being instances of greater and less magnitudes. Any two magnitudes which are of the same quality or relation are one greater, the other less; and greater and less are asymmetrical transitive relations.

Among the terms which have magnitude are not only many qualities, but also asymmetrical relations by which certain kinds of series are constituted. These may be called *distances*. When there are distances in a series, any two terms of the series have a distance, which is the same as, greater than, or less than, the distance of any two other terms in the series. Another peculiar class of magnitudes discussed in Chapter xx is constituted by the degrees of divisibility of different wholes. This, we found, is the only case in which quantities are divisible, while there is no instance of divisible magnitudes.

Numerical measurement, which was discussed in Chapter xxi, required, owing to the decision that most quantities and all magnitudes are indivisible, a somewhat unusual treatment. The problem lies, we found, in establishing a one-one relation between numbers and the magnitudes of the kind to be measured. On certain metaphysical hypotheses (which were neither accepted nor rejected), this was found to be always theoretically possible as regards existents actual or possible, though often not practically feasible or important. In regard to two classes of magnitudes, namely divisibilities and distances, measurement was found to proceed from a very natural convention, which defines what is meant by saying (what can never have the simple sense which it has in connection with finite wholes and parts) that one such magnitude is double of, or n times, another. The relation of distance to stretch was discussed, and it was found that, apart from a special axiom to that effect, there was no *à priori* reason for regarding equal distances as corresponding to equal stretches.

In Chapter xxii we discussed the definition of zero. The problem of zero was found to have no connection with that of the infinitesimal, being in fact closely related to the purely logical problem as to the nature of negation. We decided that, just as there are the distinct logical and arithmetical negations, so there is a third fundamental kind, the quantitative negation; but that this is negation of that quality or relation of which the magnitudes are, not of magnitude of that quality or relation. Hence we were able to regard zero as one among the magnitudes contained in a kind of magnitude, and to distinguish the zeroes of different kinds. We showed also that quantitative negation is connected with logical negation by the fact that there cannot be any quantities whose magnitude is zero.

In the present Chapter the problems of continuity, the infinite, and the infinitesimal, were shown to belong, not specially to the theory of quantity, but to those of number and order. It was shown that, though there are kinds of magnitude in which there is no greatest and no least magnitude, this fact does not require us to admit infinite or infinitesimal magnitudes; and that there is no contradiction in supposing a kind of magnitudes to form a series in which there is a term between any two, and in which, consequently, there is no term consecutive to a given term. The supposed contradiction was shown to result from an undue use of mathematical induction—a principle, the full discussion of which presupposes the philosophy of order.

PART IV.

ORDER.

CHAPTER XXIV.

THE GENESIS OF SERIES.

187. THE notion of order or series is one with which, in connection with distance, and with the order of magnitude, we have already had to deal. The discussion of continuity in the last chapter of Part III showed us that this is properly an ordinal notion, and prepared us for the fundamental importance of order. It is now high time to examine this concept on its own account. The importance of order, from a purely mathematical standpoint, has been immeasurably increased by many modern developments. Dedekind, Cantor, and Peano have shown how to base all Arithmetic and Analysis upon series of a certain kind—*i.e.* upon those properties of finite numbers in virtue of which they form what I shall call a *progression*. Irrationals are defined (as we shall see) entirely by the help of order; and a new class of transfinite ordinals is introduced, by which the most important and interesting results are obtained. In Geometry, von Staudt's quadrilateral construction and Pieri's work on Projective Geometry have shown how to give points, lines, and planes an order independent of metrical considerations and of quantity; while descriptive Geometry proves that a very large part of Geometry demands only the possibility of serial arrangement. Moreover the whole philosophy of space and time depends upon the view we take of order. Thus a discussion of order, which is lacking in the current philosophies, has become essential to any understanding of the foundations of mathematics.

188. The notion of order is more complex than any hitherto analyzed. Two terms cannot have an order, and even three cannot have a cyclic order. Owing to this complexity, the logical analysis of order presents considerable difficulties. I shall therefore approach the problem gradually, considering, in this chapter, the circumstances under which order arises, and reserving for the second chapter the discussion as to what order really is. This analysis will raise several fundamental points in general logic, which will demand considerable discussion of an almost purely philosophical nature. From this I shall pass to more mathematical topics, such as the types of series and the ordinal definition of numbers, thus gradually preparing the way for the discussion of infinity and continuity in the following Part.

There are two different ways in which order may arise, though we shall find in the end that the second way is reducible to the first. In the first, what may be called the ordinal element consists of three terms *a*, *b*, *c*, one of which (*b* say) is *between* the other two. This happens whenever there is a relation of *a* to *b* and of *b* to *c*, which is not a relation of *b* to *a*, of *c* to *b*, or of *c* to *a*. This is the definition, or better perhaps, the necessary and sufficient condition, of the proposition "*b* is between *a* and *c*." But there are other cases of order where, at first sight, the above conditions are not satisfied, and where *between* is not obviously applicable. These are cases where we have four terms *a*, *b*, *c*, *d*, as the ordinal element, of which we can say that *a* and *c* are separated by *b* and *d*. This relation is more complicated, but the following seems to characterize it : *a* and *c* are separated from *b* and *d*, when there is an asymmetrical relation which holds between *a* and *b*, *b* and *c*, *c* and *d*, or between *a* and *d*, *d* and *c*, *c* and *b*, or between *c* and *d*, *d* and *a*, *a* and *b*; while if we have the first case, the same relation must hold either between *d* and *a*, or else between both *a* and *c*, and *a* and *d*; with similar assumptions for the other two cases*. (No further special assumption is required as to the relatio . between *a* and *c* or between *b* and *d*; it is the absence of such an assumption which prevents our reducing this case to the former in a simple manner.) There are cases—notably where our series is closed—in which it *seems* formally impossible to reduce this second case to the first, though this appearance, as we shall see, is in part deceptive. We have to show, in the present chapter, the principal ways in which series arise from collections of such ordinal elements.

Although two terms alone cannot have an order, we must not assume that order is possible except where there are relations between two terms. In all series, we shall find, there are asymmetrical relations between two terms. But an asymmetrical relation of which there is only one instance does not constitute order. We require at least two instances for *between*, and at least three for separation of pairs. Thus although order is a relation between three or four terms, it is only possible where there are other relations which hold between pairs of terms. These relations may be of various kinds, giving different ways of generating series. I shall now enumerate the principal ways with which I am acquainted.

189. (1) The simplest method of generating a series is as follows. Let there be a collection of terms, finite or infinite, such that every term (with the possible exception of a single one) has to one and only

---

* This gives a sufficient but not a necessary condition for the separation of couples.

one other term of the collection a certain asymmetrical relation (which must of course be intransitive), and that every term (with again one possible exception, which must not be the same as the term formerly excepted) has also to one and only one other term of the collection the relation which is the converse of the former one*. Further, let it be assumed that, if a has the first relation to b, and b to c, then c does not have the first relation to a. Then every term of the collection except the two peculiar terms has one relation to a second term, and the converse relation to a third, while these terms themselves do not have to each other either of the relations in question. Consequently, by the definition of *between*, our first term is between our second and third terms. The term to which a given term has one of the two relations in question is called *next after* the given term; the term to which the given term has the converse relation is called *next before* the given term. Two terms between which the relations in question hold are called *consecutive*. The exceptional terms (when they exist) are not between any pair of terms; they are called the two ends of the series, or one is called the beginning and the other the end. The existence of the one does not imply that of the other—for example the natural numbers have a beginning but no end—and neither need exist—for example, the positive and negative integers together have neither†.

The above method may perhaps become clear by a formal exhibition. Let R be one of our relations, and let its converse be denoted by $\breve{R}$‡. Then if e be any term of our set, there are two terms d, f, such that $e \breve{R} d, e R f$, i.e. such that $d R e, e R f$. Since each term only has the relation R to one other, we cannot have $d R f$; and it was one of the initial assumptions that we were not to have $f R d$. Hence e is between d and f §. If a be a term which has only the relation R, then obviously a is not between any pair of terms. We may extend the notion of *between* by defining that, if c be between b and d, and d between c and e, then c or d will be said to be also between b and e. In this way, unless we either reach an end or come back to the term with which we started, we can find any number of terms between which and b the term c will lie. But if the total number of terms be not less than seven, we cannot show in this way that of *any* three terms one must be between the other two, since the collection may consist

---

* The converse of a relation is the relation which must hold between y and x when the given relation holds between x and y.

† The above is the only method of generating series given by Bolzano, "Paradoxien des Unendlichen," § 7.

‡ This is the notation adopted by Professor Schröder.

§ The denial of $d R f$ is only necessary to this special method, but the denial of $f R d$ is essential to the definition of *between*.

of two distinct series, of which, if the collection is finite, one at least must be closed, in order to avoid more than two ends.

This remark shows that, if the above method is to give a single series, to which any term of our collection is to belong, we need a further condition, which may be expressed by saying that the collection must be *connected*. We shall find means hereafter of expressing this condition without reference to number, but for the present we may content ourselves by saying that our collection is connected when, given any two of its terms, there is a certain finite number (not necessarily unique) of steps from one term to the next, by which we can pass from one of our two terms to the other. When this condition is fulfilled, we are assured that, of any three terms of our collection, one must be between the other two.

Assuming now that our collection is connected, and therefore forms a single series, four cases may arise: (*a*) our series may have two ends, (*b*) it may have one end, (*c*) it may have no end and be open, (*d*) it may have no end and be closed. Concerning (*a*), it is to be observed that our series must be finite. For, taking the two ends, since the collection is connected, there is some finite number n of steps which will take us from one end to the other, and hence $n + 1$ is the number of terms of the series. Every term except the two ends is between them, and neither of them is between any other pair of terms. In case (*b*), on the other hand, our collection must be infinite, and this would hold even if it were not connected. For suppose the end which exists to have the relation R, but not $\breve{R}$. Then every other term of the collection has both relations, and can never have both to the same term, since R is asymmetrical. Hence the term to which (say) e has the relation R is not that to which it had the relation $\breve{R}$, but is either some new term, or one of e's predecessors. Now it cannot be the end-term a, since a does not have the relation $\breve{R}$ to any term. Nor can it be any term which can be reached by successive steps from a without passing through e, for if it were, this term would have two predecessors, contrary to the hypothesis that R is a one-one relation. Hence, if k be any term which can be reached by successive steps from a, k has a successor which is not a or any of the terms between a and k; and hence the collection is infinite, whether it be connected or not. In case (*c*), the collection must again be infinite. For here, by hypothesis, the series is open—*i.e.*, starting from any term e, no number of steps in either direction brings us back to e. And there cannot be a finite limit to the number of possible steps, since, if there were, the series would have an end. Here again, it is not necessary to suppose the series connected. In case (*d*), on the contrary, we must assume connection. By saying that the series is closed, we mean that there exists some number n of steps by which, starting from a certain term a, we shall be brought back to a. In this case, n is the number of terms, and it makes no difference with which term we start. In this case, *between* is not definite except where three terms are consecutive, and the series contains more than three terms. Otherwise, we need the more complicated relation of separation.

190. (2) The above method, as we have seen, will give either open or closed series, but only such as have consecutive terms. The second method, which is now to be discussed, will give series in which there are no consecutive terms, but will not give closed series*. In this method we have a transitive asymmetrical relation P, and a collection of terms any two of which are such that either xPy or yPx. When these conditions are satisfied our terms necessarily form a single series. Since the relation is asymmetrical, we can distinguish xPy from yPx, and the two cannot both subsist†. Since P is transitive, xPy and yPz involve xPz. It follows that $\breve{P}$ is also asymmetrical and transitive‡. Thus with respect to any term x of our collection, all other terms of the collection fall into two classes, those for which xPy, and those for which zPx. Calling these two classes $\bar{\pi}x$ and πx respectively, we see that, owing to the transitiveness of P, if y belongs to the class $\bar{\pi}x$, $\bar{\pi}y$ is contained in $\bar{\pi}x$; and if z belongs to the class πx, πz is contained in πx. Taking now two terms x, y, for which xPy, all other terms fall into three classes: (1) Those belonging to πx, and therefore to πy; (2) those belonging to $\bar{\pi}y$, and therefore to $\bar{\pi}x$; (3) those belonging to $\bar{\pi}x$ but not to $\bar{\pi}y$. If z be of the first class, we have zPx, zPy; if v be of the second, xPv and yPv; if w be of the third, xPw and wPy. The case yPu and uPx is excluded: for xPy, yPu imply xPu, which is inconsistent with uPx. Thus we have, in the three cases, (1) x is between z and y; (2) y is between x and v; (3) w is between x and y. Hence any three terms of our collection are such that one is between the other two, and the whole collection forms a single series. If the class (3) contains no terms, x and y are said to be consecutive; but many relations P can be assigned, for which there are always terms in the class (3). If for example P be *before*, and our collection be the moments in a

---

* The following method is the only one given by Vivanti in the *Formulaire de Mathématiques*, (1895), VI, § 2, No. 7; also by Gilman, "On the properties of a one-dimensional manifold," *Mind*, N.S. Vol. I. We shall find that it is general in a sense in which none of our other methods are so.

† I use the term *asymmetrical* as the contrary, rather than the contradictory, of *symmetrical*. If xPy, and the relation is symmetrical, we have always yPx; if asymmetrical, we never have yPx. Some relations—*e.g.* logical implication—are neither symmetrical nor asymmetrical. Instead of assuming P to be asymmetrical, we may make the equivalent assumption that it is what Professor Peirce calls an *aliorelative*, *i.e.* a relation which no term has to itself. (This assumption is not equivalent to asymmetry in general, but only when combined with transitiveness.)

‡ P may be read *precedes*, and $\breve{P}$ may be read *follows*, provided no temporal or spatial ideas are allowed to intrude themselves.

certain interval, or in all time, there is a moment between any two of our collection. Similarly in the case of the magnitudes which, in the last chapter of Part III, we called continuous. There is nothing in the present method, as there was in the first, to show that there *must* be consecutive terms, unless the total number of terms in our collection be finite. On the other hand, the present method will not allow closed series ; for owing to the transitiveness of the relation P, if the series were closed, and x were any one of its terms, we should have xPx, which is impossible because P is asymmetrical. Thus in a closed series, the generating relation can never be transitive*. As in the former method, the series may have two ends, or one, or none. In the first case only, it may be finite ; but even in this case it *may* be infinite, and in the other two cases it must be so.

191. (3) A series may be generated by means of distances, as was already partially explained in Part III, and as we shall see more fully hereafter. In this case, starting with a certain term x, we are to have relations, which are magnitudes, between x and a number of other terms y, z.... According as these relations are greater or less, we can order the corresponding terms. If there are no similar relations between the remaining terms $y, z, ...$, we require nothing further. But if these have relations which are magnitudes of the same kind, certain axioms are necessary to insure that the order may be independent of the particular term from which we start. Denoting by xz the distance of x and z, if xz is less than xw, we must have yz less than yw. A consequence, which did not follow when x was the only term that had a distance, is that the distances must be asymmetrical relations, and those which have one sense must be considered less than zero. For "xz is less than xw" must involve "wz is less than ww," *i.e.* wz is less than 0. In this way the present case is practically reduced to the second ; for every pair of terms x, y will be such that xy is less than 0 or else xy is greater than 0 ; and we may put in the first case yPx, in the second xPy. But we require one further axiom in order that the arrangement may be thus effected unambiguously. If $xz = yw$, and $zw' = xy$, w and w' must be the same point. With this further axiom, the reduction to case (2) becomes complete.

192. (4) Cases of triangular relations are capable of giving rise to order. Let there be a relation R which holds between y and (x, z), between z and (y, u), between u and (z, w), and so on. *Between* is itself such a relation, and this might therefore seem the most direct and natural way of generating order. We should say, in such a case, that y is between x and z, when the relation R holds between y and the couple x, z. We should need assumptions concerning R which should show that, if y is between x and z, and z between y and w, then y and z are

---

* For more precise statements, see Chap. xxviii.

each between x and w. That is, if we have $yR(x, z)$, $zR(y, w)$, we must have $yR(x, w)$ and $zR(x, w)$. This is a kind of three-term transitiveness. Also if y be between x and w, and z between y and w, then z must be between x and w, and y between x and z: that is, if $yR(x, w)$ and $zR(y, w)$, then $zR(x, w)$ and $yR(x, z)$. Also $yR(x, z)$ must be equivalent to $yR(z, x)$*. With these assumptions, an unambiguous order will be generated among any number of terms such that any triad has the relation R. Whether such a state of things can ever be incapable of further analysis, is a question which I leave for the next chapter.

193. (5) We have found hitherto no way of generating closed continuous series. There are, however, instances of such series, *e.g.* angles, the elliptic straight line, the complex numbers with a given modulus. It is therefore necessary to have some theory which allows of their possibility. In the case where our terms are asymmetrical relations, as straight lines are, or are correlated uniquely and reciprocally with such relations, the following theory will effect this object. In other cases, the sixth method (below) seems adequate to the end in view.

Let $x, y, z...$ be a set of asymmetrical relations, and let R be an asymmetrical relation which holds between any two x, y or y, x except when y is the converse relation to x. Also let R be such that, if it holds between x and y, it holds between y and the converse of x; and if x be any term of the collection, let all the terms to which x has either of the relations R, $\breve{R}$ be terms of the collection. All these conditions are satisfied by angles, and whenever they are satisfied, the resulting series is closed. For xRy implies $yR\breve{x}$, and hence $\breve{x}R\breve{y}$, and thence $\breve{y}Rx$; so that by means of relations R it is possible to travel from x back to x. Also there is nothing in the definition to show that our series cannot be continuous. Since it is closed, we cannot apply universally the notion of *between*; but the notion of separation can be always applied. The reason why it is necessary to suppose that our terms either are, or are correlated with, asymmetrical relations, is, that such series often have antipodes, *opposite* terms as they may be called; and that the notion of *opposite* seems to be essentially bound up with that of the converse of an asymmetrical relation.

194. (6) In the same way in which, in (4), we showed how to construct a series by relations of *between*, we can construct a series directly by four-term relations of separation. For this purpose, as before, certain axioms are necessary. The following five axioms have been shown by Vailati[†] to be sufficient, and by Padoa to possess ordered independence, *i.e.* to be such that none can be deduced from its predecessors[‡]. Denoting " a and b separate c from d " by $ab \parallel cd$, we must have :

---

* See Peano, *I Principii di Geometria*, Turin, 1889, Axioms VIII, IX, X, XI.

† *Rivista di Matematica*, v, pp. 76, 183. ‡ *Ibid.* p. 185.

(α)　　$ab \| cd$ is equivalent to $cd \| ab$;

(β)　　$ab \| cd$ is equivalent to $ab \| dc$;

(γ)　　$ab \| cd$ excludes $ac \| bd$;

(δ)　　For any four terms of our collection, we must have $ab \| cd$, or $ac \| bd$, or $ad \| bc$;

(ϵ)　　If $ab \| cd$, and $ac \| be$, then $ac \| de$.

By means of these five assumptions, our terms $a, b, c, d, e \ldots$ acquire an unambiguous order, in which we start from a relation between two pairs of terms, which is undefined except to the extent to which the above assumptions define it. The further consideration of this case, as generally of the relation of separation, I postpone to a later stage.

The above six methods of generating series are the principal ones with which I am acquainted, and all other methods, so far as I know, are reducible to one of these six. The last alone gives a method of generating closed continuous series whose terms neither are, nor are correlated with, asymmetrical relations *. This last method should therefore be applied in projective and elliptic Geometry, where the correlation of the points on a line with the lines through a point appears to be logically subsequent to the order of the points on a line. But before we can decide whether these six methods (especially the fourth and sixth) are irreducible and independent, we must discuss (what has not hitherto been analyzed) the meaning of order, and the logical constituents (if any) of which this meaning is compounded. This will be done in the following chapter.

* See Chap. XXVIII.

CHAPTER XXV.

THE MEANING OF ORDER.

195. WE have now seen under what circumstances there is an order among a set of terms, and by this means we have acquired a certain inductive familiarity with the nature of order. But we have not yet faced the question: What *is* order? This is a difficult question, and one upon which, so far as I know, nothing at all has been written. All the authors with whom I am acquainted are content to exhibit the genesis of order; and since most of them give only one of the six methods enumerated in Chapter xxiv, it is easy for them to confound the genesis of order with its nature. This confusion is rendered evident to us by the multiplicity of the above methods; for it is evident that we mean by *order* something perfectly definite, which, being generated equally in all our six cases, is clearly distinct from each and all of the ways in which it may be generated, unless one of these ways should turn out to be fundamental, and the others to be reducible to it. To elicit this common element in all series, and to broach the logical discussions connected with it, is the purpose of the present chapter. This discussion is of purely philosophical interest, and might be wholly omitted in a mathematical treatment of the subject.

In order to approach the subject gradually, let us separate the discussion of *between* from that of separation of couples. When we have decided upon the nature of each of these separately, it will be time to combine them, and examine what it is that both have in common. I shall begin with *between*, as being the simpler of the two.

196. *Between* may be characterized (as in Chapter xxiv) as a relation of one term y to two others x and z, which holds whenever x has to y, and y has to z, some relation which y does not have to x, nor z to y, nor z to x*.

---

* The condition that z does not have to x the relation in question is comparatively inessential, being only required in order that, if y be between x and z, we may not have x between y and z, or z between x and y. If we are willing to allow that in such cases, for example, as the angles of a triangle, each is between the other two, we may drop the condition in question altogether. The other four conditions, on the contrary, seem more essential.

These conditions are undoubtedly *sufficient* for betweenness, but it may be questioned whether they are *necessary*. Several possible opinions must be distinguished in this respect. (1) We may hold that the above conditions give the very *meaning* of between, that they constitute an actual analysis of it, and not merely a set of conditions insuring its presence. (2) We may hold that *between* is not a relation of the terms x, y, z at all, but a relation of the relation of y to x to that of y to z, namely the relation of difference of sense. (3) We may hold that *between* is an indefinable notion, like *greater* and *less*; that the above conditions allow us to infer that y is between x and z, but that there may be other circumstances under which this occurs, and even that it may occur without involving any relation except diversity among the pairs (x, y), (y, z), (x, z). In order to decide between these theories, it will be well to develop each in turn.

197. (1) In this theory, we define "y is between x and z" to mean: "There is a relation R such that xRy, yRz but not yRx, zRy"; and it remains a question whether we are to add "not zRx." We will suppose to begin with that this addition is not made. The following propositions will be generally admitted to be self-evident: (α) If y be between x and z, and z between y and w, then y is between x and w; (β) if y be between x and z, and w between x and y, then y is between w and z. For brevity, let us express "y is between x and z" by the symbol xyz. Then our two propositions are: (α) xyz and yzw imply xyw; (β) xyz and xwy imply wyz. We must add that the relation of *between* is symmetrical so far as the extremes are concerned: *i.e.* xyz implies zyx. This condition follows directly from our definition. With regard to the axioms (α) and (β), it is to be observed that *between*, on our present view, is always relative to some relation R, and that the axioms are only assumed to hold when it is the same relation R that is in question in both the premisses. Let us see whether these axioms are consequences of our definition. For this purpose, let us write $\bar{R}$ for not -R.

xyz means xRy, yRz, $y\bar{R}x$, $z\bar{R}y$.

yzw means yRz, zRw, $z\bar{R}y$, $w\bar{R}z$.

Thus yzw only adds to xyz the two conditions zRw, $w\bar{R}z$. If R is transitive, these conditions insure xyw; if not, not. Now we have seen that some series are generated by one-one relations R, which are not transitive. In these cases, however, denoting by R^2 the relation between x and z implied by xRy, yRz, and so on for higher powers, we can substitute a transitive relation R' for R, where R' means "some positive power of R." In this way, if xyz holds for a relation which is some definite power of R, then xyz holds for R', provided only that no positive power of R is equivalent to $\breve{R}$. For, in this latter event, we should have $yR'x$ whenever $xR'y$, and R' could not be substituted for R in the explanation of xyz. Now this condition, that the converse of R is not to be a positive power of R, is equivalent to the condition that our series is not to be closed. For if $\breve{R} = R^n$, then $R\breve{R} = R^{n+1}$; but since R is a one-one relation, $R\breve{R}$ implies the relation of identity. Thus $n+1$ steps bring us back from x to x, and our series is a closed series of $n+1$ terms. Now we have agreed already that *between* is not properly applicable to closed series. Hence this condition, that $\breve{R}$ is not to be a power of R, imposes only such restrictions upon our axiom (α) as we should expect it to be subject to.

With regard to (β), we have

$$xyz = xRy \,.\, yRz \,.\, y\bar{R}x \,.\, z\bar{R}y.$$
$$xwy = xRw \,.\, wRy \,.\, w\bar{R}x \,.\, y\bar{R}w.$$

The case contemplated by this axiom is only possible if R be not a one-one relation, since we have xRy and xRw. The deduction wyz is here an immediate consequence of the definition, without the need of any further conditions.

It remains to examine whether we can dispense with the condition $z\bar{R}x$ in the definition of *between*. If we suppose R to be a one-one relation, and zRx to be satisfied, we shall have

$$xyz = xRy \,.\, yRz \,.\, z\bar{R}y \,.\, y\bar{R}x,$$

and we have further by hypothesis zRx, and since R is one-one, and xRy, we have $x\bar{R}z$. Hence, in virtue of the definition, we have yzx; and similarly we shall obtain zxy. If we now adhere to our axiom (α), we shall have xzx, which is impossible; for it is certainly part of the meaning of *between* that the three terms in the relation should be different, and it is impossible that a term should be between x and x. Thus we must either insert our condition $z\bar{R}x$, or we must set up the new condition in the definition, that x and z are to be different. (It should be observed that our definition implies that x is different from y and y from z; for if not, xRy would involve yRx, and yRz would involve zRy.) It would seem preferable to insert the condition that x and z are to be different: for this is in any case necessary, and is not implied by $z\bar{R}x$. This condition must then be added to our axiom (α); xyz and yzw are to imply xyw, unless x and w are identical. In axiom (β), this addition is not necessary, since it is implied in the premisses. Thus the condition $z\bar{R}x$ is not necessary, if we are willing to admit that xyz is compatible with yzx—an admission which such cases as the angles of a triangle render possible. Or we may insert, in place of $z\bar{R}x$, the condition which we found necessary before to the universal validity of our axiom (α), namely that no power of R is to be equivalent to the converse of R: for if we have both xyz and yzx, we shall have (so far at least as x, y, z are concerned) $R^2 = \breve{R}$, *i.e.* if xRy and yRz, then

zRx. This last course seems to be the best. Hence in all cases where our first instance of *between* is defined by a one-one relation R, we shall substitute the relation R', which means "some positive power of R." The relation R' is then transitive, and the condition that no positive power of R is to be equivalent to $\breve{R}$ is equivalent to the condition that R' is to be asymmetrical. Hence, finally, the whole matter is simplified into the following:

To say that y is between x and z is equivalent to saying that there is some transitive asymmetrical relation which relates both x and y, and y and z.

This short and simple statement, as the above lengthy argument shows, contains neither more nor less than our original definition, together with the emendations which we gradually found to be necessary. The question remains, however: Is this the *meaning* of *between*?

198. A negative instance can be at once established if we allow the phrase: R is a relation *between* x and y. The phrase, as the reader will have observed, has been with difficulty excluded from the definitions of *between*, which its introduction would have rendered at least verbally circular. The phrase may have none but a linguistic importance, or again it may point to a real insufficiency in the above definition. Let us examine the relation of a relation R to its terms x and y. In the first place, there certainly is such a relation. To be a term which has the relation R to some other term is certainly to have a relation to R, a relation which we may express as "belonging to the domain of R." Thus if xRy, x will belong to the domain of R, and y to that of $\breve{R}$. If we express this relation between x and R, or between y and $\breve{R}$, by E, we shall have $xER, yE\breve{R}$. If further we express the relation of R to $\breve{R}$ by I, we shall have $\breve{R}IR$ and $RI\breve{R}$. Thus we have $xER, yEIR$. Now EI is by no means the converse of E, and thus the above definition of *between*, if for this reason only, does not apply; also neither E nor EI is transitive. Thus our definition of *between* is wholly inapplicable to such a case. Now it may well be doubted whether *between*, in this case, has at all the same meaning as in other cases. Certainly we do not in this way obtain series: x and y are not, in the same sense as R, between R and other terms. Moreover, if we admit relations of a term to itself, we shall have to admit that such relations are *between* a term and itself, which we agreed to be impossible. Hence we may be tempted to regard the use of *between* in this case as due to the linguistic accident that the relation is usually mentioned between the subject and the object, as in "A is the father of B." On the other hand, it may be urged that a relation does have a very peculiar relation to the pair of terms which it relates, and that *between* should denote a relation of one term to two others. To the objection concerning relations of a term to itself, it may be answered that such relations, in any system, constitute a grave logical difficulty; that they would, if possible, be denied philosophic validity; and that even where the relation asserted is identity, there must be *two* identical terms, which are therefore not quite identical. As this raises a fundamental difficulty, which we cannot discuss here, it will be prudent to allow the answer to pass *. And it may be further urged that use of the same word in two connections points always to some analogy, the extent of which should be carefully indicated by those who deny that the meaning is the same in both cases; and that the analogy here is certainly profounder than the mere order of words in a sentence, which is, in any case, far more variable in this respect than the phrase that a relation is between its terms. To these remarks, however, it may be retorted that the objector has himself indicated the precise extent of the analogy: the relation of a relation to its terms is a relation of one term to two others, just as *between* is, and this is what makes the two cases similar. This last retort is, I think, valid, and we may allow that the relation of a relation to its terms, though involving a most important logical problem, is not the same as the relation of *between* by which order is to be constituted.

Nevertheless, the above definition of *between*, though we shall be ultimately forced to accept it, seems, at first sight, scarcely adequate from a philosophical point of view. The reference to *some* asymmetrical relation is vague, and seems to require to be replaced by some phrase in which no such undefined relation appears, but only the terms and the betweenness. This brings us to the second of the above opinions concerning *between*.

199. (2) *Between*, it may be said, is not a relation of three terms at all, but a relation of two relations, namely difference of sense. Now if we take this view, the first point to be observed is, that we require the two opposite relations, not merely in general, but as particularized by belonging to one and the same term. This distinction is already familiar from the case of magnitudes and quantities. *Before* and *after* in the abstract do not constitute *between*: it is only when one and the same term is both before and after that *between* arises: this term is then between what it is before and what it is after. Hence there is a difficulty in the reduction of *between* to difference of sense. The particularized relation is a logically puzzling entity, which in Part I (§ 55) we found it necessary to deny; and it is not quite easy to distinguish a relation of two relations, particularized as belonging to the same term, from a relation of the term in question to two others. At the same time, great advantages are secured by this reduction. We get rid of the necessity for a triangular relation, to which many philosophers may object, and we assign a common element to all cases of *between*, namely difference of sense, *i.e.* the difference between an asymmetrical relation and its converse.

* Cf. § 95.

200. The question whether there can be an ultimate triangular relation is one whose actual solution is both difficult and unimportant, but whose precise statement is of very great importance. Philosophers seem usually to assume—though not, so far as I know, explicitly—that relations never have more than two terms; and even such relations they reduce, by force or guile, to predications. Mathematicians, on the other hand, almost invariably speak of relations of many terms. We cannot, however, settle the question by a simple appeal to mathematical instances, for it remains a question whether these are, or are not, susceptible of analysis. Suppose, for example, that the projective plane has been defined as a relation of three points: the philosopher may always say that it should have been defined as a relation of a point and a line, or of two intersecting lines—a change which makes little or no mathematical difference. Let us see what is the precise meaning of the question. There are among terms two radically different kinds, whose difference constitutes the truth underlying the doctrine of substance and attribute. There are terms which can never occur except as terms; such are points, instants, colours, sounds, bits of matter, and generally terms of the kind of which existents consist. There are, on the other hand, terms which can occur otherwise than as terms; such are being, adjectives generally, and relations. Such terms we agreed to call concepts*. It is the presence of concepts not occurring as terms which distinguishes propositions from mere concepts; in every proposition there is at least one more concept than there are terms. The traditional view—which may be called the subject-predicate theory—holds that in every proposition there is one term, the subject, and one concept which is not a term, the predicate. This view, for many reasons, must be abandoned†. The smallest departure from the traditional opinion lies in holding that, where propositions are not reducible to the subject-predicate form, there are always two terms only, and one concept which is not a term. (The two terms may, of course, be complex, and may each contain concepts which are not terms.) This gives the opinion that relations are always between only two terms; for a relation may be defined as any concept which occurs in a proposition containing more than one term. But there seems no *à priori* reason for limiting relations to two terms, and there are instances which lead to an opposite view. In the first place, when the concept of a number is asserted of a collection, if the collection has n terms, there are n terms, and only one concept (namely n) which is not a term. In the second place, such relations as those of an existent to the place and time of its existence are only reducible by a very cumbrous method to relations of two terms‡. If, however, the reduction be held essential, it seems to be always formally possible,

---

* See Part I, Chap. iv.

† See *The Philosophy of Leibniz*, by the present author, Cambridge, 1900; Chapter ii, § 10.

‡ See Part VII, Chap. liv.

by compounding part of the proposition into one complex term, and then asserting a relation between this part and the remainder, which can be similarly reduced to one term. There may be cases where this is not possible, but I do not know of them. The question whether such a formal reduction is to be always undertaken is not, however, so far as I have been able to discover, one of any great practical or theoretical importance.

201.   There is thus no valid *à priori* reason in favour of analyzing *between* into a relation of two relations, if a triangular relation seems otherwise preferable. The other reason in favour of the analysis of *between* is more considerable. So long as *between* is a triangular relation of the terms, it must be taken either as indefinable, or as involving a reference to *some* transitive asymmetrical relation. But if we make *between* consist essentially in the opposition of two relations belonging to one term, there seems to be no longer any undue indeterminateness. Against this view we may urge, however, that no reason now appears why the relations in question should have to be transitive, and that— what is more important—the very meaning of *between* involves the terms, for it is they, and not their relations, that have order. And if it were only the relations that were relevant, it would not be necessary, as in fact it is, to particularize them by the mention of the terms between which they hold. Thus on the whole, the opinion that *between* is not a triangular relation must be abandoned.

202.   (3) We come now to the view that *between* is an ultimate and indefinable relation. In favour of this view it might be urged that, in all our ways of generating open series, we could see that cases of *between* did arise, and that we could apply a test to suggested definitions. This seems to show that the suggested definitions were merely conditions which imply relations of *between*, and were not true definitions of this relation. The question : Do such and such conditions insure that *y* shall be between *x* and *z*? is always one which we can answer, without having to appeal (at least consciously) to any previous definition. And the unanalyzable nature of *between* may be supported by the fact that the relation is symmetrical with respect to the two extremes, which was not the case with the relations of pairs from which *between* was inferred. There is, however, a very grave difficulty in the way of such a view, and that is, that sets of terms have many different orders, so that in one we may have *y* between *x* and *z*, while in another we have *x* between *y* and *z* *. This seems to show that *between* essentially involves reference to the relations from which it is inferred. If not, we shall at least have to admit that these relations are relevant to the genesis of series ; for series require imperatively that there should be at most one relevant

* This case is illustrated by the rational numbers, which may be taken in order of magnitude, or in one of the orders (*e.g.* the logical order) in which they are denumerable. The logical order is the order 1, 2, 1/2, 3, 1/3, 2/3, 4, ...

relation of *between* among three terms. Hence we must, apparently, allow that *between* is not the sole source of series, but must always be supplemented by the mention of some transitive asymmetrical relation with respect to which the betweenness arises. The most that can be said is, that this transitive asymmetrical relation of two terms may itself be logically subsequent to, and derived from, some relation of three terms, such as those considered in Chapter XXIV, in the fourth way of generating series. When such relations fulfil the axioms which were then mentioned, they lead of themselves to relations between pairs of terms. For we may say that b precedes c when acd implies bcd, and that b follows c when abd implies cbd, where a and d are fixed terms. Though such relations are merely derivative, it is in virtue of them that *between* occurs in such cases. Hence we seem finally compelled to leave the reference to an asymmetrical relation in our definition. We shall therefore say :

A term y is between two terms x and z with reference to a transitive asymmetrical relation R when xRy and yRz. In no other case can y be said properly to be between x and z; and this definition gives not merely a criterion, but the very *meaning* of betweenness.

203. We have next to consider the meaning of separation of couples. This is a more complicated relation than *between*, and was but little considered until elliptic Geometry brought it into prominence. It has been shown by Vailati* that this relation, like *between*, always involves a transitive asymmetrical relation of two terms; but this relation of a pair of terms is itself relative to three other fixed terms of the set, as, in the case of *between*, it was relative to two fixed terms. It is further sufficiently evident that wherever there is a transitive asymmetrical relation, which relates every pair of terms in a collection of not less than four terms, there there are pairs of couples having the relation of separation. Thus we shall find it possible to express separation, as well as *between*, by means of transitive asymmetrical relations and their terms. But let us first examine directly the meaning of separation.

We may denote the fact that a and c are separated by b and d by the symbol $abcd$. If, then, a, b, c, d, e be any five terms of the set we require the following properties to hold of the relation of separation (of which, it will be observed, only the last involves five terms) :

1. $abcd = badc$.
2. $abcd = adcb$.
3. $abcd$ excludes $acbd$.
4. We must have $abcd$ or $acdb$ or $adbc$.
5. $abcd$ and $acde$ together imply $abde$†.

---

* *Rivista di Matematica*, v, pp. 75—78. See also Pieri, *I Principii della Geometria di Posizione*, Turin, 1898, § 7.

† These five properties are taken from Vailati, *loc. cit.* and *ib.* p. 183.

These properties may be illustrated by the consideration of five points on a circle, as in the accompanying figure. Whatever relation of two pairs of terms possesses these properties we shall call a relation of separation between the pairs. It will be seen that the relation is symmetrical, but not in general transitive.

204. Wherever we have a transitive asymmetrical relation R between any two terms of a set of not less than four terms, the relation of separation necessarily arises. For in any series, if four terms have the order *abcd*, then a and c are separated by b and d; and every transitive asymmetrical relation, as we have seen, provided there are at least two consecutive instances of it, gives rise to a series. Thus in this case, separation is a mere extension of *between*: if R be asymmetrical and transitive, and aRb, bRc, cRd, then a and c are separated by b and d. The existence of such a relation is therefore a sufficient condition of separation.

It is also a necessary condition. For, suppose a relation of separation to exist, and let a, b, c, d, e be five terms of the set to which the relation applies. Then, considering a, b, c as fixed, and d and e as variable, twelve cases may arise. In virtue of the five fundamental properties, we may introduce the symbol *abcde* to denote that, striking out any one of these five letters, the remaining four have the relation of separation which is indicated by the resulting symbol. Thus by the fifth property, *abcd* and *acde* imply *abcde**. Thus the twelve cases arise from permuting d and e, while keeping a, b, c fixed. (It should be observed that it makes no difference whether a letter appears at the end or the beginning: *i.e. abcde* is the same case as *eabcd*. We may therefore decide not to put either d or e before a.) Of these twelve cases, six will have d before e, and six will have e before d. In the first six cases, we say that, with respect to the sense *abc*, d precedes e; in the other six cases, we say that e precedes d. In order to deal with limiting cases, we shall say further that a precedes every other term, and that b precedes c†. We shall then find that the relation of preceding is asymmetrical and transitive, and that every pair of terms of our set is such that one precedes and the other follows. In this way our relation of separation is reduced, formally at least, to the combination of "a precedes b," "b precedes c," and "c precedes d."

The above reduction is for many reasons highly interesting. In the first place, it shows the distinction between open and closed series to be somewhat superficial. For although our series may initially be of the sort which is called closed, it becomes, by the introduction of the above transitive relation, an open series, having a for its beginning, but having

---

* The argument is somewhat tedious, and I therefore omit it. It will be found in Vailati, *loc. cit.*

† Pieri, *op. cit.* p. 32.

possibly no last term, and not in any sense returning to *a*. Again it is
of the highest importance in Geometry, since it shows how order may
arise on the elliptic straight line, by purely projective considerations,
in a manner which is far more satisfactory than that obtained from
von Staudt's construction*. And finally, it is of great importance as
unifying the two sources of order, *between* and separation; since it
shows that transitive asymmetrical relations are always present where
either occurs, and that either implies the other. For, by the relation of
preceding, we can say that one term is between two others, although we
started solely from separation of pairs.

205. At the same time, the above reduction (and also, it would
seem, the corresponding reduction in the case of *between*) cannot be
allowed to be more than formal. That is, the three terms *a*, *b*, *c* by
relation to which our transitive asymmetrical relation was defined, are
essential to the definition, and cannot be omitted. The reduction shows
no reason for supposing that there is any transitive asymmetrical relation
independent of *all* other terms than those related, though it is arbitrary
what other terms we choose. And the fact that the term *a*, which is
not essentially peculiar, appears as the beginning of the series, illustrates
this fact. Where there are transitive asymmetrical relations independent
of all outside reference, our series cannot have an arbitrary beginning,
though it may have none at all. Thus the four-term relation of separation remains logically prior to the resulting two-term relation, and
cannot be analyzed into the latter.

206. But when we have said that the reduction is formal, we have
not said that it is irrelevant to the genesis of order. On the contrary,
it is just because such a reduction is possible that the four-term relation
leads to order. The resulting asymmetrical transitive relation is in
reality a relation of five terms; but when three of these are kept fixed,
it becomes asymmetrical and transitive as regards the other two. Thus
although *between* applies to such series, and although the essence of
order consists, here as elsewhere, in the fact that one term has, to two
others, converse relations which are asymmetrical and transitive, yet
such an order can only arise in a collection containing at least five terms,
because five terms are needed for the characteristic relation. And it
should be observed that *all* series, when thus explained, are open series,
in the sense that there is some relation between pairs of terms, no power
of which is equal to its converse, or to identity.

207. Thus finally, to sum up this long and complicated discussion:
The six methods of generating series enumerated in Chapter xxiv are all
genuinely distinct; but the second is the only one which is fundamental,

---

* The advantages of this method are evident from Pieri's work quoted above,
where many things which seemed incapable of projective proof are rigidly deduced
from projective premisses. See Part VI, Chap. xlv.

and the other five agree in this, that they are all reducible to the second.
Moreover, it is solely in virtue of their reducibility to the second that
they give rise to order. The minimum ordinal proposition, which can
always be made wherever there is an order at all, is of the form : "y is
between x and z"; and this proposition means; "There is some
asymmetrical transitive relation which holds between x and y and
between y and z." This very simple conclusion might have been guessed
from the beginning; but it was only by discussing all the apparently
exceptional cases that the conclusion could be solidly established.

CHAPTER XXVI.

ASYMMETRICAL RELATIONS.

208. WE have now seen that all order depends upon transitive asymmetrical relations. As such relations are of a kind which traditional logic is unwilling to admit, and as the refusal to admit them is one of the main sources of the contradictions which the Critical Philosophy has found in mathematics, it will be desirable, before proceeding further, to make an excursion into pure logic, and to set forth the grounds which make the admission of such relations necessary. At a later stage (in Part VI, Chap. LI), I shall endeavour to answer the general objections of philosophers to relations; for the present, I am concerned only with asymmetrical relations.

Relations may be divided into four classes, according as they do or do not possess either of two attributes, transitiveness* and symmetry. Relations such that xRy always implies yRx are called *symmetrical*; relations such that xRy, yRz together always imply xRz are called *transitive*. Relations which do not possess the first property I shall call *not symmetrical*; relations which do possess the opposite property, *i.e.* for which xRy always excludes yRx, I shall call *asymmetrical*. Relations which do not possess the second property I shall call *not transitive*; those which possess the property that xRy, yRz always exclude xRz I shall call *intransitive*. All these cases may be illustrated from human relationships. The relation *brother or sister* is symmetrical, and is transitive if we allow that a man may be his own brother, and a woman her own sister. The relation *brother* is not symmetrical, but is transitive. *Half-brother or half-sister* is symmetrical but not transitive. *Spouse* is symmetrical but intransitive; *descendant* is asymmetrical but transitive. *Half-brother* is not symmetrical and not transitive; if third marriages were forbidden, it would be intransitive. *Son-in-law* is asymmetrical and not transitive; if second marriages were forbidden, it would be intransitive. *Brother-in-law* is not symmetrical

---

* This term appears to have been first used in the present sense by De Morgan; see *Camb. Phil. Trans.* IX, p. 104; x, p. 346. The term is now in general use.

and not transitive. Finally, *father* is both asymmetrical and intransitive. Of not-transitive but not intransitive relations there is, so far as I know, only one *important* instance, namely *diversity*; of not-symmetrical but not asymmetrical relations there seems to be similarly only one important instance, namely *implication*. In other cases, of the kind that usually occur, relations are either transitive or intransitive, and either symmetrical or asymmetrical.

209. Relations which are both symmetrical and transitive are formally of the nature of equality. Any term of the field of such a relation has the relation in question to itself, though it may not have the relation to any other term. For denoting the relation by the sign of equality, if *a* be of the field of the relation, there is some term *b* such that $a = b$. If *a* and *b* be identical, then $a = a$. But if not, then, since the relation is symmetrical, $b = a$; since it is transitive, and we have $a = b$, $b = a$, it follows that $a = a$. The property of a relation which insures that it holds between a term and itself is called by Peano *reflexiveness*, and he has shown, contrary to what was previously believed, that this property cannot be inferred from symmetry and transitiveness. For neither of these properties asserts that there is a *b* such that $a = b$, but only what follows in case there is such a *b*; and if there is no such *b*, then the proof of $a = a$ fails*. This property of reflexiveness, however, introduces some difficulty. There is only one relation of which it is true without limitation, and that is identity. In all other cases, it holds only of the terms of a certain class. Quantitative equality, for example, is only reflexive as applied to quantities; of other terms, it is absurd to assert that they have quantitative equality with themselves. Logical equality, again, is only reflexive for classes, or propositions, or relations. Simultaneity is only reflexive for events, and so on. Thus, with any given symmetrical transitive relation, other than identity, we can only assert reflexiveness within a certain class; and of this class, apart from the principle of abstraction (already mentioned in Part III, Chap. XIX, and shortly to be discussed at length), there need be no definition except as the extension of the transitive symmetrical relation in question. And when the class is so defined, reflexiveness within that class, as we have seen, follows from transitiveness and symmetry.

210. By introducing what I have called the principle of abstraction†, a somewhat better account of reflexiveness becomes possible. Peano has defined‡ a process which he calls definition by abstraction, of which, as he shows, frequent use is made in Mathematics. This process is as

---

* See *e.g. Revue de Mathématiques*, T. VII, p. 22 ; *Notations de Logique Mathématique*, Turin, 1894, p. 45, F. 1901, p. 193.

† An axiom virtually identical with this principle, but not stated with the necessary precision, and not demonstrated, will be found in De Morgan, *Camb. Phil. Trans.* Vol. x, p. 345.

‡ *Notations de Logique Mathématique*, p. 45.

follows: when there is any relation which is transitive, symmetrical and (within its field) reflexive, then, if this relation holds between u and v, we define a new entity $\phi(u)$, which is to be identical with $\phi(v)$. Thus our relation is analyzed into sameness of relation to the new term $\phi(u)$ or $\phi(v)$. Now the legitimacy of this process, as set forth by Peano, requires an axiom, namely the axiom that, if there is any instance of the relation in question, then there is such an entity as $\phi(u)$ or $\phi(v)$. This axiom is my principle of abstraction, which, precisely stated, is as follows: " Every transitive symmetrical relation, of which there is at least one instance, is analyzable into joint possession of a new relation to a new term, the new relation being such that no term can have this relation to more than one term, but that its converse does not have this property." This principle amounts, in common language, to the assertion that transitive symmetrical relations arise from a common property, with the addition that this property stands, to the terms which have it, in a relation in which nothing else stands to those terms. It gives the precise statement of the principle, often applied by philosophers, that symmetrical transitive relations always spring from identity of content. Identity of content is, however, an extremely vague phrase, to which the above proposition gives, in the present case, a precise signification, but one which in no way answers the purpose of the phrase, which is, apparently, the reduction of relations to adjectives of the related terms.

It is now possible to give a clearer account of the reflexive property. Let R be our symmetrical relation, and let S be the asymmetrical relation which two terms having the relation R must have to some third term. Then the proposition xRy is equivalent to this: " There is some term a such that xSa and ySa." Hence it follows that, if x belongs to what we have called the domain of S, *i.e.* if there is any term a such that xSa, then xRx; for xRx is merely xSa and xSa. It does not of course follow that there is any other term y such that xRy, and thus Peano's objections to the usual proof of reflexiveness are valid. But by means of the analysis of symmetrical transitive relations, we obtain the proof of the reflexive property, together with the exact limitation to which it is subject.

211. We can now see the reason for excluding from our accounts of the methods of generating series a seventh method, which some readers may have expected to find. This is the method in which position is merely relative—a method which, in Chap. XIX, § 154, we rejected as regards quantity. As the whole philosophy of space and time is bound up with the question as to the legitimacy of this method, which is in fact the question as to absolute and relative position, it may be well to give an account of it here, and to show how the principle of abstraction leads to the absolute theory of position.

If we consider such a series as that of events, and if we refuse to allow absolute time, we shall have to admit three fundamental relations among events, namely, simultaneity, priority, and posteriority. Such a theory may be formally stated as follows : Let there be a class of terms, such that any two, x and y, have either an asymmetrical transitive relation P, or the converse relation $\breve{P}$, or a symmetrical transitive relation R. Also let xRy, yPz imply xPz, and let xPy, yRz imply xPz. Then all the terms can be arranged in a series, in which, however, there may be many terms which have the same place in the series. This place, according to the relational theory of position, is nothing but the transitive symmetrical relation R to a number of other terms. But it follows from the principle of abstraction that there is some relation S, such that, if xRy, there is some one entity t for which xSt, ySt. We shall then find that the different entities t, corresponding to different groups of our original terms, also form a series, but one in which any two different terms have an asymmetrical relation (formally, the product $\breve{S}RS$). These terms t will then be the absolute positions of our x's and y's, and our supposed seventh method of generating series is reduced to the fundamental second method. Thus there will be no series having only relative position, but in all series it is the positions themselves that constitute the series *.

212. We are now in a position to meet the philosophic dislike of relations. The whole account of order given above, and the present argument concerning abstraction, will be necessarily objected to by those philosophers—and they are, I fear, the major part—who hold that no relations can possess absolute and metaphysical validity. It is not my intention here to enter upon the general question, but merely to exhibit the objections to any analysis of asymmetrical relations.

It is a common opinion—often held unconsciously, and employed in argument, even by those who do not explicitly advocate it—that all propositions, ultimately, consist of a subject and a predicate. When this opinion is confronted by a relational proposition, it has two ways of dealing with it, of which the one may be called monadistic, the other monistic. Given, say, the proposition aRb, where R is some relation, the monadistic view will analyse this into two propositions, which we may call ar_1 and br_2, which give to a and b respectively adjectives supposed to be together equivalent to R. The monistic view, on the contrary, regards the relation as a property of the whole composed of a and b, and as thus equivalent to a proposition which we may denote by $(ab)r$. Of these views, the first is represented by Leibniz and (on the whole) by Lotze, the second by Spinoza and Mr Bradley. Let us examine these views successively, as applied to

---

* A formal treatment of relative position is given by Schröder, *Sur une extension de l'idée d'ordre, Congrès*, Vol. III, p. 235.

asymmetrical relations; and for the sake of definiteness, let us take
the relations of greater and less.

213. The monadistic view is stated with admirable lucidity by
Leibniz in the following passage* :

"The ratio or proportion between two lines L and M may be
conceived three several ways; as a ratio of the greater L to the
lesser M; as a ratio of the lesser M to the greater L; and lastly, as
something abstracted from both, that is, as the ratio between L and M,
without considering which is the antecedent, or which the consequent;
which the subject, and which the object....In the first way of considering
them, L the greater, in the second M the lesser, is the subject of that
accident which philosophers call *relation*. But which of them will be
the subject, in the third way of considering them? It cannot be said
that both of them, L and M together, are the subject of such an
accident; for if so, we should have an accident in two subjects, with
one leg in one, and the other in the other; which is contrary to the
notion of accidents. Therefore we must say that this relation, in this
third way of considering it, is indeed *out of* the subjects; but being
neither a substance nor an accident, it must be a mere ideal thing,
the consideration of which is nevertheless useful."

214. The third of the above ways of considering the relation of
greater and less is, roughly speaking, that which the monists advocate,
holding, as they do, that the whole composed of L and M is one subject,
so that their way of considering ratio does not compel us, as Leibniz
supposed, to place it among bipeds. For the present our concern is only
with the first two ways. In the first way of considering the matter, we
have "L is (greater than M)," the words in brackets being considered
as an adjective of L. But when we examine this adjective it is at once
evident that it is complex: it consists, at least, of the parts *greater*
and M, and both these parts are essential. To say that L is greater
does not at all convey our meaning, and it is highly probable that M is
also greater. The supposed adjective of L involves some reference to M;
but what can be meant by a reference the theory leaves unintelligible.
An adjective involving a reference to M is plainly an adjective which is
relative to M, and this is merely a cumbrous way of describing a relation.
Or, to put the matter otherwise, if L has an adjective corresponding
to the fact that it is greater than M, this adjective is logically subsequent to, and is merely derived from, the direct relation of L to M.
Apart from M, nothing appears in the analysis of L to differentiate it
from M; and yet, on the theory of relations in question, L should differ
intrinsically from M. Thus we should be forced, in all cases of asymmetrical relations, to admit a specific difference between the related
terms, although no analysis of either singly will reveal any relevant

* *Phil. Werke*, Gerhardt's ed., Vol. VII, p. 401.

property which it possesses and the other lacks. For the monadistic theory of relations, this constitutes a contradiction ; and it is a contradiction which condemns the theory from which it springs*. Let us examine further the application of the monadistic theory to quantitative relations. The proposition "A is greater than B" is to be analyzable into two propositions, one giving an adjective to A, the other giving one to B. The advocate of the opinion in question will probably hold that A and B are quantities, not magnitudes, and will say that the adjectives required are the magnitudes of A and B. But then he will have to admit a relation between the magnitudes, which will be as asymmetrical as the relation which the magnitudes were to explain. Hence the magnitudes will need new adjectives, and so on *ad infinitum* ; and the infinite process will have to be completed before any *meaning* can be assigned to our original proposition. This kind of infinite process is undoubtedly objectionable, since its sole object is to explain the meaning of a certain proposition, and yet none of its steps bring it any nearer to that meaning†. Thus we cannot take the magnitudes of A and B as the required adjectives. But further, if we take any adjectives whatever except such as have each a reference to the other term, we shall not be able, even formally, to give any account of the relation, without assuming just such a relation between the adjectives. For the mere fact that the adjectives are different will yield only a symmetrical relation. Thus if our two terms have different colours we find that A has to B the relation of differing in colour, a relation which no amount of careful handling will render asymmetrical. Or if we were to recur to magnitudes, we could merely say that A and B differ in magnitude, which gives us no indication as to which is the greater. Thus the adjectives of A and B must be, as in Leibniz's analysis, adjectives having a reference each to the other term. The adjective of A must be "greater than B," and that of B must be "less than A." Thus A and B differ, since they have different adjectives— B is not greater than B, and A is not less than A—but the adjectives are extrinsic, in the sense that A's adjective has reference to B, and B's to A. Hence the attempted analysis of the relation fails, and we

---

* See a paper on "The Relations of Number and Quantity," *Mind*, N.S. No. 23. This paper was written while I still adhered to the monadistic theory of relations : the contradiction in question, therefore, was regarded as inevitable. The following passage from Kant raises the same point : "Die rechte Hand ist der linken ähnlich und gleich, und wenn man blos auf eine derselben allein sieht, auf die Proportion der Lage der Theile unter einander und auf die Grösse des Ganzen, so muss eine vollständige Beschreibung der einen in allen Stücken auch von der andern gelten." (*Von dem ersten Grunde des Unterschiedes der Gegenden im Raume*, ed. Hart. Vol. ii, p. 389.)

† Where an infinite process of this kind is required we are necessarily dealing with a proposition which is an infinite unity, in the sense of Part II, Chap. xvii.

are forced to admit what the theory was designed to avoid, a so-called "external" relation, *i.e.* one implying no complexity in either of the related terms.

The same result may be proved of asymmetrical relations generally, since it depends solely upon the fact that both identity and diversity are symmetrical. Let a and b have an asymmetrical relation R, so that aRb and $b\breve{R}a$. Let the supposed adjectives (which, as we have seen, must each have a reference to the other term) be denoted by β and α respectively. Thus our terms become $a\beta$ and $b\alpha$. α involves a reference to a, and β to b; and α and β differ, since the relation is asymmetrical. But a and b have no intrinsic differences corresponding to the relation R, and prior to it; or, if they have, the points of difference must themselves have a relation analogous to R, so that nothing is gained. Either α or β expresses a difference between a and b, but one which, since either α or β involves reference to a term other than that whose adjective it is, so far from being prior to R, is in fact the relation R itself. And since α and β both presuppose R, the difference between α and β cannot be used to supply an intrinsic difference between a and b. Thus we have again a difference without a prior point of difference. This shows that some asymmetrical relations must be ultimate, and that at least one such ultimate asymmetrical relation must be a component in any asymmetrical relation that may be suggested.

It is easy to criticize the monadistic theory from a general standpoint, by developing the contradictions which spring from the relations of the terms to the adjectives into which our first relation has been analyzed. These considerations, which have no special connection with asymmetry, belong to general philosophy, and have been urged by advocates of the monistic theory. Thus Mr Bradley says of the monadistic theory*: "We, in brief, are led by a principle of fission which conducts us to no end. Every quality in relation has, in consequence, a diversity within its own nature, and this diversity cannot immediately be asserted of the quality. Hence the quality must exchange its unity for an internal relation. But, thus set free, the diverse aspects, because each something in relation, must each be something also beyond. This diversity is fatal to the internal unity of each; and it demands a new relation, and so on without limit." It remains to be seen whether the monistic theory, in avoiding this difficulty, does not become subject to others quite as serious.

215. The monistic theory holds that every relational proposition aRb is to be resolved into a proposition concerning the whole which a and b compose—a proposition which we may denote by $(ab)r$. This view, like the other, may be examined with special reference to asym-

---

* *Appearance and Reality*, 1st edition, p. 31.

metrical relations, or from the standpoint of general philosophy. We are told, by those who advocate this opinion, that the whole contains diversity within itself, that it synthesizes differences, and that it performs other similar 'feats. For my part, I am unable to attach any precise significance to these phrases. But let us do our best.

The proposition "*a* is greater than *b*," we are told, does not really say anything about either *a* or *b*, but about the two together. Denoting the whole which they compose by (*ab*), it says, we will suppose, "(*ab*) contains diversity of magnitude." Now to this statement—neglecting for the present all general arguments—there is a special objection in the case of asymmetry. (*ab*) is symmetrical with regard to *a* and *b*, and thus the property of the whole will be exactly the same in the case where *a* is greater than *b* as in the case where *b* is greater than *a*. Leibniz, who did not accept the monistic theory, and had therefore no reason to render it plausible, clearly perceived this fact, as appears from the above quotation. For, in his third way of regarding ratio, we do not consider which is the antecedent, which the consequent; and it is indeed sufficiently evident that, in the whole (*ab*) as such, there is neither antecedent nor consequent. In order to distinguish a whole (*ab*) from a whole (*ba*), as we must do if we are to explain asymmetry, we shall be forced back from the whole to the parts and their relation. For (*ab*) and (*ba*) consist of precisely the same parts, and differ in no respect whatever save the sense of the relation between *a* and *b*. "*a* is greater than *b*" and "*b* is greater than *a*" are propositions containing precisely the same constituents, and giving rise therefore to precisely the same whole; their difference lies solely in the fact that *greater* is, in the first case, a relation of *a* to *b*, in the second, a relation of *b* to *a*. Thus the distinction of sense, *i.e.* the distinction between an asymmetrical relation and its converse, is one which the monistic theory of relations is wholly unable to explain.

Arguments of a more general nature might be multiplied almost indefinitely, but the following argument seems peculiarly relevant. The relation of whole and part is itself an asymmetrical relation, and the whole—as monists are peculiarly fond of telling us—is distinct from all its parts, both severally and collectively. Hence when we say "*a* is part of *b*," we really mean, if the monistic theory be correct, to assert something of the whole composed of *a* and *b*, which is not to be confounded with *b*. If the proposition concerning this new whole be not one of whole and part there will be no true judgments of whole and part, and it will therefore be false to say that a relation between the parts is really an adjective of the whole. If the new proposition is one of whole and part, it will require a new one for its meaning, and so on. If, as a desperate measure, the monist asserts that the whole composed of *a* and *b* is not distinct from *b*, he is compelled to admit that a whole is the sum (in the sense of Symbolic Logic) of its parts, which, besides being an abandonment of his whole position, renders it inevitable that the whole should be symmetrical as regards its parts—a view which we have already seen to be fatal. And hence we find monists driven to the view that the only true whole, the Absolute, has no parts at all, and that no propositions in regard to it or anything else are quite true—a view which, in the mere statement, unavoidably contradicts itself. And surely an opinion which holds all propositions to be in the end self-contradictory is sufficiently condemned by the fact that, if it be accepted, it also must be self-contradictory.

216. We have now seen that asymmetrical relations are unintelligible on both the usual theories of relation*. Hence, since such relations are involved in Number, Quantity, Order, Space, Time, and Motion, we can hardly hope for a satisfactory philosophy of Mathematics so long as we adhere to the view that no relation can be "purely external." As soon, however, as we adopt a different theory, the logical puzzles, which have hitherto obstructed philosophers, are seen to be artificial. Among the terms commonly regarded as relational, those that are symmetrical and transitive—such as equality and simultaneity—*are* capable of reduction to what has been vaguely called identity of content, but this in turn must be analyzed into sameness of relation to some other term. For the so-called properties of a term are, in fact, only other terms to which it stands in some relation ; and a common property of two terms is a term to which both stand in the same relation.

The present long digression into the realm of logic is necessitated by the fundamental importance of order, and by the total impossibility of explaining order without abandoning the most cherished and widespread of philosophic dogmas. Everything depends, where order is concerned, upon asymmetry and difference of sense, but these two concepts are unintelligible to the traditional logic. In the next chapter we shall have to examine the connection of difference of sense with what appears in Mathematics as difference of sign. In this examination, though some pure logic will still be requisite, we shall approach again to mathematical topics; and these will occupy us wholly throughout the succeeding chapters of this Part.

---

* The grounds of these theories will be examined from a more general point of view in Part VI, Chap. LI.

CHAPTER XXVII.

DIFFERENCE OF SENSE AND DIFFERENCE OF SIGN.

217. WE have now seen that order depends upon asymmetrical relations, and that these always have two senses, as before and after, greater and less, east and west, etc. The difference of sense is closely connected (though not identical) with the mathematical difference of sign. It is a notion of fundamental importance in Mathematics, and is, so far as I can see, not explicable in terms of any other notions. The first philosopher who realized its importance would seem to be Kant. In the *Versuch den Begriff der negativen Grösse in die Weltweisheit einzuführen* (1763), we find him aware of the difference between logical opposition and the opposition of positive and negative. In the discussion *Von dem ersten Grunde des Unterschiedes der Gegenden im Raume* (1768), we find a full realization of the importance of asymmetry in spatial relations, and a proof, based on this fact, that space cannot be wholly relational*. But it seems doubtful whether he realized the connection of this asymmetry with difference of sign. In 1763 he certainly was not aware of the connection, since he regarded pain as a negative amount of pleasure, and supposed that a great pleasure and a small pain can be added to give a less pleasure†—a view which seems both logically and psychologically false. In the *Prolegomena* (§ 13), as is well known, he made the asymmetry of spatial relations a ground for regarding space as a mere form of intuition, perceiving, as appears from the discussion of 1768, that space could not consist, as Leibniz supposed, of mere relations among objects, and being unable, owing to his adherence to the logical objection to relations discussed in the preceding chapter, to free from contradiction the notion of absolute space with asymmetrical relations between its points. Although I cannot regard this later and more distinctively Kantian theory as an advance upon that of 1768, yet credit is undoubtedly due to Kant for having first called attention to the logical importance of asymmetrical relations.

---

* See especially ed. Hart, Vol. II, pp. 386, 391.
† Ed. Hart, Vol. II, p. 83.

218. By difference of sense I mean, in the present discussion at least, the difference between an asymmetrical relation and its converse. It is a fundamental logical fact that, given any relation R, and any two terms a, b, there are two propositions to be formed of these elements, the one relating a to b (which I call aRb), the other (bRa) relating b to a. These two propositions are always different, though sometimes (as in the case of diversity) either implies the other. In other cases, such as logical implication, the one does not imply either the other or its negation; while in a third set of cases, the one implies the negation of the other. It is only in cases of the third kind that I shall speak of difference of sense. In these cases, aRb excludes bRa. But here another fundamental logical fact becomes relevant. In all cases where aRb does not imply bRa there is another relation, related to R, which must hold between b and a. That is, there is a relation $\breve{R}$ such that aRb implies bRa; and further, $b\breve{R}a$ implies aRb. The relation of R to $\breve{R}$ is difference of sense. This relation is one-one, symmetrical, and intransitive. Its existence is the source of series, of the distinction of signs, and indeed of the greater part of mathematics.

219. A question of considerable importance to logic, and especially to the theory of inference, may be raised with regard to difference of sense. Are aRb and $b\breve{R}a$ really different propositions, or do they only differ linguistically? It may be held that there is only one relation R, and that all necessary distinctions can be obtained from that between aRb and bRa. It may be said that, owing to the exigencies of speech and writing, we are compelled to mention either a or b first, and that this gives a seeming difference between "a is greater than b" and "b is less than a"; but that, in reality, these two propositions are identical. But if we take this view we shall find it hard to explain the indubitable distinction between *greater* and *less*. These two words have certainly each a meaning, even when no terms are mentioned as related by them. And they certainly have different meanings, and are certainly relations. Hence if we are to hold that "a is greater than b" and "b is less than a" are the same proposition, we shall have to maintain that both *greater* and *less* enter into each of these propositions, which seems obviously false; or else we shall have to hold that what really occurs is neither of the two, but that third abstract relation mentioned by Leibniz in the passage quoted above. In this case the difference between *greater* and *less* would be one essentially involving a reference to the terms a and b. But this view cannot be maintained without circularity; for neither the greater nor the less is inherently the antecedent, and we can only say that, when the greater is the antecedent, the relation is *greater*; when the less, the relation is *less*. Hence, it would seem, we must admit that R and $\breve{R}$ are distinct relations. We cannot escape this conclusion by the analysis into adjectives attempted in the last chapter. We there analyzed aRb into $a\beta$ and ba. But, corresponding to every b, there will be two adjectives, β and $\breve{\beta}$, and corresponding to every a there will also be two, a and $\breve{a}$. Thus if R be *greater*, a will be " greater than A " and $\breve{a}$ " less than A," or *vice versâ*. But the difference between a and $\breve{a}$ presupposes that between greater and less, between R and $\breve{R}$, and therefore cannot explain it. Hence R and $\breve{R}$ must be distinct, and "aRb implies $b\breve{R}a$" must be a genuine inference.

I come now to the connection between difference of sense and difference of sign. We shall find that the latter is derivative from the former, being a difference which only exists between terms which either are, or are correlated with, asymmetrical relations. But in certain cases we shall find some complications of detail which will demand discussion.

The difference of signs belongs, traditionally, only to numbers and magnitudes, and is intimately associated with addition. It may be allowed that the notation cannot be usefully employed where there is no addition, and even that, where distinction of sign is possible, addition in some sense is in general also possible. But we shall find that the difference of sign has no very intimate connection with addition and subtraction. To make this clear, we must, in the first place, clearly realize that numbers and magnitudes which have no sign are radically different from such as are positive. Confusion on this point is quite fatal to any just theory of signs.

220. Taking first finite numbers, the positive and negative numbers arise as follows*. Denoting by R the relation between two integers in virtue of which the second is next after the first, the proposition mRn is equivalent to what is usually expressed by $m + 1 = n$. But the present theory will apply to progressions generally and does not depend upon the logical theory of cardinals developed in Part II. In the proposition mRn, the integers m and n are considered, as when they result from the logical definition, to be wholly destitute of sign. If now mRn and nRp, we put mR^2p; and so on for higher powers. Every power of R is an asymmetrical relation, and its converse is easily shown to be the same power of $\breve{R}$ as it is itself of R. Thus mR^aq is equivalent to $q\breve{R}^am$. These are the two propositions which are commonly written $m + a = q$ and $q - a = m$. Thus the relations R^a, $\breve{R}^a$ are the true positive and negative integers; and these, though associated with a, are both wholly distinct from it. Thus in this case the connection with difference of sense is obvious and straightforward.

221. As regards magnitudes, several cases must be distinguished. We have (1) magnitudes which are not either relations or stretches, (2) stretches, (3) magnitudes which are relations.

---

* I give the theory briefly here, as it will be dealt with more fully and generally in the chapter on Progressions, § 233.

(1) Magnitudes of this class are themselves neither positive nor negative. But two such magnitudes, as explained in Part III, determine either a distance or a stretch, and these are always positive or negative. These are moreover always capable of addition. But since our original magnitudes are neither relations nor stretches, the new magnitudes thus obtained are of a different kind from the original set. Thus the difference of two pleasures, or the collection of pleasures intermediate between two pleasures, is not a pleasure, but in the one case a relation, in the other a class.

(2) Magnitudes of divisibility in general have no sign, but when they are magnitudes of stretches they acquire sign by correlation. A stretch is distinguished from other collections by the fact that it consists of all the terms of a series intermediate between two given terms. By combining the stretch with one sense of the asymmetrical relation which must exist between its end-terms, the stretch itself acquires sense, and becomes asymmetrical. That is, we can distinguish (1) the collection of terms between a and b without regard to order, (2) the terms from a to b, (3) the terms from b to a. Here (2) and (3) are complex, being compounded of (1) and one sense of the constitutive relation. Of these two, one must be called positive, the other negative. Where our series consists of magnitudes, usage and the connection with addition have decided that, if a is less than b, (2) is positive and (3) is negative. But where, as in Geometry, our series is not composed of magnitudes, it becomes wholly arbitrary which is to be positive and which negative. In either case, we have the same relation to addition, which is as follows. Any pair of collections can be added to form a new collection, but not any pair of stretches can be added to form a new stretch. For this to be possible the end of one stretch must be consecutive to the beginning of the other. In this way, the stretches ab, bc can be added to form the stretch ac. If ab, bc have the same sense, ac is greater than either; if they have different senses, ac is less than one of them. In this second case the addition of ab and bc is regarded as the subtraction of ab and cb, bc and cb being negative and positive respectively. If our stretches are numerically measurable, addition or subtraction of their measures will give the measure of the result of adding or subtracting the stretches, where these are such as to allow addition or subtraction. But the whole opposition of positive and negative, as is evident, depends upon the fundamental fact that our series is generated by an asymmetrical relation.

(3) Magnitudes which are relations may be either symmetrical or asymmetrical relations. In the former case, if a be a term of the field of one of them, the other terms of the various fields, if certain conditions are fulfilled*, may be arranged in series according as their relations to a are greater or smaller. This arrangement may be different when we choose

* Cf. § 245.

some term other than a; for the present, therefore, we shall suppose a to be chosen once for all. When the terms have been arranged in a series, it may happen that some or all places in the series are occupied by more than one term; but in any case the assemblage of terms between a and some other term m is definite, and leads to a stretch with two senses. We may then combine the magnitude of the relation of a to m with one or other of these two senses, and so obtain an asymmetrical relation of a to m, which, like the original relation, will have magnitude. Thus the case of symmetrical relations may be reduced to that of asymmetrical relations. These latter lead to signs, and to addition and subtraction, in exactly the same way as stretches with sense; the only difference being that the addition and subtraction are now of the kind which, in Part III, we called relational. Thus in all cases of magnitudes having sign, the difference between the two senses of an asymmetrical relation is the source of the difference of sign.

The case which we discussed in connection with stretches is of fundamental importance in Geometry. We have here a magnitude without sign, an asymmetrical relation without magnitude, and some intimate connection between the two. The combination of both then gives a magnitude which has sign. All geometrical magnitudes having sign arise in this way. But there is a curious complication in the case of volumes. Volumes are, in the first instance, signless quantities; but in analytical Geometry they always appear as positive or negative. Here the asymmetrical relations (for there are two) appear as terms, between which there is a symmetrical relation, but one which yet has an opposite of a kind very similar to the converse of an asymmetrical relation. This relation, as an exceptional case, must be here briefly discussed.

222. The descriptive straight line is a serial relation in virtue of which the points of the line form a series*. Either sense of the descriptive straight line may be called a ray, the sense being indicated by an

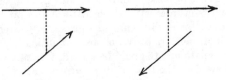

arrow. Any two non-coplanar rays have one or other of two relations, which may be called right and left-handedness respectively †. This relation is symmetrical but not transitive, and is the essence of the usual distinction of right and left. Thus the relation of the upward vertical to a line from north to east is right-handed, and to a line from south to

---

* See Part VI.

† The two cases are illustrated in the figure. The difference is the same as that between the two sorts of coordinate axes.

east is left-handed. But though the relation is symmetrical, it is changed into its opposite by changing either of the terms of the relation into its converse. That is, denoting right-handedness by R, left-handedness by L (which is not $\breve{R}$), if A and B be two rays which are mutually right-handed, we shall have

$$ARB,\ \breve{A}LB,\ AL\breve{B},\ \breve{A}R\breve{B},\ BRA,\ \breve{B}LA,\ BL\breve{A},\ \breve{B}R\breve{A}.$$

That is, every pair of non-coplanar straight lines gives rise to eight such relations, of which four are right-handed, and four left-handed. The difference between L and R, though not, as it stands, a difference of sense, is, nevertheless, the difference of positive and negative, and is the reason why the volumes of tetrahedra, as given by determinants, always have signs. But there is no difficulty in following the plain man's reduction of right and left to asymmetrical relations. The plain man takes one of the rays (say A) as fixed—when he is sober, he takes A to be the upward vertical—and then regards right and left as properties of the single ray B, or, what comes to the same thing, as relations of any two points which determine B. In this way, right and left become asymmetrical relations, and even have a limited degree of transitiveness, of the kind explained in the fifth way of generating series (in Chapter XXIV). It is to be observed that what is fixed must be a ray, not a mere straight line. For example, two planes which are not mutually perpendicular are not one right and the other left with regard to their line of intersection, but only with regard to either of the rays belonging to this line.* But when this is borne in mind, and when we consider, not semi-planes, but complete planes, through the ray in question, right and left become asymmetrical and each other's converses. Thus the signs associated with right and left, like all other signs, depend upon the asymmetry of relations. This conclusion, therefore, may now be allowed to be general.

223. Difference of sense is, of course, more general than difference of sign, since it exists in cases with which mathematics (at least at present) is unable to deal. And difference of sign seems scarcely applicable to relations which are not transitive, or are not intimately connected with some transitive relation. It would be absurd, for example, to regard the relation of an event to the time of its occurrence, or of a quantity to its magnitude, as conferring a difference of sign. These relations are what Professor Schröder calls *erschöpft*†, *i.e.* if they hold between a and b, they can never hold between b and some third term. Mathematically, their square is null. These relations, then, do not give rise to difference of sign.

---

* This requires that the passage from the one plane to the other should be made *via* one of the acute angles made by their intersection.

† *Algebra der Logik*, Vol. III, p. 328. Professor Peirce calls such relations *non-repeating* (reference in Schröder, *ib.*).

All magnitudes with sign, so the above account has led us to believe, are either relations or compound concepts into which relations enter. But what are we to say of the usual instances of opposites: good and evil, pleasure and pain, beauty and ugliness, desire and aversion? The last pair are very complex, and if I were to attempt an analysis of them, I should emit some universally condemned opinions. With regard to the others, they seem to me to have an opposition of a very different kind from that of two mutually converse asymmetrical relations, and analogous rather to the opposition of red and blue, or of two different magnitudes of the same kind. From these oppositions, which are constituted by what may be called synthetic incompatibility *, the oppositions above mentioned differ only in the fact that there are only two incompatible terms, instead of a whole series. The incompatibility consists in the fact that two terms which are thus incompatible cannot coexist in the same spatio-temporal place, or cannot be predicates of the same existent, or, more generally, cannot both enter into true propositions of a certain form, which differ only in the fact that one contains one of the incompatibles while the other contains the other. This kind of incompatibility (which usually belongs, with respect to some class of propositions, to the terms of a given series) is a most important notion in general logic, but is by no means to be identified with the difference between mutually converse relations. This latter is, in fact, a special case of such incompatibility; but it is the special case only that gives rise to the difference of sign. All difference of sign—so we may conclude our argument—is primarily derived from transitive asymmetrical relations, from which it may be extended by correlation to terms variously related to such relations†; but such extensions are always subsequent to the original opposition derived from difference of sense.

---

* See *The Philosophy of Leibniz*, by the present author (Cambridge 1900), pp. 19, 20.

† Thus in mathematical Economics, pleasure and pain may be taken as positive and negative without logical error, by the theory (whose psychological correctness we need not examine) that a man must be paid to endure pain, and must pay to obtain pleasure. The opposition of pleasure and pain is thus correlated with that of money paid and money received, which is an opposition of positive and negative in the sense of elementary Arithmetic.

CHAPTER XXVIII.

ON THE DIFFERENCE BETWEEN OPEN AND CLOSED SERIES.

224. WE have now come to the end of the purely logical discussions concerned with order, and can turn our attention with a free mind to the more mathematical aspects of the subject. As the solution of the most ancient and respectable contradictions in the notion of infinity depends mainly upon a correct philosophy of order, it has been necessary to go into philosophical questions at some length—not so much because they *are* relevant, as because most philosophers think them so. But we shall reap our reward throughout the remainder of this work.

The question to be discussed in this chapter is this: Can we ultimately distinguish open from closed series, and if so, in what does the distinction consist? We have seen that, mathematically, *all* series are open, in the sense that all are generated by an asymmetrical transitive relation. But philosophically, we must distinguish the different ways in which this relation may arise, and especially we must not confound the case where this relation involves no reference to other terms with that where such terms are essential. And practically, it is plain that there is some difference between open and closed series—between, for instance, a straight line and a circle, or a pedigree and a mutual admiration society. But it is not quite easy to express the difference precisely.

225. Where the number of terms in the series is finite, and the series is generated in the first of the ways explained in Chapter XXIV, the method of obtaining a transitive relation out of the intransitive relation with which we start is radically different according as the series is open or closed. If R be the generating relation, and n be the number of terms in our series, two cases may arise. Denoting the relation of any term to the next but one by R^2, and so on for higher powers, the relation R^n can have only one of two values, zero and identity. (It is assumed that R is a one-one relation.) For starting with the first term, if there be one, R^{n-1} brings us to the last term; and thus R^n gives no new term, and there is no instance of the relation R^n. On the other hand, it may happen that, starting with any term,

R^n brings us back to that term again. These two are the only possible alternatives. In the first case, we call the series open; in the second, we call it closed. In the first case, the series has a definite beginning and end; in the second case, like the angles of a polygon, it has no peculiar terms. In the first case, our transitive asymmetrical relation is the disjunctive relation "a power of R not greater than the $(n-1)$th." By substituting this relation, which we may call R', for R, our series becomes of the second of the six types. But in the second case no such simple reduction to the second type is possible. For now, the relation of any two terms a and m of our series may be just as well taken to be a power of $\breve{R}$ as a power of R, and the question which of any three terms is between the other two becomes wholly arbitrary. We might now introduce, first the relation of separation of four terms, and then the resulting five-term relation explained in Chapter xxv. We should then regard three of the terms in the five-term relation as fixed, and find that the resulting relation of the other two is transitive and asymmetrical. But here the first term of our series is wholly arbitrary, which was not the case before; and the generating relation is, in reality, one of five terms, not one of two. There is, however, in the case contemplated, a simpler method. This may be illustrated as follows: In an open series, any two terms a and m define two senses in which the series may be described, the one in which a comes before m, and the other in which m comes before a. We can then say of any two other terms c and g that the sense of the order from c to g is the same as that of the order from a to m, or different, as the case may be. In this way, considering a and m fixed, and c and g variable, we get a transitive asymmetrical relation between c and g, obtained from a transitive symmetrical relation of the pair c, g to the pair a, m (or m, a, as the case may be). But this transitive symmetrical relation can, by the principle of abstraction, be analyzed into possession of a common property, which is, in this case, the fact that a, m and c, g have the generating relation with the same sense. Thus the four-term relation is, in this case, not essential. But in a closed series, a and m do not define a sense of the series, even when we are told that a is to precede m: we can start from a and get to m in either direction. But if now we take a third term d, and decide that we are to start from a and reach m taking d on the way, then a sense of the series is defined. The stretch adm includes one portion of the series, but not the other. Thus we may go from England to New Zealand either by the east or by the west; but if we are to take India on the way, we must go by the east. If now we consider any other term, say k, this will have some definite position in the series which starts with a and reaches m by way of d. In this series, k will come either between a and d, or between d and m, or after m. Thus the three-term relation of a, d, m seems in this case sufficient to generate a perfectly definite series. Vailati's five-term relation will then consist in this, that with regard to the order adm, k comes before (or after) any other term l of the collection. But it is not necessary to call in this relation in the present case, since the three-term relation suffices. This three-term relation may be formally defined as follows. There is between any two terms of our collection a relation which is a power of R less than the nth. Let the relation between a and d be R^x, that between a and m R^y. Then if x is less than y, we assign one sense to adm; if x is greater than y, we assign the other. There will be also between a and d the relation $\breve{R}^{n-x}$, and between a and m the relation $\breve{R}^{n-y}$. If x is less than y, then $n - x$ is greater than $n - y$; hence the asymmetry of the two cases corresponds to that of R and $\breve{R}$. The terms of the series are simply ordered by correlation with their numbers x and y, those with smaller numbers preceding those with larger ones. Thus there is here no need of the five-term relation, everything being effected by the three-term relation, which is itself reduced to an asymmetrical transitive relation of two numbers. But the closed series is still distinguished from the open one by the fact that its first term is arbitrary.

226. A very similar discussion will apply to the case where our series is generated by relations of three terms. To keep the analogy with the one-one relation of the above case, we will make the following assumptions. Let there be a relation B of one term to two others, and let the one term be called the mean, the two others the extremes. Let the mean be uniquely determined when the extremes are given, and let one extreme be uniquely determined by the mean and the other extreme. Further let each term that occurs as mean occur also as extreme, and each term that occurs as extreme (with at most two exceptions) occur also as mean. Finally, if there be a relation in which c is mean, and b and d are extremes, let there be always (except when b or d is one of the two possible exceptional terms) a relation in which b is the mean and c one of the extremes, and another in which d is the mean and c one of the extremes. Then b and c will occur together in only two relations. This fact constitutes a relation between b and c, and only one other term besides b will have this new relation to c. By means of this relation, if there are two exceptional terms, or if, our collection being infinite, there is only one, we can construct an open series. If our two-term relation be asymmetrical, this is sufficiently evident; but the same result can be proved if our two-term relation is symmetrical. For there will be at either end, say a, an asymmetrical relation of a to the only term which is the mean between a and some other term. This relation multiplied by the nth power of our two-term relation, where $n + 1$ is any integer less than the number of terms in our collection, will give a relation which holds between a and a number (not exceeding $n + 1$) of terms of our collection, of which terms one and only one is such that no number less than n gives a relation of a to this term. Thus we obtain a correlation of our terms with the natural numbers, which generates an open series with a for one of its ends. If, on the other hand, our collection has no exceptional terms, but is finite, then we shall obtain a closed series. Let our two-term relation be P, and first suppose it symmetrical. (It will be symmetrical if our original three-term relation was symmetrical with regard to the extremes.) Then every term c of our collection will have the relation P to two others, which will have to each other the relation P^2. Of all the relations of the form P^m which hold between two given terms, there will be one in which m is least: this may be called the principal relation of our two terms. Let the number of terms of the collection be n. Then every term of our collection will have to every other a principal relation P^x, where x is some integer not greater than $n/2$. Given any two terms c and g of the collection, provided we do not have $cP^{n/2}g$ (a case which will not arise if n be odd), let us have cP^xg, where x is less than $n/2$. This assumption defines a sense of the series, which may be shown as follows. If cP^yk, where y is also less than $n/2$, three cases may arise, assuming y is greater than x. We may have $gP^{y-x}k$, or, if $x+y$ is less than $n/2$, we may have $gP^{x+y}k$, or, if $x+y$ is greater than $n/2$, we may have $gP^{\frac{n}{2}-x-y}k$. (We choose always the principal relation.) These three cases are illustrated in the accompanying figure. We shall say, in these three cases,

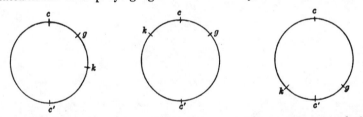

that, with regard to the sense cg, (1) k comes after c and g, (2) and (3) k comes before c and g. If y is less than x, and $kP^{x-y}g$, we shall say that k is between c and g in the sense cg. If n is odd, this covers all possible cases. But if n is even, we have to consider the term c', which is such that $cP^{n/2}c'$. This term is, in a certain sense, antipodal to c; we may define it as the first term in the series when the above method of definition is adopted. If n is odd, the first term will be that term of class (3) for which $cP^{(n-1)/2}k$. Thus the series acquires a definite order, but one in which, as in all closed series, the first term is arbitrary.

227. The only remaining case is that where we start from four-term relations, and the generating relation has, strictly speaking, five terms. This is the case of projective Geometry. Here the series is necessarily closed; that is, in choosing our three fixed terms for the five-term relation, there is never any restriction upon our choice; and any one of these three may be defined to be the first.

228. Thus, to sum up : Every series being generated by a transitive asymmetrical relation between any two terms of the series, a series is open when it has either no beginning, or a beginning which is not arbitrary ; it is closed when it has an arbitrary beginning. Now if R be the constitutive relation, the beginning of the series is a term having the relation R but not the relation $\breve{R}$. Whenever R is genuinely a two-term relation, the beginning, if it exists, must be perfectly definite. It is only when R involves some other term (which may be considered fixed) besides the two with regard to which it is transitive and asymmetrical (which are to be regarded as variable), that the beginning can be arbitrary. Hence in all cases of closed series, though there may be an asymmetrical one-one relation if the series is discrete, the *transitive* asymmetrical relation must be one involving one or more fixed terms in addition to the two variable terms with regard to which it generates the series. Thus although, mathematically, every closed series can be rendered open, and every open series closed, yet there is, in regard to the nature of the generating relation, a genuine distinction between them—a distinction, however, which is of philosophical rather than mathematical importance.

CHAPTER XXIX.

PROGRESSIONS AND ORDINAL NUMBERS.

229. It is now time to consider the simplest type of infinite series, namely that to which the natural numbers themselves belong. I shall postpone to the next Part all the supposed difficulties arising out of the infinity of such series, and concern myself here only to give the elementary theory of them in a form not presupposing numbers*.

The series now to be considered are those which can be correlated, term for term, with the natural numbers, without requiring any change in the order of the terms. But since the natural numbers are a particular case of such series, and since the whole of Arithmetic and Analysis can be developed out of any one such series, without any appeal to number, it is better to give a definition of progressions which involves no appeal to number.

A progression is a discrete series having consecutive terms, and a beginning but no end, and being also *connected*. The meaning of connection was explained in Chapter xxiv by means of number, but this explanation cannot be given now. Speaking popularly, when a series is not connected it falls into two or more parts, each being a series for itself. Thus numbers and instants together form a series which is not connected, and so do two parallel straight lines. Whenever a series is originally given by means of a transitive asymmetrical relation, we can express connection by the condition that any two terms of our series are to have the generating relation. But progressions are series of the kind that may be generated in the first of our six ways, namely, by an asymmetrical one-one relation. In order to pass from this to a transitive relation, we before employed numbers, defining the transitive relation as any power of the one-one relation. This definition will not serve now, since numbers are to be excluded. It is one of the triumphs of modern mathematics to have adapted an ancient principle to the needs of this case.

* The present chapter closely follows Peano's Arithmetic. See *Formulaire de Mathématiques*, Vol. II, § 2. I have given a mathematical treatment of the subject in RdM, Vols. VII and VIII. The subject is due, in the main, to Dedekind and Georg Cantor.

The definition which we want is to be obtained from mathematical induction. This principle, which used to be regarded as a mere subterfuge for eliciting results of which no other proof was forthcoming, has gradually grown in importance as the foundations of mathematics have been more closely investigated. It is now seen to be the principle upon which depend, so far as ordinals are concerned, the commutative law and one form of the distributive law*. This principle, which gives the widest possible extension to the finite, is the distinguishing mark of progressions. It may be stated as follows:

Given any class of terms *s*, to which belongs the first term of any progression, and to which belongs the term of the progression next after any term of the progression belonging to *s*, then every term of the progression belongs to *s*.

We may state the same principle in another form. Let $\phi(x)$ be a propositional function, which is a determinate proposition as soon as *x* is given. Then $\phi(x)$ is a function of *x*, and will in general be true or false according to the value of *x*. If *x* be a member of a progression, let seq *x* denote the term next after *x*. Let $\phi(x)$ be true when *x* is the first term of a certain progression, and let $\phi(\text{seq } x)$ be true whenever $\phi(x)$ is true, where *x* is any term of the progression. It then follows, by the principle of mathematical induction, that $\phi(x)$ is always true if *x* be any term of the progression in question.

The complete definition of a progression is as follows. Let *R* be any asymmetrical one-one relation, and *u* a class such that every term of *u* has the relation of *R* to some term also belonging to the class *u*. Let there be at least one term of the class *u* which does not have the relation $\breve{R}$ to any term of *u*. Let *s* be any class to which belongs at least one of the terms of *u* which do not have the relation $\breve{R}$ to any term of *u*, and to which belongs also every term of *u* which has the relation $\breve{R}$ to some term belonging to both *u* and *s*; and let *u* be such as to be wholly contained in any class *s* satisfying the above conditions. Then *u*, considered as ordered by the relation *R*, is a progression†.

230. Of such progressions, everything relevant to finite Arithmetic can be proved. In the first place, we show that there can only be one term of *u* which does not have the relation $\breve{R}$ to any term of *u*. We then define the term to which *x* has the relation *R* as the successor of *x* (*x* being a *u*), which may be written seq *x*. The definitions and properties of addition, subtraction, multiplication, division, positive and

---

* Namely $(a + \beta)\gamma = a\beta + a\gamma$. The other form, $a(\beta + \gamma) = a\beta + a\gamma$, holds also for infinite ordinal numbers, and is thus independent of mathematical induction.

† It should be observed that a discrete open series generated by a transitive relation can always be reduced, as we saw in the preceding chapter, to one generated by an asymmetrical one-one relation, provided only that the series is finite or a progression.

negative terms, and rational fractions are easily given; and it is easily shown that between any two rational fractions there is always a third. From this point it is easy to advance to irrationals and the real numbers*.

Apart from the principle of mathematical induction, what is chiefly interesting about this process is, that it shows that only the serial or ordinal properties of finite numbers are used by ordinary mathematics, what may be called the logical properties being wholly irrelevant. By the logical properties of numbers, I mean their definition by means of purely logical ideas. This process, which has been explained in Part II, may be here briefly recapitulated. We show, to begin with, that a one-one correlation can be effected between any two null classes, or between any two classes u, v which are such that, if x is a u, and x' differs from x, then x' cannot be a u, with a like condition for v. The possibility of such one-one correlation we call similarity of the two classes u, v. Similarity, being symmetrical and transitive, must be analyzable (by the principle of abstraction) into possession of a common property. This we define as the *number* of either of the classes. When the two classes u, v have the above-defined property, we say their number is *one*; and so on for higher numbers; the general definition of finite numbers demanding mathematical induction, or the non-similarity of whole and part, but being always given in purely logical terms.

It is numbers so defined that are used in daily life, and that are essential to any assertion of numbers. It is the fact that numbers have these logical properties that makes them important. But it is not these properties that ordinary mathematics employs, and numbers might be bereft of them without any injury to the truth of Arithmetic and Analysis. What is relevant to mathematics is solely the fact that finite numbers form a progression. This is the reason why mathematicians—*e.g.* Helmholtz, Dedekind, and Kronecker—have maintained that ordinal numbers are prior to cardinals; for it is solely the ordinal properties of number that are relevant. But the conclusion that ordinals are prior to cardinals seems to have resulted from a confusion. Ordinals and cardinals alike form a progression, and have exactly the same ordinal properties. Of either, all Arithmetic can be proved without any appeal to the other, the propositions being symbolically identical, but different in meaning. In order to prove that ordinals are prior to cardinals, it would be necessary to show that the cardinals can only be defined in terms of the ordinals. But this is false, for the logical definition of the cardinals is wholly independent of the ordinals†. There seems, in fact, to be nothing to choose, as regards logical priority, between ordinals and cardinals, except that the existence of the ordinals

---

* See my article on the Logic of Relations, RdM, VII.

† Professor Peano, who has a rare immunity from error, has recognized this fact. See *Formulaire*, 1898, 210, note (p. 39).

is inferred from the series of cardinals. The ordinals, as we shall see in the next paragraph, can be defined without any appeal to the cardinals; but when defined, they are seen to imply the cardinals. Similarly, the cardinals can be defined without any appeal to the ordinals; but they essentially form a progression, and all progressions, as I shall now show, necessarily imply the ordinals.

231. The correct analysis of ordinals has been prevented hitherto by the prevailing prejudice against relations. People speak of a series as consisting of certain terms *taken* in a certain order, and in this idea there is commonly a psychological element. All sets of terms have, apart from psychological considerations, all orders of which they are capable; that is, there are serial relations, whose fields are a given set of terms, which arrange those terms in any possible order. In some cases, one or more serial relations are specially prominent, either on account of their simplicity, or of their importance. Thus the order of magnitude among numbers, or of before and after among instants, seems emphatically the *natural* order, and any other seems to be artificially introduced by our arbitrary choice. But this is a sheer error. Omnipotence itself cannot give terms an order which they do not possess already: all that is psychological is the *consideration* of such and such an order. Thus when it is said that we can arrange a set of terms in any order we please, what is really meant is, that we can consider any of the serial relations whose field is the given set, and that these serial relations will give between them any combinations of before and after that are compatible with transitiveness and connection. From this it results that an order is not, properly speaking, a property of a given set of terms, but of a serial relation whose field is the given set. Given the relation, its field is given with it; but given the field, the relation is by no means given. The notion of a set of terms in a given order is the notion of a set of terms considered as the field of a given serial relation; but the consideration of the terms is superfluous, and that of the relation alone is quite sufficient.

We may, then, regard an ordinal number as a common property of sets of serial relations which generate ordinally similar series. Such relations have what I shall call *likeness*, *i.e.* if P, Q be two such relations, their fields can be so correlated term for term that two terms of which the first has to the second the relation P will always be correlated with two terms of which the first has to the second the relation Q, and *vice versâ*. As in the case of cardinal numbers*, so here, we may, in virtue of the principle of abstraction, define the ordinal number of a given finite serial relation as the class of like relations. It is easy to show that the generating relations of progressions are all alike; the class of such relations will be the ordinal number of the finite integers in order of magnitude. When a class is finite, all series that can be

* Cf. § 111.

formed of its terms are ordinally similar, and are ordinally different from series having a different cardinal number of terms. Hence there is a one-one correlation of finite ordinals and cardinals, for which, as we shall see in Part V, there is no analogy in respect of infinite numbers. We may therefore define the ordinal number n as the class of serial relations whose *domains* have n terms, where n is a finite cardinal. It is necessary, unless 1 is to be excluded, to take domains instead of' fields here, for no relation which implies diversity can have one term in its field, though it may have none. This has a practical inconvenience, owing to the fact that $n+1$ must be obtained by adding *one* term to the field; but the point involved is one for conventions as to notation, and is quite destitute of philosophical importance.

232. The above definition of ordinal numbers is direct and simple, but does not yield the notion of "nth," which would usually be regarded as *the* ordinal number. This notion is far more complex : a term is not intrinsically the nth, and does not become so by the mere specification of $n-1$ other terms. A term is the nth in respect of a certain serial relation, when, in respect of that relation, the term in question has $n-1$ predecessors. This is the definition of "nth," showing that this notion is relative, not merely to predecessors, but also to a specified serial relation. By induction, the various finite ordinals can be defined without mentioning the cardinals. A finite serial relation is one which is not like (in the above sense) any relation implying it but not equivalent to it; and a finite ordinal is one consisting of finite serial relations. If n be a finite ordinal, $n+1$ is an ordinal such that, if the last term* of a series of the type $n+1$ be cut off, the remainder, in the same order, is of the type n. In more technical language, a serial relation of the type $n+1$ is one which, when confined to its domain instead of its field, becomes of the type n. This gives by induction a definition of every particular finite ordinal, in which cardinals are never mentioned. Thus we cannot say that ordinals presuppose cardinals, though they are more complex, since they presuppose both serial and one-one relations, whereas cardinals only presuppose one-one relations.

Of the ordinal number of the finite ordinals in order of magnitude, several equivalent definitions may be given. One of the simplest is, that this number belongs to any serial relation, which is such that any class contained in its field and not null has a first term, while every term of the series has an immediate successor, and every term except the first has an immediate predecessor. Here, again, cardinal numbers are in no way presupposed.

Throughout the above discussions our serial relations are taken to be transitive, not one-one. The one-one relations are easily derived from

---

* The last term of a series (if it exists) is the term belonging to the converse domain but not to the domain of the generating relation, *i.e.* the term which is after but not before other terms.

the transitive ones, while the converse derivation is somewhat complicated. Moreover the one-one relations are only adequate to define finite series, and thus their use cannot be extended to the study of infinite series unless they are taken as derivative from the transitive ones.

233. A few words concerning positive and negative ordinals seem to be here in place. If the first n terms of a progression be taken away (n being any finite number), the remainder still form a progression. With regard to the new progression, negative ordinals may be assigned to the terms that have been abstracted; but for this purpose it is convenient to regard the beginning of the smaller progression as the 0th term. In order to have a series giving *any* positive or negative ordinal, we need what may be called a double progression. This is a series such that, choosing any term x out of it, two progressions start from x, the one generated by a serial relation R, the other by $\breve{R}$. To x we shall then assign the ordinal 0, and to the other terms we shall assign positive or negative ordinals according as they belong to the one or the other of the two progressions starting from x. The positive and negative ordinals themselves form such a double progression. They express essentially a relation to the arbitrarily chosen origin of the two progressions, and $+n$ and $-n$ express mutually converse relations. Thus they have all the properties which we recognize in Chapter XXVII as characterizing terms which have signs.

CHAPTER XXX.

DEDEKIND'S THEORY OF NUMBER.

234. THE theory of progressions and of ordinal numbers, with which we have been occupied in the last chapter, is due in the main to two men—Dedekind and Cantor. Cantor's contributions, being specially concerned with infinity, need not be considered at present; and Dedekind's theory of irrationals is also to be postponed. It is his theory of integers of which I wish now to give an account—the theory, that is to say, which is contained in his " *Was sind und was sollen die Zahlen?*"* In reviewing this work, I shall not adhere strictly to Dedekind's phraseology. He appears to have been, at the time of writing, unacquainted with symbolic logic ; and although he invented as much of this subject as was relevant to his purpose, he naturally adopted phrases which were not usual, and were not always so convenient as their conventional equivalents.

The fundamental ideas of the pamphlet in question are these† : (1) the representation (*Abbildung*) of a system (21) ; (2) the notion of a *chain* (37) ; (3) the chain of an element (44) ; (4) the generalized form of mathematical induction (59) ; (5) the definition of a singly infinite system (71). From these five notions Dedekind deduces numbers and ordinary Arithmetic. Let us first explain the notions, and then examine the deduction.

235. (1) A *representation* of a class u is any law by which, to every term of u, say x, corresponds some one and only one term $\phi(x)$. No assumption is made, to begin with, as to whether $\phi(x)$ belongs to the class u, or as to whether $\phi(x)$ may be the same as $\phi(y)$, when x and y are different terms of u. The definition thus amounts to this :

A *representation* of a class u is a many-one relation, whose domain contains u, by which terms, which may or may not also belong to u, are

---

* 2nd ed. Brunswick, 1893 (1st ed. 1887). The principal contents of this book, expressed by the Algebra of Relations, will be found in my article in RdM, VII, 2, 3.

† The numbers in brackets refer, not to pages, but to the small sections into which the work is divided.

correlated one with each of the terms of u*. The representation is similar when, if x differs from y, both being u's, then $\phi(x)$ differs from $\phi(y)$; that is, when the relation in question is one-one. He shows that similarity between classes is reflexive, symmetrical and transitive, and remarks (34) that classes can be classified by similarity to a given class—a suggestion of an idea which is fundamental in Cantor's work.

236. (2) If there exists a relation, whether one-one or many-one, which correlates with a class u only terms belonging to that class, then this relation is said to constitute a representation of u in itself (36), and with respect to this relation u is called a chain (37). That is to say, any class u is, with respect to any many-one relation, a chain, if u is contained in the domain of the relation, and the correlate of a u is always itself a u. The collection of correlates of a class is called the image (*Bild*) of the class. Thus a chain is a class whose image is part or the whole of itself. For the benefit of the non-mathematical reader, it may be not superfluous to remark that a chain with regard to a one-one relation, provided it has any term not belonging to the image of the chain, cannot be finite, for such a chain must contain the same number of terms as a proper part of itself †.

237. (3) If a be any term or collection of terms, there may be, with respect to a given many-one relation, many chains in which a is contained. The common part of all these chains, which is denoted by a_o, is what Dedekind calls the *chain of a* (44). For example, if a be the number n, or any set of numbers of which n is the least, the chain of a with regard to the relation "less by 1" will be all numbers not less than n.

238. (4) Dedekind now proceeds (59) to a theorem which is a generalized form of mathematical induction. This theorem is as follows : Let a be any term or set of terms contained in a class s, and let the image of the common part of s and the chain of a be also contained in s; then it follows that the chain of a is contained in s. This somewhat complicated theorem may become clearer by being put in other language. Let us call the relation by which the chain is generated (or rather the converse of this relation) succession, so that the correlate or image of a term will be its successor. Let a be a term which has a successor, or a collection of such terms. A chain in general (with regard to succession) will be any set of terms such that the successor of any one of them also belongs to the set. The chain of a will be the common

---

* A many-one relation is one in which, as in the relation of a quantity to its magnitude, the right-hand term, *to* which the relation is, is uniquely determined when the left-hand term is given. Whether the converse holds is left undecided. Thus a one-one relation is a particular case of a many-one relation.

† A *proper part* (Echter Theil) is a phrase analogous to "proper fraction"; it means a part not the whole.

part of all the chains containing *a*. Then the data of the theorem inform us that *a* is contained in *s*, and, if any term of the chain of *a* be an *s*, so is its successor; and the conclusion is, that every term in the chain of *a* is an *s*. This theorem, as is evident, is very similar to mathematical induction, from which it differs, first by the fact that *a* need not be a single term, secondly by the fact that the constitutive relation need not be one-one, but may be many-one. It is a most remarkable fact that Dedekind's previous assumptions suffice to demonstrate this theorem.

239. (5) I come next to the definition of a singly infinite system or class (71). This is defined as a class which can be represented in itself by means of a one-one relation, and which is further such as to be the chain, with regard to this one-one relation, of a single term of the class not contained in the image of the class. Calling the class *N*, and the one-one relation *R*, there are, as Dedekind remarks, four points in this definition. (1) The image of *N* is contained in *N*; that is, every term to which an *N* has the relation *R* is an *N*. (2) *N* is the chain of one of its terms. (3) This one term is such that no *N* has the relation *R* to it, *i.e.* it is not the image of any other term of *N*. (4) The relation *R* is one-one, in other words, the representation is similar. The abstract system, defined simply as possessing these properties, is defined by Dedekind as the ordinal numbers (73). It is evident that his singly infinite system is the same as what we called a *progression*, and he proceeds to deduce the various properties of progressions, in particular mathematical induction (80), which follows from the above generalized form. One number *m* is said to be less than another *n*, when the chain of *n* is contained in the image of the chain of *m* (89); and it is shown (88, 90) that of two different numbers, one must be the less. From this point everything proceeds simply.

240. The only further point that seems important for our present purpose is the definition of cardinals. It is shown (132) that all singly infinite systems are similar to each other and to the ordinals, and that conversely (133) any system which is similar to a singly infinite system is singly infinite. When a system is finite, it is similar to some system Z_n, where Z_n means all the numbers from 1 to *n* both inclusive; and *vice versâ* (160). There is only one number *n* which has this property in regard to any given finite system, and when considered in relation to this property it is called a *cardinal number*, and is said to be the number of elements of which the said system consists (161). Here at last we reach the cardinal numbers. Their dependence on ordinals, if I may venture to interpret Dedekind, is as follows: owing to the order of the ordinals, every ordinal *n* defines a class of ordinals Z_n, consisting of all that do not succeed it. They may be defined as all that are not contained in the image of the chain of *n*. This class of ordinals may be similar to another class, which is then said to have the

cardinal number *n*. But it is only because of the order of the ordinals
that each of them defines a class, and thus this order is presupposed in
obtaining cardinals.

241. Of the merits of the above deduction it is not necessary for
me to speak, for they are universally acknowledged. But some points
call for discussion. In the first case, Dedekind *proves* mathematical
induction, while Peano regards it as an axiom. This gives Dedekind
an apparent superiority, which must be examined. In the second place,
there is no reason, merely because the numbers which Dedekind obtains
have an order, to hold that they *are* ordinal numbers; in the third
place, his definition of cardinals is unnecessarily complicated, and the
dependence of cardinals upon order is only apparent. I shall take these
points in turn.

As regards the proof of mathematical induction, it is to be observed
that it makes the practically equivalent assumption that numbers form
the chain of one of them. Either can be deduced from the other, and
the choice as to which is to be an axiom, which a theorem, is mainly
a matter of taste. On the whole, though the consideration of chains
is most ingenious, it is somewhat difficult, and has the disadvantage
that theorems concerning the finite class of numbers not greater than *n*
as a rule have to be deduced from corresponding theorems concerning
the infinite class of numbers greater than *n*. For these reasons, and
not because of any logical superiority, it seems simpler to begin with
mathematical induction. And it should be observed that, in Peano's
method, it is only when theorems are to be proved concerning *any*
number that mathematical induction is required. The elementary
Arithmetic of our childhood, which discusses only particular numbers,
is wholly independent of mathematical induction; though to prove that
this is so for *every* particular number would itself require mathematical
induction. In Dedekind's method, on the other hand, propositions
concerning particular numbers, like general propositions, demand the
consideration of chains. Thus there is, in Peano's method, a distinct
advantage of simplicity, and a clearer separation between the particular
and the general propositions of Arithmetic. But from a purely logical
point of view, the two methods seem equally sound; and it is to be
remembered that, with the logical theory of cardinals, both Peano's and
Dedekind's axioms become demonstrable*.

242. On the second point, there is some deficiency of clearness in
what Dedekind says. His words are (73): "If in the contemplation
of a singly infinite system *N*, ordered by a representation ϕ, we disregard
entirely the peculiar nature of the elements, retaining only the possibility
of distinguishing them, and considering only the relations in which they
are placed by the ordering representation ϕ, then these elements are
called *natural numbers* or *ordinal numbers* or simply *numbers*." Now

* Cf. Chap. xiii.

it is impossible that this account should be quite correct. For it implies that the terms of all progressions other than the ordinals are complex, and that the ordinals are elements in all such terms, obtainable by abstraction. But this is plainly not the case. A progression can be formed of points or instants, or of transfinite ordinals, or of cardinals, in which, as we shall shortly see, the ordinals are not elements. Moreover it is impossible that the ordinals should be, as Dedekind suggests, nothing but the terms of such relations as constitute a progression. If they are to be anything at all, they must be intrinsically something; they must differ from other entities as points from instants, or colours from sounds. What Dedekind intended to indicate was probably a definition by means of the principle of abstraction, such as we attempted to give in the preceding chapter. But a definition so made always indicates some class of entities having (or being) a genuine nature of their own, and not logically dependent upon the manner in which they have been defined. The entities defined should be visible, at least to the mind's eye; what the principle asserts is that, under certain conditions, there are such entities, if only we knew where to look for them. But whether, when we have found them, they will be ordinals or cardinals, or even something quite different, is not to be decided off-hand. And in any case, Dedekind does not show us what it is that all progressions have in common, nor give any reason for supposing it to be the ordinal numbers, except that all progressions obey the same laws as ordinals do, which would prove equally that *any* assigned progression is what all progressions have in common.

243. This brings us to the third point, namely the definition of cardinals by means of ordinals. Dedekind remarks in his preface (p. ix) that many will not recognize their old friends the natural numbers in the shadowy shapes which he introduces to them. In this, it seems to me, the supposed persons are in the right—in other words, I am one among them. What Dedekind presents to us is not the numbers, but any progression: what he says is true of all progressions alike, and his demonstrations nowhere—not even where he comes to cardinals—involve any property distinguishing numbers from other progressions. No evidence is brought forward to show that numbers are prior to other progressions. We are told, indeed, that they are what all progressions have in common; but no reason is given for thinking that progressions have anything in common beyond the properties assigned in the definition, which do not themselves constitute a new progression. The fact is that all depends upon one-one relations, which Dedekind has been using throughout without perceiving that they alone suffice for the definition of cardinals. The relation of similarity between classes, which he employs consciously, combined with the principle of abstraction, which he implicitly assumes, suffice for the definition of cardinals; for the definition of ordinals these do not suffice; we require, as we saw in the preceding chapter, the relation of likeness between well-ordered serial relations. The definition of particular finite ordinals is effected explicitly in terms of the corresponding cardinals: if n be a finite cardinal number, the ordinal number n is the class of serial relations which have n terms in their domain (or in their field, if we prefer this definition). In order to define the notion of "nth," we need, besides the ordinal number n, the notion of powers of a relation, *i.e.* of the relative product of a relation multiplied into itself a finite number of times. Thus if R be any one-one serial relation, generating a finite series or a progression, the first term of the field of R (which field we will call r) is the term belonging to the domain, but not to the converse domain, *i.e.*, having the relation R but not the relation $\breve{R}$. If r has n or more terms, where n is a finite number, the nth term of r is the term to which the first term has the relation R^{n-1}, or, again, it is the term having the relation $\breve{R}^{n-1}$ but not the relation $\breve{R}^{n}$. Through the notion of powers of a relation, the introduction of cardinals is here unavoidable; and as powers are defined by mathematical induction, the notion of nth, according to the above definition, cannot be extended beyond finite numbers. We can however extend the notion by the following definition: If P be a transitive aliorelative generating a well-ordered series p, the nth term of p is the term x such that, if P' be the relation P limited to x and its predecessors, then P' has the ordinal number n. Here the dependence upon cardinals results from the fact that the ordinal n can, in general, only be defined by means of the cardinal n.

It is important to observe that no set of terms has inherently one order rather than another, and that no term is the nth of a set except in relation to a particular generating relation whose field is the set or part of the set. For example, since in any progression, any finite number of consecutive terms including the first may be taken away, and the remainder will still form a progression, the ordinal number of a term in a progression may be diminished to any smaller number we choose. Thus the ordinal number of a term is relative to the series to which it belongs. This may be reduced to a relation to the first term of the series; and lest a vicious circle should be suspected, it may be explained that the *first* term can always be defined non-numerically. It is, in Dedekind's singly infinite system, the only term not contained in the image of the system; and generally, in any series, it is the only term which has the constitutive relation with one sense, but not with the other*. Thus the relation expressed by nth is not only a relation to n, but also to the first term of the series; and *first* itself depends

---

* Though when the series has two ends, we have to make an arbitrary selection as to which we will call first, which last. The obviously non-numerical nature of *last* illustrates that of its correlative, *first*.

upon the terms included in the series, and upon the relation by which they are ordered, so that what was first may cease to be so, and what was not first may become so. Thus the first term of a series must be assigned, as is done in Dedekind's view of a progression as the chain of its first term. Hence *n*th expresses a four-cornered relation, between the term which is *n*th, an assigned term (the first), a generating serial relation, and the cardinal number *n*. Thus it is plain that ordinals, either as classes of like serial relations, or as notions like "*n*th," are more complex than cardinals; that the logical theory of cardinals is wholly independent of the general theory of progressions, requiring independent development in order to show that the cardinals form a progression; and that Dedekind's ordinals are not essentially either ordinals or cardinals, but the members of any progression whatever. I have dwelt on this point, as it is important, and my opinion is at variance with that of most of the best authorities. If Dedekind's view were correct, it would have been a logical error to begin, as this work does, with the theory of cardinal numbers rather than with order. For my part, I do not hold. it an absolute error to begin with order, since the properties of progressions, and even most of the properties of series in general, seem to be largely independent of number. But the properties of number must be capable of proof without appeal to the general properties of progressions, since cardinal numbers can be independently defined, and must be seen to form a progression before theorems concerning progressions can be applied to them. Hence the question, whether to begin with order or with numbers, resolves itself into one of convenience and simplicity; and from this point of view, the cardinal numbers seem naturally to precede the very difficult considerations as to series which have occupied us in the present Part.

CHAPTER XXXI.

DISTANCE.

244. THE notion of distance is one which is often supposed essential to series*, but which seldom receives precise definition. An emphasis on distance characterizes, generally speaking, those who believe in relative position. Thus Leibniz, in the course of his controversy with Clarke, remarks:

"As for the objection, that space and time are quantities, or rather things endowed with quantity, and that situation and order are not so: I answer, that order also has its quantity; there is that in it which goes before, and that which follows; there is distance or interval. Relative things have their quantity, as well as absolute ones. For instance, ratios or proportions in mathematics have their quantity, and are measured by logarithms; and yet they are relations. And therefore, though time and space consist in relations, yet they have their quantity†."

In this passage, the remark: "There is that which goes before, and that which follows; there is distance or interval," if considered as an inference, is a *non sequitur*; the mere fact of order does *not* prove that there is distance or interval. It proves, as we have seen, that there are stretches, that these are capable of a special form of addition closely analogous to what I have called relational addition, that they have sign, and that (theoretically at least) stretches which fulfil the axioms of Archimedes and of linearity are always capable of numerical measurement. But the idea, as Meinong rightly points out, is entirely distinct from that of stretch. Whether any particular series does or does not contain distances, will be, in most compact series (*i.e.* such as have a term between any two), a question not to be decided by argument. In discrete series there must be distance; in others, there may be—unless, indeed, they are series obtained from progressions as the rationals or the real numbers are obtained from the integers, in which

---

* *E.g.* by Meinong, *op. cit.* § 17.
† *Phil. Werke*, Gerhardt's ed. Vol. VII, p. 404.

case there must be distance. But we shall find that stretches are mathematically sufficient, and that distances are complicated and unimportant.

245. The definition of distance, to begin with, is no easy matter. What has been done hitherto towards this end is chiefly due to non-Euclidean Geometry*; something also has been done towards settling the definition by Meinong†. But in both these cases, there is more concern for numerical measurement of distance than for its actual definition. Nevertheless, distance is by no means indefinable. Let us endeavour to generalize the notion as much as possible. In the first place, distance need not be asymmetrical; but the other properties of distance always allow us to render it so, and we may therefore take it to be so. Secondly, a distance need not be a quantity or a magnitude; although it is usually taken to be such, we shall find the taking it so to be irrelevant to its other properties, and in particular to its numerical measurement. Thirdly, when distance is taken asymmetrically, there must be only one term to which a given term has a given distance, and the converse relation to the given distance must be a distance of the same kind. (It will be observed that we must first define a *kind* of distance, and proceed thence to the general definition of distance.) Thus every distance is a one-one relation; and in respect to such relations it is convenient to respect the converse of a relation as its –1th power. Further the relative product of two distances of a kind must be a distance of the same kind. When the two distances are mutually converse, their product will be identity, which is thus one among distances (their zero, in fact), and must be the only one which is not asymmetrical. Again the product of two distances of a kind must be commutative‡. If the distances of a kind be magnitudes, they must form a kind of magnitude—*i.e.* any two must be equal or unequal. If they are not magnitudes, they must still form a series generated in the second of our six ways, *i.e.* every pair of different distances must have a certain asymmetrical relation, the same for all pairs except as regards sense. And finally, if Q be this relation, and $R_1 QR_2$ (R_1, R_2, being distances of the kind), then if R_3 be any other distance of the kind, we must have $R_1 R_3 QR_2 R_3$. All these properties, so far as I can discover, are independent; and we ought to add a property of the field, namely this: any two terms, each of which belongs to the field of some distance of the kind (not necessarily the same for both), have a relation which is a distance of the kind. Having now defined a kind of distance, a distance is any relation belonging to some kind of distance; and thus the work of definition seems completed.

The notion of distance, it will be seen, is enormously complex. The properties of distances are analogous to those of stretches with sign, but

* See *e.g.* Whitehead, *Universal Algebra*, Cambridge, 1898, Book vi, Chap. i.
† *Op. cit.* Section iv.
‡ This is an independent property; consider for instance the difference between "maternal grandfather" and "paternal grandmother."

are far less capable of mutual deduction. The properties of stretches corresponding to many of the above properties of distances are capable of proof. The difference is largely due to the fact that stretches can be added in the elementary logical (not arithmetical) way, whereas distances require what I have called *relational* addition, which is much the same as relative multiplication.

246. The numerical measurement of distances has already been partially explained in Part III. It requires, as we saw, for its full application, two further postulates, which, however, do not belong to the definition of distances, but to certain kinds of distances only. These are, the postulate of Archimedes: given any two distances of a kind, there exists a finite integer n such that the nth power of the first distance is greater than the second distance; and Du Bois Reymond's postulate of linearity: Any distance has an nth root, where n is any integer (or any prime, whence the result follows for any integer). When these two postulates are satisfied, we can find a meaning for R^x, where R is a distance of the kind other than identity, and x is any real number*. Moreover, any distance of the kind is of the form R^x, for some value of x. And x is, of course, the numerical measure of the distance.

In the case of series generated in the first of our six ways, the various powers of the generating relation R give the distances of terms. These various powers, as the reader can see for himself, verify all the above characteristics of distances. In the case of series generated from progressions as rationals or real numbers from integers, there are always distances; thus in the case of the rationals themselves, which are one-one relations, their differences, which are again rationals, measure or indicate relations between them, and these relations are of the nature of distances. And we shall see, in Part V, that these distances have some importance in connection with limits. For numerical measurement in some form is essential to certain theorems about limits, and the numerical measurement of distances is apt to be more practically feasible than that of stretches.

247. On the general question, however, whether series unconnected with number—for instance spatial and temporal series—are such as to contain distances, it is difficult to speak positively. Some things may be said against this view. In the first place, there must be stretches, and these must be magnitudes. It then becomes a sheer assumption—which must be set up as an axiom—that equal stretches correspond to equal distances. This may, of course, be denied, and we might even seek an

---

* The powers of distances are here understood in the sense resulting from relative multiplication; thus if a and b have the same distance as b and c, this distance is the square root of the distance of a and c. The postulate of linearity, whose expression in ordinary language is: "every linear quantity can be divided into n equal parts, where n is any integer," will be found in Du Bois Reymond's *Allgemeine Functionentheorie* (Tübingen, 1882), p. 46.

interpretation of non-Euclidean Geometry in the denial. We might regard the usual coordinates as expressing stretches, and the logarithms of their anharmonic ratios as expressing distances; hyperbolic Geometry, at least, might thus find a somewhat curious interpretation. Herr Meinong, who regards all series as containing distances, maintains an analogous principle with regard to distance and stretch in general. The distance, he thinks, increases only as the logarithm of the stretch. It may be observed that, where the distance itself is a rational number (which is possible, since rationals are one-one relations), the opposite theory can be made formally convenient by the following fact. The square of a distance, as we saw generally, is said to be twice as great as the distance whose square it is. We might, where the distance is a rational, say instead that the *stretch* is twice as great, but that the *distance* is truly the square of the former distance. For where the distance is already numerical, the usual interpretation of numerical measurement conflicts with the notation R^2. Thus we shall be compelled to regard the stretch as proportional to the logarithm of the distance. But since, outside the theory of progressions, it is usually doubtful whether there are distances, and since, in almost all other series, stretches seem adequate for all the results that are obtainable, the retention of distance adds a complication for which, as a rule, no necessity appears. It is therefore generally better, at least in a philosophy of mathematics, to eschew distances except in the theory of progressions, and to measure them, in that theory, merely by the indices of the powers of the generating relation. There is no logical reason, so far as I know, to suppose that there are distances elsewhere, except in a finite space of two dimensions and in a projective space; and if there are, they are not mathematically important. We shall see in Part VI how the theory of space and time may be developed without presupposing distance; the distances which appear in projective Geometry are derivative relations, not required in defining the properties of our space; and in Part V we shall see how few are the functions of distance with regard to series in general. And as against distance it may be remarked that, if every series must contain distances, an endless regress becomes unavoidable, since every kind of distance is itself a series. This is not, I think, a logical objection, since the regress is of the logically permissible kind; but it shows that great complications are introduced by regarding distances as essential to every series. On the whole, then, it seems doubtful whether distances in general exist; and if they do, their existence seems unimportant and a source of very great complications.

248. We have now completed our review of order, in so far as is possible without introducing the difficulties of continuity and infinity. We have seen that all order involves asymmetrical transitive relations, and that every series as such is open. But closed series, we found, could be distinguished by the mode of their generation, and by the fact that, though they always have a first term, this term may always be selected arbitrarily. We saw that asymmetrical relations must be sometimes unanalyzable, and that when analyzable, other asymmetrical relations must appear in the analysis. The difference of sign, we found, depends always upon the difference between an asymmetrical relation and its converse. In discussing the particular type of series which we called progressions, we saw how all Arithmetic applies to every such series, and how finite ordinals may be defined by means of them. But though we found this theory to be to a certain extent independent of the cardinals, we saw no reason to agree with Dedekind in regarding cardinals as logically subsequent to ordinals. Finally, we agreed that distance is a notion which is not essential to series, and of little importance outside Arithmetic. With this equipment, we shall be able, I hope, to dispose of all the difficulties which philosophers have usually found in infinity and continuity. If this can be accomplished, one of the greatest of philosophical problems will have been solved. To this problem Part V is to be devoted.

PART V.

INFINITY AND CONTINUITY.

CHAPTER XXXII.

THE CORRELATION OF SERIES.

249. WE come now to what has been generally considered the fundamental problem of mathematical philosophy—I mean, the problem of infinity and continuity. This problem has undergone, through the labours of Weierstrass and Cantor, a complete transformation. Since the time of Newton and Leibniz, the nature of infinity and continuity had been sought in discussions of the so-called Infinitesimal Calculus. But it has been shown that this Calculus is not, as a matter of fact, in any way concerned with the infinitesimal, and that a large and most important branch of mathematics is logically prior to it. The problem of continuity, moreover, has been to a great extent separated from that of infinity. It was formerly supposed—and herein lay the real strength of Kant's mathematical philosophy—that continuity had an essential reference to space and time, and that the Calculus (as the word *fluxion* suggests) in some way presupposed motion or at least change. In this view, the philosophy of space and time was prior to that of continuity, the Transcendental Aesthetic preceded the Transcendental Dialectic, and the antinomies (at least the mathematical ones) were essentially spatio-temporal. All this has been changed by modern mathematics. What is called the arithmetization of mathematics has shown that all the problems presented, in this respect, by space and time, are already present in pure arithmetic. The theory of infinity has two forms, cardinal and ordinal, of which the former springs from the logical theory of number; the theory of continuity is purely ordinal. In the theory of continuity and the ordinal theory of infinity, the problems that arise are not specially concerned with numbers, but with all series of certain types which occur in arithmetic and geometry alike. What makes the problems in question peculiarly easy to deal with in the case of numbers is, that the series of rationals, which is what I shall call a *compact* series, arises from a progression, namely that of the integers, and that this fact enables us to give a proper name to *every* term of the series of rationals—a point in which this series differs from others of the same type. But theorems of the kind which will occupy us in most of the following chapters, though obtained in arithmetic, have a far wider application, since they are purely ordinal, and involve none of the logical properties of numbers. That is to say, the idea which the Germans call *Anzahl*, the idea of the number of terms in some class, is irrelevant, save only in the theory of transfinite cardinals—an important but very distinct part of Cantor's contributions to the theory of infinity. We shall find it possible to give a general definition of continuity, in which no appeal is made to the mass of unanalyzed prejudice which Kantians call "intuition"; and in Part VI we shall find that no other continuity is involved in space and time. And we shall find that, by a strict adherence to the doctrine of limits, it is possible to dispense entirely with the infinitesimal, even in the definition of continuity and the foundations of the Calculus.

250. It is a singular fact that, in proportion as the infinitesimal has been extruded from mathematics, the infinite has been allowed a freer development. From Cantor's work it appears that there are two respects in which infinite numbers differ from those that are finite. The first, which applies to both cardinals and ordinals, is, that they do not obey mathematical induction—or rather, they do not form part of a series of numbers beginning with 1 or 0, proceeding in order of magnitude, containing all numbers intermediate in magnitude between any two of its terms, and obeying mathematical induction. The second, which applies only to cardinals, is, that a whole of an infinite number of terms always contains a part consisting of the same number of terms. The first respect constitutes the true definition of an infinite series, or rather of what we may call an infinite term in a series: it gives the essence of the ordinal infinite. The second gives the definition of an infinite collection, and will doubtless be pronounced by the philosopher to be plainly self-contradictory. But if he will condescend to attempt to exhibit the contradiction, he will find that it can only be proved by admitting mathematical induction, so that he has merely established a connection with the ordinal infinite. Thus he will be compelled to maintain that the denial of mathematical induction is self-contradictory; and as he has probably reflected little, if at all, on this subject, he will do well to examine the matter before pronouncing judgment. And when it is admitted that mathematical induction may be denied without contradiction, the supposed antinomies of infinity and continuity one and all disappear. This I shall endeavour to prove in detail in the following chapters.

251. Throughout this Part we shall often have occasion for a notion which has hitherto been scarcely mentioned, namely the correlation of series. In the preceding Part we examined the nature of isolated series, but we scarcely considered the relations between different series. These relations, however, are of an importance which philosophers have wholly overlooked, and mathematicians have but lately realized. It has long been known how much could be done in Geometry by means of homography, which is an example of correlation; and it has been shown by Cantor how important it is to know whether a series is denumerable, and how similar two series capable of correlation are. But it is not usually pointed out that a dependent variable and its independent variable are, in most mathematical cases, merely correlated series, nor has the general idea of correlation been adequately dealt with. In the present work only the philosophical aspects of the subject are relevant.

Two *series* s, s' are said to be correlated when there is a one-one relation R coupling every term of s with a term of s', and *vice versâ*, and when, if x, y be terms of s, and x precedes y, then their correlates x', y' in s' are such that x' precedes y'. Two *classes* or collections are correlated whenever there is a one-one relation between the terms of the one and the terms of the other, none being left over. Thus two series may be correlated as classes without being correlated as series; for correlation as classes involves only the same cardinal number, whereas correlation as series involves also the same ordinal type—a distinction whose importance will be explained hereafter. In order to distinguish these cases, it will be well to speak of the correlation of classes as correlation simply, and of the correlation of series as ordinal correlation. Thus whenever correlation is mentioned without an adjective, it is to be understood as being not necessarily ordinal. Correlated classes will be called *similar*; correlated series will be called *ordinally similar*; and their generating relations will be said to have the relation of *likeness*.

Correlation is a method by which, when one series is given, others may be generated. If there be any series whose generating relation is P, and any one-one relation which holds between any term x of the series and some term which we may call x_R, then the class of terms x_R will form a series of the same type as the class of terms x. For suppose y to be any other term of our original series, and assume xPy. Then we have $x_R \breve{R} x$, xPy, and yRy_R. Hence $x_R \breve{R} PR y_R$. Now it may be shown* that, if P be transitive and asymmetrical, so is $\breve{R}PR$; hence the correlates of terms of the P-series form a series whose generating relation is $\breve{R}PR$. Between these two series there is ordinal correlation, and the series have complete ordinal similarity. In this way a new series, similar to the original one, is generated by any one-one relation whose field includes the original series. It can also be shown that, conversely, if P, P' be the generating relations of two similar series, there is a one-one relation R, whose domain is the field of P, which is such that $P' = \breve{R}PR$.

* See my article in RdM, Vol. VIII, No. 2.

252. We can now understand a distinction of great importance, namely that between self-sufficient or independent series, and series by correlation. In the case just explained there is perfect mathematical symmetry between the original series and the series by correlation; for, if we denote by Q the relation $\breve{R}PR$, we shall find $P = RQ\breve{R}$. Thus we may take either the Q-series or the P-series as the original, and regard the other as derivative. But if it should happen that R, instead of being one-one, is many-one, the terms of the field of Q, which we will call q, will have an order in which there is repetition, the same term occurring in different positions corresponding to its different correlates in the field of P, which we will call p. This is the ordinary case of mathematical functions which are not linear. It is owing to preoccupation with such series that most mathematicians fail to realize the impossibility, in an independent series, of any recurrence of the same term. In every sentence of print, for example, the letters acquire an order by correlation with the points of space, and the same letter will be repeated in different positions. Here the series of letters is essentially derivative, for we cannot order the points of space by relation to the letters: this would give us several points in the same position, instead of one letter in several positions. In fact, if P be a serial relation, and R be a many-one relation whose domain is the field of P, and $Q = \breve{R}PR$, then Q has all the characteristics of a serial relation except that of implying diversity; but $RQ\breve{R}$ is not equivalent to P, and thus there is a lack of symmetry. It is for this reason that inverse functions in mathematics, such as $\sin^{-1} x$, are genuinely distinct from direct functions, and require some device or convention before they become unambiguous. Series obtained from a many-one correlation as q was obtained above will be called series by correlation. They are not genuine series, and it is highly important to eliminate them from discussions of fundamental points.

253. The notion of *likeness* corresponds, among relations, to similarity among classes. It is defined as follows: Two relations P, Q are like when there is a one-one relation S such that the domain of S is the field of P, and $Q = \breve{S}PS$. This notion is not confined to serial relations, but may be extended to all relations. We may define the *relation-number* of a relation P as the class of all relations that are like P; and we can proceed to a very general subject which may be called relation-arithmetic. Concerning relation-numbers we can prove those of the formal laws of addition and multiplication that hold for transfinite ordinals, and thus obtain an extension of a part of ordinal arithmetic to relations in general. By means of likeness we can define a finite relation as one which is not like any proper part of itself—a proper part of a relation being a relation which implies it but is not equivalent to it. In this way we can completely emancipate ourselves from cardinal arithmetic. Moreover the properties of likeness are in themselves interesting and important. One curious property is that, if S be one-one and have the field of P for its domain, the above equation $Q = \breve{S}PS$ is equivalent to $SQ = PS$ or to $Q\breve{S} = \breve{S}P$*.

254. Since the correlation of series constitutes most of the mathematical examples of functions, and since function is a notion which is not often clearly explained, it will be well at this point to say something concerning the nature of this notion. In its most general form, functionality does not differ from relation. For the present purpose it will be well to recall two technical terms, which were defined in Part I. If x has a certain relation to y, I shall call x the *referent*, and y the *relatum*, with regard to the relation in question. If now x be defined as belonging to some class contained in the domain of the relation, then the relation defines y as a function of x. That is to say, an independent variable is constituted by a collection of terms, each of which can be referent in regard to a certain relation. Then each of these terms has one or more relata, and any one of these is a certain function of its referent, the function being defined by the relation. Thus *father* defines a function, provided the independent variable be a class contained in that of male animals who have or will have propagated their kind; and if A be the father of B, B is said to be a function of A. What is essential is an independent variable, *i.e.* any term of some class, and a relation whose extension includes the variable. Then the referent is the independent variable, and its function is any one of the corresponding relata.

But this most general idea of a function is of little use in mathematics. There are two principal ways of particularizing the function: first, we may confine the relations to be considered to such as are one-one or many-one, *i.e.* such as give to every referent a unique relatum; secondly, we may confine the independent variable to series. The second particularization is very important, and is specially relevant to our present topics. But as it almost wholly excludes functions from Symbolic Logic, where series have little importance, we may as well postpone it for a moment while we consider the first particularization alone.

The idea of function is so important, and has been so often considered with exclusive reference to numbers, that it is well to fill our minds with instances of non-numerical functions. Thus a very important class of functions are propositions containing a variable†. Let there be some proposition in which the phrase "any a" occurs, where a is some class. Then in place of "any a" we may put x, where x is an undefined member of the class a—in other words, any a. The proposition then becomes a function of x, which is unique when x is given. This proposition will, in general, be true for some values of x and false for others.

---

* On this subject see my article in RdM, Vol. viii, especially Nos. 2, 6.
† These are what in Part I we called propositional functions.

The values for which the function is true form what might be called, by analogy with Analytic Geometry, a logical curve. This general view may, in fact, be made to include that of Analytic Geometry. The equation of a plane curve, for example, is a propositional function which is a function of two variables x and y, and the curve is the assemblage of points which give to the variables values that make the proposition true. A proposition containing the word *any* is the assertion that a certain propositional function is true for all values of the variable for which it is significant. Thus "any man is mortal" asserts that "x is a man implies x is a mortal" is true for all values of x for which it is significant, which may be called the admissible values. Propositional functions, such as "x is a number," have the peculiarity that they look like propositions, and *seem* capable of implying other propositional functions, while yet they are neither true nor false. The fact is, they are propositions for all admissible values of the variable, but not while the variable remains a variable, whose value is not assigned; and although they may, for every admissible value of the variable, imply the corresponding value of some other propositional function, yet while the variable remains as a variable they can imply nothing. The question concerning the nature of a propositional function as opposed to a proposition, and generally of a function as opposed to its values, is a difficult one, which can only be solved by an analysis of the nature of the variable. It is important, however, to observe that propositional functions, as was shown in Chapter VII, are more fundamental than other functions, or even than relations. For most purposes, it is convenient to identify the function and the relation, i.e., if $y = f(x)$ is equivalent to xRy, where R is a relation, it is convenient to speak of R as the function, and this will be done in what follows; the reader, however, should remember that the idea of functionality is more fundamental than that of relation. But the investigation of these points has been already undertaken in Part I, and enough has been said to illustrate how a proposition may be a function of a variable.

Other instances of non-numerical functions are afforded by dictionaries. The French for a word is a function of the English, and *vice versâ*, and both are functions of the term which both designate. The press-mark of a book in a library catalogue is a function of the book, and a number in a cipher is a function of the word for which it stands. In all these cases there is a relation by which the relatum becomes unique (or, in the case of languages, generally unique) when the referent is given; but the terms of the independent variable do not form a series, except in the purely external order resulting from the alphabet.

255. Let us now introduce the second specification, that our independent variable is to be a series. The dependent variable is then a series by correlation, and may be also an independent series. For example, the positions occupied by a material point at a series of instants form a series by correlation with the instants, of which they are a function ; but in virtue of the continuity of motion, they also form, as a rule, a geometrical series independent of all reference to time. Thus motion affords an admirable example of the correlation of series. At the same time it illustrates a most important mark by which, when it is present, we can tell that a series is not independent. When the time is known, the position of a material particle is uniquely determined; but when the position is given, there may be several moments, or even an infinite number of them, corresponding to the given position. (There will be an infinite number of such moments if, as is commonly said, the particle has been at rest in the position in question. *Rest* is a loose and ambiguous expression, but I defer its consideration to Part VII.) Thus the relation of the time to the position is not strictly one-one, but may be many-one. This was a case considered in our general account of correlation, as giving rise to dependent series. We inferred, it will be remembered, that two correlated independent series are mathematically on the same level, because if P, Q be their generating relations, and R the correlating relation, we infer $P = RQ\breve{R}$ from $Q = \breve{R}PR$. But this inference fails as soon as R is not strictly one-one, since then we no longer have $R\breve{R}$ contained in 1', where 1' means identity. For example, my father's son need not be myself, though my son's father must be. This illustrates the fact that, if R be a many-one relation, $R\breve{R}$ and $\breve{R}R$ must be carefully distinguished : the latter is contained in identity, but not the former. Hence whenever R is a many-one relation, it may be used to form a series by correlation, but the series so formed cannot be independent. This is an important point, which is absolutely fatal to the relational theory of time*. For the present let us return to the case of motion. When a particle describes a closed curve, or one which has double points, or when the particle is sometimes at rest during a finite time, then the series of points which it occupies is essentially a series by correlation, not an independent series. But, as I remarked above, a curve is not only obtainable by motion, but is also a purely geometrical figure, which can be defined without reference to any supposed material point. When, however, a curve is so defined, it must not contain points of rest: the path of a material point which sometimes moves, but is sometimes at rest for a finite time, is different when considered kinematically and when considered geometrically; for geometrically the point in which there is rest is one, whereas kinematically it corresponds to many terms in the series.

The above discussion of motion illustrates, in a non-numerical instance, a case which normally occurs among the functions of pure mathematics. These functions (when they are functions of a real

---

* See my article "Is position in Time and Space absolute or relative?" *Mind,* July 1901.

variable) usually fulfil the following conditions: Both the independent and the dependent variable are classes of numbers, and the defining relation of the function is many-one*. This case covers rational functions, circular and elliptic functions of a real variable, and the great majority of the direct functions of pure mathematics. In all such cases, the independent variable is a series of numbers, which may be restricted in any way we please—to positive numbers, rationals, integers, primes, or any other class. The dependent variable consists also of numbers, but the order of these numbers is determined by their relation to the corresponding term of the independent variable, not by that of the numbers forming the dependent variable themselves. In a large class of functions the two orders happen to coincide; in others, again, where there are maxima and minima at finite intervals, the two orders coincide throughout a finite stretch, then they become exactly opposite throughout another finite stretch, and so on. If x be the independent variable, y the dependent variable, and the constitutive relation be many-one, the same number y will, in general, be a function of, *i.e.* correspond to, several numbers x. Hence the y-series is essentially by correlation, and cannot be taken as an independent series. If, then, we wish to consider the inverse function, which is defined by the converse relation, we need certain devices if we are still to have correlation of series. One of these, which seems the most important, consists in dividing the values of x corresponding to the same value of y into classes, so that (what may happen) we can distinguish (say) n different x's, each of which has a distinct one-one relation to y, and is therefore simply reversible. This is the usual course, for example, in distinguishing positive and negative square roots. It is possible wherever the generating relation of our original function is formally capable of exhibition as a disjunction of one-one relations. It is plain that the disjunctive relation formed of n one-one relations, each of which contains in its domain a certain class u, will, throughout the class u, be an n-one relation. Thus it may happen that the independent variable can be divided into n classes, within each of which the defining relation is one-one, *i.e.* within each of which there is only one x having the defining relation to a given y. In such cases, which are usual in pure mathematics, our many-one relation can be made into a disjunction of one-one relations, each of which separately is reversible. In the case of complex functions, this is, *mutatis mutandis*, the method of Riemann surfaces. But it must be clearly remembered that, where our function is not naturally one-one, the y which appears as dependent variable is ordinally distinct from the y which appears as independent variable in the inverse function.

The above remarks, which will receive illustration as we proceed,

---

* I omit for the present complex variables, which, by introducing dimensions, lead to complications of an entirely distinct kind.

have shown, I hope, how intimately the correlation of series is associated
with the usual mathematical employment of functions. Many other
cases of the importance of correlation will meet us as we proceed. It
may be observed that every denumerable class is related by a one-valued
function to the finite integers, and *vice versâ*. As ordered by correlation
with the integers, such a class becomes a series having the type of order
which Cantor calls ω. The fundamental importance of correlation to
Cantor's theory of transfinite numbers will appear when we come to the
definition of the transfinite ordinals.

256. In connection with functions, it seems desirable to say something concerning the necessity of a formula for definition. A function
was originally, after it had ceased to be merely a power, essentially
something that could be expressed by a formula. It was usual to start
with some expression containing a variable x, and to say nothing to
begin with as to what x was to be, beyond a usually tacit assumption
that x was some kind of number. Any further limitations upon x were
derived, if at all, from the formula itself; and it was mainly the desire
to remove such limitations which led to the various generalizations of
number. This algebraical generalization* has now been superseded by
a more ordinal treatment, in which all classes of numbers are defined by
means of the integers, and formulae are not relevant to the process.
Nevertheless, for the use of functions, where both the independent and
the dependent variables are infinite classes, the formula has a certain
importance. Let us see what is its definition.

A formula, in its most general sense, is a proposition, or more
properly a propositional function, containing one or more variables,
a variable being any term of some defined class, or even any term
without restriction. The kind of formula which is relevant in connection
with functions of a single variable is a formula containing two variables.
If both variables are defined, say one as belonging to the class u,
the other as belonging to the class v, the formula is true or false. It is
true if every u has to every v the relation expressed by the formula;
otherwise it is false. But if one of the variables, say x, be defined as
belonging to the class u, while the other, y, is only defined by the
formula, then the formula may be regarded as defining y as a function
of x. Let us call the formula P_{xy}. If in the class u there are terms x
such that there is no term y which makes P_{xy} a true proposition, then
the formula, as regards those terms, is impossible. We must therefore
assume that u is a class every term of which will, for a suitable value
of y, make the proposition P_{xy} true. If, then, for every term x of u,
there are some entities y, which make P_{xy} true, and others which do not
do so, then P_{xy} correlates to every x a certain class of terms y. In
this way y is defined as a function of x.

* Of which an excellent account will be found in Couturat, *De l'Infini Mathématique*, Paris, 1896, Part I, Book II.

But the usual meaning of *formula* in mathematics involves another element, which may also be expressed by the word *law*. It is difficult to say precisely what this element is, but it seems to consist in a certain degree of intensional simplicity of the proposition P_{xy}. In the case of two languages, for example, it would be said that there is no formula connecting them, except in such cases as Grimm's law. Apart from the dictionary, the relation which correlates words in different languages is sameness of meaning; but this gives no method by which, given a word in one language, we can infer the corresponding word in the other. What is absent is the possibility of calculation. A formula, on the other hand (say $y = 2x$), gives the means, when we know x, of discovering y. In the case of languages, only enumeration of all pairs will define the dependent variable. In the case of an algebraical formula, the independent variable and the relation enable us to know all about the dependent variable. If functions are to extend to infinite classes, this state of things is essential, for enumeration has become impossible. It is therefore essential to the correlation of infinite classes, and to the study of functions of infinite classes, that the formula P_{xy} should be one in which, given x, the class of terms y satisfying the formula should be one which we can discover. I am unable to give a logical account of this condition, and I suspect it of being purely psychological. Its practical importance is great, but its theoretical importance seems highly doubtful.

There is, however, a logical condition connected with the above, though perhaps not quite identical with it. Given any two terms, there is some relation which holds between those two terms and no others. It follows that, given any two classes of terms u, v, there is a disjunctive relation which any one term of u has to at least one term of v, and which no term not belonging to u has to any term. By this method, when two classes are both finite, we can carry out a correlation (which may be one-one, many-one, or one-many) which correlates terms of these classes and no others. In this way any set of terms is theoretically a function of any other; and it is only thus, for example, that diplomatic ciphers are made up. But if the number of terms in the class constituting the independent variable be infinite, we cannot in this way practically define a function, unless the disjunctive relation consists of relations developed one from the other by a law, in which case the formula is merely transferred to the relation. This amounts to saying that the defining relation of a function must not be infinitely complex, or, if it be so, must be itself a function defined by some relation of finite complexity. This condition, though it is itself logical, has again, I think, only psychological necessity, in virtue of which we can only master the infinite by means of a law of order. The discussion of this point, however, would involve a discussion of the relation of infinity to order—a question which will be resumed later, but which we are not yet in a position to treat intelligently. In any case, we may say that a formula containing two variables and defining a function must, if it is to be practically useful, give a relation between the two variables by which, when one of them is given, all the corresponding values of the other can be found; and this seems to constitute the mathematical essence of all formulae.

257. There remains an entirely distinct logical notion of much importance in connection with limits, namely the notion of a complete series. If R be the defining relation of a series, the series is complete when there is a term x belonging to the series, such that every other term which has to x either the relation R or the relation $\breve{R}$ belongs to the series. It is *connected* (as was explained in Part IV) when no other terms belong to the series. Thus a complete series consists of those terms, and only those terms, which have the generating relation or its converse to some one term, together with that one term. Since the generating relation is transitive, a series which fulfils this condition for one of its terms fulfils it for all of them. A series which is connected but not complete will be called incomplete or partial. Instances of complete series are the cardinal integers, the positive and negative integers and zero, the rational numbers, the moments of time, or the points on a straight line. Any selection from such a series is incomplete with respect to the generating relations of the above complete series. Thus the positive numbers are an incomplete series, and so are the rationals between 0 and 1. When a series is complete, no term can come before or after any term of the series without belonging to the series; when the series is incomplete, this is no longer the case. A series may be complete with respect to one generating relation, but not with respect to another. Thus the finite integers are a complete series when the series is defined by powers of the relation of consecutiveness, as in the discussion of progressions in Part IV; but when they are ordered by correlation with whole and part, they form only part of the series of finite and transfinite integers, as we shall see hereafter. A complete series may be regarded as the extension of a term with respect to a given relation, together with this term itself; and owing to this fact it has, as we shall find, some important differences from ordinally similar incomplete series. But it can be shown, by the Logic of Relations, that any incomplete series can be rendered complete by a change in the generating relation, and *vice versâ*. The distinction between complete and incomplete series is, therefore, essentially relative to a given generating relation.

CHAPTER XXXIII.

REAL NUMBERS.

258. THE philosopher may be surprised, after all that has already been said concerning numbers, to find that he is only now to learn about *real* numbers; and his surprise will be turned to horror when he learns that *real* is opposed to *rational*. But he will be relieved to learn that real numbers are really not numbers at all, but something quite different.

The series of real numbers, as ordinarily defined, consists of the whole assemblage of rational and irrational numbers, the irrationals being defined as the limits of such series of rationals as have neither a rational nor an infinite limit. This definition, however, introduces grave difficulties, which will be considered in the next chapter. For my part I see no reason whatever to suppose that there are any irrational numbers in the above sense; and if there are any, it seems certain that they cannot be greater or less than rational numbers. When mathematicians have effected a generalization of number they are apt to be unduly modest about it—they think that the difference between the generalized and the original notions is less than it really is. We have already seen that the finite cardinals are not to be identified with the positive integers, nor yet with the ratios of the natural numbers to 1, both of which express relations, which the natural numbers do not. In like manner there is a real number associated with every rational number, but distinct from it. A real number, so I shall contend, is nothing but a certain class of rational numbers. Thus the class of rationals less than ½ is a real number, associated with, but obviously not identical with, the rational number ½. This theory is not, so far as I know, explicitly advocated by any other author, though Peano suggests it, and Cantor comes very near to it*. My grounds in favour of this opinion are, first, that such classes of rationals have all the mathematical properties commonly assigned to real numbers, secondly, that the opposite theory presents logical difficulties which appear to me insuperable. The second point will be discussed in the next chapter; for the present I shall merely expound

---

* Cf. Cantor, *Math. Annalen*, Vol. XLVI, § 10; Peano, *Rivista di Matematica*, Vol. VI, pp. 126–140, esp. p. 133.

my own view, and endeavour to show that real numbers, so understood, have all the requisite characteristics. It will be observed that the following theory is independent of the doctrine of limits, which will only be introduced in the next chapter.

259. The rational numbers in order of magnitude form a series in which there is a term between any two. Such series, which in Part III we provisionally called continuous, must now receive another name, since we shall have to reserve the word *continuous* for the sense which Cantor has given to it. I propose to call such series *compact* *. The rational numbers, then, form a compact series. It is to be observed that, in a compact series, there are an infinite number of terms between any two, there are no consecutive terms, and the stretch between any two terms (whether these be included or not) is again a compact series. If now we consider any one rational number†, say r, we can define, by relation to r, four infinite classes of rationals: (1) those less than r, (2) those not greater than r, (3) those greater than r, (4) those not less than r. (2) and (4) differ from (1) and (3) respectively solely by the fact that the former contain r, while the latter do not. But this fact leads to curious differences of properties. (2) has a last term, while (1) has none ; (1) is identical with the class of rational numbers less than a variable term of (1), while (2) does not have this characteristic. Similar remarks apply to (3) and (4), but these two classes have less importance in the present case than in (1) and (2). Classes of rationals having the properties of (1) are called *segments*. A segment of rationals may be defined as a class of rationals which is not null, nor yet coextensive with the rationals themselves (*i.e.* which contains some but not all rationals), and which is identical with the class of rationals less than a (variable) term of itself, *i.e.* with the class of rationals x such that there is a rational y of the said class such that x is less than y‡. Now we shall find that segments are obtained by the above method, not only from single rationals, but also from finite or infinite classes of rationals, with the proviso, for infinite classes, that there must be some rational greater than any member of the class. This is very simply done as follows.

Let u be any finite or infinite class of rationals. Then four classes may be defined by relation to u§, namely (1) those less than every u, (2) those less than a variable u, (3) those greater than every u, (4) those greater than a variable u, *i.e.* those such that for each a term of u can be found which is smaller than it. If u be a finite class, it must have a maximum and a minimum term ; in this case the former alone is relevant to (2) and (3), the latter alone to (1) and (4). Thus this case is reduced to the former, in which we had only a single rational. I shall therefore

---

* Such series are called by Cantor *überall dicht*.

† I shall for simplicity confine myself entirely to rationals without sign. The extension to such as are positive or negative presents no difficulty whatever.

‡ See *Formulaire de Mathématiques*, Vol. ii, Part iii, § 61 (Turin, 1899).

§ Eight classes *may* be defined, but four are all that we need.

assume in future that u is an infinite class, and further, to prevent reduction to our former case, I shall assume, in considering (2) and (3), that u has no maximum, that is, that every term of u is less than some other term of u; and in considering (1) and (4), I shall assume that u has no minimum. For the present I confine myself to (2) and (3), and I assume, in addition to the absence of a maximum, the existence of rationals greater than any u, that is, the existence of the class (3). Under these circumstances, the class (2) will be a segment. For (2) consists of all rationals which are less than a variable u; hence, in the first place, since u has no maximum, (2) contains the whole of u. In the second place, since every term of (2) is less than some u, which in turn belongs to (2), every term of (2) is less than some other term of (2); and every term less than some term of (2) is *a fortiori* less than some u, and is therefore a term of (2). Hence (2) is identical with the class of terms less than some term of (2), and is therefore a segment.

Thus we have the following conclusion: If u be a single rational, or a class of rationals all of which are less than some fixed rational, then the rationals less than u, if u be a single term, or less than a variable term of u, if u be a class of terms, always form a segment of rationals. My contention is, that a segment of rationals *is* a real number.

260. So far, the method employed has been one which may be employed in any compact series. In what follows, some of the theorems will depend upon the fact that the rationals are a denumerable series. I leave for the present the disentangling of the theorems dependent upon this fact, and proceed to the properties of segments of rationals.

Some segments, as we have seen, consist of the rationals less than some given rational. Some, it will be found, though not so defined, are nevertheless capable of being so defined. For example, the rationals less than a variable term of the series ·9, ·99, ·999, *etc.*, are the same as the rationals less than 1. But other segments, which correspond to what are usually called irrationals, are incapable of any such definition. How this fact has led to irrationals we shall see in the next chapter. For the present I merely wish to point out the well-known fact that segments are not capable of a one-one correlation with rationals. There are classes of rationals defined as being composed of all terms less than a *variable* term of an infinite class of rationals, which are not definable as all the rationals less than some one definite rational*. Moreover there are more segments than rationals, and hence the series of segments has continuity of a higher order than the rationals. Segments form a series in virtue of the relation of whole and part, or of logical inclusion (excluding identity). Any two segments are such that one of them is wholly contained in the other, and in virtue of this fact they form a series. It can be easily shown that they form a compact series. What is more remarkable is this: if we apply the above process to the

---

* Cf. Part I, chap. v, p. 60.

series of segments, forming segments of segments by reference to classes of segments, we find that every segment of segments can be defined as all segments contained in a certain definite segment. Thus the segment of segments defined by a class of segments is always identical with the segment of segments defined by some one segment. Also every segment defines a segment of segments which can be defined by an infinite class of segments. These two properties render the series of segments *perfect*, in Cantor's language; but the explanation of this term must be left till we come to the doctrine of limits.

We might have defined our segments as all rationals greater than some term of a class u of rationals. If we had done this, and inserted the conditions that u was to have no minimum, and that there were to be rationals less than every u, we should have obtained what may be called upper segments, as distinguished from the former kind, which may be called lower segments. We should then have found that, corresponding to every upper segment, there is a lower segment which contains all rationals not contained in the upper segment, with the occasional exception of a single rational. There will be one rational not belonging to either the upper or the lower segment, when the upper segment can be defined as all rationals greater than a single rational. In this case, the corresponding lower segment will consist of all rationals less than this single rational, which will itself belong to neither segment. Since there is a rational between any two, the class of rationals not greater than a given rational cannot ever be identical with the class of rationals less than some other; and a class of rationals having a maximum can never be a segment. Hence it is impossible, in the case in question, to find a lower segment containing all the rationals not belonging to the given upper segment. But when the upper segment cannot be defined by a single rational, it will always be possible to find a lower segment containing *all* rationals not belonging to the upper segment.

Zero and infinity may be introduced as limiting cases of segments, but in the case of zero the segment must be of the kind which we called (1) above, not of the kind (2) hitherto discussed. It is easy to construct a class of rationals such that some term of the class will be less than any given rational. In this case, the class (1) will contain no terms, and will be the null-class. This is the real number zero, which, however, is not a segment, since a segment was defined as a class which is not null. In order to introduce zero as a class of the kind which we called (2), we should have to start with a null class of rationals. No rational is less than a term of a null class of rationals, and thus the class (2), in such a case, is null. Similarly the real number infinity may be introduced. This is identical with the whole class of rationals. If we have any class u of rationals such that no rational is greater than all u's, then every rational is contained in the class of rationals less than some

u. Or again, if we have a class of rationals of which a term is less than any assigned rational, the resulting class (4) (of terms greater than some u) will contain every rational, and will thus be the real number infinity. Thus both zero and infinity may be introduced as extreme terms among the real numbers, but neither is a segment according to the definition.

261. A given segment may be defined by many different classes of rationals. Two such classes u and v may be regarded as having the segment as a common property. Two infinite classes u and v will define the same lower segment if, given any u, there is a v greater than it, and given any v, there is a u greater than it. If each class has no maximum, this is also a *necessary* condition. The classes u and v are then what Cantor calls coherent (*zusammengehörig*). It can be shown, without considering segments, that the relation of being coherent is symmetrical and transitive*, whence we should infer, by the principle of abstraction, that both have to some third term a common relation which neither has to any other term. This third term, as we see from the preceding discussion, may be taken to be the segment which both define. We may extend the word *coherent* to two classes u and v, of which one defines an upper segment, the other a lower segment, which between them include all rationals with at most one exception. Similar remarks, *mutatis mutandis*, will still apply in this case.

We have now seen that the usual properties of real numbers belong to segments of rationals. There is therefore no mathematical reason for distinguishing such segments from real numbers. It remains to set forth, first the nature of a limit, then the current theories of irrationals, and then the objections which make the above theory seem preferable.

Note. The above theory is virtually contained in Professor Peano's article already referred to ("Sui Numeri Irrazionali," *Rivista di Matematica*, VI, pp. 126—140), and it was from this article, as well as from the *Formulaire de Mathématiques*, that I was led to adopt the theory. In this article, separate definitions of real numbers (§ 2, No. 5) and of segments (§ 8, ·0) are given, which makes it seem as though the two were distinguished. But after the definition of segments, we find the remark (p. 133): "Segments so defined differ only in nomenclature from real numbers." Professor Peano proceeds first to give purely technical reasons for distinguishing the two by the notation, namely that the addition, subtraction, *etc.* of real numbers is to be differently conducted from analogous operations which are to be performed on segments. Hence it would appear that the whole of the view I have advocated is contained in this article. At the same time, there is some lack of clearness, since it appears from the definition of real numbers that they are regarded as the limits of classes of rationals, whereas a segment is

---

* Cf. Cantor, *Math. Annalen*, XLVI, and *Rivista di Matematica*, V, pp. 158, 159.

in no sense a limit of a class of rationals. Also it is nowhere suggested—indeed, from the definition of real numbers the opposite is to be inferred—that no real number can be a rational, and no rational can be a real number. And this appears where he points out (p. 134) that 1 differs from the class of proper fractions (which is no longer the case as regards the real number 1, when this is distinguished both from the integer 1 and from the rational number 1 : 1), or that we say 1 is less than $\sqrt{2}$ (in which case, I should say, 1 must be interpreted as the class of proper fractions, and the assertion must be taken to mean: the proper fractions are some, but not all, of the rationals whose square is less than 2). And again he says (*ib.*): " The real number, although determined by, and determining, a segment *u*, is commonly regarded as the extremity, or end, or upper limit, of the segment"; whereas there is no reason to suppose that segments not having a rational limit have a limit at all. Thus although he confesses (*ib.*) that a complete theory of irrationals *can* be constructed by means of segments, he does not seem to perceive the reasons (which will be given in the next chapter) why this *must* be done—reasons which, in fact, are rather philosophical than mathematical.

CHAPTER XXXIV.

LIMITS AND IRRATIONAL NUMBERS.

262. THE mathematical treatment of continuity rests wholly upon the doctrine of limits. It has been thought by some mathematicians and some philosophers that this doctrine had been superseded by the Infinitesimal Calculus, and that this has shown true infinitesimals to be presupposed in limits*. But modern mathematics has shown, conclusively as it seems to me, that such a view is· erroneous. The method of limits has more and more emerged as fundamental. In this Chapter, I shall first set forth the general definition of a limit, and then examine its application to the creation of irrationals.

A compact series we defined as one in which there is a term between any two. But in such a series it is always possible to find two *classes* of terms which have no term between them, and it is always possible to reduce *one* of these classes to a single term. For example, if P be the generating relation and x any term of the series, then the class of terms having to x the relation P is one between which and x there is no term†. The class of terms so defined is one of the two segments determined by x; the idea of a segment is one which demands only a series in general, not necessarily a numerical series. In this case, if the series be compact, x is said to be the *limit* of the class; when there is such a term as x, the segment is said to be terminated, and thus every terminated segment in a compact series has its defining term as a limit. But this does not constitute a definition of a limit. To obtain the general definition of a limit, consider any class u contained in the series generated by P. Then the class u will in general, with respect to any term x not belonging to it, be divisible into two classes, that whose terms have to x the relation P (which I shall call the class of terms preceding x), and that whose terms have to x the relation $\breve{P}$ (which I shall call the class of terms following x). If x be itself a term of u, we

---

* This is the view, for instance, of Cohen, *Das Princip der Infinitesimal-Methode und seine Geschichte*, Berlin, 1883; see pp. 1, 2.

† It is perhaps superfluous to explain that a term is between two classes u, v, when it has the relation P to every term of u, and the relation $\breve{P}$ to every term of v, or *vice versâ*.

consider all the terms of u other than x, and these are still divisible into the above two classes, which we may call $\pi_u x$ and $\bar{\pi}_u x$ respectively. If, now, $\pi_u x$ be such that, if y be any term preceding x, there is a term of $\pi_u x$ following y, *i.e.* between x and y, then x is a limit of $\pi_u x$. Similarly if $\bar{\pi}_u x$ be such that, if z be any term after x, there is a term of $\bar{\pi}_u x$ between x and z, then x is a limit of $\bar{\pi}_u x$. We now define that x is a limit of u if it is a limit of either $\pi_u x$ or $\bar{\pi}_u x$. It is to be observed that u may have many limits, and that all the limits together form a new class contained in the series generated by P. This is the class (or rather this, by the help of certain further assumptions, becomes the class) which Cantor designates as the first derivative of the class u.

263. Before proceeding further, it may be well to make some general remarks of an elementary character on the subject of limits. In the first place, limits belong usually to classes contained in compact series—classes which may, as an extreme case, be identical with the compact series in question. In the second place, a limit may or may not belong to the class u of which it is a limit, but it always belongs to some series in which u is contained, and if it is a term of u, it is still a limit of the class consisting of all terms of u except itself. In the third place, no class can have a limit unless it contains an infinite number of terms. For, to revert to our former division, if u be finite, $\pi_u x$ and $\bar{\pi}_u x$ will both be finite. Hence each of them will have a term nearest to x, and between this term and x no term of u will lie. Hence x is not a limit of u; and since x is any term of the series, u will have no limits at all. It is common to add a theorem that every infinite class, provided its terms are all contained between two specified terms of the·series generated by P, must have at least one limit; but this theorem, we shall find, demands an interpretation in terms of segments, and is not true as it stands. In the fourth place, if u be co-extensive with the whole compact series generated by P, then every term of this series is a limit of u. There can be no other terms that are limits in the same sense, since limits have only been defined in relation to this compact series. To obtain other limits, we should have to regard the series generated by P as forming part of some other compact series—a case which, as we shall see, may arise. In any case, if u be any compact series, every term of u is a limit of u; whether u has also other limits, depends upon further circumstances. A limit may be defined generally as a term which immediately follows (or precedes) some class of terms belonging to an infinite series, without immediately following (or preceding, as the case may be) any one term of the series. In this way, we shall find, limits may be defined generally in all infinite series which are not progressions—as, for instance, in the series of finite and transfinite integers.

264. We may now proceed to the various arithmetical theories of irrationals, all of which depend upon limits. We shall find that, in the exact form in which they have been given by their inventors, they all involve an axiom, for which there are no arguments, either of philosophical necessity or of mathematical convenience; to which there are grave logical objections; and of which the theory of real numbers given in the preceding Chapter is wholly independent.

Arithmetical theories of irrationals could not be treated in Part II, since they depend essentially upon the notion of order. It is only by means of them that numbers become continuous in the sense now usual among mathematicians; and we shall find in Part VI that no other sense of continuity is required for space and time. It is very important to realize the logical reasons for which an arithmetical theory of irrationals is imperatively necessary. In the past, the definition of irrationals was commonly effected by geometrical considerations. This procedure was, however, highly illogical; for if the application of numbers to space is to yield anything but tautologies, the numbers applied must be independently defined; and if none but a geometrical definition were possible, there would be, properly speaking, no such arithmetical entities as the definition pretended to define. The algebraical definition, in which irrationals were introduced as the roots of algebraic equations having no rational roots, was liable to similar objections, since it remained to be shown that such equations have roots; moreover this method will only yield the so-called algebraic numbers, which are an infinitesimal proportion of the real numbers, and do not have continuity in Cantor's sense, or in the sense required by Geometry. And in any case, if it is possible, without any further assumption, to pass from Arithmetic to Analysis, from rationals to irrationals, it is a logical advance to show how this can be done. The generalizations of number—with the exception of the introduction of imaginaries, which must be independently effected—are all necessary consequences of the admission that the natural numbers form a progression. In every progression the terms have two kinds of relations, the one constituting the general analogue of positive and negative integers, the other that of rational numbers. The rational numbers form a denumerable compact series; and segments of a denumerable compact series, as we saw in the preceding Chapter, form a series which is continuous in the strictest sense. Thus all follows from the assumption of a progression. But in the present Chapter we have to examine irrationals as based on limits; and in this sense, we shall find that they do not follow without a new assumption.

There are several somewhat similar theories of irrational numbers. I will begin with that of Dedekind*.

265. Although rational numbers are such that, between any two, there is always a third, yet there are many ways of dividing *all* rational

---

* *Stetigkeit und irrationale Zahlen*, 2nd ed., Brunswick, 1892.

numbers into two classes, such that all numbers of one class come after all numbers of the other class, and no rational number lies between the two classes, while yet the first class has no first term and the second has no last term.   For example, all rational numbers, without exception, may be classified according as their squares are greater or less than 2. All the terms of both classes may be arranged in a single series, in which there exists a definite section, before which comes one of the classes, and after which comes the other.   Continuity *seems* to demand that some term should correspond to this section.   A number which lies between the two classes must be a new number, since all the old numbers are classified.   This new number, which is thus defined by its position in a series, is an *irrational* number.   When these numbers are introduced, not only is there always a number between any two numbers, but there is a number between any two classes of which one comes wholly after the other, and the first has no minimum, while the second has no maximum. Thus we can extend to numbers the axiom by which Dedekind defines the continuity of the straight line (*op. cit.* p. 11):—

  " If all the points of a line can be divided into two classes such that every point of one class is to the left of every point of the other class, then there exists one and only one point which brings about this division of all points into two classes, this section of the line into two parts."

  266.   This axiom of Dedekind's is, however, rather loosely worded, and requires an emendation suggested by the derivation of irrational numbers. If *all* the points of a line are divided into two classes, no point is left over to represent the section.   If *all* be meant to exclude the point representing the section, the axiom no longer characterizes continuous series, but applies equally to all series, *e.g.* the series of integers.   The axiom must be held to apply, as regards the division, not to all the points of the line, but to all the points forming some compact series, and distributed throughout the line, but consisting only of a portion of the points of the line.   When this emendation is made, the axiom becomes admissible.   If, from among the terms of a series, some can be chosen out to form a compact series which is distributed throughout the previous series; and if this new series can always be divided in Dedekind's manner into two portions, between which lies no term of the new series, but one and only one term of the original series, then the original series is continuous in Dedekind's sense of the word.   The emendation, however, destroys entirely the self-evidence upon which alone Dedekind relies (p. 11) for the proof of his axiom as applied to the straight line.

  Another somewhat less complicated emendation may be made, which gives, I think, what Dedekind *meant* to state in his axiom.   A series, we may say, is continuous in Dedekind's sense when, and only when, if *all* the terms of the series, without exception, be divided into two classes, such that the whole of the first class precedes the whole of the second, then, however the division be effected, either the first class has a last term, or the second class has a first term, but never both. This term, which comes at one end of one of the two classes, may then be used, in Dedekind's manner, to define the section. In discrete series, such as that of finite integers, there is both a last term of the first class and a first term of the second class*; while in compact series such as the rationals, where there is not continuity, it sometimes happens (though not for every possible division) that the first class has no last term and the last class has no first term. Both these cases are excluded by the above axiom. But I cannot see any vestige of self-evidence in such an axiom, either as applied to numbers or as applied to space.

267. Leaving aside, for the moment, the general problem of continuity, let us return to Dedekind's definition of irrational numbers. The first question that arises is this: What right have we to assume the existence of such numbers? What reason have we for supposing that there must be a position between two classes of which one is wholly to the right of the other, and of which one has no minimum and the other no maximum? This is not true of series in general, since many series are discrete. It is not demanded by the nature of order. And, as we have seen, continuity in a certain sense is possible without it. Why then should we postulate such a number at all? It must be remembered that the algebraical and geometrical problems, which irrationals are intended to solve, must not here be brought into the account. The existence of irrationals has, in the past, been inferred from such problems. The equation $x^2 - 2 = 0$ must have a root, it was argued, because, as x grows from 0 to 2, $x^2 - 2$ increases, and is first negative and then positive; if x changes continuously, so does $x - 2$; hence $x^2 - 2$ must assume the value 0 in passing from negative to positive. Or again, it was argued that the diagonal of unit square has evidently a precise and definite length x, and that this length is such that $x^2 - 2 = 0$. But such arguments were powerless to show that x is truly a number. They might equally well be regarded as showing the inadequacy of numbers to Algebra and Geometry. The present theory is designed to prove the arithmetical existence of irrationals. In its design, it is preferable to the previous theories; but the execution seems to fall short of the design.

Let us examine in detail the definition of $\sqrt{2}$ by Dedekind's method. It is a singular fact that, although a rational number lies between any two single rational numbers, two classes of rational numbers may be defined so that no rational number lies between them, though all of

---

* If the series contains a proper part which is a progression, it is only true *in general*, not without exception, that the first class must have a last term.

one class are higher than all of the other. It is evident that one at least of these classes must consist of an infinite number of terms. For if not, we could pick out the two of opposite kinds which were nearest together, and insert a new number between them. This one would be between the two classes, contrary to the hypothesis. But when one of the classes is infinite, we may arrange all or some of the terms in a series of terms continually approaching the other class, without reaching it, and without having a last term. Let us, for the moment, suppose our infinite class to be denumerable. We then obtain a denumerable series of numbers a_n, all belonging to the one class, but continually approaching the other class. Let B be a fixed number of the other class. Then between a_n and B there is always another rational number; but this may be chosen to be another of the a's, say a_{n+1}; and since the series of a's is infinite, we do not necessarily obtain, in this way, any number not belonging to the series of a's. In the definition of irrationals, the series of b's is also infinite. Moreover, if the b's also be denumerable, any rational number between a_n and b_m, for suitable values of p and q, either is a_{n+p} or b_{m+q}, or else lies between a_{n+p} and a_{n+p+1} or between b_{m+q} and b_{m+q+1}. In fact, a_{n+p} always lies between a_n and b_m. By successive steps, no term is obtained which lies between all the b's and all the a's. Nevertheless, both the a's and the b's are convergent. For, let the a's increase, while the b's diminish. Then $b_n - a_n$ and $b_n - a_{n+1}$ continually diminish, and therefore $a_{n+1} - a_n$, which is less than either, is less than a continually diminishing number. Moreover this number diminishes without limit; for if $b_n - a_n$ had a limit ϵ, the number $a_n + \epsilon/2$ would finally lie between the two classes. Hence $a_{n+1} - a_n$ becomes finally less than any assigned number. Thus the a's and b's are both convergent. Since, moreover, their difference may be made less than any assigned number ϵ, they have the same limit, if they have any. But this limit cannot be a rational number, since it lies between all the a's and all the b's. Such seems to be the argument for the existence of irrationals. For example, if

$$x = \sqrt{2} + 1,\ x^2 - 2x - 1 = 0.$$

Thus $\qquad x = 2 + 1/x = 2 + \dfrac{1}{2+}\dfrac{1}{x}$, and $x - 1 = 1 + \dfrac{1}{2+}\dfrac{1}{2+}\dfrac{1}{x} = $ etc.

The successive convergents to the continued fraction $1 + \dfrac{1}{2+}\dfrac{1}{2+}\dfrac{1}{2+}...$
are such that all the odd convergents are less than all the even convergents, while the odd convergents continually grow, and the even ones continually diminish. Moreover the difference between the odd and the next even convergent continually diminishes. Thus both series, if they have a limit, have the same limit, and this limit is defined as $\sqrt{2}$.

But the existence of a limit, in this case, is evidently a sheer assumption. In the beginning of this Chapter, we saw that the existence of a limit demands a larger series of which the limit forms part. To create the limit by means of the series whose limit is to be found would therefore be a logical error. It is essential that the distance from the limit should diminish indefinitely. But here, it is only the distance of consecutive terms which is known to diminish indefinitely. Moreover all the a's are less than b_n. Hence they continually differ less and less from b_n. But whatever n may be, b_n cannot be the limit of the a's, for b_{n+1} lies between b_n and all the a's. This cannot prove that a limit exists, but only that, *if* it existed, it would not be any one of the a's or b's, nor yet any other rational number. Thus irrationals are not proved to exist, but *may* be merely convenient fictions to describe the relations of the a's and b's.

268. The theory of Weierstrass concerning irrationals is somewhat similar to that of Dedekind. In Weierstrass's theory, we have a series of terms $a_1, a_2, \ldots, a_n, \ldots$, such that Σa_n, for all values of n, is less than some given number. This case is presented, *e.g.*, by an infinite decimal. The fraction $3\cdot14159\ldots$, however many terms we take, remains less than $3\cdot1416$. In this method, as Cantor points out*, the limit is not created by the summation, but must be supposed to exist already in order that $\overset{\infty}{\underset{1}{\Sigma}} a_n$ may be defined by means of it. This is the same state of things as we found in Dedekind's theory: series of rational numbers cannot prove the existence of irrational numbers as their limits, but can only prove that, *if* there is a limit, it must be irrational.

Thus the arithmetical theory of irrationals, in either of the above forms, is liable to the following objections. (1) No proof is obtained from it of the existence of other than rational numbers, unless we accept some axiom of continuity different from that satisfied by rational numbers; and for such an axiom we have as yet seen no ground. (2) Granting the existence of irrationals, they are merely specified, not defined, by the series of rational numbers whose limits they are. Unless they are independently postulated, the series in question cannot be known to have a limit; and a knowledge of the irrational number which is a limit is presupposed in the proof that it is a limit. Thus, although without any appeal to Geometry, any given irrational number can be *specified* by means of an infinite series of rational numbers, yet, from rational numbers alone, no proof can be obtained that there are irrational numbers at all, and their existence must be proved from a new and independent postulate.

Another objection to the above theory is that it supposes rationals and irrationals to form part of one and the same series generated by relations of greater and less. This raises the same kind of difficulties as we found to result, in Part II, from the notion that integers are greater

* *Mannichfaltigkeitslehre*, p. 22. I quote Weierstrass's theory from the account in Stolz, *Vorlesungen über allgemeine Arithmetik*, i.

or less than rationals, or that some rationals are integers. Rationals are essentially relations between integers, but irrationals are not such relations. Given an infinite series of rationals, there may be two integers whose relation is a rational which limits the series, or there may be no such pair of integers. The entity postulated as the limit, in this latter case, is no longer of the same kind as the terms of the series which it is supposed to limit; for each of them is, while the limit is not, a relation between two integers. Of such heterogeneous terms, it is difficult to suppose that they can have relations of greater and less; and in fact, the constitutive relation of greater and less, from which the series of rationals springs, has to receive a new definition for the case of two irrationals, or of a rational and an irrational. This definition is, that an irrational is greater than a rational, when the irrational limits a series containing terms greater than the given rational. But what is really given here is a relation of the given rational to a class of rationals, namely the relation of belonging to the segment defined by the series whose limit is the given irrational. And in the case of two irrationals, one is defined to be greater than the other when its defining series contains terms greater than any terms of the defining series of the other—a condition which amounts to saying that the segment corresponding to the one contains as a proper part the segment corresponding to the other. These definitions define a relation quite different from the inequality of two rationals, namely the logical relation of inclusion. Thus the irrationals cannot form part of the series of rationals, but new terms corresponding to the rationals must be found before a single series can be constructed. Such terms, as we saw in the last chapter, are found in segments; but the theories of Dedekind and Weierstrass leave them still to seek.

269. The theory of Cantor, though not expressed, philosophically speaking, with all the requisite clearness, lends itself more easily to the interpretation which I advocate, and is specially designed to *prove* the existence of limits. He remarks * that, in his theory, the existence of the limit is a strictly demonstrable proposition; and he strongly emphasizes the logical error involved in attempting to deduce the existence of the limit from the series whose limit it is (*ib.*, p. 22)†. Cantor starts by considering what he calls fundamental series (which are the same as what I have called progressions) contained in a larger series. Each of these fundamental series is to be wholly ascending or wholly descending. Two such series are called coherent (*zusammengehörig*) under the following circumstances :—

* *Op. cit.*, p. 24.

† Cantor's theory of irrationals will be found in *op. cit.*, p. 23, and in Stolz, *Vorlesungen uber allgemeine Arithmetik*, I, 7. I shall follow, to begin with, a later account, which seems to me clearer; this forms § 10 in an article contained in *Math. Annalen*, XLVI, and in *Rivista di Matematica*, V.

(1) If both are ascending, and after any term of either there is always a term of the other;

(2) If both are descending, and before any term of either there is always a term of the other;

(3) If one is ascending, the other descending, and the one wholly precedes the other, and there is *at most* one term which is between the two fundamental series.

The relation of being coherent is symmetrical, in virtue of the definition; and Cantor shows that it is transitive. In the article from which the above remarks are extracted, Cantor is dealing with more general topics than the definition of irrationals. But the above general account of coherent series will help us to understand the theory of irrationals. This theory is set forth as follows in the *Mannichfaltig-keitslehre* (p. 23 ff.):—

A fundamental series of rationals is defined as a denumerable series such that, given any number ϵ, there are at most a finite number of terms in the series the absolute values of whose differences from subsequent terms exceed ϵ. That is to say, given any number ϵ, however small, any two terms of the series which both come after a certain term have a difference which lies between $+\epsilon$ and $-\epsilon$. Such series must be of one of three kinds: (1) Any number ϵ being mentioned, the absolute values of the terms, from some term onwards, will all be less than ϵ, whatever ϵ may be; (2) from some term onwards, all the terms may be greater than a certain positive number ρ; (3) from some term onwards, all the terms may be less than a certain negative number $-\rho$. A real number b is to be defined by the fundamental series, and is said in the first case to be zero, in the second to be positive, and in the third to be negative. To define the addition, *etc.*, of these new real numbers, we observe that, if a_ν, a_ν' be the νth terms of two fundamental series, the series whose νth term is $a_\nu + a_\nu'$ or $a_\nu - a_\nu'$ or $a_\nu \times a_\nu'$ is also a fundamental series; while if the real number defined by the series (a_ν)* is not zero, (a_ν'/a_ν) also defines a fundamental series. If b, b' be the real numbers defined by the series (a_ν), (a_ν'), the real numbers defined by $(a_\nu + a_\nu')$, $(a_\nu - a_\nu')$, $(a_\nu \times a_\nu')$ and (a_ν'/a_ν) are defined to be $b + b'$, $b - b'$, $b \times b'$ and b'/b respectively. Hence we proceed to the definitions of equal, greater and less among real numbers. We define that $b = b'$ means $b - b' = 0$; $b > b'$ means that $b - b'$ is positive; and $b < b'$ means that $b - b'$ is negative—all terms which have been already defined. Cantor remarks further that in these definitions one of the numbers may be rational. This may be formally justified, in part, by the remark that a denumerable series whose terms are all one and the same rational number is a fundamental series, according to the definition; hence in constructing

---

* The symbol (a_ν) denotes the whole series whose νth term is a_ν, not this term alone.

the differences $a_\nu - a_\nu'$, by which $b - b'$ is defined, we may put some fixed rational a in place of a_ν' for all values of ν. But the consequence that we can define $b - a$ does not follow, and that for the following reason. There is absolutely nothing in the above definition of the real numbers to show that a is the real number defined by a fundamental series whose terms are all equal to a. The only reason why this seems self-evident is, that the definition by limits is unconsciously present, making us think that, since a is plainly the limit of a series whose terms are all equal to a, therefore a must be the real number defined by such a series. Since, however, Cantor insists—rightly, as I think—that his method is independent of limits, which, on the contrary, are to be deduced from it (pp. 24—5), we must not allow this preconception to weigh with us. And the preconception, if I am not mistaken, is in fact erroneous. There is nothing in the definitions above enumerated to show that a real number and a rational number can ever be either equal or unequal, and there are very strong reasons for supposing the contrary. Hence also we must reject the proposition (p. 24) that, if b be the real number defined by a fundamental series (a_ν), then

$$\operatorname*{Lim}_{\nu = \infty} a_\nu = b.$$

Cantor is proud of the supposed fact that his theory renders this proposition strictly demonstrable. But, as we have seen, there is nothing to show that a rational can be subtracted from a real number, and hence the supposed proof is fallacious. What is true, and what has all the mathematical advantages of the above theorem, is this : Connected with every rational a there is a real number, namely that defined by the fundamental series whose terms are all equal to a; if b be the real number defined by a fundamental series (a_ν) and if b_ν be the real number defined by a fundamental series whose terms are all equal to a_ν, then (b_ν) is a fundamental series of real numbers whose limit is b. But from this we cannot infer, as Cantor supposes (p. 24), that $\operatorname{Lim} a_\nu$ exists ; this will only be true in the case where (a_ν) has a rational limit. The limit of a series of rationals either does not exist, or is rational ; in no case is it a real number. But in all cases a fundamental series of rationals *defines* a real number, which is never identical with any rational.

270. Thus to sum up what has been said on Cantor's theory: By proving that two fundamental series may have the relation of being coherent, and that this relation is symmetrical and transitive, Cantor shows, by the help of the principle of abstraction (which is tacitly assumed), that two such series both have some one relation to one third term, and to no other. This term, when our series consist of rationals, we define as the real number which both determine. We can then define the rules of operation for real numbers, and the relations of equal, greater and less between them. But the principle of abstraction leaves

us in doubt as to what the real numbers really are. One thing, however, seems certain. They cannot form part of any series containing rationals, for the rationals are relations between integers, while the real numbers are not so; and the constitutive relation in virtue of which rationals form a series is defined solely by means of the integers between which they are relations, so that the same relation cannot hold between two real numbers, or between a real and a rational number. In this doubt as to what real numbers may be, we find that segments of rationals, as defined in the preceding chapter, fulfil all the requirements laid down in Cantor's definition, and also those derived from the principle of abstraction. Hence there is no logical ground for distinguishing segments of rationals from real numbers. If they are to be distinguished, it must be in virtue of some immediate intuition, or of some wholly new axiom, such as, that all series of rationals must have a limit. But this would be fatal to the uniform development of Arithmetic and Analysis from the five premisses which Peano has found sufficient, and would be wholly contrary to the spirit of those who have invented the arithmetical theory of irrationals. The above theory, on the contrary, requires no new axiom, for if there are rationals, there must be segments of rationals; and it removes what seems, mathematically, a wholly unnecessary complication, since, if segments will do all that is required of irrationals, it seems superfluous to introduce a new parallel series with precisely the same mathematical properties. I conclude, then, that an irrational actually *is* a segment of rationals which does not have a limit; while a real number which would be commonly identified with a rational is a segment which does have a rational limit; and this applies, e.g., to the real number defined by a fundamental series of rationals whose terms are all equal. This is the theory which was set forth positively in the preceding Chapter, and to which, after examining the current theories of irrationals, we are again brought back. The greater part of it applies to compact series in general; but some of the uses of fundamental series, as we shall see hereafter, presuppose either numerical measurement of distances or stretches, or that a denumerable compact series is contained in our series in a certain manner*. The whole of it, however, applies to any compact series obtained from a progression as the rationals are obtained from the integers; and hence no property of numbers is involved beyond the fact that they form a progression.

* See Chapter XXXVI.

CHAPTER XXXV.

CANTOR'S FIRST DEFINITION OF CONTINUITY.

271. THE notion of continuity has been treated by philosophers, as a rule, as though it were incapable of analysis. They have said many things about it, including the Hegelian dictum that everything discrete is also continuous and *vice versâ* *. This remark, as being an exemplification of Hegel's usual habit of combining opposites, has been tamely repeated by all his followers. But as to what they meant by continuity and discreteness, they preserved a discreet and continuous silence; only one thing was evident, that whatever they did mean could not be relevant to mathematics, or to the philosophy of space and time.

In the last chapter of Part III, we agreed provisionally to call a series continuous if it had a term between any two. This definition usually satisfied Leibniz†, and would have been generally thought sufficient until the revolutionary discoveries of Cantor. Nevertheless there was reason to surmise, before the time of Cantor, that a higher order of continuity is possible. For, ever since the discovery of incommensurables in Geometry—a discovery of which is the proof set forth in the tenth Book of Euclid—it was probable that space had continuity of a higher order than that of the rational numbers, which, nevertheless, have the kind of continuity defined in Part III. The kind which belongs to the rational numbers, and consists in having a term between any two, we have agreed to call *compactness*; and to avoid confusion, I shall never again speak of this kind as continuity. But that other kind of continuity, which was seen to belong to space, was treated, as Cantor remarks‡, as a kind of religious dogma, and was exempted from that conceptual analysis which is requisite to its comprehension. Indeed it was often held to show, especially by philosophers, that any subject-matter possessing it was not validly analyzable into elements. Cantor has shown that this view is mistaken, by a precise definition of the kind

---

* *Logic*, Wallace's Translation, p. 188; *Werke*, v. p. 201.

† *Phil. Werke*, Gerhardt's ed., Vol. II, p. 515. But cf. Cassirer, *Leibniz' System*, Berlin, 1901, p. 183.

‡ *Mannichfaltigkeitslehre*, p. 28.

of continuity which must belong to space. This definition, if it is to be explanatory of space, must, as he rightly urges*, be effected without any appeal to space. We find, accordingly, in his final definition, only ordinal notions of a general kind, which can be fully exemplified in Arithmetic. The proof that the notion so defined is precisely the kind of continuity belonging to space, must be postponed to Part VI. Cantor has given his definition in two forms, of which the earlier is not *purely* ordinal, but involves also either number or quantity. In the present chapter, I wish to translate this earlier definition into language as simple and untechnical as possible, and then to show how series which are continuous in this sense occur in Arithmetic, and generally in the theory of any progression whatever. The later definition will be given in the following Chapter.

272. In order that a series should be continuous, it must have two characteristics: it must be *perfect* and *cohesive* (zusammenhängend, bien enchaînée)†. Both these terms have a technical meaning requiring considerable explanation. I shall begin with the latter.

(1) Speaking popularly, a series is cohesive, or has cohesion, when it contains no finite gaps. The precise definition, as given by Cantor, is as follows: "We call T a *cohesive* collection of points, if for any two points t and t' of T, for a number ϵ given in advance and as small as we please, there are always, in several ways, a finite number of points $t_1, t_2, \ldots t_\nu$, belonging to T, such that the distances $tt_1, t_1t_2, t_2t_3, \ldots t_\nu t'$ are all less than ϵ."‡ This condition, it will be seen, has essential reference to distance. It is not necessary that the collection considered should consist of numbers, nor that ϵ should be a number. All that is necessary is, that the collection should be a series in which there are distances obeying the axiom of Archimedes and having no minimum, and that ϵ should be an arbitrary distance of the kind presented by the series. If the series be the whole field of some asymmetrical transitive relation, or if it be the whole of the terms having a certain asymmetrical transitive relation to a given term, we may substitute stretch for distance; and even if the series be only part of such a series, we may substitute the stretch in the complete series of which our series forms part. But we must, in order to give any meaning to cohesion, have something numerically measurable. How far this condition is necessary, and what can be done without it, I shall show at a later stage. It is through this condition that our discussions of quantity and measurement, in Part III, become relevant to the discussion of continuity.

---

* *Acta Math.* II, p. 403.

† *Acta Math.* II, pp. 405, 406; *Mannichfaltigkeitslehre*, p. 31.

‡ The words "in several ways" seem superfluous. They are omitted by Vivanti: see *Formulaire de Mathématiques*, Vol. I, VI, § 1, No. 22.

If the distances or stretches in our series do not obey the axiom of Archimedes, there are some among them that are incapable of a finite numerical measure in terms of some others among them. In this case, there is no longer an analogy of the requisite kind with either the rational or the real numbers, and the series is necessarily not cohesive. For let δ, d be two distances; let them be such that, for any finite number n, $n\delta$ is less than d. In this case, if δ be the distance ϵ, and d be the distance tt', it is plain that the condition of cohesion cannot be satisfied. Such cases actually occur, and—what seems paradoxical— they can be created by merely interpolating terms in certain cohesive series. For example, the series of segments of rationals is cohesive; and when these segments have rational limits, the limits are not contained in them. Add now to the series what may be called the *completed* segments, *i.e.* the segments having rational limits together with their limits. These are new terms, forming part of the same series, since they have the relation of whole and part to the former terms. But now the difference between a segment and the corresponding completed segment consists of a single rational, while all other differences in the series consist of an infinite number of rationals. Thus the axiom of Archimedes fails, and the new series is not cohesive.

The condition that distances in the series are to have no minimum is satisfied by real or rational numbers; and it is necessary, if cohesion is to be extended to non-numerical series, that, when any unit distance is selected, there should be distances whose numerical measure is less than ϵ, where ϵ is any rational number. For, if there be a minimum distance, we cannot make our distances tt_1, t_1t_2... less than this minimum, which is contrary to the definition of cohesion. And there must not only be no minimum to distances in general, but there must be no minimum to distances from any given term. Hence every cohesive series must be compact, *i.e.* must have a term between any two.

It must not be supposed, however, that every compact series is cohesive. Consider, for example, the series formed of 0 and $2 - m/n$, where m, n are any integers such that m is less than n. Here there is a term between any two, but the distance from 0 cannot be made less than 1. Hence the series, though compact, is not cohesive. This series, however, is not complete, being part only of the series of rationals, by means of which its distances are measured. In a complete series, the conditions are somewhat different. We must distinguish two cases, according as there are or are not distances. (*a*) If there are distances, and equal distances do not correspond to equal stretches, it may happen that, though the series is compact, the distances from some term never become less than some finite distance. This case would be presented by magnitudes, if we were to accept Meinong's opinion that the distance of any finite magnitude from zero is always infinite (*op. cit.* p. 84). It is presented by numbers, if we measure distances (as there are many reasons for doing) by $\log x/y$. Thus in this case, with regard to distances, the series is not cohesive, though it is complete and compact. (*b*) If there are no distances, but only stretches, then, assuming the axiom of Archimedes, any stretch will be less than $n\epsilon$, for a suitable value of n. Hence, dividing the stretch into n parts, one at least of these will be less than ϵ. But there is no way of proving that all can be made less than ϵ, unless we assume either the axiom of linearity (that any stretch can be divided into n equal parts), or a more complicated but more general axiom, to the effect that a stretch d can be divided into n parts, each of which is greater than $d/(n+1)$ and less than $d/(n-1)$, whatever integer n may be. With this axiom and the axiom of Archimedes, a complete compact series must be cohesive; but these two axioms together render completeness superfluous and compactness redundant. Thus we see that cohesion is in almost all cases a condition distinct from compactness. Compactness is purely serial, while cohesion has essential reference to numbers or to the conditions of numerical measurement. Cohesion implies compactness, but compactness never implies cohesion, except in the sole case of the complete series of rationals or real numbers.

273. (2) To explain what is meant by a *perfect* series is more difficult. A series is perfect when it coincides with its first derivative*. To explain this definition, we must examine the notion of the *derivatives* of a series†, and this demands an explanation of a *limiting-point* of a series. Speaking generally, the terms of a series are of two kinds, those which Cantor calls *isolated* points, and those which he calls limiting-points. A finite series has only isolated points; an infinite series must define at least one limiting-point, though this need not belong to the series. A limiting-point of a series is defined by Cantor to be a term such that, in any interval containing the term, there are an infinite number of terms of the series (*ib.* p. 343). The definition is given in terms of the points on a line, but it has no essential reference to space. The limiting-point may or may not be a term of the original series. The assemblage of all limiting-points is called the first derivative of the series. The first derivative of the first derivative is called the second derivative, and so on. Peano gives the definition of the first derivative of a class of real numbers as follows: Let u be a class of real numbers, and let x be a real number (which may or may not be a u) such that the lower limit of the absolute values of the differences of x from terms of u other than x is zero; then the class of terms x satisfying this condition is the first derivative of u‡. This definition is virtually identical with that of Cantor, but it brings out more explicitly the connection of the derivative with limits. A series, then, is perfect, when it consists of

* *Acta Math.* ii, p. 405. † *Ib.* pp. 341–4.

‡ *Formulaire*, Vol. ii, No. 3 (1899), § 71, 1·0 and 4·0.

exactly the same terms as its first derivative; *i.e.* when all its points are limiting-points, and all its limiting-points belong to it.

274. But with regard to the latter point, namely, that all limiting-points of the series must belong to it, some explanation is necessary. Take, for example, the series of rational numbers. Every rational number is the limit of some series of rational numbers, and thus the rationals are contained in their first derivative. But as regards those series of rationals which do not have a rational limit, we agreed in the preceding chapter that they do not have a limit at all. Hence all series of rationals which have a limit have a rational limit, and therefore, by the letter of the definition, the rationals should form a perfect series. But this is not the case. Cantor, as we saw in connection with irrationals, believes, what we were compelled to regard as erroneous, that every series fulfilling certain conditions, which may be called the conditions of convergency, must have a limit. Hence he regards those series of rationals which have no rational limit as having an irrational limit, and as therefore having a limit not belonging to the series of rationals; and therefore the series of rationals does not contain all the terms of its first derivative. In fact, the first derivative of the rational numbers is held to be the real numbers. But when we regard the real numbers as segments of rationals, it is impossible to take this view; and when we deny the existence-theorem for limits, it is necessary to modify Cantor's definition of perfection*. This modification we must now examine.

What we must say is, that a series is perfect when all its points are limiting-points, and when further, any series being chosen out of our first series, if this new series is of the sort which is usually regarded as defining a limit, then it actually has a limit belonging to our first series. To make this statement precise, we must examine what are the conditions usually considered as defining a limit. In the case of denumerable series, they are simple, and have already been set forth. They come to this, that, given any distance ϵ, however small, all the terms of our series after some definite term, say the mth, are such that any two of them have a difference whose absolute value is less than ϵ. This statement, it will be seen, involves either number or quantity, *i.e.* it is not purely ordinal. It is a curious fact that, though the supposed condition for the existence of a limit cannot, by our present method, be stated in purely ordinal terms, the limit of a denumerable series, if there be one, can always be defined in purely ordinal terms. I shall distinguish Cantor's fundamental series in a compact series into progressions and regressions, according as earlier terms have to later ones always the relation P, or always the relation $\breve{P}$ (where P is the generating relation of the compact series in which the said progressions and regressions are

---

* This point is ably discussed by Couturat, *Revue de Mét. et de Morale*, March, 1900, p. 167.

contained). The compact series is further assumed to be complete. A term x is then the limit of a progression, if every term of the progression has to x the relation P, while every term which has to x the relation P also has this relation to some term of the progression. This definition, it will be seen, is purely ordinal; and a similar definition will apply to a regression.

Let us examine next what are the usual conditions for the existence of a limit to a non-denumerable series. When we come to examine non-numerical series, we shall find it inconvenient to be restricted to denumerable series, and therefore it will be well to consider other series at once. Here, of course, if any denumerable series contained in our larger series fulfils the conditions for a limit, there will be a corresponding definition of a limiting-point in our larger series. And the upper or lower limit of the whole or part of our larger series, if there is one, may be defined exactly as in the case of a progression or a regression. But general conditions for the existence of a limit cannot be laid down, except by reference to denumerable series contained in our larger series. And it will be observed that Cantor's definition of a limiting-point assumes the existence of such a point, and cannot be turned into a definition of the conditions under which there are such points. This illustrates the great importance of Cantor's fundamental series.

The method of segments will, however, throw some light on this matter. We saw in Chapter XXXIII that any class of terms in a series defines a segment, and that this segment sometimes can, but sometimes cannot, be defined by a single term. When it can be so defined, this term is its upper limit; and if this term does not belong to the class by which the segment was defined, then it is also the upper limit of that class. But when the segment has no upper limit, then the class by which the segment was defined also has no upper limit. In all cases, however—and this is one of the chief virtues of segments—the segment defined by an infinite class which has no upper limit is the upper limit of the segments defined by the several members of the class. Thus, whether or not the class has an upper limit, the segments which its various terms define always have one—provided, that is, that the compact series in which the class is contained has terms coming after all terms of the class.

We can now express, without assuming the existence of limits in cases where this is not demonstrable, what is meant by a series containing its first derivative. When any class of terms is contained in a compact series, the conditions which are commonly said to insure the existence of an upper limit to the class, though they do not insure this, do insure an upper limit to the class of segments defined by the several members of the class. And as regards lower limits, the same proposition holds concerning what we called upper segments. Hence we may define: A class u of terms forming the whole or part of a series is perfect when each of the terms of u is the upper or lower limit of some class contained in u, and when, if v be any class contained in u, and the lower segments defined by the several members of v have an upper limit, or the upper segments have a lower limit, this limiting segment is one of those that can be defined by a single term of u, *i.e.* have a term of u for their upper or lower limit respectively. This definition, it must be admitted, is more complicated than Cantor's, but it is free from the unjustifiable assumption of the existence of limits.

We may repeat the definition of perfection in what is perhaps less difficult language. Given any series, and any class of terms u contained in this series, there are an upper and a lower segment corresponding to every term of u. Any infinite set of terms v being chosen out of u, there are certain conditions which are commonly said to insure that v has an upper limit, which, it is admitted, may belong neither to u, nor to the series in which u is contained. What these conditions do insure, however, is that the class of lower segments corresponding to v has an upper limit. If the series is perfect, v will have an upper limit whenever the corresponding class of segments has one, and this upper limit of v will be a term of u. The definition of perfection requires that this should hold both for upper and lower limits, and for any class v contained in u.

275. As the question concerning the existence of limits, which has necessitated the above complication, is one of some philosophical importance, I shall repeat the arguments against assuming the existence of limits in the class of series to which the rational numbers belong. Where a series is imperfect, while its first derivative is perfect, there the first derivative is logically prior to its own formation. That is to say, it is only by presupposing the perfect series that it can be shown to be the derivative of the imperfect series. We have already seen that this is the case with individual irrational numbers; it is easy to see that the principle is general. Wherever the derivative contains a term not belonging to the original series, that term is the limit of some denumerable series forming an integral part of the first series. If this series with a limit have the general term a_n, then—wording the definition so as not to apply only to series of numbers—there is always a definite number m, for any specified distance ϵ, however small, such that, if n is greater than m, the distance between a_{n+p} and a_n is less than ϵ, whatever positive integer p may be. From this it is inferred that the series (a_n) has a limit, and it is shown that, in many cases, this limit cannot belong to the series out of which the series (a_n) was chosen. But the inference that there is a limit is precarious. It may be supported either by previous knowledge of the term which is the limit, or by some axiom necessitating the existence of such a term. When the term which is the limit is independently known, it may be easily shown to be the limit. But when it is not known, it cannot be proved to exist at all, unless we introduce some axiom of continuity. Such an axiom is introduced by Dedekind, but we saw that his axiom is unsatisfactory. The principle of abstraction, which shows that two coherent series have something in common, is fully satisfied by segments. And in some cases, among which is that of the rationals, it seems that the constitutive relation of the imperfect series cannot hold between any terms not belonging to this series, so that the existence of limits not belonging to the series is wholly impossible. For a limit must have a certain position in a series of which the series which it limits forms part, and this requires some constitutive relation of which the limit, as well as the terms limited, must be capable. An independent complete series, such as the rationals, cannot, in fact, have any limiting-points not belonging to it. For, if R be the constitutive relation, and two terms a, b, have the relation R, any third term c, which has this relation or its converse to either, and therefore both, of the terms a, b, belongs to the same series as a and b. But the limit, if it exists, must have the constitutive relation to the terms which it limits; hence it must belong to the complete series to which they belong. Hence any series which has actual limiting-points not belonging to it is only part of some complete series; and a complete series which is not perfect is one in which the limits defined in the usual way, but not belonging to the series, do not exist at all. Hence, in any complete series, either some definable limits do not exist, or the series contains its first derivative.

In order to render the arbitrariness of assuming the existence of limits still more evident, let us endeavour to set up an axiom of continuity more irreproachable than Dedekind's. We shall find that it can still be denied with perfect impunity.

When a number of positions in a series continually differ less and less from each other, and are known to be all on one side of some given position, there must exist (so our axiom might run) *some* position to which they approximate indefinitely, so that no distance can be specified so small that they will not approach nearer than by this distance. If this axiom be admitted, it will follow that all imperfect series, whose first derivatives are perfect, presuppose these first derivatives, and are to be regarded as selections from them. Let us examine the consequences of denying our axiom in the case of a series of numbers. In this case, the unwary might suppose, the position next to all the terms a_n, but not belonging to them, would be (say) p, where $p - a_n$ is greater than ϵ, for a suitable value of ϵ, whatever n may be. But if our series is compact, there is a term between p and $p - \epsilon$, say p'. Thus $p' - a_n$ is less than $p - a_n$, whatever n may be. Thus p' is nearer all the a's than p is, contrary to the hypothesis. But the above denial was not direct, and the fact that it seemed correct illustrates the fallacies which in this subject are hard to avoid. The axiom is: There is a term to which the a's approach as near as we like.

The denial was: There is a term nearest to the a's, but at a finite distance. The denial should have been: There is no term to which the a's approach as near as we like. In other words, whatever term we specify, say p, there is some finite distance ϵ, such that $p - a_n$ is greater than ϵ, whatever a_n may be. This is true in the case of series of rational numbers which have no rational limit. In this case, there is no term nearest to the a's, but at a finite distance, while also, whatever term beyond all the a's we specify (except where our series has a rational limit), none of the a's approach nearer to this term than by a certain finite distance ϵ. Every term beyond the a's is at more than some finite distance from all of them, but there is no finite distance which every term beyond the a's exceeds. The introduction of irrationals introduces symmetry into this odd state of things, so that there is a term to which the a's approach indefinitely, as well as a series of terms indefinitely approaching the a's. When irrationals are not admitted, if we have a term p after all the a's, and a small distance ϵ, then, if ϵ be specified, p can be chosen so that $p - a_n$ is less than ϵ, whatever n may be; but if p be specified, an ϵ can always be found (except when the limit is rational) so that $p - a_n$ is greater than ϵ, whatever n may be. This state of things, though curious, is not self-contradictory. The admission of irrationals, as opposed to segments, is thus logically unnecessary; as it is also mathematically superfluous, and fatal to the theory of rationals, there are no reasons in its favour, and strong reasons against it. Hence, finally, any axiom designed to show the existence of limits in cases where they cannot otherwise be shown to exist, is to be rejected; and Cantor's definition of perfection must be modified as above. This conclusion will, in future, be regarded as established.

Having now analyzed Cantor's earlier definition of continuity, I shall proceed to examine his later ordinal definition, and the application of its various portions to series more general than those of numbers, showing, if possible, the exact points in which these various portions are required.

CHAPTER XXXVI.

ORDINAL CONTINUITY*.

276. THE definition of continuity which we examined in the preceding chapter was, as we saw, not purely ordinal; it demanded, in at least two points, some reference to either numbers, or numerically measurable magnitudes. Nevertheless continuity seems like a purely ordinal notion; and this has led Cantor to construct a definition which is free from all elements extraneous to order†. I shall now examine this definition, as well as others which may be suggested. We shall find that, so long as all references to number and quantity are excluded, there are theorems of great importance, especially as regards fundamental series, which, with any suggested ordinal definition except that of Cantor, remain indemonstrable, and are presumably sometimes false‡— a fact from which the merits of Cantor's definition, now to be given, are apparent.

277. Cantor's definition of the continuum in his later article§ is as follows. We start (§ 9) from the type of series presented by the rational numbers greater than 0 and less than 1, in their order of magnitude. This type we call η. A series of this type we define by the following marks. (1) It is denumerable, that is, by taking its terms in a suitable order (which, however, must be different from that in which they are given), we can give them a one-one correspondence with the finite integers. (2) The series has no first or last term. (3) There is a term between any two, *i.e.* the series is compact (*überall dicht*). It is then proved that these three characteristics completely define the type of order presented by the rationals, that is to say, there is a one-one correspondence, between any two series having these three properties, in which earlier terms correspond to earlier terms, and later ones to later ones. This is established by the use of mathematical induction, which is applicable in virtue of the fact that series of this type are denumer-

---

* The present chapter deals with the same subject as M. Couturat's article, "Sur la définition du Continu," *Revue de Métaphysique et de Morale*, March, 1900. I agree in the main with this article, in which much of what I said in the preceding chapter, and shall say in this, will be found.

† *Math. Annalen*, XLVI.

‡ Mathematical proofs of such theorems as are not already well known will be found in RdM, VII, 3.

§ *Math. Annalen*, XCVI, § 11.

able. Thus all series which are denumerable, endless*, and compact, are ordinally similar. We now proceed (§ 10) to the consideration of fundamental series contained in any one-dimensional series M. We show (as has been already explained) what is meant by calling two fundamental series *coherent*, and we give an ordinal definition of the limit of a fundamental series, namely, in the case of a progression, the limit comes after the whole progression, but every term before the limit comes before some term of the progression; with a corresponding definition for the limit of a regression. We prove that no fundamental series can have more than one limit, and that, if a fundamental series has a limit, this is also the limit of all coherent series; also that two fundamental series, of which one is part of the other, are coherent. Any term of M which is the limit of some fundamental series in M is called a *principal* term of M. If all the terms of M are principal terms, M is called *condensed in itself* (insichdicht). If every fundamental series in M has a limit in M, M is called *closed* (abgeschlossen)†. If M is both closed and condensed in itself, it is *perfect*. All these properties, if they belong to M, belong to any series which is ordinally similar to M. With these preparations, we advance at last to the definition of the continuum (§ 11). Let θ be the type of the series to which belong the real numbers from 0 to 1, both inclusive. Then θ, as we know, is a perfect type. But this alone does not characterize θ. It has further the property of containing within itself a series of the type η, to which the rationals belong, in such a way that between any two terms of the θ-series there are terms of the η-series. Hence the following definition of the continuum:

A one-dimensional continuum M is a series which (1) is perfect, (2) contains within itself a denumerable series S of which there are terms between any two terms of M.

In this definition, it is not necessary to add the other properties which are required to show that S is of the type η. For if S had a first or last term, this would be also the first or last term of M; hence we could take it away from S, and the remaining series would still satisfy the condition (2), but would have no first or last term; and the condition (2) together with (1) insures that S is a compact series. Cantor proves that any series M satisfying the above conditions is ordinally similar to the number-continuum, *i.e.* the real numbers from 0 to 1, both inclusive; and hence it follows that the above definition includes precisely the same class of series as those that were included in his former definition. He does not assert that his new definition is purely ordinal, and it might be doubted, at first sight, whether it is so. Let us see for ourselves whether any extra-ordinal notions are contained in it.

---

* *I.e.* having neither a beginning nor an end.

† Not to be confounded with the elementary sense of a closed series discussed in Part IV.

278. The only point as to which any doubt could arise is with regard to the condition of being denumerable. To be a denumerable collection is to be a collection whose terms are all the terms of some progression. This notion, so far, is purely ordinal. But in the case supposed, that of the rationals or of any ordinally similar series, the terms forming the series must be capable of two orders, in one of which they form a compact series, while in the other they form a progression. To discover whether or not a given set of terms is capable of these two orders, will in general demand other than ordinal conditions; nevertheless, the notion itself is purely ordinal. Now we know, from the similarity of all such series to the series of rationals (which involves only ordinal ideas), that no such series is perfect. But it remains to be seen whether we can prove this without appealing to the special properties of the rationals which result from there being a series in which there is distance. We know, as a matter of fact, that no denumerable series can be perfect*, but we want here a purely ordinal proof of this theorem. Such a proof, however, is easily given. For take the terms of our denumerable compact series S in the order in which they form a progression, and in this order call them u. Starting with the first in this order, which we will call x_0, there must be one which, in the other order S, follows this term. Take the first such term, x_1, as the second in a fundamental series v. This term has a finite number of predecessors in the progression u, and therefore has successors in S which are also successors in u; for the number of successors in S is always infinite. Take the first of these common successors, say x_2, as the third term of our fundamental series v. Proceeding in this way, we can construct an ascending fundamental series in S, the terms of which have the same order in u as in S. This series cannot have a limit in S, for each term x_n succeeds, in S, every term which precedes it in u. Hence any term of S will be surpassed by some term x_n of our fundamental series v, and hence this fundamental series has no limit in S. The theorem that a denumerable endless series cannot be perfect is, therefore, purely ordinal. From this point onwards there is no difficulty, and our former theory of segments enables us to state the matter simply. Given a denumerable, endless, compact series S, construct all the segments defined by fundamental series in S. These form a perfect series, and between any two terms of the series of segments there is a segment whose upper (or lower) limit is a term of S. Segments of this kind, which may be called rational segments, are a series of the same type as S, and are contained in the whole series of segments in the required manner. Hence the ordinal definition of the continuum is complete.

279. It must not be supposed that continuity as above defined can only be exemplified, in Arithmetic, by the devious course from integers to rationals, and thence to real numbers. On the contrary, the integers

* *Acta Mathematica*, ii, p. 409.

themselves can be made to illustrate continuity. Consider all possible infinite classes of integers, and let them be arranged on the following plan. Of two classes u, v, of which the smallest number in u is less than the smallest in v, u comes first. If the first n terms of u and v are identical, but the $(n+1)^{th}$ terms are different, that which has the smaller $(n+1)^{th}$ term is to come first. This series has a first term, namely, the whole class of the integers, but no last term. Any completed segment of the series, however, is a continuous series, as the reader can easily see for himself. The denumerable compact series contained in it is composed of those infinite classes which contain all numbers greater than some number, *i.e.* those containing all but a finite number of numbers. Thus classes of finite integers alone suffice to generate continuous series.

280. The above definition, it will be observed, depends upon progressions. As progressions are the very essence of discreteness, it seems paradoxical that we should require them in defining continuity*. And, after all, as it is certain that people have not in the past associated any precise idea with the word *continuity*, the definition we adopt is, in some degree, arbitrary. Series having the properties enumerated in Cantor's definition would generally be called continuous, but so would many others which his definition excludes. In any case it will be a valuable inquiry to ask what can be done by compact series without progressions.

Let u be any endless compact series, whose generating relation is P, and concerning which nothing further is known. Then, by means of any term or any class of terms in u, we can define a segment of u. Let us denote by U the class of all lower segments of u. A lower segment, it may be well to repeat, is a class v of terms contained in u, not null, and not coextensive with u, and such that v has no last term, and every term preceding a v is a v. In the converse case, when v has no first term, and every term following a v is a v, v is called an upper segment. It is then easy to prove that every segment consists of all the terms preceding (or following) either some single term of u, or a variable term of some class of terms of u; and that every single term, and every class of terms, defines an upper and a lower segment in this manner. Then, if V denote the class of upper segments, it is easy to prove that both U and V are again endless compact series, whose generating relation is that of whole or part; while if u has one or two ends, so have U and V, though the end-terms are not segments according to the definition. If we now proceed to the consideration of segments

---

* Mr Whitehead has shown that the following simpler definition is equivalent to Cantor's. A series is continuous when (1) every segment, upper or lower, has a limit, and the series has a first and a last term; (2) a denumerable compact series is contained in it in such a way that there are terms of this latter series between any two terms of our original series. In this definition, progressions are relevant only in defining a denumerable series.

in U or V (U say), we shall find that the segment of U's defined by any class whatever of U's can always be defined by a single U, which, if the class is infinite and has no last term, is the upper limit of the class, and which, in all cases, is the logical sum of all the members of the class—members which, it must be remembered, are all themselves classes contained in u^*. Hence all classes contained in U and having no last term have an upper limit in U; and also (what is a distinct proposition) all classes contained in U and having no first term have a lower limit in U, except in the case where the lower limit is the logical zero or null-class; and the lower limit is always the logical product of all the classes composing the class which it limits. Thus by adding to U the null-class, we insure that U shall be a closed series. There is a sense in which U is condensed in itself, namely, this: every term of U is the upper limit of a suitably chosen class contained in U, for every term is the upper limit of the segment of U's which it defines; and every term of U is a lower limit of the class of those U's of which it is a proper part. But there is absolutely no proof, so far at least as I have been able to discover, that every term of U is the upper or lower limit of a *fundamental* series. There is no *à priori* reason why, in any series, the limit of any class should always be also the limit of a fundamental series; this seems, in fact, to be a prerogative of series of the types to which rationals and real numbers respectively belong. In our present case, at least, though our series is, in the above general sense, condensed in itself, there seems no reason for supposing its terms to be all of them limits of fundamental series, and in this special sense the series may not be condensed in itself.

281. It is instructive to examine the result of confining the terms of U to such segments as can be defined by fundamental series. In this case it is well to consider, in addition to upper and lower segments, their supplements, as they may be called, of which I shall shortly give the definition. Let a compact series v be given, generated by a transitive asymmetrical relation P, and let u be any fundamental series in v. If earlier terms of u have to later ones the relation P, I shall call u a *progression*; if the relation $\breve{P}$, I shall call u a *regression*. If now w be any class whatever contained in v, w defines, as we have already seen, four other classes in v, namely (1) the class of terms before every w, which I shall call $w\pi$; (2) the class of terms after every w, which I shall call $w\bar{\pi}$; (3) the class of terms before some w, which I shall call πw; (4) the class of terms after some w, which I shall call $\bar{\pi}w$. The classes (3) and (4) are lower and upper segments respectively; the classes (1) and

---

* The definition of the logical sum of the members of a class of classes, in a form not involving finitude, is, I believe, due to Peano. It is as follows: Let w be a class of classes; then the logical sum of the members of w is the class of terms x such that there is some class belonging to w, to which x belongs. See *Formulaire*, Vol. II, Part I (1897), No. 461.

(2) are supplements to (4) and (3) respectively, and I shall call them supplemental segments. When w has an upper limit, this is the first term of $w\overline{\pi}$, and thus $w\overline{\pi}$ is not a segment, since no upper segment has a first term. But when w has no upper limit, then, whether w be finite or infinite, $w\overline{\pi}$ is a segment. Similar remarks apply to lower limits. If w has a last term, this belongs neither to πw nor to $w\overline{\pi}$, but all other terms of v belong to one or other class; if w has no last term, all terms of v belong to πw or $w\overline{\pi}$. Similar remarks apply to $w\pi$ and $\overline{\pi}w$. Applying these general definitions to the cases of progressions and regressions, we shall find that, for a progression, only the classes (2) and (3) are important; for a regression, only the classes (1) and (4). The question where a progression begins or a regression ends is quite unimportant. Since a progression has no last term, and a regression no first term, the segment defined by either, together with its supplement, contains every term of v. Whether progressions and regressions in v have limits always, sometimes, or never, there seems no way of deciding from the given premisses. I have not been able to discover an instance of a compact series where they never have limits, but I cannot find any proof that such an instance is impossible.

Proceeding now to classes of segments, as we proceeded before to our class U, we have here four such classes to consider, namely : (1) The class $v\pi$, each of whose terms is the class $u\pi$ defined by some *regression* u, i.e., the terms of v which come before all the terms of some regression in v ; (2) the class $v\overline{\pi}$, consisting of all the classes $u\overline{\pi}$ defined by *progressions* u ; (3) the class πv, whose terms are πu, where u is some *progression* ; (4) the class $v\pi$, whose terms are $u\pi$, where u is some *regression*. Each of these four classes is a class of classes, for its terms are classes contained in v. Each of the four is itself a compact series. There is no way of proving, so far as I know, that (1) and (3), or (2) and (4), have any common terms. Each pair would have a common term if v contained a progression and a regression which were coherent, and had no limit in v, but there is no way of discovering whether this case ever arises in the given series v.

When we come to examine whether the four classes thus defined are condensed in themselves, we obtain the most curious results. Every fundamental series in. any one of the four classes has a limit, but not necessarily in the series of which its terms are composed, and conversely, every term of each of our four classes is the limit of a fundamental series, but not necessarily of a series contained in the same class to which the limiting term belongs. The state of things, in fact, is as follows :

Every progression in $v\pi$ or πv has a limit in πv.
Every progression in $v\overline{\pi}$ or $\overline{\pi}v$ has a limit in $\overline{\pi}v$.
Every regression in $v\pi$ or πv has a limit in $v\pi$.
Every regression in $v\overline{\pi}$ or $\overline{\pi}v$ has a limit in $v\overline{\pi}$.
Every term of $v\pi$ is the limit of a regression in $v\pi$ and of one in πv.

Every term of $v\breve{\pi}$ is the limit of a regression in $v\breve{\pi}$ and of one in $\breve{\pi}v$.

Every term of πv is the limit of a progression in $v\pi$ and of one in πv.

Every term of $\breve{\pi}v$ is the limit of a progression in $v\breve{\pi}$ and of one in $\breve{\pi}v$.

Hence $v\pi$ is identical with the class of limits of regressions in $v\pi$ or πv;
$v\breve{\pi}$ is identical with the class of limits of regressions in $v\breve{\pi}$ or $\breve{\pi}v$;
πv is identical with the class of limits of progressions in $v\pi$ or πv;
$\breve{\pi}v$ is identical with the class of limits of progressions in $\breve{\pi}v$ or $v\breve{\pi}$.

Thus each of our four classes has a kind of one-sided perfection; two of the four are perfect on one side, the other two on the other. But I cannot prove of any one of the four classes that it is wholly perfect. We might attempt the combination of $v\pi$ and πv, and also of $v\breve{\pi}$ and $\breve{\pi}v$. For $v\pi$ and πv together form one series, whose generating relation is still whole and part. This series will be perfect, and will contain the limits alike of progressions and of regressions in itself. But this series may not be compact; for if there be any progression u and regression u' in v, which both have the same limit in v (a case which, as we know, occurs in some compact series), then πu and $u'\pi$ will be consecutive terms of the series formed of πv and $v\pi$ together, for $u'\pi$ will contain the common limit, while πu will not contain it, but all other terms of v will belong to both or to neither. Hence when our series is compact, we cannot show that it is perfect; and when we have made it perfect, we can show that it may not be compact. And a series which is not compact can hardly be called continuous.

Although we can prove that, in our original compact series v, there are an infinite number of progressions coherent with a given progression and having no term in common with it, we cannot prove that there is even one regression coherent with a given progression; nor can we prove that any progression or regression in v has a limit, or that any term of v is a limit of a progression or regression. We cannot prove that any progression u and regression u' are such that $\pi u = u'\pi$, nor yet that πu and $u'\pi$ may differ by only a single term of v. Nor, finally, can we prove that any single progression in $v\pi$ has a limit in $v\pi$, with similar propositions concerning the other three classes $v\breve{\pi}$, πv, $\breve{\pi}v$. At least, I am unable to discover any way of proving any of these theorems, though in the absence of instances of the falsity of some of them it seems not improbable that these may be demonstrable.

If it is the fact—as it seems to be—that, starting only from a compact series, so many of the usual theorems are indemonstrable, we see how fundamental is the dependence of Cantor's ordinal theory upon the condition that the compact series from which we start is to be denumerable. As soon as this assumption is made, it becomes easy to prove all those of the above propositions which hold concerning the types η and θ respectively. This is a fact which is obviously of considerable philosophical importance; and it is with a view of bringing it out clearly that I have dwelt so long upon compact series which are not assumed to be denumerable.

282. The remark which we made just now, that two compact series may be combined to form one which sometimes has consecutive terms, is rather curious, and applies equally to continuity as defined by Cantor. Segments of rationals form a continuous series, and so do completed segments (*i.e.* segments together with their limits); but the two together form a series which is not compact, and therefore not continuous. It is certainly contrary to the usual idea of continuity that a continuous series should cease to be so merely by the interpolation of new terms between the old ones. This should, according to the usual notions, make our series still more continuous. It might be suggested that, philosophically speaking, a series cannot be called continuous unless it is *complete*, *i.e.* contains a certain term together with all the terms having to the given term a specified asymmetrical transitive relation or its converse. If we add this condition, the series of segments of rationals is not complete with regard to the relation by which we have hitherto regarded it as generated, since it does not consist of all classes of rationals to which a given segment has the relation of whole and part, and each of which contains all terms less than any one of its terms—this condition is also satisfied by completed segments. But every series is complete with regard to some relation, simple or complex. This is the reason why completeness need not, from a mathematical standpoint, be mentioned in the definition of continuity, since it can always be insured by a suitable choice of the generating relation.

We have now seen in what Cantor's definition of continuity consists, and we have seen that, while instances fulfilling the definition may be found in Arithmetic, the definition itself is purely ordinal—the only datum required is a denumerable compact series. Whether or not the kind of series which Cantor defines as continuous is thought to be the most similar to what has hitherto been vaguely denoted by the word, the definition itself, and the steps leading to it, must be acknowledged to be a triumph of analysis and generalization.

Before entering upon the philosophical questions raised by the continuum, it will be well to continue our review of Cantor's most remarkable theorems, by examining next his transfinite cardinal and ordinal numbers. Of the two problems with which this Part is concerned, we have hitherto considered only continuity; it is now time to consider what mathematics has to say concerning infinity. Only when this has been accomplished, shall we be in a position adequately to discuss the closely allied philosophical problems of infinity and continuity.

CHAPTER XXXVII.

TRANSFINITE CARDINALS.

283. THE mathematical theory of infinity may almost be said to begin with Cantor. The Infinitesimal Calculus, though it cannot wholly dispense with infinity, has as few dealings with it as possible, and contrives to hide it away before facing the world. Cantor has abandoned this cowardly policy, and has brought the skeleton out of its cupboard. He has been emboldened in this course by denying that it is a skeleton. Indeed, like many other skeletons, it was wholly dependent on its cupboard, and vanished in the light of day. Speaking without metaphor, Cantor has established a new branch of Mathematics, in which, by mere correctness of deduction, it is shown that the supposed contradictions of infinity all depend upon extending, to the infinite, results which, while they can be proved concerning finite numbers, are in no sense necessarily true of *all* numbers. In this theory, it is necessary to treat separately of cardinals and ordinals, which are far more diverse in their properties when they are transfinite than when they are finite. Following the same order as previously—the order which seems to me to be alone philosophically correct—I shall begin with transfinite cardinals*.

284. The transfinite cardinals, which are also called *powers*, may be defined in the first place so as to include the finite cardinals, leaving it to be investigated in what respects the finite and the transfinite are distinguished. Thus Cantor gives the following definition†.

"We call the power or cardinal number of M that general idea which, by means of our active faculty of thought, is deduced from the collection M, by abstracting from the nature of its diverse elements and from the order in which they are given."

This, it will be seen, is merely a phrase indicating what is to be spoken of, not a true definition. It presupposes that every collection

---

* This is the order followed in *Math. Annalen*, XLVI, but not in the *Mannich-faltigkeitslehre*.

† *Math. Annalen*, XLVI, § 1.

has some such property as that indicated—a property, that is to say, independent of the nature of its terms and of their order; depending, we might feel tempted to add, only upon their number. In fact, number is taken by Cantor to be a primitive idea, and it is, in his theory, a primitive proposition that every collection has a number. He is therefore consistent in giving a specification of number which is not a formal definition.

By means, however, of the principle of abstraction, we can give, as we saw in Part II, a formal definition of cardinal numbers. This method, in essentials, is given by Cantor immediately after the above informal definition. We have already seen that, if two classes be called *similar* when there is a one-one relation which couples every term of either with one and only one term of the other, then similarity is symmetrical and transitive, and is reflexive for all classes. A one-one relation, it should be observed, can be defined without any reference to number, as follows: A relation is one-one when, if x has the relation to y, and x' differs from x, y' from y, then it follows that x' does not have the relation to y, nor x to y'. In this there is no reference to number; and the definition of similarity also is therefore free from such reference. Since similarity is reflexive. transitive, and symmetrical, it can be analyzed into the product of a many-one relation and its converse, and indicates at least one common property of similar classes. This property, or, if there be several, a certain one of these properties, we may call the cardinal number of similar classes, and the many-one relation is that of a class to the number of its terms. In order to fix upon one definite entity as *the* cardinal number of a given class, we decide to identify the number of a class with the whole class of classes similar to the given class. This class, taken as a single entity, has, as the proof of the principle of abstraction shows, all the properties required of a cardinal number. The method, however, is philosophically subject to the doubt resulting from the contradiction set forth in Part I, Chapter x.*

In this way we obtain a definition of the cardinal number of a class. Since similarity is reflexive for classes, every class has a cardinal number. It might be thought that this definition would only apply to finite classes, since, to prove that *all* terms of one class are correlated with *all* of another, complete enumeration might be thought necessary. This, however, is not the case, as may be seen at once by substituting *any* for *all*—a word which is generally preferable where infinite classes are concerned. Two classes u, v are similar when there is some one-one relation R such that, if x be any u, there is some term y of v such that xRy; and if y' be any v, there is some term x' of u such that $x'Ry'$. Here there is no need whatever of complete enumeration, but only of propositions concerning *any* u and *any* v. For example, the points on a given line are similar to the lines through a given point and meeting

* See Appendix.

the given line; for *any* point on the given line determines one and only one line through the given point, and *any* line through the given point meeting the given line determines one and only one point on the given line. Thus where our classes are infinite, we need some general proposition about *any* term of either class to establish similarity, but we do not need enumeration. And in order to prove that every (or any) class has a cardinal number, we need only the observation that any term of any class is identical with itself. No other general proposition about the terms of a class is requisite for the reflexive property of similarity.

285. Let us now examine the chief properties of cardinal numbers. I shall not give proofs of any of these properties, since I should merely repeat what has been said by Cantor. Considering first their relations to classes, we may observe that, if there be two sets of classes which are similar in pairs, and no two of the one set have any common part, nor yet any two of the other set, then the logical sum of all the classes of one set is similar to the logical sum of all the classes of the other set. This proposition, familiar in the case of finite classes, holds also of infinite classes. Again, the cardinal number of a class *u* is said to be greater than that of a class *v*, when no part of *v* is similar to *u*, but there is a part of *u* which is similar to *v*. In this case, also, the number of *v* is said to be less than that of *u*. It can be proved that, if there is a part of *u* which is similar to *v*, and a part of *v* which is similar to *u*, then *u* and *v* are similar*. Thus equal, greater, and less are all incompatible with each other, all transitive, and the last two asymmetrical. We cannot prove at all simply—and it seems more or less doubtful whether we can prove at all—that of two different cardinal numbers one must be greater and the other less†. It is to be observed that the definition of *greater* contains a condition not required in the case of finite cardinals. If the number of *v* be finite, it is sufficient that a proper part of *u* should be similar to *v*. But among transfinite cardinals this is not sufficient. For the general definition of *greater*, therefore, both parts are necessary. This difference between finite and transfinite cardinals results from the defining difference of finite and infinite, namely that when the number of a class is not finite, it always has a proper part which is similar to the whole; that is, every infinite class contains a part (and therefore an infinite number of parts) having the same number as itself. Certain particular cases of this proposition have long been known, and have been regarded as constituting a contradiction in the notion of infinite number. Leibniz, for example, points out‡ that, since every number can be doubled, the

* Bernstein and Schröder's theorem; for proofs see Borel, *Leçons sur la théorie des fonctions*, Paris, 1898, Note I, and Zermelo, *Göttinger Nachrichten*, 1901, pp. 34—38.

† Cantor's grounds for holding that this is so are vague, and do not appear to me to be valid. They depend upon the postulate that every class is the field of some well-ordered relation. See Cantor, *Math. Annalen*, XLVI, note to § 2.

‡ Gerhardt's ed. I, p. 338.

number of numbers is the same as the number of even numbers, whence he deduces that there is no such thing as infinite number. The first to generalize this property of infinite collections, and to treat it as not contradictory, was, so far as I know, Bolzano*. But the strict proof of the proposition, when the finite cardinals are defined by means of mathematical induction, as well as the demonstration that it is not contradictory, are due to Cantor and Dedekind. The proposition itself may be taken as the definition of the transfinite among cardinal numbers, for it is a property belonging to all of them, and to none of the finite cardinals†. Before examining this property further, however, we must acquire a more intimate acquaintance with the other properties of cardinal numbers.

286. I come now to the strictly arithmetical properties of cardinals, *i.e.* their addition, multiplication, etc.‡. The *addition* of numbers is defined, when they are transfinite, exactly as it was defined in the case of finite numbers, namely by means of logical addition. The number of the logical sum of two classes which have no common term is the sum of the numbers of the two classes. This can be extended by successive steps to any finite number of classes; for an infinite number of classes, forming a class of classes, the sum of their numbers, if no two have any common term, is still the number of their logical sum—and the logical sum of *any* class of classes, finite or infinite, is logically definable. For sums of two or three numbers, so defined, the commutative and associative laws still hold, *i.e.* we still have

$$a + b = b + a \text{ and } a + (b + c) = (a + b) + c.$$

The *multiplication* of two numbers is thus defined by Cantor: If M and N be two classes, we can combine any element of M with any element of N to form a couple (m, n); the number of all such couples is the product of the numbers of M and N. If we wish to avoid the notion of a couple in the definition, we may substitute the following§: Let u be a class of classes, a in number; let each of these classes belonging to u contain b terms; and let no two of these classes have any common term; then ab is the number of the logical sum of all these classes. This definition is still purely logical, and avoids the notion of a couple. Multiplication so defined obeys the commutative, associative, and distributive laws, *i.e.* we have

$$ab = ba, \ a\,(bc) = (ab)\,c, \ a\,(b + c) = ab + ac.$$

Hence addition and multiplication of cardinals, even when these are transfinite, satisfy all the elementary rules of Arithmetic.

* *Paradoxien des Unendlichen*, § 21.
† See Dedekind, *Was sind und was sollen die Zahlen?* No. 64.
‡ Cantor, *Math. Annalen*, XLVI, § 3; Whitehead, *American Journal of Math.* Vol. XXIV, No. 4.
§ Vivanti, *Théorie des Ensembles, Formulaire de Mathématiques*, Vol. I, Part VI, § 2, No. 4.

The definition of powers of a number (a^b) is also effected logically (*ib.* § 4). For this purpose, Cantor first defines what he calls a covering (*Belegung*) of one class N by another M. This is a law by which, to every element n of N is joined one and only one element m of M, but the same element m may be joined to many elements of N. That is, a *Belegung* is a many-one relation, whose domain includes N, and which correlates with the terms of N always terms of M. If a be the number of terms in M, b the number in N, then the number of all such many-one relations is defined to be a^b. It is easy to see that, for finite numbers, this definition agrees with the usual one. For transfinite numbers, indices have still the usual properties, *i.e.*

$$a^b a^c = a^{b+c}, \; a^c b^c = (ab)^c, \; (a^b)^c = a^{bc}.$$

In the case where $a = 2$, a^b is capable of a simpler definition, deduced from the above. If $a = 2$, 2^b will be the number of ways in which b terms can be related each to one of two terms. Now when those which are related to one of the two are given, the rest are related to the other. Hence it is enough, in each case, to specify the class of terms related to one of the two. Hence we get in each case a class chosen out of the b terms, and in all cases we get all such classes. Hence 2^b is the number of classes that can be formed out of b terms, or the number of combinations of b things any number at a time—a familiar theorem when b is finite, but still true when b is transfinite. Cantor has a proof that 2^b is always greater than b—a proof which, however, leads to difficulties when b is the number of all classes, or, more generally, when there is some collection of b terms in which all the sets chosen out of the b terms are themselves single terms of b*.

The definitions of multiplication given by Cantor and Vivanti require that the number of factors in a product should be finite; and this makes it necessary to give a new and independent definition of powers, if the exponent is allowed to be infinite. Mr A. N. Whitehead† has given a definition of multiplication which is free from this restriction, and therefore allows powers to be defined in the ordinary way as products. He has also found proofs of the formal laws when the number of summands, brackets, or factors is infinite. The definition of a product is as follows : Let k be a class of classes, no two of which have any terms in common. Choose out, in every possible way, one term and only one from each of the classes composing k. By doing this in all possible ways, we get a class of classes, called the multiplicative class of k. The number of terms in this class is defined to be the product of the numbers of terms in the various classes which are members of k. Where k has a finite number of members, it is easy to see that this agrees with the usual definition. Let u, v, w be the members of k, and let them have respectively a, β, γ terms. Then one term can be chosen out of u in

---

* See Chapter XLIII, *infra*.　　† *American Journal of Mathematics, loc. cit.*

a ways: for every way there are β ways of choosing one term out of v, and for every way of choosing one term out of u and one out of v, there are γ ways of choosing one out of w. Hence there are $a\,\beta\,\gamma$ ways of choosing one term out of each, when multiplication is understood in its usual sense. The multiplicative class is an important notion, by means of which transfinite cardinal Arithmetic can be carried a good deal further than Cantor has carried it.

287. All the above definitions apply to finite and transfinite integers alike, and, as we see, the formal laws of Arithmetic still hold. Transfinite integers differ from finite ones, however, both in the properties of their relation to the classes of which they are the numbers, and also in regard to the properties of classes of the integers themselves. Classes of numbers have, in fact, very different properties according as the numbers are all finite or are in part at least transfinite.

Among transfinite cardinals, some are particularly important, especially the number of finite numbers, and the number of the continuum. The number of finite numbers, it is plain, is not itself a finite number; for the class *finite number* is similar to the class *even finite number*, which is a part of itself. Or again the same conclusion may be proved by mathematical induction—a principle which also serves to define finite numbers, but which, being of a more ordinal nature, I shall not consider until the next chapter. The number of finite numbers, then, is transfinite. This number Cantor denotes by the Hebrew Aleph with the suffix 0; for us it will be more convenient to denote it by a_0. Cantor proves that this is the least of all the transfinite cardinals. This results from the following theorems (*loc. cit.* § 6):

(*A*) Every transfinite collection contains others as parts whose number is a_0.

(*B*) Every transfinite collection which is part of one whose number is a_0, also has the number a_0.

(*C*) No finite collection is similar to any proper part of itself.

(*D*) Every transfinite collection is similar to some proper part of itself[*].

From these theorems it follows that no transfinite number is less than the number of finite numbers. Collections which have this number are said to be denumerable, because it is always possible to *count* such collections, in the sense that, given any term of such a collection, there is some finite number n such that the given term is the nth. This is merely another way of saying that all the terms of a denumerable collection have a one-one correlation with the finite numbers, which again is equivalent to saying that the number of the collection is the same as that of the finite numbers. It is easy to see that the even numbers, the primes, the perfect squares, or any other class of finite numbers

* Theorems C and D require that the finite should be defined by mathematical induction, or else they become tautologous.

having no maximum, will form a denumerable series. For, arranging any such class in order of magnitude, there will be a finite number of terms, say n, before any given term, which will thus be the $(n+1)$th term. What is more remarkable is, that all the rationals, and even all real roots of equations of a finite degree and with rational co-efficients (*i.e.* all algebraic numbers), form a denumerable series. And even an n-dimensional series of such terms, where n is a finite number, or the smallest transfinite ordinal, is still denumerable*. That the rational numbers are denumerable can be easily seen, by arranging them in the order in which those with smaller sum of numerator and denominator precede those with larger sum, and of those with equal sums, those with the smaller numerators precede those with larger ones. Thus we get the series

$$1, 1/2, 2, 1/3, 3, 1/4, 2/3, 3/2, 4, 1/5 \ldots$$

This is a discrete series, with a beginning and no end; *every* rational number will occur in this series, and will have a finite number of predecessors. In the other cases the proof is rather more difficult.

All denumerable series have the same cardinal number a_0, however different they may appear. But it must not be supposed that there is no number greater than a_0. On the contrary, there is an infinite series of such numbers†. The transfinite cardinals are asserted by Cantor to be well-ordered, that is, such that every one of them except the last of all (if there be a last) has an immediate successor, and so has every class of them which has any numbers at all after it. But they do not all have an immediate predecessor; for example, a_0 itself has no immediate predecessor. For if it had one, this would have to be the last of the finite numbers; but we know that there is no last finite number. But Cantor's grounds for his assertion that the cardinals are well-ordered seem insufficient, so that for the present this must remain an open question.

288. Of the transfinite numbers other than a_0, the most important is the number of the continuum. Cantor has proved that this number is not a_0‡, and hopes to prove that it is a_1§—a hope which, though he has long cherished it, remains unfulfilled. He has shown that the number of the continuum is 2^{a_0}‖—a most curious theorem : but it must still remain doubtful whether this number is a_1, though there are reasons which rendered this probable¶. As to the definition of a_1

---

* See *Acta Mathematica*, II, pp. 306, 313, 326.

† See *Jahresbericht der deutschen Mathematiker-Vereinigung* I, 1892 ; *Rivista di Matematica*, II, pp. 165–7. Cantor's assertion that there is no greatest transfinite cardinal is open to question. See Chap. XLIII, *infra*.

‡ *Acta Math.* II, p. 308.   § *Ib.* p. 404. a_1 is the number next after a_0.

‖ *Math. Annalen*, XLVI, § 4, note.

¶ See Couturat, *De l'Infini Mathématique*, Paris, 1896, p. 655. The ground alleged by Cantor for identifying the second power with that of the continuum is, that every infinite linear collection of points has either the first power, or that of the continuum, whence it would seem to follow that the power of the continuum must be the next after the first. (*Math. Annalen*, 23, p. 488; see also *Acta Math.* VII.) But and of the whole succession of transfinite cardinals, this is a matter which is better postponed until we have discussed the transfinite ordinals. It must not be supposed that we can obtain a new transfinite cardinal by merely adding one to it, or even by adding any finite number or α_0. On the contrary, such puny weapons cannot disturb the transfinite cardinals. It is known that in the case of α_0 and a certain class of transfinite cardinals, a number is equal to its double; also that in the case of α_0 and a presumably different class of transfinite cardinals, a number is equal to its square. The sum of two numbers belonging to the former of these classes is equal to the greater of the two numbers. It is not known whether all transfinite cardinals belong to one or both of these classes *.

289. It may be asked: In what respect do the finite and transfinite cardinals together form a single series? Is not the series of finite numbers complete in itself, without the possibility of extending its generating relation? If we define the series of integers by means of the generating relation of differing by one—the method which is most natural when the series is to be considered as a progression— then, it must be confessed, the finite integers form a complete series, and there is no possibility of adding terms to them. But if, as is appropriate in the theory of cardinals, we consider the series as arising by correlation with that of whole and part among classes of which the integers can be asserted, then we see that this relation does extend beyond finite numbers. There are an infinite number of infinite classes in which any given finite class is contained; and thus, by correlation with these, the number of the given finite class precedes that of any one of the infinite classes. Whether there is any other sense in which all integers, finite and transfinite, form a single series, I leave undecided; the above sense would be sufficient to show that there is no logical error in regarding them as a single series, if it were known that of any two cardinals one must be the greater. But it is now time to turn our attention to the transfinite ordinals.

the inference seems somewhat precarious. Consider, for example, the following analogy: in a compact series, the stretch determined by two terms consists either of an infinite number of terms, or, when the two terms coincide, of one term only, and never of a finite number of terms other than one. But finite stretches are presented by other types of series, *e.g.* progressions.

The theorem that the number of the continuum is 2^{α_0} results very simply from the proposition of Chapter xxxvi, that infinite classes of finite integers form a continuous series. The number of all classes of finite integers is 2^{α_0} (*vide supra*), and the number of finite classes is α_0. Hence the number of all infinite classes of finite integers is 2^{α_0} for the subtraction of α_0 does not diminish any number greater than α_0; 2^{α_0} is therefore the number of the continuum. To prove that this number is α_1 it would therefore be sufficient to show that the number of infinite classes of finite integers is the same as the number of types of series that can be formed of all the finite integers; for the latter number, as we shall see in the next chapter, is α_1.

* Cf. Whitehead, *loc. cit.* pp. 392–4.

CHAPTER XXXVIII.

TRANSFINITE ORDINALS.

290. THE transfinite ordinals are, if possible, even more interesting and remarkable than the transfinite cardinals. Unlike the latter, they do not obey the commutative law, and their arithmetic is therefore quite different from elementary arithmetic. For every transfinite cardinal, or at any rate for any one of a certain class, there is an infinite collection of transfinite ordinals, although the cardinal number of all ordinals is the same as or less than that of all cardinals. The ordinals which belong to series whose cardinal number is a_0 are called the second class of ordinals; those corresponding to a_1 are called the third class, and so on. The ordinal numbers are essentially classes of series, or better still, classes of generating relations of series; they are defined, for the most part, by some relation to mathematical induction. The finite ordinals, also, may be conceived as types of series: for example, the ordinal number n may be taken to mean "a serial relation of n terms;" or, in popular language, n terms in a row. This is an ordinal notion, distinct from "nth," and logically prior to it[*]. In this sense, n is the name of a class of serial relations. It is this sense, not that expressed by "nth," which is generalized by Cantor so as to apply to infinite series.

291. Let us begin with Cantor's definition of the second class of ordinal numbers[†].

"It is now to be shown," he says, "how we are led to the definitions of the new numbers, and in what way are obtained the natural sections, which I call *classes of numbers*, in the absolutely endless series of real integers.... The series (1) of positive real whole numbers 1, 2, 3, ... ν,... arises from repeated positing and combination of units which are pre-supposed and regarded as equal; the number ν is the expression both for a certain finite amount (*Anzahl*) of such successive positings, and for the combination of the units posited into a whole. Thus the formation of finite

---

[*] Cf. *supra* Part IV, Chap. xxix, §§ 231, 232.
[†] *Mannichfaltigkeitslehre*, § 11, pp. 32, 33.

real whole numbers rests on the addition of a unit to a number which
has already been formed; I call this moment, which, as we shall see
immediately, also plays an essential part in the formation of the higher
integers, the *first principle of formation*. The amount (*Anzahl*) of
possible numbers ν of the class (1) is infinite, and there is no greatest
among them. Thus however contradictory it would be to speak of
a greatest number of the class (1), there is yet nothing objectionable
in imagining a new number, which we will call ω, which is to express
that the whole collection (1) is given by its law in its natural order
of succession. (In the same way as ν expresses the combination of a
certain finite amount of units into a whole.) It is even permissible to
think of the newly created number ω as a limit, towards which the
numbers ν tend, if by this nothing else is understood but that ω is
the first integer which follows all the numbers ν, *i.e.* is to be called
greater than each of the numbers ν. By allowing further additions
of units to follow the positing of the number ω we obtain, by the help
of the *first* principle of formation, the further numbers

$$\omega + 1, \quad \omega + 2, \ldots\ldots\ldots \quad \omega + \nu, \ldots\ldots\ldots;$$

Since here again we come to no greatest number, we imagine a new one,
which we may call 2ω, and which is to be the first after all previous
numbers ν and $\omega + \nu$.

"The logical function which has given us the two numbers ω and 2ω
is evidently different from the *first* principle of formation; I call it the
second principle of formation of real integers, and define it more exactly
as follows: If we have any determinate succession of defined real integers,
among which there is no greatest number, by means of this second
principle of formation a new number is created, which is regarded as
the *limit* of those numbers, *i.e.* is defined as the next number greater
than all of them."

The two principles of formation will be made clearer by considering
that an ordinal number is merely a type or class of series, or rather
of their generating relations. Thus if we have any series which has
no last term, every part of such a series which can be defined as all
the terms up to and including a certain term of the series will have
a last term. But since the series itself has no last term, it is of a
different type from any such part or segment of itself. Hence the
ordinal number representing the series as a whole must be different
from that representing any such segment of itself, and must be a
number having no immediate predecessor, since the series has no last
term. Thus ω is simply the name of the class *progression*, or of the
generating relations of series of this class. The second principle of
formation, in short, is that by which we define a certain type of series
having no last term. Considering the ordinals preceding any ordinal
α which is obtained by the second principle as representing segments

of a series represented by α, the ordinal α itself represents the limit of such segments; and as we saw before, the segments always have a limit (provided they have no maximum), even when the original series has none *.

In order to define a class among transfinite ordinals (of which, as is evident, the succession is infinite), Cantor introduces what he calls a principle of limitation (*Hemmungsprincip*)†. According to this principle, the *second class* of ordinals is to consist only of those whose predecessors, from 1 upwards, form a series of the first power, *i.e.* a series whose cardinal number is α_0, or one whose terms, in a suitable order, have a one-one relation to the finite integers. It is then shown that the power, or cardinal number, of the second class of ordinals as a whole, is different from α_0 (p. 35), and is further the very next cardinal number after α_0 (p. 37). What is meant by the next cardinal number to α_0 results clearly from the following proposition (p. 38): " If M be any well-defined collection of the power of the second class of numbers, and if any infinite portion M of M be taken, then either the collection M' can be considered as a simply infinite series, or it is possible to establish a unique and reciprocal correspondence between M and M'." That is to say, any part of a collection of the second power is either finite, or of the first power, or of the second; and hence there is no power between the first and second.

292. Before proceeding to the addition, multiplication, *etc.*, of ordinals, it will be well to take the above propositions, as far as possible, out of their mathematical dress, and to state, in ordinary language, exactly what it is they mean. As for the ordinal ω, this is simply the name for the class of generating relations of progressions. We have seen how a progression is defined: it is a series which has a first term, and a term next after each term, and which obeys mathematical induction. By mathematical induction itself we can show that every part of a progression, if it has a last term, has some finite ordinal number n, where n denotes the class of series consisting of n terms in order; while every part which has no last term is itself a progression; also we can show (what is indeed obvious) that no finite ordinal will represent a progression. Now progressions are a perfectly definite class of series, and the principle of abstraction shows that there is some entity to which all of them have a relation which they have to nothing else— for all progressions are ordinally similar (*i.e.* have a one-one relation

---

* On the segments of well-ordered series see Cantor's article in *Math. Annalen*, XLIX, § 13. It is important to observe that the ordinals above explained are analogous, in their genesis, to the real numbers considered as segments (*vide* Chap. XXXIII, *supra*). Here, as there, the existence of ω is not open to question when the segment-theory is adopted, whereas on any other theory the existence-theorem is indemonstrable and unplausible.

† *Mannichfaltigkeitslehre*, p. 34.

such that earlier terms are correlated with earlier ones, and later with later), and ordinal similarity is symmetrical, transitive, and (among series) reflexive. This entity, to which the principle of abstraction points, may be taken to be the type or class of serial relations, since no series can belong to more than one type of series. The type to which progressions belong, then, is what Cantor calls ω. Mathematical induction, starting from any finite ordinal, can never reach ω, since ω is not a member of the class of finite ordinals. Indeed, we may define the finite ordinals or cardinals—and where series are concerned, this seems the best definition—as those which, starting from 0 or 1, can be reached by mathematical induction. This principle, therefore, is not to be taken as an axiom or postulate, but as the definition of finitude. It is to be observed that, in virtue of the principle that every number has an immediate successor, we can *prove* that any assigned number, say, 10,937, is finite—provided, of course, that the number assigned is a finite number. That is to say, every proposition concerning 10,937 can be proved without the use of mathematical induction, which, as most of us can remember, was not mentioned in the Arithmetic of our childhood. There is therefore no kind of logical error in using the principle as a definition of the class of finite numbers, nor is there a shadow of a reason for supposing that the principle applies to *all* ordinal or *all* cardinal numbers.

At this point, a word to the philosophers may be in season. Most of them seem to suppose that the distinction between the finite and the infinite is one whose meaning is immediately evident, and they reason on the subject as though no precise definitions were needed. But the fact is, that the distinction of the finite from the infinite is by no means easy, and has only been brought to light by modern mathematicians. The numbers 0 and 1 are capable of logical definition, and it can be shown logically that every number has a successor. We can now define finite numbers either by the fact that mathematical induction can reach them, starting from 0 or 1—in Dedekind's language, that they form the chain of 0 or 1—or by the fact that they are the numbers of collections such that no proper part of them has the same number as the whole. These two conditions may be easily shown to be equivalent. But they alone precisely distinguish the finite from the infinite, and any discussion of infinity which neglects them must be more or less frivolous.

293. With regard to numbers of the second class other than ω, we may make the following remark. A collection of two or more terms is always, except possibly for some very large infinite collections, the field of more than one serial relation. Men may be arranged by their rank, age, wealth, or in alphabetical order: all these relations among men generate series, and each places mankind in a different order. But when a collection is finite, all possible orders give one and

the same ordinal number, namely that corresponding to the cardinal number of the collection. That is to say, all series which can be formed of a certain finite number of terms are ordinally similar. With infinite series, this is quite different. An infinite collection of terms which is capable of different orders may belong, in its various orders, to quite different types. We have already seen that the rationals, in one order, form a compact series with no beginning or end, while in another order they form a progression. These are series of entirely different types; and the same possibility extends to all infinite series. The ordinal type of a series is not changed by the interchange of two consecutive terms, nor, consequently, in virtue of mathematical induction, by any finite number of such interchanges. The general principle is, that the type of a series is not changed by what may be called a *permutation*. That is, if P be a serial relation by which the terms of u are ordered, R a one-one relation whose domain and whose converse domain are both u, then $\breve{R}PR$ is a serial relation of the same type as P; and all serial relations whose field is u, and which are of the same type as P, are of the above form $\breve{R}PR$. But by a rearrangement not reducible to a permutation, the type, in general, is changed. Consider, for example, the natural numbers, first in their natural order, and then in the order in which 2 comes first, then all the higher numbers in their natural order, and last of all 1. In the first order, the natural numbers form a progression; in the second, they form a progression together with a last term. In the second form, mathematical induction no longer applies; there are propositions which hold of 2, and of every subsequent finite number, but not of 1. The first form is the type of any fundamental series of the kind we considered in Chapter XXXVI; the second is the type of any such series together with its limit. Cantor has shown that every denumerable collection can be given an order which corresponds to any assigned ordinal number of the second class[*]. Hence the second class of ordinal numbers may be defined as all the types of well-ordered series in which any one given denumerable collection can be arranged by means of different generating relations. The possibility of such different types depends upon the fundamental property of infinite collections, that an infinite part of an infinite collection can always be found, which will have a one-one correlation with the whole. If the original collection was a series, the part, by this correlation, becomes a series ordinally similar to the whole: the remaining terms, if added after all the terms of the infinite part, will then make the whole ordinally different from what it was[†].

---

[*] *Acta Math.* II, p. 394.

[†] The remaining terms, if they be finite in number, will often not alter the type if added at the beginning; but if they be infinite, they will in general alter it even then. This will soon be more fully explained.

We may assimilate the theory of ordinals to that of cardinals as follows. Two relations will be said to be *like* when there is a one-one relation S, whose domain is the field of one of them (P), and which is such that the other relation is $\breve{S}PS$. If P be a well-ordered relation, *i.e.* one which generates a well-ordered series, the class of relations like P may be defined as the ordinal number of P. Thus ordinal numbers result from likeness among relations as cardinals from similarity among classes.

294. We can now understand the rules for the addition and multiplication of transfinite ordinals. Both operations obey the associative, but not the commutative law. The distributive law is true, in general, only in the form
$$\gamma(\alpha + \beta) = \gamma\alpha + \gamma\beta, \quad \cdot$$
where $\alpha + \beta$, α, β are multipliers*. That addition does not obey the commutative law may be easily seen. Take for example $\omega + 1$ and $1 + \omega$. The first denotes a progression followed by a single term: this is the type presented by a progression and its limit, which is different from a simple progression. Hence $\omega + 1$ is a different ordinal from ω. But $1 + \omega$ denotes a progression *preceded* by a single term, and this is again a progression. Hence $1 + \omega = \omega$, but $1 + \omega$ does not equal $\omega + 1$†. The numbers of the second class are, in fact, of two kinds, (1) those which have an immediate predecessor, (2) those which have none. Numbers such as ω, $\omega.2$, $\omega.3,\ldots\omega^2$, $\omega^3\ldots\omega^\omega\ldots$ have no immediate predecessor. If any of these numbers be added to a finite number, the same transfinite number reappears; but if a finite number be added to any of these numbers, we get a new number. The numbers with no predecessor represent series which have no end, while those which have a predecessor represent series which have an end. It is plain that terms added at the beginning of a series with no end leave it endless; but the addition of a terminating series after an endless one produces a terminating series, and therefore a new type of order. Thus there is nothing mysterious about these rules of addition, which simply express the type of series resulting from the combination of two given series.

Hence it is easy to obtain the rules of subtraction‡. If α is less than β, the equation
$$\alpha + \xi = \beta$$
has always one and only one solution in ξ, which we may represent by $\beta - \alpha$. This gives the type of series that must be added after α to produce β. But the equation
$$\xi + \alpha = \beta$$

---

* *Mannichfaltigkeitslehre*, p. 39; $\alpha + \beta$ will be the type of a series consisting of two parts, namely a part of the type α followed by a part of the type β; $\gamma\alpha$ will be the type of a series consisting of a series of the type α of series of the type γ. Thus a series composed of two progressions is of the type $\omega.2$.

† *Math. Annalen*, XLVI, § 8.

‡ *Mannichfaltigkeitslehre*, p. 39.

will sometimes have no solution, and at other times an infinite number of solutions. Thus the equation

$$\xi + \omega = \omega + 1$$

has no solution at all : no number of terms added at the beginning of a progression will produce a progression together with a last term. In fact, in the equation $\xi + \alpha = \beta$, if α represents an endless type, while β represents a terminating type, it is sufficiently evident that terms added before α can never produce a terminating type, and therefore can never produce the type β. On the other hand, if we consider the equation

$$\xi + \omega = \omega . 2$$

this will be satisfied by $\xi = \omega + n$, where n is zero or any finite number. For n before the second ω will coalesce with this to form ω, and thus $\omega + n + \omega = \omega . 2$. In this case, therefore, ξ has an infinite number of values. In all such cases, however, the possible values of ξ have a minimum, which is a sort of principal value of the difference between β and α. Thus subtraction is of two kinds, according as we seek a number which, added to α, will give β, or a number to which α may be added so as to give β. In the first case, provided α is less than β, there is always a unique solution ; in the second case, there may be no solution, and there may be an infinite number of solutions.

295. The multiplication of ordinals is defined as follows*. Let M and N be two series of the types α and β. In N, in place of each element n, substitute a series M_n of the type α ; and let S be the series formed of all the terms of all series M_n, taken in the following order : (1) any two elements of S which belong to the same series M_n are to preserve the order they had in M_n ; two elements which belong to different series M_n, $M_{n'}$ are to have the order which n and n' have in N. Then the type of S depends only upon α and β, and is defined to be their product $\alpha\beta$, where α is the multiplicand, and β the multiplicator. It is easy to see that products do not always obey the commutative law. For example, $2 . \omega$ is the type of series presented by

$$e_1, f_1; \quad e_2, f_2; \quad e_3, f_3; \quad \ldots e_\nu, f_\nu; \quad \ldots\ldots$$

which is a progression, so that $2 . \omega = \omega$. But $\omega . 2$ is the type

$$e_1, e_2, e_3 \ldots e_\nu, \ldots; \quad f_1, f_2, f_3, \ldots f_\nu, \ldots$$

which is a combination of two progressions, but not a single progression. In the former series, there is only one term, e_1, which has no immediate predecessor ; in the latter there are two, e_1 and f_1.

Of division, as of subtraction, two kinds must be distinguished†. If there are three ordinals α, β, γ, such that $\beta = \alpha\gamma$, then the equation $\beta = \alpha\xi$ has no other solution than $\xi = \gamma$, and we may therefore denote

---

* *Math. Annalen*, XLVI, § 8.
† *Mannichfaltigkeitslehre*, p. 40.

γ by β/α^*. But the equation $\beta = \xi\alpha$, if soluble at all, may have several or even an infinity of roots; of which, however, one is always the smallest. This smallest root is denoted by $\beta//\alpha$.

Multiplication of ordinals is the process of representing a series of series as a single series, each series being taken as a whole, and preserving its place in the series of series. Division, on the other hand, is the process of splitting up a single series into a series of series, without altering the order of its terms. Both these processes have some importance in connection with dimensions. Division, as is plain, is only possible with some types of series; those with which it is not possible may be called primes. The theory of primes is interesting, but it is not necessary for us to go into it †.

296. Every rational integral or exponential function of ω is a number of the second class, even when such numbers as ω^ω, ω^{ω^2}, *etc.*, occur‡. But it must not be supposed that all types of denumerable series are capable of such a form. For example, the type η, which represents the rationals in order of magnitude§, is wholly incapable of expression in terms of ω. Such a type is not called by Cantor an ordinal *number*. The term *ordinal number* is reserved for *well-ordered* series, *i.e.* such as have the following two properties‖:

I. There is in the series F a first term.

II. If F' is a part of F, and if F possesses one or more terms which come after all the terms of F', then there is a term f' of F which immediately follows F', so that there is no term of F before f' and after all terms of F'.

All possible functions of ω and finite ordinals only, to the exclusion of other types such as that of the rationals, represent well-ordered series, though the converse does not hold. In every well-ordered series, there is a term next after any given term, except the last term if there be one; and provided the series is infinite, it always contains parts which are progressions. A term which comes next after a progression has no immediate predecessor, and the type of the segment formed of its predecessors is of what is called the second species. The other terms have immediate predecessors, and the types of the segments formed of their predecessors are said to be of the first species.

---

* Cantor has changed his notation in regard to multiplication: formerly, in $a \cdot \beta$, a was the multiplicator, and β the multiplicand; now, the opposite order is adopted. In following older works, except in actual quotations, I have altered the order to that now adopted.

† See *Mannichfaltigkeitslehre*, p. 40.

‡ On the exponential function, see *Math. Annalen*, XLIX, §§ 18–20.

§ *Math. Annalen*, XLVI, § 9.

‖ *Math. Annalen*, XLIX, § 12. The definition may be replaced by the following, which is equivalent to it: a series is well-ordered if every class contained in the series (except of course the null-class) has a first term.

297. The consideration of series which are not well-ordered is important, though the results have far less affinity to Arithmetic than in the case of well-ordered series. Thus the type η is not expressible as a function of ω, since all functions of ω represent series with a first term, whereas η has no first term, and all functions of ω represent series in which every term has an immediate successor, which again is not the case with η. Even the series of negative and positive integers and zero cannot be expressed in terms of ω, since this series has no beginning. Cantor defines for this purpose a serial type $*\omega$, which may be taken as the type of a *regression* (*ib.* § 7). The definition of a progression, as we have seen, is relative to some one-one aliorelative $P\dagger$. When $\overset{\smile}{P}$ generates a progression, this progression with respect to $\overset{\smile}{P}$ is a regression with respect to P, and its type, considered as generated by P, is denoted by $*\omega$. Thus the whole series of negative and positive integers is of the type $*\omega + \omega$. Such a series can be divided anywhere into two progressions, generated by converse relations; but in regard to one relation, it is not reducible to any combination of progressions. Such a series is completely defined, by the methods of Part IV, as follows : P is a one-one aliorelative ; the field of P is identical with that of $\overset{\smile}{P}$; the disjunctive relation " some finite positive power of P " is transitive and asymmetrical ; and the series consists of all terms having this relation or its converse to a given term together with the given term. The class of series corresponding to any transfinite ordinal type may always be thus defined by the methods of Part IV ; but where a type cannot be expressed as a function of ω or $*\omega$ or both, it will usually be necessary, if we are to define our type completely, either to bring in a reference to some other relation, in regard to which the terms of our series form a progression, or to specify the behaviour of our series with respect to limits. Thus the type of the series of rationals is not defined by specifying that it is compact, and has no beginning or end ; this definition applies also, for example, to what Cantor calls the semi-continuum, *i.e.* the continuum with its ends cut off. We must add that the rationals are denumerable, *i.e.* that, with respect to another relation, they form a progression. I doubt whether, in this case, the behaviour of the rationals with regard to limits can be used for definition. Their chief characteristics in this respect are : (1) that they are condensed in themselves, *i.e.* every term of them is the limit of certain progressions and regressions ; (2) in any interval, a progression or a regression which has no limit is contained. But both these characteristics belong to the series of irrational numbers, *i.e.* to the series obtained by omitting all rationals from the series of real numbers ; yet this series is not denumerable.

$\dagger$ An aliorelative is a relation which no term can have to itself. This term is due to Pierce. See Schröder, *Algebra u. Logik der Relative*, p. 131.

Thus it would seem that we cannot define the type η, to which the rationals belong, without reference to two generating relations. The type η is that of endless compact series whose terms, with reference to another relation, form a progression.

From the last remark, we see clearly the importance of the correlation of series, with which we began the discussions of Part V. For it is only by means of correlation that the type of the rationals, and hence the continuum, can be defined. Until we bring in some other relation than that by which the order of magnitude among rationals arises, there is ·..thing to distinguish the type of the rationals from that of the irrationals.

298. The consideration of ordinals not expressible as functions of ω shows clearly that ordinals in general are to be considered—as I suggested at the beginning of this chapter—as classes or types of serial relations, and to this view Cantor himself now apparently adheres; for in the article in the *Mathematische Annalen*, Vol. XLVI, he speaks of them always as types of order, not as numbers, and in the following article (*Math. Annalen*, XLIX, § 12), he definitely restricts ordinal numbers to well-ordered series. In his earlier writings, he confined himself more to functions of ω, which bear many analogies to more familiar kinds of numbers. These are, in fact, types of order which may be presented by series of finite and transfinite cardinals which begin with some cardinal. But other types of order, as we have now seen, have very little resemblance to numbers.

299. It is worth while to repeat the definitions of general notions involved in terms of what may be called relation-arithmetic*. If P, Q be two relations such that there is a one-one relation S whose domain is the field of P and which is such that $Q = \tilde{S}PS$, then P and Q are said to be *like*. The class of relations like P, which I denote by λP, is called P's *relation-number*. If the fields of P and Q have no common terms, $P + Q$ is defined to be P or Q or the relation which holds between any term of the field of P and any term of the field of Q, and between no other terms. Thus $P + Q$ is not equal to $Q + P$. Again $\lambda P + \lambda Q$ is defined as $\lambda (P + Q)$. For the summation of an infinite number of relations, we require an aliorelative whose field is composed of relations whose fields are mutually exclusive. Let P be such a relation, p its field, so that p is a class of relations. Then $\Sigma_P p$ is to denote either one of the relations of the class p or the relation of any term belonging to the field of some relation Q of the class p to a term belonging to the field of another relation R (of the class p) to which Q has the relation P. (If P be a serial relation, and p a class of serial relations, $\Sigma_P p$ will be the generating relation of the sum of the various series generated by terms of p taken in the order generated by P.) We may define the sum

* Cf. Part IV, Chap. XXIX, § 231.

of the relation-numbers of the various terms of p as the relation-number of $\Sigma_P p$. If all the terms of p have the same relation-number, say α, and if β be the relation-number of P, $\alpha \times \beta$ will be defined to be the relation-number of $\Sigma_P p$. Proceeding in this way, it is easy to prove generally the three formal laws which hold of well-ordered series, namely :

$$(\alpha + \beta) + \gamma = \alpha + (\beta + \gamma)$$

$$\alpha\,(\beta + \gamma) = \alpha\beta + \alpha\gamma$$

$$(\alpha\beta)\,\gamma = \alpha\,(\beta\gamma).$$

The proofs are very closely analogous to those discovered by Mr Whitehead for cardinal numbers (*Amer. Journal of Math.*, Vol. XXIV); but they differ by the fact that no method has yet been discovered for defining an infinite product of relation-numbers, or even of ordinal numbers.

300. It is to be observed that the merit of the above method is that it allows no doubt as to existence-theorems—a point in which Cantor's work leaves something to be desired. As this is an important matter, and one in which philosophers are apt to be sceptical, I shall here repeat the argument in outline. It may be shown, to begin with, that no finite class embraces all terms : this results, with a little care, from the fact that, since 0 is a cardinal number, the number of numbers up to and including a finite number n is $n + 1$. Further, if n be a finite number, $n + 1$ is a new finite number different from all its predecessors. Hence finite cardinals form a progression, and therefore the ordinal number ω and the cardinal number α_0 exist (in the mathematical sense). Hence, by mere rearrangements of the series of finite cardinals, we obtain all ordinals of Cantor's second class. We may now define the ordinal number ω_1 as the class of serial relations such that, if u be a class contained in the field of one of them, to say that u has successors implies and is implied by saying that u has α_0 terms or a finite number of terms ; and it is easy to show that the series of ordinals of the first and second classes in order of magnitude is of this type. Hence the existence of ω_1 is proved; and α_1 is defined to be the number of terms in a series whose generating relation is of the type ω_1. Hence we can advance to ω_2 and α_2 and so on, and even to ω_ω and α_ω, whose existence can be similarly proved : ω_ω will be the type of generating relation of a series such that, if u be a class contained in the series, to say that u has successors is equivalent to saying that u is finite or has, for a suitable finite value of n, α_n terms. This process gives us a one-one correlation of ordinals with cardinals : it is evident that, by extending the process, we can make each cardinal which can belong to a well-ordered series correspond to one and only one ordinal. Cantor assumes as an axiom that *every* class is the field of some well-ordered series, and deduces that *all* cardinals can be correlated with ordinals by the above method. This assumption seems to me unwarranted, especially in view of the fact that no one has yet succeeded in arranging a class of 2^{α_0} terms in a well-ordered series. We do not know that of any two different cardinal numbers one must be the greater, and it may be that 2^{α_0} is neither greater nor less than α_1 and α_2 and their successors, which may be called well-ordered cardinals because they apply to well-ordered classes.

301. There is a difficulty as regards the type of the whole series of ordinal numbers. It is easy to prove that every segment of this series is well-ordered, and it is natural to suppose that the whole series is also well-ordered. If so, its type would have to be the greatest of all ordinal numbers, for the ordinals less than a given ordinal form, in order of magnitude, a series whose type is the given ordinal. But there cannot be a greatest ordinal number, because every ordinal is increased by the addition of 1. From this contradiction, M. Burali-Forti, who discovered it*, infers that of two different ordinals, as of two different cardinals, it is not necessary that one should be greater and the other less. In this, however, he consciously contradicts a theorem of Cantor's which affirms the opposite†. I have examined this theorem with all possible care, and have failed to find any flaw in the proof‡. But there is another premiss in M. Burali-Forti's argument, which appears to me more capable of denial, and that is, that the series of all ordinal numbers is well-ordered. This does not follow from the fact that all its segments are well-ordered, and must, I think, be rejected, since, so far as I know, it is incapable of proof. In this way, it would seem, the contradiction in question can be avoided.

302. We may now return to the subject of the successive derivatives of a series, already briefly discussed in Chapter xxxvi. This forms one of the most interesting applications of those ordinals which are functions of ω, and may even be used as an independent method of defining them. We have already seen how, from a series P, its first derivative is obtained§. The first derivative of P, which is denoted by P', is the class of its limiting points. P'', the second derivative of P', consists of the limiting-points of P', and so on. Every infinite collection has at least one limiting-point: for example, ω is the limit of the finite ordinals. By induction we can define any derivative of finite order P^ν. If P^ν consists of a finite number of points, $P^{\nu+1}$ vanishes; if this happens for any finite number ν, P is said to be of the 1st genus and the νth

---

* "Una questione sui numeri transfiniti," *Rendiconti del circolo Matematico di Palermo*, Vol. xi (1897).

† Theorem к of § 13 of Cantor's article in *Math. Annalen*, Vol. xlix.

‡ I have reproduced the proof in symbolic form, in which errors are more easily detected, in RdM, Vol. viii, Prop. 5.47 of my article.

§ What follows is extracted from *Acta Math.* ii, pp. 341–360. I shall assume for simplicity that all definable limits exist, *i.e.* that a series has a limit whenever the corresponding segments have one. I have shown in Chapter xxxvi how to state results so as to avoid this assumption; but the necessary circumlocution is tiresome.

species. But it may happen that no P^ν vanishes, and in this case all finite derivatives may have common points. The points which all have in common form a collection which is defined as P^ω. It is to be observed that P^ω is thus defined without requiring the definition of ω. A term x belongs to P^ω if, whatever finite integer ν may be, x belongs to P^ν. It is to be observed that, though P' may contain points not belonging to P, yet subsequent derivatives introduce no new points. This illustrates the creative nature of the method of limits, or rather of segments : when it is first applied, it may yield new terms, but later applications give no further terms. That is, there is an intrinsic difference between a series which has been, or may have been, obtained as the derivative of some other series, and one not so obtainable. Every series which contains its first derivative is itself the derivative of an infinite number of other series*. The successive derivatives, like the segments determined by the various terms of a regression, form a series in which each term is part of each of its predecessors ; hence P^ω, if it exists, is the lower limit of all the derivatives of finite order. From P^ω it is easy to go on to $P^{\omega+\nu}$, $P^{\omega.2}$, etc. Series can be actually constructed in which any assigned derivative, finite or transfinite of the second class, is the first to vanish. When none of the finite derivatives vanishes, P is said to be of the second genus. It must not be inferred, however, that P is not denumerable. On the contrary, the first derivative of the rationals is the number-continuum, which is perfect, so that all its derivatives are identical with itself ; yet the rationals, as we know, are denumerable. But when P^ν vanishes, P is always denumerable, if ν be finite or of the second class.

The theory of derivatives is of great importance to the theory of real functions†, where it practically enables us to extend mathematical induction to any ordinal of the second class. But for philosophy, it seems unnecessary to say more of it than is contained in the above remarks and in those of Chapter xxxvi. Popularly speaking, the first derivative consists of all points in whose neighbourhood an infinite number of terms of the collection are heaped up ; and subsequent derivatives give, as it were, different degrees of concentration in any neighbourhood. Thus it is easy to see why derivatives are relevant to continuity : to be continuous, a collection must be as concentrated as possible in every neighbourhood containing any terms of the collection. But such popular modes of expression are incapable of the precision which belongs to Cantor's terminology.

---

* *Formulaire de Mathématiques*, Vol. ii, Part iii, § 71, 4–8.

† See Dini, *Theorie der Functionen*, Leipzig, 1892; esp. Chap. xiii and Translator's preface.

CHAPTER XXXIX.

THE INFINITESIMAL CALCULUS.

303. The Infinitesimal Calculus is the traditional name for the differential and integral calculus together, and as such I have retained it; although, as we shall shortly see, there is no allusion to, or implication of, the infinitesimal in any part of this branch of mathematics.

The philosophical theory of the Calculus has been, ever since the subject was invented, in a somewhat disgraceful condition. Leibniz himself—who, one would have supposed, should have been competent to give a correct account of his own invention—had ideas, upon this topic, which can only be described as extremely crude. He appears to have held that, if metaphysical subtleties are left aside, the Calculus is only approximate, but is justified practically by the fact that the errors to which it gives rise are less than those of observation*. When he was thinking of Dynamics, his belief in the actual infinitesimal hindered him from discovering that the Calculus rests on the doctrine of limits, and made him regard his dx and dy as neither zero, nor finite, nor mathematical fictions, but as really representing the units to which, in his philosophy, infinite division was supposed to lead†. And in his mathematical expositions of the subject, he avoided giving careful proofs, contenting himself with the enumeration of rules‡. At other times, it is true, he definitely rejects infinitesimals as philosophically valid§; but he failed to show how, without the use of infinitesimals, the results obtained by means of the Calculus could yet be exact, and not approximate. In this respect, Newton is preferable to Leibniz: his Lemmas‖ give the true foundation of the Calculus in the doctrine of limits, and, assuming the continuity of space and time in Cantor's sense, they give valid proofs

---

* Cf. *Mathematical Works*, Gerhardt's ed. iv, pp. 91–93; *Phil. Works*, Gerhardt's ed. ii, p. 282.

† See *Math. Works*, Gerhardt's ed. vi, pp. 235, 247, 252.

‡ See *Math. Works*, Gerhardt's ed., Vol. v, pp. 220 ff.

§ E.g. *Phil. Works*, Gerhardt's ed., ii, p. 305. Cf. Cassirer, *Leibniz' System* (Marburg, 1902), pp. 206–7.

‖ *Principia*, Part I, Section i.

of its rules so far as spatio-temporal magnitudes are concerned. But Newton was, of course, entirely ignorant of the fact that his Lemmas depend upon the modern theory of continuity; moreover, the appeal to time and change, which appears in the word fluxion, and to space, which appears in the Lemmas, was wholly unnecessary, and served merely to hide the fact that no definition of continuity had been given. Whether Leibniz avoided this error, seems highly doubtful : it is at any rate certain that, in his first published account of the Calculus, he defined the differential coefficient by means of the tangent to a curve. And by his emphasis on the infinitesimal, he gave a wrong direction to speculation as to the Calculus, which misled all mathematicians before Weierstrass (with the exception, perhaps, of De Morgan), and all philosophers down to the present day. It is only in the last thirty or forty years that mathematicians have provided the requisite mathematical foundations for a philosophy of the Calculus; and these foundations, as is natural, are as yet little known among philosophers, except in France*. Philosophical works on the subject, such as Cohen's *Princip der Infinitesimalmethode und seine Geschichte*†, are vitiated, as regards the constructive theory, by an undue mysticism, inherited from Kant, and leading to such results as the identification of intensive magnitude with the extensive infinitesimal‡. I shall examine in the next chapter the conception of the infinitesimal, which is essential to all philosophical theories of the Calculus hitherto propounded. For the present, I am only concerned to give the constructive theory as it results from modern mathematics.

304. The differential coefficient depends essentially upon the notion of a continuous function of a continuous variable. The notion to be defined is not purely ordinal; on the contrary, it is applicable, in the first instance, only to series of numbers, and thence, by extension, to series in which distances or stretches are numerically measureable. But first of all we must define a continuous function.

We have already seen (Chap. xxxii.) what is meant by a function of a variable, and what is meant by a continuous variable (Chap. xxxvi.). If the function is one-valued, and is only ordered by correlation with the variable, then, when the variable is continuous, there is no sense in asking whether the function is continuous; for such a series by correlation is always ordinally similar to its prototype. But when, as where the variable and the field of the function are both classes of numbers, the function has an order independent of correlation, it may or may not happen that the values of the function, in the order obtained by correlation, form a continuous series in the independent order. When they do so in any interval, the function is said to be continuous in that interval. The

---

* See Couturat, *De l'Infini Mathématique*, passim.
† Berlin, 1883. The historical part of this work, it should be said, is admirable.
‡ *Op. cit.* p. 15.

precise definitions of continuous and discontinuous functions, where both x and $f(x)$ are numerical, are given by Dini* as follows. The independent variable x is considered to consist of the real numbers, or of all the real numbers in a certain interval; $f(x)$, in the interval considered, is to be one-valued, even at the end-points of the interval, and is to be also composed of real numbers. We then have the following definitions, the function being defined for the interval between α and β, and a being some real number in this interval.

" We call $f(x)$ *continuous* for $x = a$, or in the point a, in which it has the value $f(a)$, if for every positive number σ, different from 0, but as small as we please, there exists a positive number ϵ, different from 0, such that, for all values of δ which are numerically less than ϵ, the difference $f(a + \delta) - f(a)$ is numerically less than σ. In other words, $f(x)$ is continuous in the point $x = a$, where it has the value $f(a)$, if the limit of its values to the right and left of a is the same, and equal to $f(a)$."

" Again, $f(x)$ is *discontinuous* for $x = a$, if, for any† positive value of σ, there is no corresponding positive value of ϵ such that, for all values of δ which are numerically less than ϵ, $f(a + \delta) - f(a)$ is always less than σ; in other words, $f(x)$ is discontinuous for $x = a$, when the values $f(a + h)$ of $f(x)$ to the right of a, and the values $f(a - h)$ of $f(x)$ to the left of a, the one and the other, have no determinate limits, or, if they have such, these are different on the two sides of a; or, if they are the same, they differ from the value $f(a)$, which the function has in the point a."

These definitions of the continuity and discontinuity of a function, it must be confessed, are somewhat complicated; but it seems impossible to introduce any simplification without loss of rigour. Roughly, we may say that a function is continuous in the neighbourhood of a, when its values as it approaches a approach the value $f(a)$, and have $f(a)$ for their limit both to left and right. But the notion of the limit of a function is a somewhat more complicated notion than that of a limit in general, with which we have been hitherto concerned. A function of a perfectly general kind will have no limit as it approaches any given point. In order that it should have a limit as x approaches a from the left, it is necessary and sufficient that, if any number ϵ be mentioned, any two values of $f(x)$, when x is sufficiently near to a, but less than a, will differ by less than ϵ; in popular language, the value of the function does not make any sudden jumps as x approaches a from the left. Under similar circumstances, $f(x)$ will have a limit as it approaches a from the right. But these two limits, even when both exist, need not be equal either to each other, or to $f(a)$, the *value* of the function when

* *Op. cit.* § 30, pp. 50, 51.
† The German (not the Italian) has *every* instead of *any*, but this is a slip.

$x = a$. The precise condition for a determinate finite limit may be thus stated * :

" In order that the values of y to the right or left of a finite number a (for instance to the right) should have a determinate and finite limit, it is necessary and sufficient that, for every arbitrarily small positive number σ, there should be a positive number ϵ, such that the difference $y_{a+\epsilon} - y_{a+\delta}$ between the value $y_{a+\epsilon}$ of y for $x = a + \epsilon$, and the value $y_{a+\delta}$, which corresponds to the value $a + \delta$ of x, should be numerically less than σ, for every δ which is greater than 0 and less than ϵ."

It is possible, instead of thus defining *the* limit of a function, and then discussing whether it exists, to define generally a whole class of limits†. In this method, a number z belongs to the class of limits of y for $x = a$, if, within any interval containing a, however small, y will approach nearer to z than by any given difference. Thus, for example, $\sin 1/x$, as x approaches zero, will take every value from -1 to $+1$ (both inclusive) in every finite interval containing zero, however small. Thus the interval from -1 to $+1$ forms, in this case, the class of limits for $x = 0$. This method has the advantage that the *class* of limits always exists. It is then easy to define *the* limit as the only member of the class of limits, in case this class should happen to have only one member. This method seems at once simpler and more general.

305. Being now agreed as to the meaning of a continuous function, and of the limit of a function, we can attack the question of the derivative of a function, or differential coefficient. It was formerly supposed that all continuous functions could be differentiated, but this is now known to be erroneous. Some can be differentiated everywhere, others everywhere except in one point, others have everywhere a differential on the right, but sometimes none on the left, others contain an infinite number of points, in any finite interval, in which they cannot be differentiated, though in an infinitely greater number of points they can be differentiated, others lastly—and these are properly the most general class—cannot be differentiated anywhere at all‡. But the conditions under which a function may be differentiated, though they are of some importance to the philosophy of space and of motion, need not greatly concern us here; and in any case, we must first know what a differential is.

If $f(x)$ be a function which is finite and continuous at the point x, then it may happen that the fraction

$$\{f(x + \delta) - f(x)\}/\delta$$

has a definite limit as δ approaches to zero. If this does happen, the

* Dini, *op. cit.* p. 38.

† See Peano, *Rivista di Matematica*, II, pp. 77-79; *Formulaire*, Part III, § 73, 1.0.

‡ See Dini, *op. cit.* Chaps. x, xi, xii; *Encyklopädie der math. Wissenschaften*, Band II, Heft I (Leipzig, 1899), esp. pp. 20-22.

limit is denoted by $f'(x)$, and is called the derivative or differential of $f(x)$ in the point x. If, that is to say, there be some number z such that, given any number ϵ however small, if δ be any number less than some number η, but positive, then $\{f(x \pm \delta) - f(x)\}/ \pm \delta$ differs from z by less than ϵ, then z is the derivative of $f(x)$ in the point x. If the limit in question does not exist, then $f(x)$ has no derivative at the point x. If $f(x)$ be not continuous at this point, the limit does not exist; if $f(x)$ be continuous, the limit may or may not exist.

306. The only point which it is important to notice at present is, that there is no implication of the infinitesimal in this definition. The number δ is always finite, and in the definition of the limit there is nothing to imply the contrary. In fact, $\{f(x + \delta) - f(x)\}/\delta$, regarded as a function of δ, is wholly indeterminate when $\delta = 0$. The limit of a function for a given value of the independent variable is, as we have seen, an entirely different notion from its value for the said value of the independent variable, and the two may or may not be the same number. In the present case, the limit may be definite, but the value for $\delta = 0$ can have no meaning. Thus it is the doctrine of limits that underlies the Calculus, and not any pretended use of the infinitesimal. This is the only point of philosophic importance in the present subject, and it is only to elicit this point that I have dragged the reader through so much mathematics.

307. Before examining the infinitesimal on its own account, it remains to define the definite integral, and to show that this, too, does not involve the infinitesimal. The indefinite integral, which is the mere converse of the differential, is of no importance to us ; but the definite integral has an independent definition, which must be briefly examined.

Just as the derivative of a function is the limit of a fraction, so the definite integral is the limit of a sum[*]. The definite integral may be defined as follows : Let $f(x)$ be a function which is one-valued and finite in the interval a to β (both inclusive). Divide this interval into any n portions by means of the $(n-1)$ points $x_1, x_2, \ldots x_{n-1}$, and denote by $\delta_1, \delta_2, \ldots \delta_n$ the n intervals $x_1 - a, x_1 - x_2, \ldots \beta - x_{n-1}$. In each of these intervals, δ_s, take any one of the values, say $f(\zeta_s)$, which $f(x)$ assumes in this interval, and multiply this value by the interval δ_s.

Now form the sum $\sum_1^n f(\zeta_s) \delta_s$. This sum will always be finite. If now, as n increases, this sum tends to one definite limit, however $f(\zeta_s)$

---

[*] The definition of the definite integral differs little in different modern works. Cp. Dini, *op. cit.* §§ 178–181; Jordan, *Cours d'Analyse*, Vol. i (Paris, 1893), Chap. i, §§ 41–58; *Encyklopädie der mathematischen Wissenschaften*, ii, A. 2, § 31. The definition as the limit of a sum is more consonant with Leibniz's views than that as the inverse of a derivative, but was banished by Bernoulli and Euler, and only brought back by Cauchy. See references in the last-mentioned place.

may be chosen in its interval, and however the intervals be chosen (provided only that all are less than any assigned number for sufficiently great values of n)—then this one limit is called the definite integral of $f(x)$ from α to β. If there is no such limit, $f(x)$ is not integrable from α to β.

308. As in the case of the derivative, there is only one important remark to make about this definition. The definite integral involves neither the infinite nor the infinitesimal, and is itself not a sum, but only and strictly the limit of a sum. All the terms which occur in the sum whose limit is the definite integral are finite, and the sum itself is finite. If we were to suppose the limit actually attained, it is true, the number of intervals would be infinite, and the magnitude of each would be infinitesimal; but in this case, the sum becomes meaningless. Thus the sum must not be regarded as actually attaining its limit. But this is a respect in which series in general agree. Any series which always ascends or always descends and has no last term cannot reach its limit; other infinite series *may* have a term equal to their limit, but if so, this is a mere accident. The general rule is, that the limit does not belong to the series which it limits; and in the definition of the derivative and the definite integral we have merely another instance of this fact. The so-called infinitesimal calculus, therefore, has nothing to do with the infinitesimal, and has only indirectly to do with the infinite—its connection with the infinite being, that it involves limits, and only infinite series have limits.

The above definitions, since they involve multiplication and division, are essentially arithmetical. Unlike the definitions of limits and continuity, they cannot be rendered purely ordinal. But it is evident that they may be at once extended to any numerically measurable magnitudes, and therefore to all series in which stretches or distances can be measured. Since spaces, times, and motions are included under this head, the Calculus is applicable to Geometry and Dynamics. As to the axioms involved in the assumption that geometrical and dynamical functions can be differentiated and integrated, I shall have something to say at a later stage. For the present, it is time to make a critical examination of the infinitesimal on its own account.

CHAPTER XL.

THE INFINITESIMAL AND THE IMPROPER INFINITE.

309. UNTIL recent times, it was universally believed that continuity, the derivative, and the definite integral, all involved actual infinitesimals, *i.e.* that even if the definitions of these notions could be formally freed from explicit mention of the infinitesimal, yet, where the definitions applied, the actual infinitesimal must always be found. This belief is now generally abandoned. The definitions which have been given in previous chapters do not in any way imply the infinitesimal, and this notion appears to have become mathematically useless. In the present chapter, I shall first give a definition of the infinitesimal, and then examine the cases where this notion arises. I shall end by a critical discussion of the belief that continuity implies the infinitesimal.

The infinitesimal has, in general, been very vaguely defined. It has been regarded as a number or magnitude which, though not zero, is less than any finite number or magnitude. It has been the dx or dy of the Calculus, the time during which a ball thrown vertically upwards is at rest at the highest point of its course, the distance between a point on a line and the next point, etc., etc. But none of these notions are at all precise. The dx and dy, as we saw in the last chapter, are nothing at all: dy/dx is the limit of a fraction whose numerator and denominator are finite, but is not itself a fraction at all. The time during which a ball is at rest at its highest point is a very complex notion, involving the whole philosophic theory of motion: in Part VII we shall find, when this theory has been developed, that there is no such time. The distance between consecutive points presupposes that there are consecutive points—a view which there is every reason to deny. And so with most instances—they afford no precise definition of what is meant by the infinitesimal.

310. There is, so far as I know, only one precise definition, which renders the infinitesimal a purely relative notion, correlative to something arbitrarily assumed to be finite. When, instead, we regard what had been taken to be infinitesimal as finite, the correlative notion is what Cantor calls the improper infinite (*Uneigentlich-Unendliches*). The

definition of the relation in question is obtained by denying the axiom
of Archimedes, just as the transfinite was obtained by denying mathematical induction.   If P, Q be any two numbers, or any two measurable
magnitudes, they are said to be finite with respect to each other when,
if P be the lesser, there exists a finite integer n such that nP is greater
than Q.   The existence of such an integer constitutes the axiom of
Archimedes and the definition of relative finitude.   It will be observed
that it presupposes the definition of absolute finitude among numbers—
a definition which, as we have seen, depends upon two points, (1) the
connection of 1 with the logical notion of simplicity, or of 0 with
the logical notion of the null-class; (2) the principle of mathematical
induction.   The notion of relative finitude is plainly distinct from that
of absolute finitude.   The latter applies only to numbers, classes and
divisibilities, whereas the former applies to any kind of measurable
magnitude.   Any two numbers, classes, or divisibilities, which are both
absolutely finite are also relatively finite; but the converse does not
hold.   For example, ω and $\omega \cdot 2$, an inch and a foot, a day and a year,
are relatively finite pairs, though all three consist of terms which are
absolutely infinite.

The definition of the infinitesimal and the improper infinite is then
as follows.   If P, Q be two numbers, or two measurable magnitudes of
the same kind, and if, n being any finite integer whatever, nP is always
less than Q, then P is infinitesimal with respect to Q, and Q is infinite
with respect to P.   With regard to numbers, these relative terms are
not required; for if, in the case supposed, P is absolutely finite, then Q
is absolutely infinite; while if it were possible for Q to be absolutely
finite, P would be absolutely infinitesimal—a case, however, which we
shall see reason to regard as impossible.   Hence I shall assume in future
that P and Q are not numbers, but are magnitudes of a kind of which
some, at least, are numerically measurable.   It should be observed that,
as regards magnitudes, the axiom of Archimedes is the only way of
defining, not only the infinitesimal, but the infinite also.   Of a magnitude not numerically measurable, there is nothing to be said except that
it is greater than some of its kind, and less than others; but from such
propositions infinity cannot be obtained.   Even if there be a magnitude
greater than all others of its kind, there is no reason for regarding it as
infinite.   Finitude and infinity are essentially numerical notions, and
it is only by relation to numbers that these terms can be applied to
other entities.

311.   The next question to be discussed is, What instances of infinitesimals are to be found?   Although there are far fewer instances
than was formerly supposed, there are yet some that are important.   To
begin with, if we have been right in regarding divisibility as a magnitude, it is plain that the divisibility of any whole containing a finite
number of simple parts is infinitesimal as compared with one containing

an infinite number. The number of parts being taken as the measure, every infinite whole will be greater than n times every finite whole, whatever finite number n may be. This is therefore a perfectly clear instance. But it must not be supposed that the ratio of the divisibilities of two wholes, of which one at least is transfinite, can be measured by the ratio of the cardinal numbers of their simple parts. There are two reasons why this cannot be done. The first is, that two transfinite cardinals do not have any relation strictly analogous to ratio; indeed, the definition of ratio is effected by means of mathematical induction. The relation of two transfinite cardinals α, γ expressed by the equation $\alpha\beta = \gamma$ bears a certain resemblance to integral ratios, and $\alpha\beta = \gamma\delta$ may be used to define other ratios. But ratios so defined are not very similar to finite ratios. The other reason why infinite divisibilities must not be measured by transfinite numbers is, that the whole must always have more divisibility than the part (provided the remaining part is not relatively infinitesimal), though it may have the same transfinite number. In short, divisibilities, like ordinals, are equal, so long as the wholes are finite, when and only when the cardinal numbers of the wholes are the same; but the notion of magnitude of divisibility is distinct from that of cardinal number, and separates itself visibly as soon as we come to infinite wholes.

Two infinite wholes may be such that one is infinitely less divisible than the other. Consider, for example, the length of a finite straight line and the area of the square upon that straight line; or the length of a finite straight line and the length of the whole straight line of which it forms part (except in finite spaces); or an area and a volume; or the rational numbers and the real numbers; or the collection of points on a finite part of a line obtainable by von Staudt's quadrilateral construction, and the total collection of points on the said finite part*. All these are magnitudes of one and the same kind, namely divisibilities, and all are infinite divisibilities; but they are of many different orders. The points on a limited portion of a line obtainable by the quadrilateral construction form a collection which is infinitesimal with respect to the said portion; this portion is ordinally infinitesimal† with respect to any bounded area; any bounded area is ordinally infinitesimal with respect to any bounded volume; and any bounded volume (except in finite spaces) is ordinally infinitesimal with respect to all space. In all these cases, the word *infinitesimal* is used strictly according to the above definition, obtained from the axiom of Archimedes. What makes these various infinitesimals somewhat unimportant, from a mathematical standpoint, is, that measurement essentially depends upon the axiom of Archimedes, and cannot, in general, be extended by means of transfinite numbers, for the reasons which have just been explained. Hence two divisibilities, of

---

* See Part VI, Chap. xlv.   † See Part VI, Chap. xlvii, § 397.

which one is infinitesimal with respect to the other, are regarded usually as different kinds of magnitude ; and to regard them as of the same kind gives no advantage save philosophic correctness. All of them, however, are strictly instances of infinitesimals, and the series of them well illustrates the relativity of the term *infinitesimal*.

An interesting method of comparing certain magnitudes, analogous to the divisibilities of any infinite collections of points, with those of continuous stretches is given by Stolz[*], and a very similar but more general method is given by Cantor[†]. These methods are too mathematical to be fully explained here, but the gist of Stolz's method may be briefly explained. Let a collection of points x' be contained in some finite interval a to b. Divide the interval into any number n of parts, and divide each of these parts again into any number of parts, and so on ; and let the successive divisions be so effected that all parts become in time less than any assigned number δ. At each stage, add together all the parts that contain points of x'. At the mth stage, let the resulting sum be S_m. Then subsequent divisions may diminish this sum, but cannot increase it. Hence as the number of divisions increases, S_m must approach a limit L. If x' is compact throughout the interval, we shall have $L = b - a$; if any finite derivative of x' vanishes, $L = 0$. L obviously bears an analogy to a definite integral ; but no conditions are required for the existence of L. But L cannot be identified with the divisibility ; for some compact series, *e.g.* that of rationals, are less divisible than others, *e.g.* the continuum, but give the same value of L.

312. The case in which infinitesimals were formerly supposed to be peculiarly evident is that of compact series. In this case, however, it is possible to prove that there can be no infinitesimal segments[‡], provided numerical measurement be possible at all—and if it be not possible, the infinitesimal, as we have seen, is not definable. In the first place, it is evident that the segment contained between two different terms is always infinitely divisible ; for since there is a term c between any two a and b, there is another d between a and c, and so on. Thus no terminated segment can contain a finite number of terms. But segments defined by a class of terms may (as we saw in Chapter XXXIV) have no limiting term. In this case, however, provided the segment does not consist of a single term a, it will contain some other term b, and therefore an infinite number of terms. Thus all segments are infinitely divisible. The next point is to define multiples of segments. Two terminated segments can be added by placing a segment equal to the one at the end of the other to form a new segment ; and if the two were equal, the new one is said

   [*] *Math. Annalen*, 23, "Ueber einen zu einer unendlichen Punktmenge gehörigen Grenzwerth."

   [†] *Ib.* "Ueber unendliche lineare Punktmannigfaltigkeiten," No. 6.

   [‡] See Peano, *Rivista di Matematica*, Vol. II, pp. 58–62.

to be double of each of them. But if the two segments are not terminated, this process cannot be employed. Their sum, in this case, is defined by Professor Peano as the logical sum of all the segments obtained by adding two terminated segments contained respectively in the two segments to be added *. Having defined this sum, we can define any finite multiple of a segment. Hence we can define the class of terms contained in *some* finite multiple of our segment, *i.e.* the logical sum of all its finite multiples. If, with respect to all greater segments, our segment obeys the axiom of Archimedes, then this new class will contain all terms that come after the origin of our segment. But if our segment be infinitesimal with respect to any other segment, then the class in question will fail to contain some points of this other segment. In this case, it is shown that all transfinite multiples of our segment are equal to each other. Hence it follows that the class formed by the logical sum of all finite multiples of our segment, which may be called the infinite multiple of our segment, must be a non-terminated segment, for a terminated segment is always increased by being doubled. " Each of these results," so Professor Peano concludes, " is in contradiction with the usual notion of a segment. And from the fact that the infinitesimal segment cannot be rendered finite by means of any actually infinite multiplication, I conclude, with Cantor, that it cannot be an element in finite magnitudes " (p. 62). But I think an even stronger conclusion is warranted. For we have seen that, in compact series, there is, corresponding to every segment, a segment of segments, and that this is always terminated by its defining segment; further that the numerical measurement of segments of segments is exactly the same as that of simple segments; whence, by applying the above result to segments of segments, we obtain a definite contradiction, since none of them can be unterminated, and an infinitesimal one cannot be terminated.

In the case of the rational or the real numbers, the complete knowledge which we possess concerning them renders the non-existence of infinitesimals demonstrable. A rational number is the ratio of two finite integers, and any such ratio is finite. A real number other than zero is a segment of the series of rationals; hence if x be a real number other than zero, there is a class u, not null, of rationals such that, if y is a u, and z is less than y, z is an x, *i.e.* belongs to the segment which is x. Hence every real number other than zero is a class containing rationals, and all rationals are finite; consequently every real number is finite. Consequently if it were possible, in any sense, to speak of infinitesimal numbers, it would have to be in some radically new sense.

313. I come now to a very difficult question, on which I would gladly say nothing—I mean, the question of the orders of infinity and infinitesimality of functions. On this question the greatest authorities

---

* *Loc. cit.* p. 61, No. 9.

are divided : Du Bois Reymond, Stolz, and many others, maintaining that these form a special class of magnitudes, in which actual infinitesimals occur, while Cantor holds strongly that the whole theory is erroneous *. To put the matter as simply as possible, consider a function $f(x)$ whose limit, as x approaches zero, is zero. It may happen that, for some finite real number α, the ratio $f(x)/x^a$ has a finite limit as x approaches zero. There can be only one such number, but there may be none. Then α, if there is such a number, may be called the order to which $f(x)$ becomes infinitesimal, or the order of smallness of $f(x)$ as x approaches zero. But for some functions, *e.g.* $1/\log x$, there is no such number α. If α be any finite real number, the limit of $1/x^a \log x$, as x approaches zero, is infinite. That is, when x is sufficiently small, $1/x^a \log x$ is very large, and may be made larger than any assigned number by making x sufficiently small—and this whatever finite number α may be. Hence, to express the order of smallness of $1/\log x$, it is necessary to invent a new infinitesimal number, which may be denoted by $1/g$. Similarly we shall need infinitely great numbers to express the order of smallness of (say) $e^{-1/x}$ as x approaches zero. And there is no end to the succession of these orders of smallness : that of $1/\log (\log x)$, for example, is infinitely smaller than that of $1/\log x$, and so on. Thus we have a whole hierarchy of magnitudes, of which all in any one class are infinitesimal with respect to all in any higher class, and of which one class only is formed of all the finite real numbers.

In this development, Cantor finds a vicious circle ; and though the question is difficult, it would seem that Cantor is in the right. He objects (*loc. cit.*) that such magnitudes cannot be introduced unless we have reason to think that there are such magnitudes. The point is similar to that concerning limits ; and Cantor maintains that, in the present case, definite contradictions may be proved concerning the supposed infinitesimals. If there were infinitesimal numbers j, then even for them we should have

$$\mathrm{Lim}_{x=0} 1/ (\log x . x^j) = 0$$

since x^j must ultimately exceed $\frac{1}{2}$. And he shows that even continuous, differentiable, and uniformly growing functions may have an entirely ambiguous order of smallness or infinity : that, in fact, for some such functions, this order oscillates between infinite and infinitesimal values, according to the manner in which the limit is approached. Hence we may, I think, conclude that these infinitesimals are mathematical fictions. And this may be reinforced by the consideration that, if there were infinitesimal numbers, there would be infinitesimal segments of the number-continuum, which we have just seen to be impossible.

* See Du Bois Reymond, *Allgemeine Functionentheorie* (1882), p. 279 ff. ; Stolz, *Allgemeine Arithmetik*, Part I (Leipzig, 1885), Section IX, Anhang ; Cantor, *Rivista di Matematica*, v, pp. 104–8.

314. Thus to sum up what has been said concerning the infinitesimal, we see, to begin with, that it is a relative term, and that, as regards magnitudes other than divisibilities, or divisibilities of wholes which are infinite in the absolute sense, it is not capable of being other than a relative term. But where it has an absolute meaning, there this meaning is indistinguishable from finitude. We saw that the infinitesimal, though completely useless in mathematics, does occur in certain instances—for example, lengths of bounded straight lines are infinitesimal as compared to areas of polygons, and these again as compared to volumes of polyhedra. But such genuine cases of infinitesimals, as we saw, are always regarded by mathematics as magnitudes of another kind, because no numerical comparison is possible, even by means of transfinite numbers, between an area and a length, or a volume and an area. Numerical measurement, in fact, is wholly dependent upon the axiom of Archimedes, and cannot be extended as Cantor has extended numbers. And finally we saw that there are no infinitesimal segments in compact series, and—what is closely connected—that orders of smallness of functions are not to be regarded as genuine infinitesimals. The infinitesimal, therefore—so we may conclude—is a very restricted and mathematically very unimportant conception, of which infinity and continuity are alike independent.

CHAPTER XLI.

PHILOSOPHICAL ARGUMENTS CONCERNING THE INFINITESIMAL.

315. WE have now completed our summary review of what mathematics has to say concerning the continuous, the infinite, and the infinitesimal. And here, if no previous philosophers had treated of these topics, we might leave the discussion, and apply our doctrines to space and time. For I hold the paradoxical opinion that what can be mathematically demonstrated is true. As, however, almost all philosophers disagree with this opinion, and as many have written elaborate arguments in favour of views different from those above expounded, it will be necessary to examine controversially the principal types of opposing theories, and to defend, as far as possible, the points in which I differ from standard writers. For this purpose, the work of Cohen already referred to will be specially useful, not only because it deals explicitly with our present theme, but also because, largely owing to its historical excellence, certain very important mathematical errors, which it appears to me to contain, have led astray other philosophers who have not an acquaintance with modern mathematics at first hand *.

316. In the above exposition, the differential appeared as a philosophically unimportant application of the doctrine of limits. Indeed, but for its traditional importance, it would scarcely have deserved even mention. And we saw that its definition nowhere involves the infinitesimal. The dx and dy of a differential are nothing in themselves, and dy/dx is not a fraction. Hence, in modern works on the Calculus, the notation $f'(x)$ has replaced dy/dx, since the latter form suggests erroneous notions. The notation $f'(x)$, it may be observed, is more similar to Newton's $\dot{y}$, and its similarity is due to the fact that, on this point, modern mathematics is more in harmony with Newton than with Leibniz. Leibniz employed the form dy/dx because he believed in infinitesimals; Newton, on the other hand, definitely asserts that his fluxion is not a fraction. "Those ultimate ratios," he says, "with

---

* For example, Mr Latta, in his article "On the Relations of the Philosophy of Spinoza and that of Leibniz," *Mind*, N. S. No 31.

which quantities vanish are not truly the ratios of ultimate quantities, but limits towards which the ratios of quantities decreasing without limit do always converge, and to which they approach nearer than by any given difference*."

But when we turn to such works as Cohen's, we find the dx and the dy treated as separate entities, as real infinitesimals, as the intensively real elements of which the continuum is composed (pp. 14, 28, 144, 147). The view that the Calculus requires infinitesimals is apparently not thought open to question; at any rate, no arguments whatever are brought up to support it. This view is certainly assumed as self-evident by most philosophers who discuss the Calculus. Let us see for ourselves what kind of grounds can be urged in its favour.

317. Many arguments in favour of the view in question are derived by most writers from space and motion—arguments which Cohen to some extent countenances (pp. 34, 37), though he admits that the differential can be obtained from numbers alone, which however, following Kant, he regards as implying time (pp. 20, 21). Since the analysis of space and motion is still to come, I shall confine myself for the present to such arguments as can be derived from purely numerical instances. For the sake of definiteness, I shall as far as possible extract the opinions to be controverted from Cohen.

318. Cohen begins (p. 1) by asserting that the problem of the infinitesimal is not purely logical: it belongs rather to Epistemology, which is distinguished, I imagine, by the fact that it depends upon the pure intuitions as well as the categories. This Kantian opinion is wholly opposed to the philosophy which underlies the present work; but it would take us too far from our theme to discuss it here, and I mention it chiefly to explain the phraseology of the work we are examining. Cohen proceeds at once to reject the view that the infinitesimal calculus can be independently derived by mathematics from the method of limits. This method, he says (p. 1), "consists in the notion that the elementary conception of equality must be completed by the exact notion of the limit. Thus in the first place the conception of equality is presupposed.... Again, in the second place, the method of limits presupposes the conception of magnitude.... But in the presupposed conception of magnitude the limiting magnitude is at the same time presupposed. The equality which is defined in the elementary doctrine of magnitude pays no attention to these limiting magnitudes. For it, magnitudes count as equal if and although their difference consists in a limiting magnitude. Hence the elementary conception of equality must be— this is the notion of the method of limits—not so much *completed* as

---

* *Principia*, Bk i, Section i, Lemma xi, Scholium. The whole Scholium is highly important, though portions of it are less free from error than the passage quoted in the text.

corrected by the exact conception of the limit. Equality is to be regarded as an *earlier stage* of the limiting relation*."

319. I have quoted this passage in full, because its errors are typical of those to which non-mathematicians are liable in this question. In the first place, equality has no relevance to limits. I imagine that Cohen has in mind such cases as a circle and the inscribed polygon, where we cannot say that the circle is equal to any of the polygons, but only that it is their limit; or, to take an arithmetical instance, a convergent series whose sum is π or $\sqrt{2}$. But in all such instances there is much that is irrelevant and adventitious, and there are many unnecessary complications. The absolutely simplest instance of a limit is ω considered as the limit of the ordinal numbers. There is here certainly no kind of equality. Yet in all cases where limits are defined by progressions—and these are the usual cases—we have a series of the type presented by the finite ordinals together with ω. Consider, for example, the series $2 - \dfrac{1}{n}$ together with 2, the n being capable of all positive integral finite values. Here the series is of the same type as before, and here, as before, 2 is the limit of the series. But here—and this is what has misled Cohen—the difference between 2 and the successive terms of the series becomes less than any assigned magnitude, and thus we seem to have a sort of extended quality between 2 and the late terms of the series $2 - \dfrac{1}{n}$. But let us examine this. In the first place, it depends upon the fact that rationals are a series in which we have distances which are again rationals. But we know that distances are unnecessary to limits, and that stretches are equally effective. Now considering stretches, 2 is the limit of $2 - \dfrac{1}{n}$ because no rational comes between 2 and all terms of the series $2 - \dfrac{1}{n}$ —precisely the sense in which ω is the limit of the finite integers. And it is only because $2 - \dfrac{1}{n}$ forms a progression, *i.e.* is similar to the series of finite integers, that we know its limit to be 2. The fact that the terms, as we advance, differ little from 2, depends either upon our having a series in which there is distance, which is a fortuitous and irrelevant circumstance, or upon the fact that the successive stretches up to 2 may be made less than any assigned stretch up to 2, which follows from the notion of a limit, but has nothing to do with equality. And whenever our series which is to have a limit is part of a series which is a function of ω, the stretch from any term to the limit is always infinite in the only sense in which such series have infinite stretches; and in a very real sense the stretch

---

* Or ratio: the German is *Grenzverhältniss*.

grows no smaller as we approach the limit, for both the ordinal and the cardinal number of its terms remain constant.

We have seen so fully already in what sense, and how far, magnitude is involved in limits, that it seems unnecessary to say much on this subject here. Magnitude is certainly *not* involved in the sense, which is undoubtedly that intended by Cohen, that the limit and the terms limited must be magnitudes. Every progression which forms part of a series which is a function of ω, and in which there are terms after the progression, has a limit, whatever may be the nature of the terms. Every endless series of segments of a compact series has a limit, whatever may be the nature of the compact series. Now of course in all series we have magnitudes, namely the divisibilities of stretches; but it is not of these that we find the limit. Even in the case of segments, the limit is an actual segment, not the magnitude of a segment; and what is relevant is only that the segments are classes, not that they are quantities. But the distinction of quantities and magnitudes is, of course, wholly foreign to Cohen's order of ideas.

320. But we now come to a greater error. The conception of magnitude, Cohen says, which is presupposed in limits, in turn presupposes limiting magnitudes. By limiting magnitudes, as appears from the context, he means infinitesimals, the ultimate differences, I suppose, between the terms of a series and its limit. What he means seems to be, that the kinds of magnitude which lead to limits are compact series, and that, in compact series, we must have infinitesimals. Every point in this opinion is mistaken. Limits, we have just seen, need not be limits of magnitudes; segments of a compact series, as we saw in the preceding chapter, cannot be infinitesimal; and limits do not in any way imply that the series in which they occur are compact. These points have been so fully proved already that it is unnecessary to dwell upon them.

321. But the crowning mistake is the supposition that limits introduce a new meaning of equality. Among magnitudes, equality, as we saw in Part III, has an absolutely rigid and unique meaning: it applies only to quantities, and means that they have the *same* magnitude. There is no question of approximation here: what is meant is simply absolute logical identity of magnitude. Among numbers (which Cohen probably regards as magnitudes), there is no such thing as equality. There is identity, and there is the relation which is usually expressed by the sign of equality, as in the equation $2 \times 3 = 6$. This relation had puzzled those who endeavoured to philosophize about Arithmetic, until it was explained by Professor Peano*. When one term of the equation is a single number, while the other is an expression composed of two or more numbers, the equation expresses the fact that the class

* See *e.g. Riv. di Mat.* VII, p. 35.

defined by the expression contains only one term, which is the single number on the other side of the equation. This definition again is absolutely rigid : there is nothing whatever approximate in it, and it is incapable of any modification by infinitesimals. I imagine that what Cohen means may be expressed as follows. In forming a differential coefficient, we consider two numbers x and $x + dx$, and two others y and $y + dy$. In elementary Arithmetic, x and $x + dx$ would count as equal, but not in the Calculus. There are, in fact, two ways of defining equality. Two terms may be said to be equal when their ratio is unity, or when their difference is zero. But when we allow real infinitesimals dx, x and $x + dx$ will have the ratio unity, but will not have zero for their difference, since dx is different from absolute zero. This view, which I suggest as equivalent to Cohen's, depends upon a misunderstanding of limits and the Calculus. There are in the Calculus no such magnitudes as dx and dy. There are finite differences Δx and Δy, but no view, however elementary, will make x equal to $x + \Delta x$. There are ratios of finite differences, $\Delta y / \Delta x$, and in cases where the derivative of y exists, there is one real number to which $\Delta y / \Delta x$ can be made to approach as near as we like by diminishing Δx and Δy. This single real number we choose to denote by dy/dx; but it is not a fraction, and dx and dy are nothing but typographical parts of one symbol. There is no correction whatever of the notion of equality by the doctrine of limits; the only new element introduced is the consideration of infinite classes of terms chosen out of a series.

322. As regards the nature of the infinitesimal, we are told (p. 15) that the differential, or the inextensive, is to be identified with the intensive, and the differential is regarded as the embodiment of Kant's category of reality. This view (in so far as it is independent of Kant) is quoted with approval from Leibniz ; but to me, I must confess, it seems destitute of all justification. It is to be observed that dx and dy, if we allow that they are entities at all, are not to be identified with single terms of our series, nor yet with differences between consecutive terms, but must be always stretches containing an infinite number of terms, or distances corresponding to such stretches. Here a distinction must be made between series of numbers and series in which we have only measurable distances or stretches. The latter is the case of space and time. Here dx and dy are not points or instants, which alone would be truly inextensive ; they are primarily numbers, and hence must correspond to infinitesimal stretches or distances—for it would be preposterous to assign a numerical ratio to two points, or—as in the case of velocity—to a point and an instant. But dx and dy cannot represent the distances of consecutive points, nor yet the stretch formed by two consecutive points. Against this we have, in the first place, the general ground that our series must be regarded as compact, which precludes the idea of consecutive terms. To evade this, if we are dealing with a series in which there are only stretches, not distances, would be impossible: for to say that there are always an infinite number of intermediate points except when the *stretch* consists of a finite number of terms would be a mere tautology. But when there is distance, it might be said that the distance of two terms may be finite or infinitesimal, and that, as regards infinitesimal distances, the stretch is not compact, but consists of a finite number of terms. This being allowed for the moment, our dx and dy may be made to be the distances of consecutive points, or else the stretches composed of consecutive points. But now the distance of consecutive points, supposing for example that both are on one straight line, would seem to be a constant, which would give $dy/dx = \pm 1$. We cannot suppose, in cases where x and y are both continuous, and the function y is one-valued, as the Calculus requires, that x and $x + dx$ are consecutive, but not y and $y + dy$; for every value of y will be correlated with one and only one value of x, and *vice versâ*; thus y cannot skip any supposed intermediate values between y and $y + dy$. Hence, given the values of x and y, even supposing the distances of consecutive terms to differ from place to place, the value of dy/dx will be determinate; and any other function y' which, for some value of x, is equal to y, will, for that value, have an equal derivative, which is an absurd conclusion. And leaving these mathematical arguments, it is evident, from the fact that dy and dx are to have a numerical ratio, that if they be intensive magnitudes, as is suggested, they must be numerically measurable ones: but how this measurement is effected, it is certainly not easy to see. This point may be made clearer by confining ourselves to the fundamental case in which both x and y are numbers. If we regard x and $x + dx$ as consecutive, we must suppose either that y and $y + dy$ are consecutive, or that they are identical, or that there are a finite number of terms between them, or that there are an infinite number. If we take stretches to measure dx and dy, it will follow that dy/dx must be always zero, or integral, or infinite, which is absurd. It will even follow that, if y is not constant, dy/dx must be ± 1. Take for example $y = x^2$, where x and y are positive real numbers. As x passes from one number to the next, y must do so likewise; for to every value of y corresponds one of x, and y grows as x grows. Hence if y skipped the number next to any one of its values, it could never come back to pick it up; but we know that every real number is among the values of y. Hence y and $y + dy$ must be consecutive, and $dy/dx = 1$. If we measure by distances, not stretches, the distance dy must be fixed when y is given, and the distance dx when x is given. Now if $x = 1$, $y = 1$, $dy/dx = 2$; but, since x and y are the same number, dx and dy must be equal, since each is the distance to the next number: therefore $dy/dx = 1$, which is absurd. Similarly, if we take for y a decreasing function, we shall find $dy/dx = -1$. Hence the admission of consecutive numbers is fatal to the

Calculus; and since the Calculus must be maintained, the Calculus is fatal to consecutive numbers.

323. The notion that there must be consecutive numbers is reinforced by the idea of continuous change, which is embodied in calling x and y "variables." Change in time is a topic which we shall have to discuss at a later stage, but which has, undoubtedly, greatly influenced the philosophy of the Calculus. People picture a variable to themselves—often unconsciously—as successively assuming a series of values, as might happen in a dynamical problem. Thus they might say: How can x pass from x_1 to x_2, without passing through all intermediate values? And in this passage, must there not be a next value, which it assumes on first leaving the value x_1? Everything is conceived on the analogy of motion, in which a point is supposed to pass through all intermediate positions in its path. Whether or not this view of motion is correct, I do not now decide: at any rate it is irrelevant where a fundamental point in the theory of continuous series is concerned, since time and the path of motion must both be continuous series, and the properties of such series must be decided before appealing to motion to confirm our views. For my part, to return to Cohen, I must confess, it seems evident that intensive magnitude is something wholly different from infinitesimal extensive magnitude: for the latter must always be smaller than finite extensive magnitudes, and must therefore be of the same kind with them; while intensive magnitudes seem never in any sense smaller than any extensive magnitudes. Thus the metaphysical theory by which infinitesimals are to be rescued seems, both mathematically and philosophically, destitute of grounds in its favour.

324. We cannot, then, agree with the following summary of Cohen's theory (p. 28): "That I may be able to posit an element *in and for itself*, is the *desideratum*, to which corresponds the *instrument of thought* reality. This instrument of thought must first be set up, in order to be able to enter into that combination with intuition, with the *consciousness of being given*, which is completed in the *principle of intensive magnitude*. This presupposition of intensive reality is latent in all principles, and must therefore be made independent. *This presupposition is the meaning of reality and the secret of the concept of the differential.*" What we can agree to, and what, I believe, confusedly underlies the above statement, is, that every continuum must consist of elements or terms; but these, as we have just seen, will not fulfil the function of the dx and dy which occur in old-fashioned accounts of the Calculus. Nor can we agree that "this finite" (*i.e.* that which is the object of physical science) "can be thought as a sum of those infinitesimal intensive realities, as a *definite integral*" (p. 144). The definite integral is not a sum of elements of a continuum, although there are such elements: for example, the length of a curve, as obtained by integration, is not the sum of its points, but strictly and only the limit of the lengths of inscribed polygons. The only sense which can be given to the sum of the points of the curve is the logical class to which they all belong, *i.e.* the curve itself, not its length. All lengths are magnitudes of divisibility of stretches, and all stretches consist of an infinite number of points; and any two terminated stretches have a finite ratio to each other. There is no such thing as an infinitesimal stretch; if there were, it would not be an element of the continuum; the Calculus does not require it, and to suppose its existence leads to contradictions. And as for the notion that in every series there must be consecutive terms, that was shown, in the last Chapter of Part III, to involve an illegitimate use of mathematical induction. Hence infinitesimals as explaining continuity must be regarded as unnecessary, erroneous, and self-contradictory.

CHAPTER XLII.

THE PHILOSOPHY OF THE CONTINUUM.

325. THE word *continuity* has borne among philosophers, especially since the time of Hegel, a meaning totally unlike that given to it by Cantor. Thus Hegel says*: " Quantity, as we saw, has two sources : the exclusive unit, and the identification or equalization of these units. When we look, therefore, at its immediate relation to self, or at the characteristic of selfsameness made explicit by abstraction, quantity is *Continuous* magnitude ; but when we look at the other characteristic, the One implied in it, it is *Discrete* magnitude." When we remember that quantity and magnitude, in Hegel, both mean " cardinal number," we may conjecture that this assertion amounts to the following : " Many terms, considered as having a cardinal number, must all be members of one class ; in so far as they are each merely an instance of the class-concept, they are indistinguishable one from another, and in this aspect the whole which they compose is called *continuous* ; but in order to their maniness, they must be *different* instances of the class-concept, and in this aspect the whole which they compose is called *discrete*." Now I am far from denying—indeed I strongly hold—that this opposition of identity and diversity in a collection constitutes a fundamental problem of Logic—perhaps even *the* fundamental problem of philosophy. And being fundamental, it is certainly relevant to the study of the mathematical continuum as to everything else. But beyond this general connection, it has no special relation to the mathematical meaning of continuity, as may be seen at once from the fact that it has no reference whatever to order. In this chapter, it is the mathematical meaning that is to be discussed. I have quoted the philosophic meaning only in order to state definitely that this is *not* here in question ; and since disputes about words are futile, I must ask philosophers to divest themselves, for the time, of their habitual associations with the word, and allow it no signification but that obtained from Cantor's definition.

326. In confining ourselves to the arithmetical continuum, we conflict in another way with common preconceptions. Of the arithmetical con-

---

* *Smaller Logic*, § 100, Wallace's Translation, p. 188.

tinuum, M. Poincaré justly remarks*: "The continuum thus conceived is nothing but a collection of individuals arranged in a certain order, infinite in number, it is true, but external to each other. This is not the ordinary conception, in which there is supposed to be, between the elements of the continuum, a sort of intimate bond which makes a whole of them, in which the point is not prior to the line, but the line to the point. Of the famous formula, the continuum is unity in multiplicity, the multiplicity alone subsists, the unity has disappeared."

It has always been held to be an open question whether the continuum is composed of elements; and even when it has been allowed to contain elements, it has been often alleged to be not *composed* of these. This latter view was maintained even by so stout a supporter of elements in everything as Leibniz†. But all these views are only possible in regard to such continua as those of space and time. The arithmetical continuum is an object selected by definition, consisting of elements in virtue of the definition, and known to be embodied in at least one instance, namely the segments of the rational numbers. I shall maintain in Part VI that spaces afford other instances of the arithmetical continuum. The chief reason for the elaborate and para-doxical theories of space and time and their continuity, which have been constructed by philosophers, has been the supposed contradictions in a continuum composed of elements. The thesis of the present chapter is, that Cantor's continuum is free from contradictions. This thesis, as is evident, must be firmly established, before we can allow the possibility that spatio-temporal continuity may be of Cantor's kind. In this argument, I shall assume as proved the thesis of the preceding chapter, that the continuity to be discussed does not involve the admission of actual infinitesimals.

327. In this capricious world, nothing is more capricious than posthumous fame. One of the most notable victims of posterity's lack of judgment is the Eleatic Zeno. Having invented four arguments, all immeasurably subtle and profound, the grossness of subsequent philosophers pronounced him to be a mere ingenious juggler, and his arguments to be one and all sophisms. After two thousand years of continual refutation, these sophisms were reinstated, and made the foundation of a mathematical renaissance, by a German professor, who probably never dreamed of any connection between himself and Zeno. Weierstrass, by strictly banishing all infinitesimals, has at last shown that we live in an unchanging world, and that the arrow, at every moment of its flight, is truly at rest. The only point where Zeno probably erred was in inferring (if he did infer) that, because there is no change, therefore the world must be in the same state at one time as at another. This consequence by no means follows, and in

* *Revue de Métaphysique et de Morale*, Vol. i, p. 26.
† See *The Philosophy of Leibniz*, by the present author, Chap. ix.

this point the German professor is more constructive than the ingenious Greek. Weierstrass, being able to embody his opinions in mathematics, where familiarity with truth eliminates the vulgar prejudices of common sense, has been able to give to his propositions the respectable air of platitudes; and if the result is less delightful to the lover of reason than Zeno's bold defiance, it is at any rate more calculated to appease the mass of academic mankind.

Zeno's arguments are specially concerned with motion, and are not therefore, as they stand, relevant to our present purpose. But it is instructive to translate them, so far as possible, into arithmetical language *.

328. The first argument, that of dichotomy, asserts: "There is no motion, for what moves must reach the middle of its course before it reaches the end." That is to say, whatever motion we assume to have taken place, this presupposes another motion, and this in turn another, and so on *ad infinitum*. Hence there is an endless regress in the mere idea of any assigned motion. This argument can be put into an arithmetical form, but it appears then far less plausible. Consider a variable x which is capable of all real (or rational) values between two assigned limits, say 0 and 1. The class of its values is an infinite whole, whose parts are logically prior to it: for it has parts, and it cannot subsist if any of the parts are lacking. Thus the numbers from 0 to 1 presuppose those from 0 to 1/2, these presuppose the numbers from 0 to 1/4, and so on. Hence, it would seem, there is an infinite regress in the notion of any infinite whole; but without such infinite wholes, real numbers cannot be defined, and arithmetical continuity, which applies to an infinite series, breaks down.

This argument may be met in two ways, either of which, at first sight, might seem sufficient, but both of which are really necessary. First, we may distinguish two kinds of infinite regresses, of which one is harmless. Secondly, we may distinguish two kinds of whole, the collective and the distributive, and assert that, in the latter kind, parts of equal complexity with the whole are not logically prior to it. These two points must be separately explained.

329. An infinite regress may be of two kinds. In the objectionable kind, two or more propositions join to constitute the *meaning* of some proposition; of these constituents, there is one at least whose meaning is similarly compounded; and so on *ad infinitum*. This form of regress commonly results from circular definitions. Such definitions may be

---

* Not being a Greek scholar, I pretend to no first-hand authority as to what Zeno really did say or mean. The form of his four arguments which I shall employ is derived from the interesting article of M. Noël, "Le mouvement et les arguments de Zénon d'Elée," *Revue de Métaphysique et de Morale*, Vol. I, pp. 107–125. These arguments are in any case well worthy of consideration, and as they are, to me, merely a text for discussion, their historical correctness is of little importance.

expanded in a manner analogous to that in which continued fractions are developed from quadratic equations. But at every stage the term to be defined will reappear, and no definition will result. Take for example the following: "Two people are said to have the *same* idea when they have ideas which are similar; and ideas are similar when they contain an identical part." If an idea may have a part which is not an idea, such a definition is not logically objectionable; but if part of an idea is an idea, then, in the second place where identity of ideas occurs, the definition must be substituted; and so on. Thus wherever the *meaning* of a proposition is in question, an infinite regress is objectionable, since we never reach a proposition which has a definite meaning. But many infinite regresses are not of this form. If A be a proposition whose meaning is perfectly definite, and A implies B, B implies C, and so on, we have an infinite regress of a quite unobjectionable kind. This depends upon the fact that implication is a synthetic relation, and that, although, if A be an aggregate of propositions, A implies any proposition which is part of A, it by no means follows that any proposition which A implies is part of A. Thus there is no logical necessity, as there was in the previous case, to complete the infinite regress before A acquires a meaning. If, then, it can be shown that the implication of the parts in the whole, when the whole is an infinite class of numbers, is of this latter kind, the regress suggested by Zeno's argument of dichotomy will have lost its sting.

330. In order to show that this is the case, we must distinguish wholes which are defined extensionally, *i.e.* by enumerating their terms, from such as are defined intensionally, *i.e.* as the class of terms having some given relation to some given term, or, more simply, as a class of terms. (For a class of terms, when it forms a whole, is merely all terms having the class-relation to a class-concept*.) Now an extensional whole—at least so far as human powers extend—is necessarily finite: we cannot enumerate more than a finite number of parts belonging to a whole, and if the number of parts be infinite, this must be known otherwise than by enumeration. But this is precisely what a class-concept effects: a whole whose parts are the terms of a class is completely defined when the class-concept is specified; and any definite individual either belongs, or does not belong, to the class in question. An individual of the class is part of the whole extension of the class, and is logically prior to this extension taken collectively; but the extension itself is definable without any reference to any specified individual, and subsists as a genuine entity even when the class contains no terms. And to say, of such a class, that it is infinite, is to say that, though it has terms, the number of these terms is not any finite number— a proposition which, again, may be established without the impossible

---

* For precise statements, *v. supra*, Part I, Chaps. VI and X.

process of enumerating *all* finite numbers. And this is precisely the case of the real numbers between 0 and 1. They form a definite class, whose meaning is known as soon as we know what is meant by *real number*, 0, 1, and *between*. The particular members of the class, and the smaller classes contained in it, are not logically prior to the class. Thus the infinite regress consists merely in the fact that every segment of real or rational numbers has parts which are again segments; but these parts are not logically prior to it, and the infinite regress is perfectly harmless. Thus the solution of the difficulty lies in the theory of denoting and the intensional definition of a class. With this an answer is made to Zeno's first argument as it appears in Arithmetic.

331. The second of Zeno's arguments is the most famous: it is the one which concerns Achilles and the tortoise. "The slower," it says, "will never be overtaken by the swifter, for the pursuer must first reach the point whence the fugitive is departed, so that the slower must always necessarily remain ahead." When this argument is translated into arithmetical language, it is seen to be concerned with the one-one correlation of two infinite classes. If Achilles were to overtake the tortoise, then the course of the tortoise would be part of that of Achilles; but, since each is at each moment at some point of his course, simultaneity establishes a one-one correlation between the positions of Achilles and those of the tortoise. Now it follows from this that the tortoise, in any given time, visits just as many places as Achilles does; hence—so it is hoped we shall conclude—it is impossible that the tortoise's path should be part of that of Achilles. This point is purely ordinal, and may be illustrated by Arithmetic. Consider, for example, $1 + 2x$ and $2 + x$, and let x lie between 0 and 1, both inclusive. For each value of $1 + 2x$ there is one and only one value of $2 + x$, and *vice versâ*. Hence as x grows from 0 to 1, the number of values assumed by $1 + 2x$ will be the same as the number assumed by $2 + x$. But $1 + 2x$ started from 1 and ends at 3, while $2 + x$ started from 2 and ends at 3. Thus there should be half as many values of $2 + x$ as of $1 + 2x$. This very serious difficulty has been resolved, as we have seen, by Cantor; but as it belongs rather to the philosophy of the infinite than to that of the continuum, I leave its further discussion to the next chapter.

332. The third argument is concerned with the arrow. "If everything is in rest or in motion in a space equal to itself, and if what moves is always in the instant, the arrow in its flight is immovable." This has usually been thought so monstrous a paradox as scarcely to deserve serious discussion. To my mind, I must confess, it seems a very plain statement of a very elementary fact, and its neglect has, I think, caused the quagmire in which the philosophy of change has long been immersed. In Part VII, I shall set forth a theory of change which may be called *static*, since it allows the justice of Zeno's remark. For the present,.I wish to divest the remark of all reference to change. We shall then find that it is a very important and very widely applicable platitude, namely : " Every possible value of a variable is a constant." If x be a variable which can take all values from 0 to 1, all the values it can take are definite numbers, such as 1/2 or 1/3, which are all absolute constants. And here a few words may be inserted concerning variables. A variable is a fundamental concept of logic, as of daily life. Though it is always connected with some class, it is not the class, nor a particular member of the class, nor yet the whole class, but *any* member of the class. On the other hand, it is not the *concept* " any member of the class," but it is that (or those) which this concept denotes. On the logical difficulties of this conception, I need not now enlarge ; enough has been said on this subject in Part I. The usual x in Algebra, for example, does not stand for a particular number, nor for all numbers, nor yet for the class *number*. This may be easily seen by considering some identity, say

$$(x + 1)^2 = x^2 + 2x + 1.$$

This certainly does not mean what it would become if, say, 391 were substituted for x, though it implies that the result of such a substitution would be a true proposition. Nor does it mean what results from substituting for x the class-concept *number*, for we cannot add 1 to this concept. For the same reason, x does not denote the concept *any number* : to this, too, 1 cannot be added. It denotes the disjunction formed by the various numbers ; or at least this view may be taken as roughly correct*. The values of x are then the terms of the disjunction ; and each of these is a constant. This simple logical fact seems to constitute the essence of Zeno's contention that the arrow is always at rest.

333. But Zeno's argument contains an element which is specially applicable to continua. In the case of motion, it denies that there is such a thing as a *state* of motion. In the general case of a continuous variable, it may be taken as denying actual infinitesimals. For infinitesimals are an attempt to extend to the *values* of a variable the variability which belongs to it alone. When once it is firmly realized that all the values of a variable are constants, it becomes easy to see, by taking *any* two such values, that their difference is always finite, and hence that there are no infinitesimal differences. If x be a variable which may take all real values from 0 to 1, then, taking any two of these values, we see that their difference is finite, although x is a continuous variable. It is true the difference might have been less than the one we chose ; but if it had been, it would still have been finite. The lower limit to possible differences is zero, but all possible differences are finite ; and in this there is no shadow of contradiction. This static

---

* See Chap. VIII, esp. § 93.

theory of the variable is due to the mathematicians, and its absence in Zeno's day led him to suppose that continuous change was impossible without a state of change, which involves infinitesimals and the contradiction of a body's being where it is not.

334. The last of Zeno's arguments is that of the measure. This is closely analogous to one which I employed in the preceding chapter, against those who regard dx and dy as distances of consecutive terms. It is only applicable, as M. Noël points out (*loc. cit.* p. 116), against those who hold to indivisibles among stretches, the previous arguments being held to have sufficiently refuted the partisans of infinite divisibility. We are now to suppose a set of discrete moments and discrete places, motion consisting in the fact that at one moment a body is in one of these discrete places, in another at another.

Imagine three parallel lines composed of the points a, b, c, d; a', b', c', d'; a'', b'', c'', d'' respectively. Suppose the second line, in one instant, to move all its points to the left by one place, while the third moves them all one place to the right. Then although the instant is indivisible, c', which was over c'', and is now over a'', must have passed b'' during the instant; hence the instant is divisible, *contra hyp.* This argument is virtually that by which I proved, in the preceding chapter, that, if there are consecutive terms, then $dy/dx = \pm 1$ always; or rather, it is this argument together with an instance in which $dy/dx = 2$. It may be put thus: Let y, z be two functions of x, and let $dy/dx = 1$, $dz/dx = -1$. Then $\frac{d}{dx}(y-z) = 2$, which contradicts the principle that the value of every derivative must be ± 1. To the argument in Zeno's form, M. Evellin, who is an advocate of indivisible stretches, replies that a'' and b' do not cross each other at all*. For if instants are indivisible—and this is the hypothesis—all we can say is, that at one instant a' is over a'', in the next, c' is over a''. Nothing has happened between the instants, and to suppose that a'' and b' have crossed is to beg the question by a covert appeal to the continuity of motion. This reply is valid, I think, in the case of motion; both time and space may, without positive contradiction, be held to be discrete, by adhering strictly to distances in addition to stretches. Geometry, Kinematics, and Dynamics become false; but there is no very good reason to think them true. In the case of Arithmetic, the matter is otherwise, since no empirical question of existence is involved. And in this case, as we see from the above

a b c d

a' b' c' d'

a'' b'' c'' d''

a b c d

a' b' c' d'

a'' b'' c'' d''

---

* *Revue de Métaphysique et de Morale*, Vol. I, p. 386.

argument concerning derivatives, Zeno's argument is absolutely sound. Numbers are entities whose nature can be established beyond question; and among numbers, the various forms of continuity which occur cannot be denied without positive contradiction. For this reason the problem of continuity is better discussed in connection with numbers than in connection with space, time, or motion.

335. We have now seen that Zeno's arguments, though they prove a very great deal, do not prove that the continuum, as we have become acquainted with it, contains any contradictions whatever. Since his day the attacks on the continuum have not, so far as I know, been conducted with any new or more powerful weapons. It only remains, therefore, to make a few general remarks.

The notion to which Cantor gives the name of *continuum* may, of course, be called by any other name in or out of the dictionary, and it is open to every one to assert that he himself means something quite different by the continuum. But these verbal questions are purely frivolous. Cantor's merit lies, not in meaning what other people mean, but in telling us what he means himself—an almost unique merit, where continuity is concerned. He has defined, accurately and generally, a purely ordinal notion, free, as we now see, from contradictions, and sufficient for all Analysis, Geometry, and Dynamics. This notion was presupposed in existing mathematics, though it was not known exactly what it was that was presupposed. And Cantor, by his almost unexampled lucidity, has successfully analyzed the extremely complex nature of spatial series, by which, as we shall see in Part VI, he has rendered possible a revolution in the philosophy of space and motion. The salient points in the definition of the continuum are (1) the connection with the doctrine of limits, (2) the denial of infinitesimal segments. These two points being borne in mind, the whole philosophy of the subject becomes illuminated.

336. The denial of infinitesimal segments resolves an antinomy which had long been an open scandal, I mean the antinomy that the continuum both does and does not consist of elements. We see now that both may be said, though in different senses. Every continuum is a series consisting of terms, and the terms, if not indivisible, at any rate are not divisible into new terms of the continuum. In this sense there are elements. But if we take consecutive terms together with their asymmetrical relation as constituting what may be called (though not in the sense of Part IV) an *ordinal* element, then, in this sense, our continuum has no elements. If we take a stretch to be essentially serial, so that it must consist of at least two terms, then there are no elementary stretches; and if our continuum be one in which there is distance, then likewise there are no elementary distances. But in neither of these cases is there the slightest logical ground for elements. The demand for consecutive terms springs, as we saw in Part III, from an illegitimate use of mathematical induction. And as regards distance, small distances are no simpler than large ones, but all, as we saw in Part III, are alike simple. And large distances do not presuppose small ones: being intensive magnitudes, they may exist where there are no smaller ones at all. Thus the infinite regress from greater to smaller distances or stretches is of the harmless kind, and the lack of elements need not cause any logical inconvenience. Hence the antinomy is resolved, and the continuum, so far at least as I am able to discover, is wholly free from contradictions.

It only remains to inquire whether the same conclusion holds concerning the infinite—an inquiry with which this Fifth Part will come to a close.

CHAPTER XLIII.

THE PHILOSOPHY OF THE INFINITE.

337. In our previous discussions of the infinite we have been compelled to go into so many mathematical points that there has been no adequate opportunity for purely philosophical treatment of the question. In the present chapter, I wish, leaving mathematics aside, to inquire whether any contradiction can be found in the notion of the infinite.

Those who have objected to infinity have not, as a rule, thought it worth while to exhibit precise contradictions in it. To have done so is one of the great merits of Kant. Of the mathematical antinomies, the second, which is concerned, essentially, with the question whether or not the continuum has elements, was resolved in the preceding chapter, on the supposition that there may be an actual infinite—that is, it was reduced to the question of infinite number. The first antinomy is concerned with the infinite, but in an essentially temporal form; for Arithmetic, therefore, this antinomy is irrelevant, except on the Kantian view that numbers must be schematized in time. This view is supported by the argument that it takes time to count, and therefore without time we could not know the number of anything. By this argument we can prove that battles always happen near telegraph wires, because if they did not we should not hear of them. In fact, we can prove generally that we know what we know. But it remains conceivable that we don't know what we don't know; and hence the necessity of time remains unproved.

Of other philosophers, Zeno has already been examined in connection with the continuum; and the paradox which underlies Achilles and the tortoise will be examined shortly. Plato's *Parmenides*—which is perhaps the best collection of antinomies ever made—is scarcely relevant here, being concerned with difficulties more fundamental than any that have to do with infinity. And as for Hegel, he cries *wolf* so often that when he gives the alarm of a contradiction we finally cease to be disturbed. Leibniz, as we have seen, gives as a contradiction the one-one correlation of whole and part, which underlies the Achilles. This is, in fact, the

only point on which most arguments against infinity turn. In what
follows I shall put the arguments in a form adapted to our present
mathematical knowledge; and this will prevent me from quoting them
from any classic opponents of infinity.

338. Let us first recapitulate briefly the positive theory of the infinite to which we have been led. Accepting as indefinable the notion
proposition and the notion *constituent of a proposition*, we may denote
by $\phi(a)$ a proposition in which a is a constituent. We can then transform a into a variable x, and consider $\phi(x)$, where $\phi(x)$ is any proposition
differing from $\phi(a)$, if at all, only by the fact that some other object
appears in the place of a; $\phi(x)$ is what we called a *propositional function*.
It will happen, in general, that $\phi(x)$ is true for some values of x and
false for others. All the values of x, for which $\phi(x)$ is true, form what
we called the *class* defined by $\phi(x)$; thus every propositional function
defines a class, and the actual enumeration of the members of a class
is not necessary for its definition. Again, without enumeration we can
define the similarity of two classes: two classes u, v are similar when
there is a one-one relation R such that "x is a u" always implies "there
is a v to which x has the relation R," and "y is a v" always implies
"there is a u which has the relation R to y." Further, R is a one-one
relation if xRy, xRz together always imply that y is identical with z,
and xRz, yRz together always imply that x is identical with y; and
"x is identical with y" is defined as meaning "every propositional
function which holds of x also holds of y." We now define the cardinal
number of a class u as the class of all classes which are similar to u;
and every class has a cardinal number, since "u is similar to v" is a
propositional function of v, if v be variable. Moreover u itself is
a member of its cardinal number, since every class is similar to itself.
The above definition of a cardinal number, it should be observed, is
based upon the notion of propositional functions, and nowhere involves
enumeration; consequently there is no reason to suppose that there
will be any difficulty as regards the numbers of classes whose terms
cannot be counted in the usual elementary fashion. Classes can be
divided into two kinds, according as they are or are not similar to
proper parts of themselves. In the former case they are called *infinite*,
in the latter *finite*. Again, the number of a class defined by a propositional function which is always false is called 0; 1 is defined as
the number of a class u such that there is a term x, belonging to u,
such that "y is a u and y differs from x" is always false; and if n
be any number, $n + 1$ is defined as the number of a class u which has
a member x such that the propositional function "y is a u and y
differs from x" defines a class whose number is n. If n is finite,
$n + 1$ differs from n; if not, not. In this way, starting from 0, we
obtain a progression of numbers, since any number n leads to a new
number $n + 1$. It is easily proved that all the numbers belonging to

the progression which starts from 1 and is generated in this way are different; that is to say, if n belongs to this progression, and m be any one of its predecessors, a class of n terms cannot have a one-one correlation with one of m terms. The progression so defined is the series of *finite numbers*. But there is no reason to think that all numbers can be so obtained; indeed it is capable of formal proof that the number of the finite numbers themselves cannot be a term in the progression of finite numbers. A number not belonging to this progression is called *infinite*. The proof that n and $n+1$ are different numbers proceeds from the fact that 0 and 1, or 1 and 2, are different numbers, by means of mathematical induction; if n- and $n+1$ be not terms of this progression, the proof fails; and what is more, there is direct proof of the contrary. But since the previous proof depended upon mathematical induction, there is not the slightest reason why the theorem should extend to infinite numbers. Infinite numbers cannot be expressed, like finite ones, by the decimal system of notation, but they can be distinguished by the classes to which they apply. The finite numbers being all defined by the above progression, if a class u has terms, but not any finite number of terms, then it has an infinite number. This is the positive theory of infinity.

339. That there are infinite classes is so evident that it will scarcely be denied. Since, however, it is capable of formal proof, it may be as well to prove it. A very simple proof is that suggested in the *Parmenides*, which is as follows. Let it be granted that there is a number 1. Then 1 is, or has Being, and therefore there is Being. But 1 and Being are two: hence there is a number 2; and so on. Formally, we have proved that 1 is not the number of numbers; we prove that n is the number of numbers from 1 to n, and that these numbers together with Being form a class which has a new finite number, so that n is not the number of finite numbers. Thus 1 is not the number of finite numbers; and if $n-1$ is not the number of finite numbers, no more is n. Hence the finite numbers, by mathematical induction, are all contained in the class of things which are not the number of finite numbers. Since the relation of similarity is reflexive for classes, every class has a number; therefore the class of finite numbers has a number which, not being finite, is infinite. A better proof, analogous to the above, is derived from the fact that, if n be any finite number, the number of numbers from 0 up to and including n is $n+1$, whence it follows that n is not the number of numbers. Again, it may be proved directly, by the correlation of whole and part, that the number of propositions or concepts is infinite[*]. For of every term or concept there is an idea, different from that of which it is the idea, but again a term or concept. On the other hand, not every term or concept is an idea. There are tables, and ideas of tables;

[*] Cf. Bolzano, *Paradoxien des Unendlichen*, § 13; Dedekind, *Was sind und was sollen die Zahlen?* No. 66.

numbers, and ideas of numbers; and so on. Thus there is a one-one relation between terms and ideas, but ideas are only some among terms. Hence there is an infinite number of terms and of ideas*.

340. The possibility that whole and part may have the same number of terms is, it must be confessed, shocking to common-sense. Zeno's Achilles ingeniously shows that the opposite view also has shocking consequences; for if whole and part cannot be correlated term for term, it does strictly follow that, if two material points travel along the same path, the one following the other, the one which is behind can never catch up: if it did, we should have, correlating simultaneous positions, a unique and reciprocal correspondence of all the terms of a whole with all the terms of a part. Commonsense, therefore, is here in a very sorry plight; it must choose between the paradox of Zeno and the paradox of Cantor. I do not propose to help it, since I consider that, in the face of proofs, it ought to commit suicide in despair. But I will give the paradox of Cantor a form resembling that of Zeno. Tristram Shandy, as we know, took two years writing the history of the first two days of his life, and lamented that, at this rate, material would accumulate faster than he could deal with it, so that he could never come to an end. Now I maintain that, if he had lived for ever, and not wearied of his task, then, even if his life had continued as eventfully as it began, no part of his biography would have remained unwritten. This paradox, which, as I shall show, is strictly correlative to the Achilles, may be called for convenience the Tristram Shandy.

In cases of this kind, no care is superfluous in rendering our arguments formal. I shall therefore set forth both the Achilles and the Tristram Shandy in strict logical shape.

I. (1) For every position of the tortoise there is one and only one of Achilles; for every position of Achilles there is one and only one of the tortoise.

(2) Hence the series of positions occupied by Achilles has the same number of terms as the series of positions occupied by the tortoise.

(3) A part has fewer terms than a whole in which it is contained and with which it is not coextensive.

(4) Hence the series of positions occupied by the tortoise is not a proper part of the series of positions occupied by Achilles.

II. (1) Tristram Shandy writes in a year the events of a day.

(2) The series of days and years has no last term.

(3) The events of the nth day are written in the nth year.

(4) Any assigned day is the nth, for a suitable value of n.

(5) Hence any assigned day will be written about.

---

* It is not necessary to suppose that the ideas of all terms *exist*, or form part of some mind; it is enough that they are entities.

(6) Hence no part of the biography will remain unwritten.

(7) Since there is a one-one correlation between the times of happening and the times of writing, and the former are part of the latter, the whole and the part have the same number of terms.

Let us express both these paradoxes as abstractly as possible. For this purpose, let u be a compact series of any kind, and let x be a variable which can take all values in u after a certain value, which we will call 0. Let $f(x)$ be a one-valued function of x, and x a one-valued function of $f(x)$; also let all the values of $f(x)$ belong to u. Then the arguments are the following.

I. Let $f(0)$ be a term preceding 0; let $f(x)$ grow as x grows, *i.e.* if $x\,P\,x'$ (where P is the generating relation), let $f(x)\,P\,f(x')$. Further let $f(x)$ take all values in u intermediate between any two values of $f(x)$. If, then, for some value a of x, such that $0\,P\,a$, we have $f(a) = a$, then the series of values of $f(x)$ will be all terms from $f(0)$ to a, while that of x will be only the terms from 0 to a, which are a part of those from $f(0)$ to a. Thus to suppose $f(a) = a$ is to suppose a one-one correlation, term for term, of whole and part, which Zeno and common-sense pronounce impossible.

II. Let $f(x)$ be a function which is 0 when x is 0, and which grows uniformly as x grows, our series being one in which there is measurement. Then if x takes all values after 0, so does $f(x)$; and if $f(x)$ takes all such values, so does x. The class of values of the one is therefore identical with that of the other. But if at any time the value of x is greater than that of $f(x)$, since $f(x)$ grows at a uniform rate, x will always be greater than $f(x)$. Hence for any assigned value of x, the class of values of $f(x)$ from 0 to $f(x)$ is a proper part of the values of x from 0 to x. Hence we might infer that all the values of $f(x)$ were a proper part of all the values of x; but this, as we have seen, is fallacious.

These two paradoxes are correlative. Both, by reference to segments, may be stated in terms of limits. The Achilles proves that two variables in a continuous series, which approach equality from the same side, cannot ever have a common limit; the Tristram Shandy proves that two variables which start from a common term, and proceed in the same direction, but diverge more and more, may yet determine the same limiting class (which, however, is not necessarily a segment, because segments were defined as having terms beyond them). The Achilles assumes that whole and part cannot be similar, and deduces a paradox; the other, starting from a platitude, deduces that whole and part may be similar. For common-sense, it must be confessed, this is a most unfortunate state of things.

341. There is no doubt which is the correct course. The Achilles must be rejected, being directly contradicted by Arithmetic. The Tristram Shandy must be accepted, since it does not involve the axiom that the whole cannot be similar to the part. This axiom, as we have seen, is essential to the proof of the Achilles; and it is an axiom doubtless very agreeable to common-sense. But there is no evidence for the axiom except supposed self-evidence, and its admission leads to perfectly precise contradictions. The axiom is not only useless, but positively destructive, in mathematics, and against its rejection there is nothing to be set except prejudice. It is one of the chief merits of proofs that they instil a certain scepticism as to the result proved. As soon as it was found that the similarity of whole and part could be *proved* to be impossible for every *finite* whole*, it became not unplausible to suppose that for infinite wholes, where the impossibility could not be proved, there was in fact no such impossibility. In fact, as regards the numbers dealt with in daily life—in engineering, astronomy, or accounts, even those of Rockefeller and the Chancellor of the Exchequer— the similarity of whole and part *is* impossible ; and hence the supposition that it is always impossible is easily explained. But the supposition rests on no better foundation than that formerly entertained by the inductive philosophers of Central Africa, that all men are black.

342. It may be worth while, as helping to explain the difference between finite and infinite wholes, to point out that whole and part are terms capable of two definitions where the whole is finite, but of only one of these, at least practically, where the whole is infinite†. A finite whole may be taken collectively, as such and such individuals, *A, B, C, D, E* say. A part of this whole may be obtained by enumerating some, but not all, of the terms composing the whole ; and in this way a single individual is part of the whole. Neither the whole nor its parts need be taken as classes, but each may be defined by extension, *i.e.* by enumeration of individuals. On the other hand, the whole and the parts may be both defined by intension, *i.e.* by class-concepts. Thus we know without enumeration that Englishmen are part of Europeans ; for whoever is an Englishman is a European, but not *vice versâ*. Though this *might* be established by enumeration, it need not be so established. When we come to infinite wholes, this twofold definition disappears, and we have only the definition by intension. The whole and the part must both be classes, and the definition of whole and part is effected by means of the notions of a variable and of logical implication. If *a* be a class-concept, an individual of *a* is a term having to *a* that specific relation which we call the class-relation. If now *b* be another class such that, for all values of *x*, "*x* is an *a*" implies "*x* is a *b*," then the extension of *a* (*i.e.* the variable *x*) is said to be *part* of the extension of *b*‡. Here no enumeration of individuals is required, and the relation of whole and part has no longer

---

* The finite being here defined by mathematical induction, to avoid tautology.
† Cf. § 330.
‡ See Peano, *Rivista di Matematica*, VII, or *Formulaire*, Vol. II, Part I.

that simple meaning which it had where finite parts were concerned. To say now that a and b are similar, is to say that there exists some one-one relation R fulfilling the following conditions: if x be an a, there is a term y of the class b such that xRy; if y' be a b, there is a term x' of the class a such that $x'Ry$. Although a is part of b, such a state of things cannot be proved impossible, for the impossibility could only be proved by enumeration, and there is no reason to suppose enumeration possible. The definition of whole and part without enumeration is the key to the whole mystery. The above definition, which is due to Professor Peano, is that which is naturally and necessarily applied to infinite wholes. For example, the primes are a proper part of the integers, but this cannot be proved by enumeration. It is deduced from "if x be a prime, x is a number," and "if x be a number, it does not follow that x is a prime." That the class of primes should be similar to the class of numbers only seems impossible because we imagine whole and part defined by enumeration. As soon as we rid ourselves of this idea the supposed contradiction vanishes.

343. It is very important to realize, as regards ω or α_0, that neither has a number immediately preceding it. This characteristic they share with all limits, for the limit of a series is never immediately preceded by any term of the series which it limits. But ω is in some sense logically prior to other limits, for the finite ordinal numbers together with ω present the formal type of a progression together with its limit. When it is forgotten that ω has no immediate predecessor, all sorts of contradictions emerge. For suppose n to be the last number before ω; then n is a finite number, and the number of finite numbers is $n + 1$. In fact, to say that ω has no predecessor is merely to say that the finite numbers have no last term. Though ω is preceded by all finite numbers, it is not preceded immediately by any of them: there is none next to ω. Cantor's transfinite numbers have the peculiarity that, although there is one next after any assigned number, there is not always one next before. Thus there seem to be gaps in the series. We have the series $1, 2, 3, \ldots \nu, \ldots$, which is infinite and has no last term. We have another series $\omega, \omega + 1, \omega + 2, \ldots \omega + \nu, \ldots$ which equally is infinite and has no last term. This second series comes wholly after the first, though there is no one term of the first which ω immediately succeeds. This state of things may, however, be paralleled by very elementary series, such as the series whose general terms are $1 - 1/\nu$ and $2 - 1/\nu$, where ν may be any finite integer. The second series comes wholly after the first, and has a definite first term, namely 1. But there is no term of the first series which immediately precedes 1. What is necessary, in order that the second series should come after the first, is that there should be some series in which both are contained. If we call an *ordinal part* of a series any series which can be obtained by omitting some of the terms of our series without changing the order of the remaining terms, then the finite and transfinite ordinals all form one series, whose generating relation is that of ordinal whole and part among the series to which the various ordinals apply. If ν be any finite ordinal, series of the type ν are ordinal parts of progressions; similarly every series of the type $\omega + 1$ contains a progression as an ordinal part. The relation *ordinal part* is transitive and asymmetrical, and thus the finite and transfinite ordinals all belong to one series. The existence of ω (in the mathematical sense of existence) is not open to question, since ω is the type of order presented by the natural numbers themselves. To deny ω would be to affirm that there is a last finite number—a view which, as we have seen, leads at once to definite contradictions. And when this is admitted, $\omega + 1$ is the type of the series of ordinals including ω, *i.e.* of the series whose terms are all series of integers from 1 up to any finite number together with the whole series of integers. Hence all the infinite hierarchy of transfinite numbers easily follows.

344. The usual objections to infinite numbers, and classes, and series, and the notion that the infinite as such is self-contradictory, may thus be dismissed as groundless. There remains, however, a very grave difficulty, connected with the contradiction discussed in Chapter x. This difficulty does not concern the infinite as such, but only certain very large infinite classes. Briefly, the difficulty may be stated as follows. Cantor has given a proof* that there can be no greatest cardinal number, and when this proof is examined, it is found to state that, if u be a class, the number of classes contained in u is greater than the number of terms of u, or (what is equivalent), if α be any number, 2^{α} is greater than α. But there are certain classes concerning which it is easy to give an apparently valid proof that they have as many terms as possible. Such are the class of all terms, the class of all classes, or the class of all propositions. Thus it would seem as though Cantor's proof must contain some assumption which is not verified in the case of such classes. But when we apply the reasoning of his proof to the cases in question, we find ourselves met by definite contradictions, of which the one discussed in Chapter x is an example†. The difficulty arises whenever we try to deal with the class of all entities absolutely, or with any equally numerous class; but for the difficulty of such a view, one would be tempted to say that the conception of the totality of things, or of the whole universe of entities and existents, is in some way illegitimate and inherently contrary to logic. But it is undesirable to adopt so desperate a measure as long as hope remains of some less heroic solution.

It may be observed, to begin with, that the class of numbers is not,

---

* He has, as a matter of fact, offered two proofs, but we shall find that one of them is not cogent.

† It was in this way that I discovered this contradiction; a similar one is given at the end of Appendix B.

as might be supposed, one of those in regard to which difficulties occur. Among finite numbers, if n were the number of numbers, we should have to infer that $n-1$ was the greatest of numbers, so that there would be no number n at all. But this is a peculiarity of finite numbers. The number of numbers up to and including α_0 is α_0, but this is also the number of numbers up to and including α_β, where β is any finite ordinal or any ordinal applicable to a denumerable well-ordered series. Thus the number of numbers up to and including α, where α is infinite, is usually less than α, and there is no reason to suppose that the number of all numbers is the greatest number. The number of numbers may be less than the greatest number, and no contradiction arises from the fact (if it be a fact) that the number of individuals is greater than the number of numbers.

But although the class of numbers causes no difficulty, there are other classes with which it is very hard to deal. Let us first examine Cantor's proofs that there is no greatest cardinal number, and then discuss the cases in which contradictions arise.

345. In the first of Cantor's proofs*, the argument depends upon the supposed fact that there is a one-one correspondence between the ordinals and the cardinals†. We saw that, when we consider the cardinal number of the series of the type represented by any ordinal, an infinite number of ordinals correspond to one cardinal—for example, all ordinals of the second class, which form a non-denumerable collection, correspond to the single cardinal α_0. But there is another method of correlation, in which only one ordinal corresponds to each cardinal. This method results from considering the series of cardinals itself. In this series, α_0 corresponds to ω, α_1 to $\omega+1$, and so on: there is always one and only one ordinal to describe the type of series presented by the cardinals from 0 up to any one of them. It seems to be assumed that there is a cardinal for every ordinal, and that no class can have so many terms that no well-ordered series can have a greater number of terms. For my part I do not see any grounds for either supposition, and I do see definite grounds against the latter. For every term of a series must be an individual, and must be a different individual (a point often overlooked) from every other term of the series. It must be different, because there are no instances of an individual: each individual is absolutely unique, and in the nature of the case only one. But two terms in a series are two, and are therefore not one and the same individual. This most important point is obscured by the fact that we do not, as a rule, fully describe the terms of our series. When we say: Consider a series $a, b, c, d, b, d, e, a, \ldots$, where terms are repeated at intervals—such a series, for example, as is presented by the digits in a decimal—we forget the theorem that where there is repetition our series is only obtainable by correlation; that is, the

---

* *Mannichfaltigkeitslehre*, p. 44.
† Cf. *supra*, Chap. xxxviii, § 300.

terms do not themselves have an order, but they have a one-many (not one-one) relation to terms which have an order*. Hence if we wish for a genuine series we must either go back to the series with which our terms are correlated, or we must form the complex terms compounded of those of the original series and those of the correlated series in pairs. But in either of these series there is no repetition. Hence every ordinal number must correspond to a series of individuals, each of which differs from each other. Now it may be doubted whether all individuals form a series at all: for my part I cannot discover any transitive asymmetrical relation which holds between *every* pair of terms. Cantor, it is true, regards it as a law of thought that every definite aggregate can be well-ordered; but I see no ground for this opinion. But allowing this view, the ordinals will have a perfectly definite maximum, namely that ordinal which represents the type of series formed by all terms without exception†. If the collection of all terms does not form a series it is impossible to prove that there must be a maximum ordinal, which in any case there are reasons for denying‡. But in this case we may legitimately doubt whether there are as many ordinals as there are cardinals. Of course, if all cardinals form a well-ordered series, then there must be an ordinal for each cardinal. But although Cantor professes that he has a proof that of two different cardinals one must be the greater (*Math. Annalen*, XLVI, § 2), I cannot persuade myself that he does more than prove that there is a series, whose terms are cardinals of which any one is greater or less than any other. That all cardinals are in this series I see no reason to think. There may be two classes such that it is not possible to correlate either with a part of the other; in this case the cardinal number of the one will be neither equal to, greater than, nor less than, that of the other. If all terms belong to a single well-ordered series, this is impossible; but if not, I cannot see any way of showing that such a case cannot arise. Thus the first proof that there is no cardinal which cannot be increased seems to break down.

346. The second of the proofs above referred to§ is quite different, and is far more definite. The proof is interesting and important on its own account, and will be produced in outline. The article in which it occurs consists of three points: (1) a simple proof that there are powers higher than the first, (2) the remark that this method of proof can be applied to any power, (3) the application of the method to prove that there are powers higher than that of the continuum‖. Let us

---

* See Chap. XXXII, *supra*.

† On the maximum ordinal, see Burali-Forti, "Una questione sui numeri transfiniti," *Rendiconti del circolo matematico di Palermo*, 1897. Also my article in *RdM*, Vol. VIII, p. 43 *note*.

‡ Cf. Chap. XXXVIII, § 301.

§ *Jahresbericht der deutschen Mathematiker-Vereinigung*, I. (1892), p. 77.

‖ *Power* is synonymous with *cardinal number*: the first power is that of the finite integers, *i.e.* a_0.

examine the first of the above points, and then see whether the method is really general.

Let m and w, Cantor says, be two mutually exclusive characters, and consider a collection M of elements E, where each element E is a denumerable collection, $x_1, x_2, \ldots x_n, \ldots$, and each x is either an m or a w. (The two characters m and w may be considered respectively as greater and less than some fixed term. Thus the x's may be rational numbers, each of which is an m when it is greater than 1, and a w when it is less than 1. These remarks are logically irrelevant, but they make the argument easier to follow.) The collection M is to consist of all possible elements E of the above description. Then M is not denumerable, *i.e.* is of a power higher than the first. For let us take any denumerable collection of E's, which are defined as follows:

$$E_1 = (a_{11}, \ a_{12}, \ \ldots \ a_{1n}, \ \ldots)$$
$$E_2 = (a_{21}, \ a_{22}, \ \ldots \ a_{2n}, \ \ldots)$$
$$\ldots\ldots\ldots\ldots\ldots\ldots\ldots\ldots\ldots\ldots$$
$$E_p = (a_{p1}, \ a_{p2}, \ \ldots \ a_{pn}, \ \ldots)$$
$$\ldots\ldots\ldots\ldots\ldots\ldots\ldots\ldots\ldots\ldots$$

where the a's are each an m or a w in some determinate manner. (For example, the first p terms of E_p might be m's, the rest all w's. Or any other law might be suggested, which insures that the E's of our series are all different.) Then however our series of E's be chosen, we can always find a term E_o, belonging to the collection M, but not to the denumerable series of E's. For let E_o be the series $(b_1, \ b_2, \ldots b_n \ldots)$, where, for every n, b_n is different from a_{nn}—*i.e.* if a_{nn} is an m, b_n is a w, and *vice versâ*. Then every one of our denumerable series of E's contains at least one term not identical with the corresponding term of E_o, and hence E_o is not any one of the terms of our denumerable series of E's. Hence no such series can contain all the E's, and therefore the E's are not denumerable, *i.e.* M has a power higher than the first.

We need not stop to examine the proof that there is a power higher than that of the continuum, which is easily obtained from the above proof. We may proceed at once to the general proof that, given any collection whatever, there is a collection of a higher power. This proof is quite as simple as the proof of the particular case. It proceeds as follows. Let u be any class, and consider the class K of relations such that, if R be a relation of the class, every term of the class u has the relation R either to 0 or to 1. (Any other pair of terms will do as well as 0 and 1.) Then the class K has a higher power than the class u. To prove this, observe in the first place that K has certainly not a lower power; for, if x be any u, there will be a relation R of the class K such that every u except x has the relation R to 0, but x has this relation to 1. Relations of this kind, for the various values of x, form a class having a one-one correlation with the terms of u, and contained in the class K. Hence K has at least the same power as u. To prove that K has a greater power, consider any class contained in K and having a one-one correlation with u. Then any relation of this class may be called R_x, where x is some u—the suffix x denoting correlation with x. Let us now define a relation R' by the following conditions: For every term x of u for which x has the relation R_x to 0, let x have the relation R' to 1; and for every term y of u for which y has the relation R_y to 1, let y have the relation R' to 0. Then R' is defined for all terms of u, and is a relation of the class K; but it is not any one of the relations R_x. Hence, whatever class contained in K and of the same power as u we may take, there is always a term of K not belonging to this class; and therefore K has a higher power than u.

347. We may, to begin with, somewhat simplify this argument, by eliminating the mention of 0 and 1 and relations to them. Each of the relations of the class K is defined when we know which of the terms of u have this relation to 0, that is, it is defined by means of a class contained in u (including the null-class and u itself). Thus there is one relation of the class K for every class contained in u, and the number of K is the same as that of classes contained in u. Thus if k be any class whatever, the logical product ku is a class contained in u, and the number of K is that of ku, where k is a variable which may be any class. Thus the argument is reduced to this: that the number of classes contained in any class exceeds the number of terms belonging to the class*.

Another form of the same argument is the following. Take any relation R which has the two properties (1) that its domain, which we will call ρ, is equal to its converse domain, (2) that no two terms of the domain have exactly the same set of relata. Then by means of R, any term of ρ is correlated with a class contained in ρ, namely the class of relata to which the said term is referent; and this correlation is one-one. We have to show that at least one class contained in ρ is omitted in this correlation. The class omitted is the class w which consists of all terms of the domain which do not have the relation R to themselves, *i.e.* the class w which is the domain of the logical product of R and diversity. For, if y be any term of the domain, and therefore of the converse domain, y belongs to w if it does not belong to the class correlated with y, and does not belong to w in the contrary case. Hence w is not the same class as the correlate of y; and this applies to whatever term y we select. Hence the class w is necessarily omitted in the correlation.

348. The above argument, it must be confessed, appears to contain no dubitable assumption. Yet there are certain cases in which the conclusion seems plainly false. To begin with the class of all terms. If we assume, as was done in § 47, that every constituent of every

* The number of classes contained in a class which has a members is 2^a; thus the argument shows that 2^a is always greater than a.

proposition is a term, then classes will be only some among terms. And conversely, since there is, for every term, a class consisting of that term only, there is a one-one correlation of all terms with some classes. Hence the number of classes should be the same as the number of terms*. This case is adequately met by the doctrine of types†, and so is the exactly analogous case of classes and classes of classes. But if we admit the notion of all objects‡ of every kind, it becomes evident that classes of objects must be only some among objects, while yet Cantor's argument would show that there are more of them than there are objects. Or again, take the class of propositions. Every object can occur in some proposition, and it seems indubitable that there are at least as many propositions as there are objects. For, if u be a fixed class, "x is a u" will be a different proposition for every different value of x; if, according to the doctrine of types, we hold that, for a given u, x has a restricted range if "x is a u" is to remain significant, we only have to vary u suitably in order to obtain propositions of this form for every possible x, and thus the number of propositions must be at least as great as that of objects. But classes of propositions are only some among objects, yet Cantor's argument shows that there are more of them than there are propositions. Again, we can easily prove that there are more propositional functions than objects. For suppose a correlation of all objects and some propositional functions to have been affected, and let ϕ_x be the correlate of x. Then "not-$\phi_x(x)$," *i.e.* "ϕ_x does not hold of x," is a propositional function not contained in the correlation; for it is true or false of x according as ϕ_x is false or true of x, and therefore it differs from ϕ_x for every value of x. But this case may perhaps be more or less met by the doctrine of types.

349. It is instructive to examine in detail the application of Cantor's argument to such cases by means of an actual attempted correlation. In the case of terms and classes, for example, if x be not a class, let us correlate it with ιx, *i.e.* the class whose only member is x, but if x be a class, let us correlate it with itself. (This is not a one-one, but a many-one correlation, for x and ιx are both correlated with ιx; but it will serve to illustrate the point in question.) Then the class which, according to Cantor's argument, should be omitted from the correlation, is the class w of those classes which are not members of themselves; yet this, being a class, should be correlated with itself. But w, as we saw in Chapter x, is a self-contradictory class, which both is and is not a member of itself. The contradiction, in this case, can be solved by the doctrine of types; but the case of propositions is more difficult. In this case, let us correlate every class of propositions with the

---

* This results from the theorem of Schröder and Bernstein, according to which, if u be similar to a part of v, and v to a part of u, then u and v must be similar. See Borel, *Leçons sur la Théorie des Fonctions* (Paris, 1898), p. 102 ff.

† See Chapter x. and Appendix B.

‡ For the use of the word *object* see p. 55, *note*.

proposition which is its logical product; by this means we appear to have
a one-one relation of all classes of propositions to some propositions.
But applying Cantor's argument, we find that we have omitted the
class *w* of those propositions which are logical products, but are not
members of the classes of propositions whose logical products they
are. This class, according to the definition of our correlation, should
be correlated with its own logical product, but on examining this
logical product, we find that it both is and is not a member of the
class *w* whose logical product it is.

Thus the application of Cantor's argument to the doubtful cases
yields contradictions, though I have been unable to find any point in
which the argument appears faulty. The only solution I can suggest
is, to accept the conclusion that there is no greatest number and the
doctrine of types, and to deny that there are any true propositions
concerning all objects or all propositions. Yet the latter, at least,
seems plainly false, since all propositions are at any rate true or false,
even if they had no other common properties. In this unsatisfactory
state, I reluctantly leave the problem to the ingenuity of the reader.

350. To sum up the discussions of this Part: We saw, to begin
with, that irrationals are to be defined as those segments of rationals
which have no limit, and that in this way analysis is able to dispense
with any special axiom of continuity. We saw that it is possible to
define, in a purely ordinal manner, the kind of continuity which belongs
to real numbers, and that continuity so defined is not self-contradictory.
We found that the differential and integral calculus has no need of the
infinitesimal, and that, though some forms of infinitesimal are admissible,
the most usual form, that of infinitesimal segments in a compact series,
is not implied by either compactness or continuity, and is in fact
self-contradictory. Finally we discussed the philosophical questions
concerning continuity and infinity, and found that the arguments of
Zeno, though largely valid, raise no sort of serious difficulty. Having
grasped clearly the twofold definition of the infinite, as that which
cannot be reached by mathematical induction starting from 1, and as
that which has parts which have the same number of terms as itself—
definitions which may be distinguished as ordinal and cardinal respectively—we found that all the usual arguments, both as to infinity
and as to continuity, are fallacious, and that no definite contradiction
can be proved concerning either, although certain special infinite classes
do give rise to hitherto unsolved contradictions.

It remains to apply to space, time, and motion, the three chief results of this discussion, which are (1) the impossibility of infinitesimal
segments, (2) the definition of continuity, and (3) the definition and
the consistent doctrine of the infinite. These applications will, I
hope, persuade the reader that the above somewhat lengthy discussions
have not been superfluous.

PART VI.

SPACE.

CHAPTER XLIV.

DIMENSIONS AND COMPLEX NUMBERS.

351. THE discussions of the preceding Parts have been concerned with two main themes, the logical theory of numbers and the theory of one-dimensional series. In the first two Parts, it was shown how, from the indispensable apparatus of general logical notions, the theory of finite integers and of rational numbers without sign could be developed. In the third Part, a particular case of order, namely the order of magnitude, was examined on its own account, and it was found that most of the problems arising in the theory of quantity are purely ordinal. In the fourth Part, the general nature of one-dimensional series was set forth, and it was shown that all the arithmetical propositions obtained by means of the logical theory of finite numbers could also be proved by assuming that the finite integers form a series of the kind which we called a progression. In the fifth Part, we examined the problems raised by endless series and by compact series—problems which, under the names of infinity and continuity, have defied philosophers ever since the dawn of abstract thought. The discussion of these problems led to a combination of the logical and ordinal theories of Arithmetic, and to the rejection, as universally valid, of two connected principles which, following Cantor, we regarded as definitions of the finite, not as applicable to all collections or series. These two principles were: (1) If one class be wholly contained in, but not coextensive with, another, then the one has not the same number of terms as the other; (2) mathematical induction, which is purely ordinal, and may be stated as follows: A series generated by a one-one relation, and having a first term, is such that any property, belonging to the first term and to the successor of any possessor of the property, belongs to every term of the series. These two principles we regarded as definitions of finite classes and of progressions or finite series respectively, but as inapplicable to some classes and some series. This view, we found, resolves all the difficulties of infinity and continuity, except a purely logical difficulty as to the notion of *all* classes. With this result, we completed the philosophical theory of one-dimensional series.

352. But in all our previous discussions, large branches of mathematics have remained unmentioned. One of the generalizations of number, namely complex numbers, has been excluded completely, and no mention has been made of the imaginary. The whole of Geometry, also, has been hitherto foreign to our thoughts. These two omissions were connected. Not that we are to accept a geometrical, *i.e.* spatial, theory of complex numbers: this would be as much out of place as a geometrical theory of irrationals. Although this Part is called *Space*, we are to remain in the region of pure mathematics: the mathematical entities discussed will have certain affinities to the space of the actual world, but they will be discussed without any logical dependence upon these affinities. Geometry may be considered as a pure *à priori* science, or as the study of actual space. In the latter sense, I hold it to be an experimental science, to be conducted by means of careful measurements. But it is not in this latter sense that I wish to discuss it. As a branch of pure mathematics, Geometry is strictly deductive, indifferent to the choice of its premisses and to the question whether there exist (in the strict sense) such entities as its premisses define. Many different and even inconsistent sets of premisses lead to propositions which would be called geometrical, but all such sets have a common element. This element is wholly summed up by the statement that Geometry deals with series of more than one dimension. The question what may be the actual terms of such series is indifferent to Geometry, which examines only the consequences of the relations which it postulates among the terms. These relations are always such as to generate a series of more than one dimension, but have, so far as I can see, no other general point of agreement. Series of more than one dimension I shall call *multiple* series: those of one dimension will be called *simple*. What is meant by dimensions I shall endeavour to explain in the course of the present chapter. At present, I shall set up, by anticipation, the following definition: *Geometry is the study of series of two or more dimensions.* This definition, it will be seen, causes complex numbers to form part of the subject-matter of Geometry, since they constitute a two-dimensional series; but it does not show that complex numbers have any logical dependence upon actual space.

The above definition of Geometry is, no doubt, somewhat unusual, and will produce, especially upon Kantian philosophers, an appearance of wilful misuse of words. I believe, however, that it represents correctly the present usage of mathematicians, though it is not necessary for them to give an explicit definition of their subject. How it has come to bear this meaning, may be explained by a brief historical retrospect, which will illustrate also the difference between pure and applied mathematics.

353. Until the nineteenth century, Geometry meant Euclidean Geometry, *i.e.* a certain system of propositions deduced from premisses which were supposed to describe the space in which we live. The subject was pursued very largely because (what is no doubt important to the engineer) its results were practically applicable in the existent world, and embodied in themselves scientific truths. But in order to be sure that this was so, one of two things was necessary. Either we must be certain of the truth of the premisses on their own account, or we must be able to show that no other set of premisses would give results consistent with experience. The first of these alternatives was adopted by the idealists and was especially advocated by Kant. The second alternative represents, roughly, the position of empiricists before the non-Euclidean period (among whom we must include Mill). But objections were raised to both alternatives. For the Kantian view, it was necessary to maintain that all the axioms are self-evident—a view which honest people found it hard to extend to the axiom of parallels. Hence arose a search for more plausible axioms, which might be declared *à priori* truths. But, though many such axioms were suggested, all could sanely be doubted, and the search only led to scepticism. The second alternative—the view that no other axioms would give results consistent with experience—could only be tested by a greater mathematical ability than falls to the lot of most philosophers. Accordingly the test was wanting until Lobatchewsky and Bolyai developed their non-Euclidean system. It was then proved, with all the cogency of mathematical demonstration, that premisses other than Euclid's could give results empirically indistinguishable, within the limits of observation, from those of the orthodox system. Hence the empirical argument for Euclid was also destroyed. But the investigation produced a new spirit among Geometers. Having found that the denial of Euclid's axiom of parallels led to a different system, which was self-consistent, and possibly true of the actual world, mathematicians became interested in the development of the consequences flowing from other sets of axioms more or less resembling Euclid's. Hence arose a large number of Geometries, inconsistent, as a rule, with each other, but each internally self-consistent. The resemblance to Euclid required in a suggested set of axioms has gradually grown less, and possible deductive systems have been more and more investigated on their own account. In this way, Geometry has become (what it was formerly mistakenly called) a branch of pure mathematics, that is to say, a subject in which the assertions are that such and such consequences follow from such and such premisses, not that entities such as the premisses describe actually exist. That is to say, if Euclid's axioms be called A, and P be any proposition implied by A, then, in the Geometry which preceded Lobatchewsky, P itself would be asserted, since A was asserted. But now-a-days, the geometer would only assert that A implies P, leaving A and P themselves doubtful. And he would have other sets of axioms, A_1, A_2... implying P_1, P_2... respectively: the *implications* would belong to Geometry, but not A_1 or P_1 or any of the other actual axioms and propositions. Thus Geometry no longer throws any direct light on the nature of actual space. But indirectly, the increased analysis and knowledge of possibilities, resulting from modern Geometry, has thrown immense light upon our actual space. Moreover it is now proved (what is fatal to the Kantian philosophy) that every Geometry is rigidly deductive, and does not employ any forms of reasoning but such as apply to Arithmetic and all other deductive sciences. My aim, in what follows, will be to set forth first, in brief outlines, what is philosophically important in the deductions which constitute modern Geometry, and then to proceed to those questions, in the philosophy of space, upon which mathematics throws light. In the first section of this Part, though I shall be discussing Geometries as branches of pure mathematics, I shall select for discussion only those which throw the most light either upon actual space, or upon the nature of mathematical reasoning. A treatise on non-Euclidean Geometry is neither necessary nor desirable in a general work such as the present, and will therefore not be found in the following chapters.

354. Geometry, we said, is the study of series which have more than one dimension. It is now time to define dimensions, and to explain what is meant by a multiple series. The relevance of our definition to Geometry will appear from the fact that the mere definition of dimensions leads to a duality closely analogous to that of projective Geometry.

Let us begin with two dimensions. A series of two dimensions arises as follows. Let there be some asymmetrical transitive relation P, which generates a series u_1. Let every term of u_1 be itself an asymmetrical transitive relation, which generates a series. Let all the field of P form a simple series of asymmetrical relations, and let each of these have a simple series of terms for its field. Then the class u_2 of terms forming the fields of all the relations in the series generated by P is a two-dimensional series. In other words, the total field of a class of asymmetrical transitive relations forming a simple series is a double series. But instead of starting from the asymmetrical relation P, we may start from the terms. Let there be a class of terms u_2, of which any given one (with possibly one exception) belongs to the field of one and only one of a certain class u_1 of serial relations. That is if x be a term of u_2, x is also a term of the field of some relation of the class u_1. Now further let u_1 be a series. Then u_2 will be a double series. This seems to constitute the definition of two-dimensional series.

To obtain three dimensions, we have only to suppose that u_2 itself consists of series, or of asymmetrical transitive relations. Or, starting with the terms of the three-dimensional series, let any term of a certain class u_3 belong to one and only one series (again with one possible exception, which may belong to many series) of a certain class u_2. Let every term of u_2 be a term of some series belonging to a class u_1 of series, and let u_1 itself be a simple series. Then u_3 is a triple series, or a series of three dimensions. Proceeding in this way, we obtain the definition of n dimensions, which may be given as follows: Let there be some series u_1 whose terms are all themselves serial relations. If x_1 be any term of u_1, and x_2 any term of the field of x_1, let x_2 be again a serial relation, and so on. Proceeding to x_3, x_4, etc., let x_{n-1}, however obtained, be always a relation generating a simple series. Then all the terms x_n belonging to the field of any serial relation x_{n-1}, form an n-dimensional series. Or, to give the definition which starts from the terms: Let u_n be a class of terms, any one of which, x_n say, belongs to the field of some serial relation, x_{n-1} say, which itself belongs to a definite class u_{n-1} of serial relations. Let each term x_n in general belong to the field of only one serial relation x_{n-1} (with exceptions which need not be discussed at present). Let u_{n-1} lead to a new class u_{n-2} of serial relations, in exactly the way in which u_n led to u_{n-1}. Let this proceed until we reach a class u_1, and let u_1 be a simple series. Then u_n is a series of n dimensions.

355. Before proceeding further, some observations on the above definitions may be useful. In the first place, we have just seen that alternative definitions of dimensions suggest themselves, which have a relation analogous to what is called duality in projective Geometry. How far this analogy extends, is a question which we cannot discuss until we have examined projective Geometry. In the second place, every series of n dimensions involves series of all smaller numbers of dimensions, but a series of $(n-1)$ dimensions does not in general imply one of n dimensions. In the second form of the definition of n dimensions, the class u_{n-1} is a series of $(n-1)$ dimensions, and generally, if m be less than n, the class u_{n-m} is a series of $(n-m)$ dimensions. And in the other method, all possible terms x_{n-1} together form a series of $(n-1)$ dimensions, and so on. In the third place, if n be finite, a class which is an n-dimensional series is also a one-dimensional series. This may be established by the following rules: In the class u_1, which is a simple series, preserve the order unchanged. In u_2, keep the internal order of each series unchanged, and place that series before which comes before in u_1, and that after which comes after in u_1. Thus u_2 is converted into a simple series. Apply now the same process to u_3, and so on. Then by mathematical induction, if n be finite, or be any infinite ordinal number, u_n can be converted into a simple series. This remarkable fact, which was discovered, for finite numbers and ω, by Cantor[*],

---

[*] Cantor has proved, not only that a simple series can be so formed, but that, if n be not greater than ω, and the constituent series all have the same cardinal number, this is also the cardinal number of the resultant series: *i.e.* an n-dimensional space has the same cardinal number of points as a finite portion of a line. See *Acta Math.* II, p. 314 ff.

has a very important bearing on the foundations of Geometry. In the fourth place, the definition of n dimensions can be extended to the case where n is ω, the first of the transfinite ordinals. For this purpose, it is only necessary to suppose that, whatever finite number m we may take, any u_m will belong to some simple series of series u_{m+1}; and that the sequence of classes of series so obtained obeys mathematical induction, and is therefore a progression. Then the number of dimensions is ω. This case brings out, what does not appear so clearly from the case of a finite number of dimensions, that the number of dimensions is an ordinal number.

356. There are very many ways of generating multiple series, as there are of generating simple series. The discussion of these various ways is not, however, of great importance, since it would follow closely the discussion of Part IV, Chapter xxiv. Instances will meet us in the course of our examination of the various Geometries; and this examination will give opportunities of testing our definition of dimensions. For the present, it is only important to observe that dimensions, like order and continuity, are defined in purely abstract terms, without any reference to actual space. Thus when we say that space has three dimensions, we are not merely attributing to it an idea which can only be obtained from space, but we are effecting part of the actual logical analysis of space. This will appear more clearly from the applicability of dimensions to complex numbers, to which we must now turn our attention.

357. The theory of imaginaries was formerly considered a very important branch of mathematical philosophy, but it has lost its philosophical importance by ceasing to be controversial. The examination of imaginaries led, on the Continent, to the Theory of Functions—a subject which, in spite of its overwhelming mathematical importance, appears to have little interest for the philosopher. But among ourselves the same examination took a more abstract direction: it led to an examination of the principles of symbolism, the formal laws of addition and multiplication, and the general nature of a Calculus. Hence arose a freer spirit towards ordinary Algebra, and the possibility of regarding it (like ordinary Geometry) as one species of a genus. This was the guiding spirit of Sir William Hamilton, De Morgan, Jevons and Peirce—to whom, as regards the result, though not as regards the motive, we must add Boole and Grassmann. Hence the philosophy of imaginaries became merged in the far wider and more interesting problems of Universal Algebra*. These problems cannot, in my opinion, be dealt with by starting with the genus, and asking ourselves: what are the essential principles of any Calculus? It is necessary to adopt a more inductive method, and

* See Whitehead, *Universal Algebra*, Cambridge, 1898; especially Book I.

examine the various species one by one. The mathematical portion of this task has been admirably performed by Mr Whitehead: the philosophical portion is attempted in the present work. The possibility of a deductive Universal Algebra is often based upon a supposed principle of the Permanence of Form. Thus it is said, for example, that complex numbers must, in virtue of this principle, obey the same laws of addition and multiplication, as real numbers obey. But as a matter of fact there is no such principle. In Universal Algebra, our symbols of operation, such as + and ×, are variables, the hypothesis of any one Algebra being that these symbols obey certain prescribed rules. In order that such an Algebra should be important, it is necessary that there should be at least one instance in which the suggested rules of operation are verified. But even this restriction does not enable us to make any general formal statement as to all possible rules of operation. The principle of the Permanence of Form, therefore, must be regarded as simply a mistake: other operations than arithmetical addition may have some or all of its formal properties, but operations can easily be suggested which lack some or all of these properties.

358. Complex numbers first appeared in mathematics through the algebraical generalization of number. The principle of this generalization is the following: Given some class of numbers, it is required that numbers should be discovered or invented which will render soluble any equation in one variable, whose coefficients are chosen from the said class of numbers. Starting with positive integers, this method leads at once, by means of simple equations alone, to all rational numbers positive and negative. Equations of finite degrees will give all the so-called algebraic numbers, but to obtain transcendent numbers, such as e and π, we need equations which are not of any finite degree. In this respect the algebraical generalization is very inferior to the arithmetical, since the latter gives all irrationals by a uniform method, whereas the former, strictly speaking, will give only the algebraic numbers. But with regard to complex numbers, the matter is otherwise. No arithmetical problem leads to these, and they are wholly incapable of arithmetical definition. But the attempt to solve such equations as $x^2 + 1 = 0$, or $x^2 + x + 1 = 0$, at once demands a new class of numbers, since, in the whole domain of real numbers, none can be found to satisfy these equations. To meet such cases, the algebraical generalization defined new numbers by means of the equations whose roots they were. It showed that, assuming these new numbers to obey the usual laws of multiplication, each of them fell into two parts, one real, the other the product of some real number and a fixed number of the new kind. This fixed number could be chosen arbitrarily, and was always taken to be one of the square roots of -1. Numbers thus composed of two parts were called complex numbers, and it was shown that no algebraic operation upon them could lead to any new class of numbers. What is still more remarkable, it was proved that any further generalization must lead to numbers disobeying some of the formal laws of Arithmetic*. But the algebraical generalization was wholly unable (as it was, in truth, at every previous stage) to prove that there are such entities as those which it postulated. *If* the said equations have roots, then the roots have such and such properties; this is all that the algebraical method allows us to infer. There is, however, no law of nature to the effect that every equation *must* have a root; on the contrary, it is quite essential to be able to point out actual entities which do have the properties demanded by the algebraical generalization.

359. The discovery of such entities is only to be obtained by means of the theory of dimensions. Ordinary complex numbers form a series of two dimensions of a certain type, which happen to occur as roots of equations in which the coefficients are real. Complex numbers of a higher order represent a certain type of n-dimensional series, but here there is no algebraical problem concerning real numbers which they are required to solve. As a matter of fact, however, the algebraical generalization, as we have seen, does not tell us what our new entities are, nor whether they are entities at all: moreover it encourages the erroneous view that complex numbers whose imaginary part vanishes are real numbers. This error is analogous to that of supposing that some real numbers are rational, some rationals integral, and positive integers identical with signless integers. All the above errors having been exposed at length, the reader will probably be willing to admit the corresponding error in the present case. No complex number, then, is a real number, but each is a term in some multiple series. It is not worth while to examine specially the usual two-dimensional complex numbers, whose claims, as we have seen, are purely technical. I shall therefore proceed at once to systems with n units. I shall give first the usual purely formal definition†, then the logical objections to this definition, and then the definition which I propose to substitute.

Let n different entities, $e_1, e_2, \ldots\ldots e_n$, which we may call elements or units, be given; and let each be capable of association with any real number, or, in special cases, with any rational or any integer. In this way let entities $\alpha_r e_r$ arise, where α_r is a number, and $\alpha_r e_r$ differs from $\alpha_s e_s$ unless $r = s$ and $\alpha_r = \alpha_s$. That is, if either the numerical or the non-numerical parts of $\alpha_r e_r$ and $\alpha_s e_s$ be different, then the wholes are different. Further, let there be a way of combining $\alpha_1 e_1, \alpha_2 e_2, \ldots, \alpha_n e_n$, for each set of values $\alpha_1, \alpha_2, \ldots \alpha_n$, to form a new entity. (The class whose members are $\alpha_1 e_1, \alpha_2 e_2, \ldots \alpha_n e_n$ will be such an entity.) Then the combination, which may be written as

$$a = \alpha_1 e_1 + \alpha_2 e_2 + \alpha_3 e_3 + \ldots + \alpha_n e_n,$$

* See Stolz, *Allgemeine Arithmetik*, II, Section 1, § 10.
† See Stolz, *ibid.* II, Section 1, § 9.

is a complex number of the nth order. The arrangement of the component terms α_1e_1, α_2e_2, ... α_ne_n may or may not be essential to the definition; but the only thing always essential is, that the combination should be such that a difference in any one or more of the numbers α_1, α_2, ... α_n insures a difference in the resulting complex number.

360. The above definition suffers from the defect that it does not point out any one entity which is *the* complex number defined by a set of real numbers. Given two real numbers, a, b, the two complex numbers $a + ib$, $b + ia$ are determinate; and it is desirable that such determinateness should appear in the general definition of complex numbers of any order. But the e's in the above definition are variables, and the suggested complex number is only determinate when the e's are specified as well as the α's. Where, as in metrical Geometry or in the Dynamics of a finite system of particles, there are important meanings for the e's, we may find that complex numbers in the above sense are important. But no special interpretation can give us *the* complex number associated with a given set of real numbers. We might take as *the* complex number the class of all such entities as the above for all possible values of the e's; but such a class would be too general to serve our purposes. A better method seems to be the following.

We wish a complex number of the nth order to be specified by the enumeration of n real numbers in a certain order, *i.e.* by the numbers α_1, α_2, ... α_n, where the order is indicated by the suffix. But we cannot define a complex number as a *series* of n real numbers, because the same real number may recur, *i.e.* α_r and α_s need not be different whenever r and s are different. Thus what defines a real number is a one-many relation whose domain consists of real numbers and whose converse domain consists of the first n integers (or, in the case of a complex number of infinite order, of all the integers); for the suffix in α_r indicates correlation with the integer r. Such one-many relations may be defined to *be* the complex numbers, and in this way a purely arithmetical definition is obtained. The n-dimensional series of complex numbers of order n results from arranging all complex numbers which differ only as to (say) α_r in the order of the real numbers which are α_r in the various cases.

In order that complex numbers in the sense defined by Stolz should have any importance, there must be some motive for considering assemblages of terms selected out of continua. Such a motive exists in a metrical space of n dimensions, owing to a circumstance which is essential to the utility, though not to the definition, of complex numbers. Let a collection of entities (points) be given, each of which has to each of the entities c_1, c_2, ... c_n a numerically measurable relation (distance), and let each be uniquely defined by the n relations which it has to e_1, e_2, ... c_n. Then the complex number a will represent one of this collection of entities, and the elements e_1, e_2, ... e_n will themselves be terms of the collection*. Thus there is a motive for considering the numbers a, which in the general case is practically absent†. But what is essential to observe, and what applies equally to the usual complex numbers of Algebra, is this: our numbers are not purely arithmetical, but involve essential reference to a plurality of dimensions. Thus we have definitely passed beyond the domain of Arithmetic, and this was my reason for postponing the consideration of complex numbers to this late stage.

* e_1 is not identical with $1 \times e_1 + 0 \times e_2 + ...$ The former is a point, the latter a complex number.

† In geometrical applications, it is usual to consider only the *ratios* $a_1 : a_2 : ... : a_n$ as relevant. In this case, our series has only $(n-1)$ dimensions.

CHAPTER XLV.

PROJECTIVE GEOMETRY.

361. THE foundations of Geometry have been subjected, in recent times, to a threefold scrutiny. First came the work of the non-Euclideans, which showed that various axioms, long known to be sufficient for certain results, were also necessary, *i.e.* that results inconsistent with the usual results but consistent with each other followed from the denial of those axioms. Next came the work of Dedekind and Cantor on the nature of continuity, which showed the necessity of investigating carefully the prerequisites of analytical Geometry. Lastly, a great change has been introduced by the Italian work on closed series, mentioned in Part IV., in virtue of which we are able, given a certain type of relation between *four* points of a line, to introduce an order of all the points of a line. The work of the non-Euclideans has, by this time, produced probably almost all the modifications that it is likely to produce in the foundations, while the work of Dedekind and Cantor only becomes relevant at a fairly advanced stage of Geometry. The work on closed series, on the contrary, being very recent, has not yet been universally recognized, although, as we shall see in the present chapter, it has enormously increased the range of pure projective Geometry.

362. In the discussions of the present Part, I shall not divide Geometries, as a rule, into Euclidean, hyperbolic, elliptic, and so on, though I shall of course recognize this division and mention it whenever it is relevant. But this is not so fundamental a division as another, which applies, generally speaking, within each of the above kinds of Geometry, and corresponds to a greater logical difference. The above kinds differ, not in respect of the indefinables with which we start, nor yet in respect of the majority of the axioms, but only in respect of comparatively few and late axioms. The three kinds which I wish to discuss differ both in respect of the indefinables and in respect of the axioms, but unlike the three previous kinds, they are, roughly speaking, mutually compatible. That is to say, given a certain body of geometrical propositions concerning a certain number of entities, it is more or less arbitrary which of the entities we take as indefinable and which of the propositions as indemonstrable. But the logical differences which result from different selections are very great, and the systems of deductions to which different selections lead must be separately discussed. All Geometries, as commonly developed, agree in starting with points as indefinables. That is, there is a certain class-concept *point* (which need not be the same in different Geometries), of which we assume that there are at least two, or three, or four instances, according to circumstances. Further instances, *i.e.* further points, result from special assumptions in the various cases. Where the three great types of Geometry begin to diverge is as regards the straight line. *Projective* Geometry begins with the whole straight line, *i.e.* it asserts that any two points determine a certain class of points which is also determined by any two other members of the class. If this class be regarded as determined in virtue of a relation between the two points, then this relation is symmetrical. What I shall call *Descriptive* Geometry, on the contrary, begins with an asymmetrical relation, or a line with sense, which may be called a ray; or again it may begin by regarding two points as determining the stretch of points between them. *Metrical* Geometry, finally, takes the straight line in either of the above senses, and adds either a second relation between any two points, namely distance, which is a magnitude, or else the consideration of stretches as magnitudes. Thus in regard to the relations of two points, the three kinds of Geometry take different indefinables, and have corresponding differences of axioms. Any one of the three, by a suitable choice of axioms, will lead to any required Euclidean or non-Euclidean space; but the first, as we shall see, is not capable of yielding as many propositions as result from the second or the third. In the present chapter, I am going to assume that set of axioms which gives the simplest form of projective Geometry; and I shall call any collection of entities satisfying these axioms a *projective space*. We shall see in the next chapter how to obtain a set of entities forming a projective space from a set forming a Euclidean or hyperbolic space; projective space itself is, so far as it goes, indistinguishable from the polar form of elliptic space. It is defined, like all mathematical entities, solely by the formal nature of the relations between its constituents, not by what those constituents are in themselves. Thus we shall see in the following chapter that the " points " of a projective space may each be an infinite class of straight lines in a non-projective space. So long as the " points " have the requisite type of mutual relations, the definition is satisfied.

363. Projective Geometry assumes a class of entities, called *points*, to which it assigns certain properties*. In the first place, there are to

---

* In what follows, I am mainly indebted to Pieri, *I Principii della Geometria di Posizione*. Turin, 1898. This is, in my opinion, the best work on the present subject.

be at least two different points, *a* and *b* say. These two points are
to determine a certain class of points, their straight line, which we
will call *ab*. This class is determined by *b* and *a*, as well as by *a* and *b*,
i.e. there is no order of *a* and *b* involved; moreover *a* (and therefore *b*)
is itself a member of the class. Further, the class contains at least
one point other than *a* and *b*; if *c* be any such point, then *b* belongs to
the class *ac*, and every point of *ac* belongs to *ab*. With these assumptions it follows* that, if *c*, *d* be any points of *ab*, then *cd* and *ab*
coincide—*i.e.* any two points of a line determine that line, or two lines
coincide if they have two points in common.

Before proceeding further, let us consider for a moment what is
meant by saying that two points *determine* a class of points. This
expression is often thought to require no explanation, but as a matter
of fact it is not a perfectly precise statement. The precise statement
of what is meant is this: There is a certain definite relation (*K* say)
which holds between any couple of points and one and only one corresponding class of points. Without some such definite relation, there
could be no question of two points determining a class. The relation *K*
may be ultimate and indefinable, in which case we need the above
properties of the class *ab*. We may obtain, however, a derivative
relation between two points, *b* and *c* say, namely that of being both
collinear with a given point *a*. This relation will be transitive and
symmetrical, but will always involve reference to a term other than
those (*b* and *c*) which are its terms. This suggests, as a simplification,
that instead of a relation *K* between a couple of points and a class
of points, we might have a relation *R* between the two points *a* and *b*.
If *R* be a symmetrical aliorelative, transitive so far as its being an
aliorelative will permit (*i.e.* if *aRb* and *bRc* imply *aRc*, unless *a* and *c*
are identical), the above properties of the straight line will belong
to the class of terms having to *a* the relation *R* together with *a* itself.
This view seems simpler than the former, and leads to the same results.
Since the view that the straight line is derived from a relation of two
points is the simpler, I shall in general adopt it. Any two points *a*,
b have, then, a relation R_{ab}; *a*, *c* have a relation R_{ac}. If R_{ab} and R_{ac}
are identical, while *b* and *c* differ, R_{bc} is identical with both R_{ab} and
R_{ac}; if not, not. It is to be observed that the formal properties of any
such relation *R* are those belonging to the disjunction of an asymmetrical transitive relation and its converse—*e.g.* greater or less, before
or after, *etc.* These are all symmetrical aliorelatives, and are transitive so
far as their being aliorelatives will permit. But not all relations of the
type in question are analyzable into a transitive asymmetrical relation
or its converse; for diversity, which is of the above type, is not so
analyzable. Hence to assume that the straight line can be generated

---

* Pieri, *op. cit.* § 1, prop. 25.

by an asymmetrical relation and its converse is a new assumption, characteristic of what I shall call Descriptive Geometry. For the present, such an assumption would be out of place. We have, then, two indefinables, namely *point*, and the relation R or K^*. No others are required in projective space.

364. The next point is the definition of the plane. It is one of the merits of projective space that, unlike other spaces, it allows a very simple and easy definition of the plane. For this purpose, we need a new axiom, namely: If a, b be two distinct points, there is at least one point not belonging to ab. Let this be c. Then the plane is the class of points lying on any line determined by c and any point x of ab. That is, if x be any point of ab, and y any point of cx, then y is a point of the plane cab; and if y be a point of the plane cab, then there is some point x in ab such that y is a point of cx. It is to be observed that this definition will not apply to the Euclidean or hyperbolic plane, since in these two lines may fail to intersect. The exclusion of Euclidean and hyperbolic space results from the following axiom†: "If a, b, c be three non-collinear points, and a' be a point of bc other than b and c, b' a point of ac other than a and c, then there is a point common to aa' and bb'." By means of this axiom we can prove that the plane cab is the same as the plane abc or bac, and generally that, if d, e, f be any three non-collinear points of abc, the plane def coincides with the plane abc; we can also show that any two lines in a plane intersect.

365. We can now proceed to the harmonic range and von Staudt's quadrilateral construction. Given three collinear points a, b, c take any two points u, v collinear with c but not on ab. Construct the points of intersection $au . bv$ and $av . bu$; join these points, and let the line joining them meet ab in d. This construction is called the quadrilateral construction. If we now assume that outside the plane abu there is at least one point, we can prove that the point d is independent of u and v, and is uniquely determined by a, b, c. The point d is called the harmonic of c with respect to a and b, and the four points are said to form a harmonic range. The uniqueness‡ of the above construction— the proof of which, it should be observed, requires a point not in the plane of the construction∥—is the fundamental proposition of projective Geometry. It gives a relation which may hold between four points of a line, and which, when two are given, is one-one as regards the other

---

* We shall see in Chap. XLIX that these notions, which are here provisionally undefined, are themselves variable members of definable classes.

† Pieri, *op. cit*, § 3, p. 9.

‡ The proof of the uniqueness of the quadrilateral construction will be found in any text-book of Projective Geometry, *e.g.* in Cremona's (Oxford 1893), Chap. VIII.

∥ A proof that this proposition requires three dimensions is easily derivable from a theorem given by Hilbert, *Grundlagen der Geometrie*, p. 51 (Gauss-Weber Festschrift, Leipzig, 1899).

two. Denoting "c and d are harmonic with respect to a and b" by $cH_{ab}d$, the following properties of the relation are important: (1) $cH_{ab}d$ implies $dH_{ab}c$, i.e. H_{ab} is symmetrical; (2) $cH_{ab}d$ implies $aH_{cd}b$, i.e. the relation of the pairs ab, cd is symmetrical; (3) $cH_{ab}d$ implies that c and d are different points, i.e. H_{ab} is an aliorelative. This last property is independent of the others, and has to be introduced by an axiom*.

Having obtained the harmonic range, we may proceed in two different directions. We may regard the harmonic relation as a relation of two pairs of points: hence, by keeping one of the pairs fixed, we obtain what is called an involution. Or we may regard the harmonic relation, as in the symbol $cH_{ab}d$, as a relation between two points, which involves a reference to two others. In this way, regarding a, b, c, as fixed, we obtain three new points d, e, f on the line ab by the relations $cH_{ab}d$, $aH_{bc}e$, $bH_{ac}f$. Each of these may be used, with two of the previous points, to determine a fourth point, and so on. This leads to what Möbius† calls a *net*, and forms the method by which Klein‡ introduces projective coordinates. This construction gives also the method of defining an harmonic ratio. These two directions in which projective Geometry may be developed must be separately pursued to begin with. I shall take the former first.

366. By means of the harmonic relation, we define an *involution*. This consists of all pairs of points which are harmonic conjugates with respect to two fixed points‖. That is to say, if a, b be the two fixed points, an involution is composed of all pairs of points x, y such that $xH_{ab}y$. If four points x, y, x', y' be given, it may or may not happen that there exist two points a, b such that $xH_{ab}y$ and $x'H_{ab}y'$. The possibility of finding such points a, b constitutes a certain relation of x, y to x', y'. It is plain that this relation sometimes holds, for it holds when x, y are respectively identical with x', y'. It is plain also that it sometimes does not hold; for if x and y be identical, but not x' and y', then the relation is impossible. Pieri§ has shown how, by means of certain axioms, this relation of four terms may be used to divide the straight line into two segments with respect to any two of its points, and to generate an order of all the points on a line. (It must be borne

---

* See Fano, *Giornale di Matimatiche*, Vol. 30; Pieri, *op. cit.* § 4, p. 17 and Appendix. Fano has proved the necessity of the above axiom in the only conclusive manner, by constructing a system satisfying all the previous axioms, but not this one. The discovery of its necessity is due to him. A simpler but equivalent axiom is that our space contains at least one line on which there are more than three points.

† *Barycentrischer Calcul*, Section ii, Chap. vi.

‡ *Math. Annalen*, 4, 6, 7, 37; *Vorlesungen über nicht-Euklidische Geometrie*, Göttingen, 1893, Vol. i, p. 308 ff.

‖ In what follows, only involutions with real double points are in question.

§ *Op. cit.* §§ 5, 6, 7. Pieri's method was presumably suggested by von Staudt. Cf. *Geometrie der Lage*, § 16: especially No. 216.

in mind that, in projective Geometry, the points of a line do not have an order to begin with.) This projective order is obtained as follows.

367. Given any three different points a, b, c on a line, consider the class of points x such that a and c, b and x are each harmonic conjugates with respect to some pair of points y, y'—in other words, a and c, b and x are pairs in an involution whose double points are y, y'. Here y, y' are supposed variable: that is, if any such points can be found, x is to belong to the class considered. This class contains the point b, but not a or c. Let us call it the segment (abc). Let us denote the relation of b to x (a and c being fixed) by $bQ_{ac}x$. Then Q_{ac} is symmetrical, and also $bQ_{ac}x$ implies $aQ_{bx}c$. We have here a relation of four points, from which, as we saw in Part IV, Chapter XXIV, an order will result if certain further axioms are fulfilled. Three such axioms are required, and are given by Pieri as follows.

(1) If d is on the line ab, but does not belong to the segment (abc), and does not coincide with a or with c, then d must belong to the segment (bca). (If d coincides with c, we know already that d belongs to the segment (bca). This case is therefore excluded from the axiom to avoid a superfluity of assumptions.) In virtue of this axiom, if a, b, c, d be distinct points on a line, we must have either $bQ_{ac}d$ or $cQ_{ab}d$. It follows that we must have either $bQ_{ac}d$ or $aQ_{bc}d$. Thus at least two Q-relations hold between any four distinct collinear points. (2) If a, b, c be distinct collinear points, and d be a point belonging to both the segments (bca) and (cab), then d cannot belong to the segment (abc). That is, of the three segments to which d may belong, it never belongs to more than two. From this and the previous axiom it results that, if d be distinct from a, b and c, then d belongs to two and only two of the three segments defined by a, b and c. (3) If a, b, c be distinct collinear points, and d a point, other than b, of the segment (abc), and e a point of the segment (adc), then e is a point of the segment (abc). (Here again, the condition that d is to be other than b is required only to avoid superfluity, not for the truth of the axiom.) In terms of Q, this axiom states that $bQ_{ac}d$ and $dQ_{ac}e$ imply $bQ_{ac}e$; that is, Q_{ac} is transitive. We saw already that Q_{ac} is symmetrical. We can now prove that, by means of this relation, all points of the line except a and c are divided into two classes, which we may call $(ac)_1$ and $(ac)_2$. Any two points in the same class have the relation Q_{ac}, any two in different classes have not. The division into *two* classes results from the fact that, if we do not have $bQ_{ac}d$, nor yet $dQ_{ac}e$ (b, d, e being points other than a and c), then we do have $bQ_{ac}e$. That is to say, Q_{ac} has the formal properties of sameness of sign, and divides the line into two classes, just as sameness of sign divides numbers into positive and negative.

The opposite of Q_{ac}, which I shall denote by T_{ac}, corresponds in like manner to difference of sign. T_{ac} is not to denote the mere negation of Q_{ac}, but the fact of belonging to different segments.   That is, $bT_{ac}d$ means that d does not coincide with a or c, that d lies in the line ac, but not in the segment (abc).   Then $bT_{ac}d$ may be taken as meaning that b and d are separated by a and c.   It is a relation which has the formal properties of separation of couples, as enumerated in Part IV, Chapter XXIV.   If a, b, c, d, e be five distinct points in one straight line, we have the following properties of the T-relation.   (1) $bT_{ac}d$ is equivalent to $dT_{ac}b$, $aT_{bd}c$, $cT_{bd}a$, $cT_{db}a$, etc.   (2) We have one and only one of the three relations $aT_{bc}d$, $aT_{bd}c$, $aT_{cd}b$.   (3) $dT_{ac}b$ implies $dT_{ac}e$ or $eT_{ac}b$*.

By comparing the above properties of T with those of separation of couples, it will be seen that T leads to a closed series (in the sense of Part IV), *i.e.* to a series in which there is a first term, but this first term is arbitrary.   The definition of the generating relation of the series (which involves, as in the general case, three fixed points) is given by Pieri as follows.   With regard to the natural order abc, a precedes every other point of the line; c precedes every point d not belonging to (abc) and not coinciding with a or c, *i.e.* every point d such that $dT_{ac}b$; a general point d precedes a general point e if $dQ_{ac}b$ and $eQ_{ad}c$, or if $dT_{ac}b$ and $eT_{ad}c$, *i.e.* if d belongs to the segment (abc) and e to the segment (acd), or if b and d are separated by a and c, and likewise c and e by a and d.   It is then shown, that of any two points of the line, one precedes the other, and that the relation is transitive and asymmetrical; hence all the points of the line acquire an order.

Having now obtained an order among our points, we can introduce an axiom of continuity, to which Pieri† gives a form analogous to that of Dedekind's axiom, namely: If any segment (abc) be divided into two parts h and k, such that, with regard to the order abc, every point of h precedes every point of k, while h and k each contain at least one point, then there must be in (abc) at least one point x such that every point of (abc) which precedes x belongs to h, and every point of (abc) which follows x belongs to k.   It follows from this axiom that every infinite class contained in (abc) and having no last (or first) term has a limit, which is either a point of (abc) or c (or a); and it is easy to prove that, when h and k are given, there can be only one such point as x in the axiom.

By means of the projective segment, it is easy to define triangles and tetrahedra.   Three points determine four triangles, which between them

---

* This last property affords an instance (almost the only one known to me) where Peirce's relative addition occurs outside the Algebra of Relatives.   "$dT_{ac}e$ or $eT_{ac}b$" is the relative sum of T_{ac} and T_{ac}, if d, e, and b be variable.   This property results formally from regarding T_{ac} as the negation of the transitive relation Q_{ac}.

† *Op. cit.* § 9, p. 7.

contain all the points of the plane, and have no common points except the angles. Also we can define harmonic transformations, and prove their properties without any further axiom *. Only one other axiom is required to complete our Geometry, namely : A plane and a line not in the plane always have a common point. This amounts to the axiom of three dimensions. Nothing is altered, in what precedes, by denying it, and proceeding to a space of n dimensions or of an infinite number of dimensions. This last, in fact, requires fewer axioms than a space of three dimensions †.

368. Let us now resume the other direction in which projective Geometry may be developed, in which we start from three fixed points on a line, and examine all the points obtainable from these three by successive quadrilateral constructions. We do not here, as in the development we have been examining, require any new axiom ; but there is a corresponding restriction in the results obtainable. In order to give projective Geometry its fullest possible development we must combine the results of both directions.

Confining ourselves, to begin with, to one straight line, let us see how to construct a net and introduce projective coordinates. Denoting by $aH_{bc}d$, as before, the proposition " a and d are harmonic conjugates with respect to b and c," we can, by the quadrilateral construction, when a, b, c are given, determine the only point d satisfying this proposition. We next construct the point e for which $bH_{cd}e$, then f for which $dH_{ce}f$, g for which $eH_{cf}g$, and so on. In this way we obtain a progression of points on our line, such that any three consecutive points, together with c, form a harmonic range. With our former definition of a segment, all these points will belong to the segments (abc) and (bca). We may number these points, beginning with a, 0, 1, 2, ..., n, ... Since c does not belong to the progression, we may assign to it the number ∞ ‡. Consider next the points obtained as follows. Let d' be such that $d'H_{ab}c$, let $e'H_{ad'}b$, $f'H_{ae'}d'$, and so on. We have thus a new progression of points, such that any three consecutive points together with a form a harmonic range, and all belonging to the segments (abc), (cab). To these points let us assign the numbers $1/n$ in order. Similarly we can construct a progression belonging to the two segments (cab), (bca), and assign to them the negative integers. By proceeding in a similar manner with any triad of points so obtained, we can obtain more and more points. The principle adopted in assigning numbers to points (a principle which, from our present standpoint, has no motive save

---

* These developments will be found in Pieri, *loc. cit.* §§ 8, 10.

† Pieri, § 12.

‡ We must not assign to c the definite number ω, since we cannot assume, without further axioms, that c is the limit of our progression. Indeed, so long as we exclude Pieri's three axioms above mentioned, we do not know, to begin with, that c has any ordinal relation to the terms of our progression.

convenience) is the following: if p, q, r be the numbers assigned to three points already constructed, and s be the number to be assigned to the harmonic conjugate (supposed not previously constructed) of the q-point with respect to the p-point and the r-point, then we are to have $\dfrac{p-q}{r-q}\Big/\dfrac{p-s}{r-s} = -1$. In this manner, we can find one and only one point of our line for each rational number, positive or negative*. Thus we obtain a denumerable endless compact series of points on our line. Whether these are all the points of our line or not, we cannot decide without a further axiom. If our line is to be a continuous series, or a collection of the power of the continuum, we must of course assume points not obtainable by quadrilateral constructions, however often repeated, which start with three given elements. But as the definition of our space is optional, we may, if we like, content ourselves with a rational space, and introduce an axiom to the effect that all points of our line can be obtained from three given points.

369. Before proceeding further, it may be well to point out a logical error, which is very apt to be committed, and has been committed, I think, even by Klein†. So long as Pieri's three axioms above enumerated are not assumed, our points have no order but that which results from the net, whose construction has just been explained. Hence only rational points (*i.e.* such as, starting from three given points, have rational coordinates) can have an order at all. If there be any other points, there can be no sense in which these can be limits of series of rational points, nor any reason for assigning irrational coordinates to them. For a limit and the series which it limits must both belong to some one series; but in this case, the rational points form the whole of the series. Hence other points (if there be any) cannot be assigned as limits of series of rational points. The notion that this can be done springs merely from the habit of assuming that all the points of a line form a series, without explicitly stating this or its equivalent as an axiom. Indeed, just as we found that series of rationals properly have no limit except when they happen to have a rational limit, so series of points obtainable by the quadrilateral construction will not have limits, *quâ* terms of the series obtained from the quadrilateral construction, except where they happen to have a limit within this series, *i.e.* when their coordinates have a rational limit. At this point, therefore, it is highly desirable to introduce Pieri's three axioms, in virtue of which *all* the points of a line have an order. We shall find that, in the natural order *cab*, the order of the rational points, resulting from Pieri's axioms, is the same as that of their coordinates assigned on the

---

* On this subject, see Klein, *Vorlesungen über nicht-Euklidische Geometrie*, p. 338 ff., where proofs will be found.

† *e.g. Op. cit.* p. 344.

above principle*. Thus we have only to assume that all infinite series of rational points have limits, as parts of Pieri's series, and that all points are either rational or limits of rational series, in order to show that our straight line has continuity in Cantor's sense. In this case we shall assign to non-rational points the irrational numbers corresponding to the series which such points limit.

370. Returning now to the quadrilateral construction, we define as the *anharmonic ratio* of four points whose coordinates are p, q, r, s the number $\dfrac{p-q}{r-q}\Big/\dfrac{p-s}{r-s}$. It can be shown that this number is independent of the choice of our three original points a, b, c. It expresses the series of quadrilateral constructions required to obtain s when p, q, r are given, and thus expresses a purely projective relation of the four points. By the introduction of irrational points, in the manner just explained, it follows that *any* four points on a line have an anharmonic ratio. (This cannot possibly be proved without Pieri's three axioms or some equivalent to them.) The anharmonic ratio is unaltered by any linear transformation, *i.e.* by substituting for every point x the point whose coordinate is $(\alpha x + \beta)/(\gamma x + \delta)$, where α, β, γ, δ are any fixed numbers such that $\alpha\delta - \beta\gamma$ is not zero. From this point we can at last advance to what was formerly the beginning of projective Geometry, namely the operation of projection, to which it owes its name.

It can be shown that, if p, r be harmonic conjugates with respect to q, s, and p, q, r, s be joined to some point o, and if op, oq, or, os meet any line in p', q', r', s', then p, r are harmonic conjugates with respect to q', s'. Hence we can show that all anharmonic ratios are unaltered by the above operation. Similarly if l be any straight line not coplanar with $pqrs$, and the planes lp, lq, lr, ls meet any line not coplanar with l in p', q', r', s', these four points will have the same anharmonic ratio as p, q, r, s. These facts are expressed by saying that anharmonic ratio is unaltered by projection. From this point we can proceed to the assignment of coordinates to any point in space†.

371. To begin with a plane, take three points a, b, c not in one straight line, and assign coordinates in the above manner to the points of ab, ac. Let p be any point of the plane abc, but not on the line bc. Then if cp meets ab in p_1, and bp meets ac in p_2, and x, y are the coordinates of p_1, p_2 respectively, let (x, y) be the two coordinates of p. In this way all points of the plane not on bc acquire coordinates. To avoid this restriction, let us introduce three homogeneous coordinates, as follows. Take any four points a, b, c, e in a plane, no three of which are collinear; let ae meet bc in e_1, be meet ca in e_2, ce meet ab

---

* This has the one exception that c came last in the order of the quadrilateral constructions, and comes first in Pieri's order. This may be remedied by the simple device of giving c the coordinate $-\infty$ instead of ∞.

† See Pasch, *Neuere Geometrie*, § 22; Klein, *Math. Annalen*, 37.

in e_3. Assign coordinates to the points of bc, ca, ab as before, giving the coordinate 1 to e_1, e_2, e_3, and in ab giving 0 to a, and ∞ to b, and similarly for the other sides. In place of the single coordinate x of any point of bc, let us introduce the homogeneous coordinates x_2, x_3, where $x = x_2/x_3$. If now p be any point of the plane abc, let ap meet bc in p_1, bp meet ca in p_2, and cp meet ab in p_3. Let x_2, x_3 be the homogeneous coordinates of p_1, x_3, x_1 those of p_2; then x_1, x_2 will be those of p_3*. Hence we may assign x_1, x_2, x_3 as homogeneous coordinates of p. In like manner we can assign four homogeneous coordinates to any point of space. We can also assign coordinates to the lines through a point, or the planes through a line, or all the planes of space, by means of the anharmonic ratios of lines and planes†. It is easy to show that, in point-coordinates, a plane has a linear equation, and a linear equation represents a plane; and that, in plane-coordinates, a point has a linear equation, and a linear equation represents a point. Thus we secure all the advantages of analytical Geometry. From this point onwards, the subject is purely technical, and ceases to have philosophic interest.

372. It is now time to ask ourselves what portions of the Geometry to which we are accustomed are not included in projective Geometry. In the first place, the series of points on a line, being obtained from a four-term relation, is closed in the sense of Part IV. That is, the order of points requires three fixed points to be given before it can be defined. The practical effect of this is, that given only three points on a line, no one of them is between the other two. This is a definite difference between projective space and Euclidean or hyperbolic space. But it is easy to exaggerate this difference. We saw in Part IV that, wherever a series is generated by a two-term relation, there is also the four-term relation of separation of couples, by which we can generate a closed series consisting of the same terms. Hence in this respect the difference does not amount to an inconsistency. Euclidean and hyperbolic spaces contain what projective space contains, and something more besides. We saw that the relation by which the projective straight line is defined has the formal properties of "P or $\breve{P}$," where P is transitive and asymmetrical. If the said relation be actually of this form, we shall have an open series with respect to P, and of three collinear points one will be between the other two. It is to be observed that, where the straight line is taken to be essentially closed, as in elliptic space, *between* must be excluded where three points only are given. Hence elliptic space, in this respect, is not only consistent with the projective axioms, but contains nothing more than they do.

It is when we come to the plane that actual inconsistencies arise

---

* See Pasch, *loc. cit.*

† The anharmonic ratio of four lines through a point or of four planes through a line is that of the four points in which they meet any line.

between projective Geometry and Euclidean or hyperbolic Geometry. In projective space, any two lines in a plane intersect; in the Euclidean and hyperbolic Geometries, this does not occur. In elliptic Geometry, any two lines in a plane intersect; but in the antipodal form they intersect twice. Thus only the polar form wholly satisfies the projective axioms. Analogous considerations apply to the intersection of two planes, or of a line and a plane. These differences render the projective definition of a plane inapplicable to Euclidean and hyperbolic spaces, and render the theory of these spaces far more complicated than that of projective space.

Finally, in metrical Geometry it is assumed either that two points have a quantitative relation called distance, which is determined when the points are given, or that stretches satisfy axioms in virtue of which their divisibilities become numerically measurable. In this point, even elliptic space differs from projective space, though the difference is of the nature of an addition, not an inconsistency. But this matter cannot be discussed until we have examined metrical Geometry, when we shall be in a position to examine also the projective theory of distance to more advantage than is at present possible.

373. A few words may be added concerning the principle of duality. This principle states, in three dimensions, that the class of planes is also a projective space, the intersection of two planes being, as before, the straight line, and the intersection of three non-collinear planes taking the place of the point. In n dimensions, similarly, a projective space results from all sub-classes of $(n-1)$ dimensions. Such a duality, as we saw in Chapter XLIV, belongs always to n-dimensional series as such. It would seem (though this is only a conjecture) that projective Geometry employs the smallest number of axioms from which it is possible to generate a series of more than two dimensions, and that projective duality therefore flows from that of dimensions in general. Other spaces have properties additional to those required to make them n-dimensional series, and in other spaces, accordingly, duality is liable to various limitations.

CHAPTER XLVI.

DESCRIPTIVE GEOMETRY.

374. THE subject which I have called descriptive Geometry is not, as a rule, sharply distinguished from projective Geometry. These two terms, and the term " Geometry of Position," are commonly used as synonyms. But it seems improper to include in projective Geometry any property which is not unaltered by projection, and it is by the introduction of one such property that I wish to define the subject of the present chapter. We have seen that, in projective space, three points on a line are not such that a definite one of them is *between* the other two. The simplest possible proposition involving *between*, in projective Geometry, requires four points, and is as follows : " If *a*, *b*, *c* be distinct collinear points, and *d* is on *ac*, but does not belong to the segment (*abc*), nor yet coincide with *a* or *c*, then, with regard to the order *abc*, *c* is between *b* and *d*." When we reflect that the definition of the segment (*abc*) involves the quadrilateral construction— which demands, for its proof, a point outside its own plane, and four pairs of triangles in perspective—we shall admit that the projective method of generating order is somewhat complicated. But at any rate the ordinal propositions which result are unaltered by projection. The elementary sense of *between*, on the contrary, which is to be introduced in the present chapter, is in general not unaltered by projection*.

In descriptive Geometry, we start, as before, with points, and as before, any two points determine a class of points. But this class now consists only of the points *between* the two given points. What is to

---

* The present subject is admirably set forth by Pasch, *Neuere Geometrie*, Leipzig, 1882, with whose empirical pseudo-philosophical reasons for preferring it to projective Geometry, however, I by no means agree (see *Einleitung* and § 1). It is carried further, especially as regards the definition of the plane, by Peano, *I Principii di Geometria logicamente esposti*, Turin, 1889. For the definition of the whole line by means of its various segments, see Peano's note in *Rivista di Matematica*, II, pp. 58–62. See also his article "Sui fondamenti della Geometria," *ib.* IV, p. 51 ff., and Vailati, " Sui Principi fondamentali della Geometria della retta," *Riv. d. Mat.* II, pp. 71–75. Whatever, in the following pages, is not controversial, will be found in the above sources.

be understood by *between* is not explained by any writer on this subject except Vailati, in terms of a transitive asymmetrical relation of two points; and Vailati's explanation is condemned by Peano*, on the ground that *between* is a relation of *three* points, not of two only. This ground, as we know from Part IV, is inadequate and even irrelevant. But on the subject of relations, even the best mathematicians go astray, for want, I think, of familiarity with the Logic of Relations. In the present case, as in that of projective Geometry, we may start either with a relation of two points, or with a relation between a pair and a class of points: either method is equally legitimate, and leads to the same results, but the former is far simpler. Let us examine first the method of Pasch and Peano, then that of Vailati.

375. We start, in the former method, with two indefinables, *point*, and *between*. If a, b, c be three points, and c is between a and b, we say that c is an ab, or belongs to the class of points ab. Professor Peano has enumerated, with his usual care, the postulates required as regards the class ab†. In the first place, the points a and b must be distinct, and when they are so, there always is a point between them. If c is between a and b, it is also between b and a: a itself (and therefore b) is not between a and b. We now introduce a new definition. If a, b be any two distinct points, then $a'b$ is the class of all points c such that b is between a and c. Similarly $b'a$ will be the class of points d such that a is between b and d. We then proceed to new postulates. If a and b be distinct points, $a'b$ must contain at least one point. If a, b, c, d be points, and c is between a and d, b between a and c, then b is between a and d. If b and c be between a and d, b is between a and c, or identical with c, or between c and d. If c, d belong to $a'b$, then either c and d are identical, or c is between b and d, or d is between b and c. If b is between a and c, and c is between b and d, then c is between a and d. This makes in all nine postulates with regard to *between*. Peano confesses‡ that he is unable to prove that all of them are independent: hence they are only shown to be sufficient, not necessary. The complete straight line (ab) is defined as $b'a$ and a and ab and b and $a'b$; that is, (1) points between which and b the point a lies; (2) the point a; (3) points between a and b; (4) the point b; (5) points between which and a the point b lies.

Concerning this method, we may observe to begin with that it is very complicated. In the second place, we must remark, as before, that the phrase "two points *determine* a class of points" must be expanded as follows: "There is a certain specific relation K, to whose domain belongs every couple of distinct points. K is a many-one relation, and the relatum, corresponding to a couple of points as referent, is a class of points." In the third place, we may observe that the points of the

* *Riv. di Mat.* IV, p. 62. † *Ib.* IV, p. 55 ff. ‡ *Ib.* p. 62.

line only acquire order by relation to i.ie segments which they terminate, and that these acquire order by the relation of whole and part, or logical inclusion. Let a, b be any two points, and consider the class of points ab or b or $a'b$. Let c, d be any two distinct points of this class. Then either ac is a proper part of ad, or ad is a proper part of ac. Here ac and ad may be called segments, and we see that segments whose origin is a and whose limits belong to ab or b or $a'b$ form a series in virtue of the transitive asymmetrical relation of whole and part. By correlation with these segments, their extremities also acquire an order ; and it is easy to prove that this order is unchanged when we substitute for a any point of ab'. But the order still results, as it always must, from a transitive asymmetrical relation of two terms, and nothing is gained by not admitting such a relation immediately between points.

376. Passing now to what I have called Vailati's theory, we find a very great simplification. We may state the present theory (which is not in every detail identical with that of Vailati) as follows. There is a certain class, which we will call K, of transitive asymmetrical relations. Between any two points there is one and only one relation of the class K. If R be a relation of the class K, $\breve{R}$ is also a relation of this class. Every such relation R defines a straight line ; that is, if a, b be two points such that aRb, then a belongs to the straight line ρ. (I use the corresponding Greek letter to denote the domain of a relation ; thus if S be a relation, σ is the class of terms having the relation S to some term or other.) If aRb, then there is some point c such that aRc and cRb ; also there is a point d such that bRd. Further, if a, b be any two distinct points belonging to ρ, then either aRb or bRa. With this apparatus we have all that we require.

We may do well to enumerate formally the above definition of the class K, or rather the postulates concerning its members—for K itself is not defined. I may remark to begin with that I define the field of a class of relations as the logical sum of the fields of the constituent relations ; and that, if K be the class, I denote its field by k. Then the axioms we require are as follows.

(1) There is a class of relations K, whose field is defined to be the class *point*.

(2) There is at least one point.

If R be any term of K we have,

(3) R is an aliorelative.

(4) $\breve{R}$ is a term of K.

(5) $R^2 = R$.

(6) $\breve{\rho}$ (the domain of $\breve{R}$) is contained in ρ.

(7) Between any two points there is one and only one relation of the class K.

(8) If a, b be points of ρ, then either aRb or bRa.

The mutual independence of these axioms is easy to see. But let us first briefly sketch the proof that they give all the required results. Since there is, by (2), at least one point, and since by (1) this point has some relation of the class K, and since by (3) all relations of the class K are aliorelatives, it follows that there is some term, other than the one point, to which this one point has a relation R of the class K. But since $\breve{R}$, by (4), is a relation of the class K, it follows that the term to which the one point is so related is also a point. Hence there are at least two distinct points. Let a, b be two distinct points, and let R be the one relation of the class K between a and b. Thus we have aRb. But we do not have bRa, for if we did, since $R^2 = R$, by (5), we should have aRa, which contradicts (3). Thus R and $\breve{R}$ are always different, *i.e.* each is asymmetrical. Since $R^2 = R$, aRb and bRc imply aRc, *i.e.* R is transitive. Hence the points which have to a the relation R or $\breve{R}$, together with a itself, form a series. Since $R = R^2$, aRb implies that there is some point c such that aRc, cRb; *i.e.* the series generated by R is compact. Since, by (6), $\breve{\rho}$ is contained in ρ, aRb implies that there is some point c such that bRc. Applying the same argument to $\breve{R}$, there is a point d such that dRa. Thus we have $\rho = \breve{\rho}$, and the field of R has no beginning or end. By (8), the field of R is what, in Part IV, we called a connected series, that is, it does not fall apart into two or more detached portions, but of any two of its terms one is before and the other after. By (7), if there be more than one relation of the class K, the fields of two such relations cannot, unless one is the converse of the other, have more than one point in common. The field of one relation of the class K is called a *straight line*; and thus (7) assures us that two straight lines have at most one common point, while (8) assures us that, if ab, cd be the same line, so are ac and bd. Thus it is proved that our axioms are sufficient for the geometry of a line, while (7) goes beyond a single line, but is inserted here because it does not imply the *existence* of points outside a single line, or of more than one relation of the class K. It is most important to observe that, in the above enumeration of fundamentals, there is only one indefinable, namely K, not two as in Peano's system.

377. With regard to the mutual independence of the axioms, it is to be observed that (1) is not properly an axiom, but the assumption of our indefinable K. (2) may obviously be denied while all the others are maintained. If (3) be denied, and R be taken to be the symmetrical relation of projective Geometry, together with identity with some term of ρ, we obtain projective Geometry, which is different from the present system, but self-consistent. If (4) be denied, all the rest can be maintained; the only difficulty is as regards (7), for if aRb, and $\breve{R}$ is not a term of K, b will not have to a any relation of the class K, unless indeed it has one which is not $\breve{R}$, which seems to be not contradictory.

As regards (5), we may deny either that R is contained in R^2, or that R^2 is contained in R. To deny the former makes our series not compact, to which there is no logical objection. The latter, but not the former, is false as regards angles *, which can be made to satisfy all the other axioms here laid down. (6) will become false if our lines have last terms : thus the space on the left of a plane, together with this plane, will satisfy all the axioms except (6). As regards (7), it is plainly independent of all the rest ; it consists of two parts, (*a*) the assertion that between any two points there is at least one relation of the class K, (*b*) the assertion that there is not more than one such relation between two given points. If we consider a Euclidean and a hyperbolic space together, all the axioms will be true except (*a*). If we combine two different classes K_1, K_2 of relations of the above kind, such that $k_1 = k_2$, (*b*) alone will be false. Nevertheless it seems plain that (*b*) cannot be deduced from the other axioms. As regards (8), it alone is false if we take for K the class of directions in Euclidean space, in which a set of parallel lines all have the same direction. Thus the necessity of all except one of our axioms is strictly proved, and that of this one is highly probable.

378. We saw that the above method enabled us to content ourselves with one indefinable, namely the class of relations K. But we may go further, and dispense altogether with indefinables. The axioms concerning the class K were all capable of statement in terms of the logic of relations. Hence we can define a class C of classes of relations, such that every member of C is a class of relations satisfying our axioms. The axioms then become parts of a definition, and we have neither indefinables nor axioms. If K be any member of the class C, and k be the field of K, then k is a descriptive space, and every term of k is a descriptive point. Here every concept is defined in terms of general logical concepts. The same method can be applied to projective space, or to any other mathematical entity except the indefinables of logic. This is, indeed, though grammatically inconvenient, the true way, philosophically speaking, to define mathematical notions. Outside logic, indefinables and primitive propositions are not required by pure mathematics, and should therefore, strictly speaking, not be introduced. This subject will be resumed in Chapter XLIX.

379. The two ways of defining the straight line—that of Pasch and Peano, and that which I have just explained—seem equally legitimate, and lead to the same consequences. The choice between them is therefore of no mathematical importance. The two methods agree in enabling us, in terms of two points only, to define three parts of a straight line, namely the part before a ($b'a$), the part between a and b (ab), and the part after b ($a'b$). This is a point in which descriptive Geometry differs from projective Geometry : there we had, with respect

---

* See Part IV, chap. XXIV.

to *a* and *b*, only two segments of the straight line *ab*, and these could not be defined without reference to another point *c* of the line, and to the quadrilateral construction.

The straight line may be regarded either as the class of points forming the field of a relation *R*, or as this relation itself. For the sake of distinction, it will be well to call the relation *R* a *ray*, since this word suggests a sense; $\breve{R}$ will then be the opposite ray. In considering a number of lines all passing through one point *O*, it will be well to give the name of *ray* also to the class of points to which *O* has some relation *R*, *i.e.* to those points of a line through *O* which lie on one side of *O*. Those on the other side of *O* will then be the opposite ray. The context will show in which sense the word is used.

380. I come now to the *plane*. Easy as it is to define the plane in projective space, its definition when the line is not a closed series, or rather, when we wish to call coplanar some pairs of lines which do not intersect, is a matter of some difficulty. Pasch* takes the plane, or rather a finite portion of the plane, as a new indefinable. It is, however, capable of definition, as, following Peano, I shall now show.

We need, to begin with, some new axioms. First, if *ρ* be any straight line, there is at least one point not belonging to *ρ*. Next, if *a*, *b*, *c* be three points not in one straight line, and *d* be a point of *bc* between *b* and *c*, *e* a point of *ad* between *a* and *d*, then *be* will meet *ac* in a point *f* and *e* will be between *b* and *f*, *f* between *a* and *c*. Again, *a*, *b*, *c*, *d* being as before, if *f* be a point between *a* and *c*, then *ad* and *bf* will intersect in a point *e* between *a* and *d* and between *b* and *f*†. We now define what may be regarded as the product (in a geometrical sense) of a point and a figure. If *a* be any point, and *k* any figure, *ak* is to denote the points which lie on the various segments between *a* and the points of *k*. That is, if *p* be any point of *k*, and *x* any point of the segment *ap*, then *x* belongs to the class *ak*. This definition may be applied even when *a* is a point of *k*, and *k* is a straight line or part of one. The figure *ak* will then be the whole line or some continuous portion of it. Peano now proves, by purely logical transformations, that, if *a*, *b*, *c* be distinct non-collinear points, $a(bc) = b(ac)$. This figure is called the triangle *abc*, and is thus wholly determined by its three defining points. It is also shown that, if *p*, *q* be points of the segments *ab*, *ac* respectively, the segment *pq* is wholly contained in the triangle *abc*. After some more theorems, we come to a new definition. If *a* be a point, and *k* any figure (*i.e.* class of points), *a'k* is to denote all the points between which and *a* lies some point of *k*, that is, as Peano remarks, the whole shadow of *k* if *a* be an illuminated point. Thus if *a*, *b*, *c* be non-collinear points, *a'*(*bc*) will represent the class of points beyond *bc* and bounded by *ab*, *ac* produced. This

enables us to define the plane (*abc*) as consisting of the straight lines
bc, *ca*, *ab*, the triangle *abc*, and the figures *a'bc*, *b'ca*, *c'ab*, *b'c'a*, *c'a'b*,
a'b'c*. It is then easy to show that any other three points of the plane
define the same plane, and that the line joining two points of a plane
lies wholly in the plane. But in place of the proposition that any two
lines in a plane intersect, we have a more complicated proposition,
namely: If *a*, *b*, *c*, *d* be coplanar points, no three of which are
collinear, then either the lines *ab*, *cd* intersect, or *ac*, *bd* do so, or
ad, *bc* do so.

381. Having successfully defined the plane, we can now advance
to solid Geometry. For this we need, to begin with, the axiom:
Given any plane, there is at least one point outside the plane. We
can then define a tetrahedron exactly as we defined a triangle. But
in order to know that two planes, which have a point in common,
have a line in common, we need a new axiom, which shows that the
space we are dealing with has three dimensions. In projective space,
this axiom was simply that a line and a plane always have at least one
point in common. But here, no such simple axiom holds. The following
is given by Peano (*loc. cit.* p. 74): If *p* be a plane, and *a* a point
not on *p*, and *b* a point of *a'p* (*i.e.* a point such that the segment *ab*
contains a point of *p*, or, in common language, a point on the other
side of the plane from *a*), then if *x* be any point, either *x* lies on the
plane, or the segment *ax* contains a point of the plane, or else the
segment *bx* contains a point of the plane. By adding to this, finally,
an axiom of continuity, we have all the apparatus of three-dimensional
descriptive Geometry †.

382. Descriptive Geometry, as above defined, applies equally to
Euclidean and to hyperbolic space: none of the axioms mentioned
discriminate between these two. Elliptic space, on the contrary, which
was included in projective Geometry, is here excluded. It is impossible,
or rather, it has hitherto proved so, to set up a general set of axioms
which will lead to a general Geometry applying to all three spaces,
for at some point our axioms must lead to either an open or a closed
series of points on a line. Such a general Geometry can be constructed
symbolically, but this results from giving different interpretations to
our symbols, the indefinables in one interpretation being definable in
another, and *vice versâ*. This will become plain by examining the method

---

* The figure *b'* (*c'a*), or *b'c'a*, represents the angle between *ba* and *ca* both
produced, as may be seen from the definition.

† I confine myself as a rule to three dimensions, since a further extension has
little theoretic interest. Three dimensions are far more interesting than two,
because, as we have seen, the greater part of projective Geometry—*i.e.* everything
dependent upon the quadrilateral construction—is impossible with less than three
dimensions, unless the uniqueness of the quadrilateral construction be taken as an
axiom.

in which projective Geometry is made applicable to the space above defined, which, for want of a better name, I shall call descriptive space.

383. When we try to apply projective Geometry to descriptive space, we are met by the difficulty that some of the points required in a construction may not exist. Thus in the quadrilateral construction, given three points a, b, c, the fourth point d may not exist at all. We can prove as before that, if it exists, it is unique, and so with other projective propositions: they become hypothetical, since the construction indicated is not always possible. This has led to the introduction of what are called *ideal* elements (points, lines and planes), by means of which it becomes possible to state our projective theorems generally. These ideal elements have a certain analogy to complex numbers in Algebra—an analogy which in analytical Geometry becomes very close. Before explaining in detail how these elements are introduced, it may be well to state the logical nature of the process. By means of the points, lines and planes of descriptive Geometry, we define a new set of entities, some of which correspond (*i.e.* have a one-one relation) to our points, lines and planes respectively, while others do not. These new entities we call ideal points, lines and planes; and we find that they have all the properties of projective points, lines and planes. Hence they constitute a projective space, and all projective propositions apply to them. Since our ideal elements are defined by means of the elements of descriptive space, projective propositions concerning these ideal elements are theorems concerning descriptive space, though not concerning its actual points, lines and planes. Pasch, who has given the best account of the way in which ideal elements are to be defined*, has not perceived (or, at any rate, does not state) that no ideal point is an actual point, even where it has a one-one relation to an actual point, and that the same holds of lines and planes. This is exactly the same remark as we have had to make concerning rationals, positive numbers, real numbers, and complex numbers, all of which are supposed, by the mathematician, to contain the cardinals or the ordinals, whereas no one of them can ever be one of the cardinals or ordinals. So here, an ideal element is never identical with an actual point, line or plane. If this be borne in mind, the air of magic which surrounds the usual expositions disappears.

384. An ideal point is defined as follows. Consider first the class of all the lines passing through some point, called the vertex. This class of lines is called a *sheaf* of lines (*Strahlenbündel*). A sheaf so defined has certain properties which can be stated without reference to the vertex†. Such are, for example, the following: Through any point (other than the vertex) there is one and only one line of the sheaf; and any two lines of the sheaf are coplanar. All the properties of a sheaf,

---

* *Op. cit.* §§ 6–8.

† These are enumerated by Killing, *Grundlagen der Geometrie*, Vol. II (Paderborn, 1898), p. 82.

which can be stated without reference to the vertex, are found to belong to certain classes of lines having no vertex, and such that no two of the class intersect. For these a simple construction can be given, as follows*. Let l, m be any two lines in one plane, A any point not in this plane. Then the planes Al, Am have a line in common. The class of such lines, for all possible points A outside the plane lm, has the properties above alluded to, and the word *sheaf* is extended to all classes of lines so defined. It is plain that if l, m intersect, the sheaf has a vertex ; if not, it has none. Thus, in Euclidean space, all the lines parallel to a given line form a sheaf which has no vertex. When our sheaf has no vertex, we define an *ideal point* by means of the sheaf. But this must not be supposed to be really a point : it is merely another name for the sheaf itself, and so, when our sheaf has a vertex, if we are to make propositions in which ideal points occur, we must substitute the sheaf for its vertex. That is, an ideal point is simply a sheaf, and no sheaf *is* an actual point.

Concerning sheaves of lines we may observe the following points. Any two straight lines in one plane uniquely determine a sheaf. Two sheaves both having a vertex always determine a line, namely that joining the vertices, which is common to both sheaves. Three sheaves, of which one at least has a vertex, determine a plane, unless they are collinear. A line and a plane always have a common sheaf, and so have three planes of which two at least have a common point.

385. Thus sheaves of lines have some projective properties, in relation to lines and planes, which are lacking to points. In order to obtain entities with further projective properties, we must, to begin with, replace our lines by ideal lines. For this purpose we must first define pencils of planes (axial pencils, *Ebenenbüschel*). An axial pencil consists, in the first instance, of all the planes through a given straight line, called the axis. But as in the case of sheaves, it is found that such a figure has many properties independent of the axis, and that these properties all belong to certain other classes of planes, to which the name of pencil is therefore extended. These figures are defined as follows†. Let A, B be two sheaves of lines. Let D be a point not on the line (if there be one) common to the two sheaves A, B. Then A, B, D determine uniquely a plane, which we may call ABD, or P (say). This will be the plane containing those lines of A and B that pass through D. Any other point E, not in the plane P, will determine a different plane ABE, or Q. The class of planes so obtained, by varying D or E, is a pencil of planes, and has all the properties of a pencil having a real axis, except those in which the axis is explicitly mentioned. Any two planes P and Q belonging to the pencil completely determine it. Moreover, in place of A and B above,

---

* Pasch, *op. cit.* § 5. † Pasch, *op. cit.* § 7.

we may substitute any other sheaves of lines A', B', belonging to both P and Q. (A sheaf belongs to a plane when one of its lines lies in the plane.) Any two sheaves belonging to both P and Q will serve to define the pencil of planes, and will belong to every plane of the pencil. Hence if, in place of actual points, we substitute ideal points, *i.e.* sheaves of lines, every pencil of planes has an axis, consisting of a certain collection of sheaves of lines, any two of which define the pencil. This collection of sheaves is called an *ideal line**.

386. Substituting ideal points and lines for actual ones, we find that we have now made a further advance towards projective space. Two ideal points determine one and only one ideal line; a given plane is determined by any three of its ideal points which do not belong to one ideal line, but three ideal points do not always determine a plane. Two ideal lines in a plane always have a common ideal point, and so have a plane and an ideal line. Also two planes always have a common ideal line, and three planes always have either a common ideal point or a common ideal line. The only point where our space is not strictly projective is in regard to planes. There is a plane through any two ideal points and one actual point, or through an ideal point and an actual line. If there is a plane at all through three non-collinear ideal points, or through an ideal line and an ideal point not on the line, then there is only one such plane; but in some cases there is no such plane. To remedy this, we must introduce one more new class of entities, namely ideal planes.

The definition of ideal planes† is comparatively simple. If A, B, C be any three ideal points, D an ideal point on the ideal line AB, and E on AC, then the ideal line DE has an ideal point in common with BC, whether there be an actual plane determined by A, B, C or not. Thus if B, C, D be any three ideal points, and E any other ideal point such that BD, CE intersect, then BC, DE intersect, and so do BE, CD. Hence, if B, C, D be not collinear, we define the ideal plane BCD as that class of ideal points E which are such that the ideal lines BD, CE intersect.

For the sake of clearness, let us repeat this definition in terms of our original points, lines and planes, without the use of the word *ideal*. Given three sheaves of lines B, C, D, which are not all contained in a common pencil of planes, let E be another sheaf of lines such that there is a sheaf of lines common to the two pencils of planes BD, CE. Then the class of all sheaves E satisfying this condition is called the ideal plane BCD.

---

* For logical purposes, it is better to define the ideal line as the class of ideal points associated with a sheaf of planes, than as the sheaf itself, for we wish a line to be, as in projective Geometry, a class of points.

† Pasch, *op. cit.* § 8.

The usual properties of planes are easily proved concerning our new ideal planes, as that any three of their points determine them, that the ideal line joining two of their ideal points is wholly contained in them, and so forth. In fact, we find now that the new points, lines and planes constitute a projective space, with all the properties described in the preceding chapter. The elementary order of points on a line, with which we began, has disappeared, and a new order has to be generated by means of the separation of couples*. Thus all projective Geometry becomes available; and wherever our ideal points, lines and planes correspond to actual ones, we have a corresponding projective proposition concerning the latter.

387. I have explained this development at length, partly because it shows the very wide applicability of projective Geometry, partly because it affords a good instance of the emphasis which mathematics lays upon relations. To the mathematician, it is wholly irrelevant what his entities are, so long as they have relations of a specified type. It is plain, for example, that an instant is a very different thing from a point; but to the mathematician as such there is no relevant distinction between the instants of time and the points on a line. So in our present instance, the highly complex notion of a sheaf of lines—an infinite class of infinite classes—is philosophically very widely dissimilar to the simple notion of a point. But since classes of sheaves can be formed, having the same relations to their constituent sheaves that projective lines and planes have to projective points, a sheaf of lines in descriptive space *is*, for mathematical purposes, a projective point. It is not, however, even for mathematical purposes, a point of descriptive space, and the above transformation clearly shows that descriptive space is not a species of projective space, but a radically distinct entity. And this is, for philosophy, the principal result of the present chapter.

It is a remarkable fact, which the above generation of a projective space demonstrates, that if we remove from a projective space all the points of a plane, or all the points on one side of a closed quadric†, the remaining points form a descriptive space, Euclidean in the first case, hyperbolic in the second. Yet, in ordinary metrical language, the projective space is finite, while the part of it which is descriptive is infinite. This illustrates the comparatively superficial nature of metrical notions.

* See Pasch, *op. cit.* § 9.

† For the projective definition of a surface of the second order (quadric) in a projective space cf. Reye, *Geometrie der Lage* (Hanover, 1868), Part II, Lecture V. A quadric is closed if there are points not on it such that all straight lines through them cut the quadric. Such points are *within* the quadric.

CHAPTER XLVII.

METRICAL GEOMETRY.

388. THE subject of the present chapter is elementary Geometry, as treated by Euclid or by any other author prior to the nineteenth century. This subject includes the usual analytical Geometry, whether Euclidean or non-Euclidean; it is distinguished from projective and descriptive Geometry, not by any opposition corresponding to that of Euclid and non-Euclid, but by its method and its indefinables. The question whether its indefinables can, or cannot, be defined in terms of those of projective and descriptive Geometry, is a very difficult one, which I postpone to the following chapter. For the present, I shall develop the subject straightforwardly, in a manner as similar to Euclid's as is consistent with the requisite generality and with the avoidance of fallacies. Metrical Geometry is logically subsequent to the two kinds which we have examined, for it necessarily assumes one or other of these two kinds, to which it merely adds further specifications. I shall, as a rule, assume descriptive Geometry, mentioning projective Geometry only in connection with points in which it shows important metrical differences from descriptive Geometry. In the former case, all the first twenty-six propositions of Euclid will hold. In the latter, the first, seventh, sixteenth, and seventeenth require modification; for these propositions assume, in one form or another, that the straight line is not a closed series. Propositions after the twenty-sixth—or, with a suitable definition of parallels, after the twenty-eighth—depend, with few exceptions, upon the postulate of parallels, and are therefore not to be assumed generally.

389. Since Euclid still has popularly, and even with mathematicians, a reputation for rigour, in virtue of which his circumlocution and long-windedness are condoned, it may be worth while to point out, to begin with, a few of the errors in his first twenty-six propositions*. To begin with the first proposition. There is no evidence whatever that the circles which we are told to construct intersect, and if they do not, the whole proposition fails. Euclid's problems are often regarded as

---

* Cf. Killing, *op. cit.* Vol. II, Section 5.

existence-theorems, and from this point of view, it is plain, the assumption that the circles in question intersect is precisely the same as the assumption that there is an equilateral triangle on a given base. And in elliptic space, where the straight line is a closed series, the construction fails when the length of the base exceeds half the length of the whole straight line. As regards the second and third propositions, there is nothing to be said, except that they are not existence-theorems. The corresponding existence-theorem—*i.e.* on any straight line, in either direction from a given point on the line, there is a point whose distance from the given point is equal to a given distance—is equivalent to the postulate concerning the circle, and is thus prior to the second and third propositions. With regard to the fourth, there is a great deal to be said; indeed Euclid's proof is so bad that he would have done better to assume this proposition as an axiom*. As the issues raised by this proof are of great importance, both to mathematics and to philosophy, I shall set forth its fallacies at some length.

390. The fourth proposition is the first in which Euclid employs the method of superposition—a method which, since he will make any *détour* to avoid it†, he evidently dislikes, and rightly, since it has no logical validity, and strikes every intelligent child as a juggle. In the first place, to speak of motion implies that our triangles are not spatial, but material. For a point of space *is* a position, and can no more change its position than the leopard can change his spots. The motion of a point of space is a phantom directly contradictory to the law of identity: it is the supposition that a given point can be now one point and now another. Hence motion, in the ordinary sense, is only possible to matter, not to space. But in this case superposition proves no geometrical property. Suppose that the triangle *ABC* is by the window, and the side *AB* consists of the column of mercury in a thermometer; suppose also that *DEF* is by the fire. Let us apply *ABC* to *DEF* as Euclid directs, and let *AB* just cover *DE*. Then we are to conclude that *ABC* and *DEF*, before the motion, were equal in all respects. But if we had brought *DEF* to *ABC*, no such result would have followed. But how foolish! I shall be told; of course *ABC* and *DEF* are to be both rigid bodies. Well and good. But two little difficulties remain. In the first place—and for my opponent, who is an empirical philosopher, this point is serious—it is as certain as anything can be that there are no rigid bodies in the universe. In the second place—and if my opponent were not an empiricist, he would find this objection far more fatal—the meaning of rigidity presupposes a purely spatial metrical equality, logically independent of matter. For

---

* This course is actually adopted, as regards the equality of the remaining angles, by Hilbert, *Grundlagen der Geometrie* (Festschrift zur Feier der Enthüllung des Gauss-Weber Denkmals, Leipzig, 1899), p. 12.

† Cf. Killing, *loc. cit.* § 2.

what is meant by a rigid body? It is one which, throughout a continuous portion of time, preserves all its metrical properties unchanged. Hence we incur a most fatally vicious circle if we attempt to define metrical properties by rigidity. If $\alpha\beta\gamma$ be a material triangle, which occupies at one time the space ABC, at another the space $A'B'C'$, to say that $\alpha\beta\gamma$ is rigid means that, however the two times be chosen (within some assigned period), the triangles ABC, $A'B'C'$ are equal in all respects. If we are to avoid this conclusion, we must define rigidity in some wholly non-geometrical manner. We may say, for example, that a rigid body *means* one which is made of steel, or of brass. But then it becomes a logical error to regard brass eternal as slave to mortal rage; and if we define equal spaces as those which can be occupied by one and the same rigid body, the propositions of metrical Geometry will be one and all false.

The fact is that motion, as the word is used by geometers, has a meaning entirely different from that which it has in daily life, just as a variable, in mathematics, is not something which changes, but is usually, on the contrary, something incapable of change. So it is with motion. Motion is a certain class of one-one relations, each of which has every point of space for its extension, and each of which has a converse also belonging to the class. That is, a motion is a one-one relation, in which the referent and the relatum are both points, and in which every point may appear as referent and again as relatum. A motion is not this only: on the contrary, it has this further characteristic, that the metrical properties of any class of referents are identical with those of the corresponding class of relata. This characteristic, together with the other, defines a motion as used in Geometry, or rather, it defines a motion or a reflexion; but this point need not be elucidated at present. What is clear is, that a motion presupposes the existence, in different parts of space, of figures having the same metrical properties, and cannot be used to define those properties. And it is this sense of the word *motion*, not the usual material sense, which is relevant to Euclid's use of superposition.

391. Returning now to Euclid's fourth proposition, we see that the superposition of ABC on DEF involves the following assumptions. (1) On the line DE there is a point E, on either side of D, such that $DE = AB$. This is provided for by the postulate about the circle. (2) On either side of the ray DE, there is a ray DF such that the angle EDF is equal to the angle BAC. This is required for the possibility of a triangle DEF such as the enunciation demands, but no axiom from which this follows can be found in Euclid. The problem, to construct an angle EDF equal to BAC, does not occur till I. 23, and there I. 4 is used in the proof. Hence the present assumption must be added to Euclid's axioms. It now follows that on DF there is a point F such that $DF = AC$. Hence the possibility of two such triangles as the enunciation demands is established. But in order to prove that *DEF* is equal in all respects to *ABC*, we need a further axiom, namely: With one angle at *D*, one side along the ray *DE*, and the other side to the right (or left) of *DE*, there exists a triangle which is equal in all respects to the triangle *ABC**. This is, in fact, the exact assumption which is concealed in the method of superposition. With this assumption, it finally becomes possible to prove that *DEF* is the triangle satisfying the above conditions and equal in all respects to *ABC*.

The next remark concerns I. 6. Here Euclid first employs an axiom of which he is wholly unconscious, though it is very essential to his system, namely: If *OA*, *OB*, *OC* be three rays which meet a straight line not passing through *O* in *A*, *B*, *C* respectively, and if *B* be between *A* and *C*, then the angle *AOB* is less than the angle *AOC*. This axiom, it will be seen, is not applicable in projective space, since it presupposes that the line is not a closed series. In I. 7, if this proposition is to apply to hyperbolic space, we require further the axiom: If three non-intersecting lines in one plane meet two lines in *A*, *B*, *C*; *A'*, *B'*, *C'*, respectively; and if *B* be between *A* and *C*; then *B'* is between *A'* and *C'*. Also it may be observed that Euclid gives no definition of the two sides of a line, a notion which again presupposes that the straight line is not a closed series. And with regard to angles, I. 7 requires sufficient axioms to show that they are a series of the kind explained in Part IV, Chapter xxiv; or else we must assume the descriptive axiom of the last chapter, to the effect that, if *A*, *B*, *C*, *D* be coplanar points, no three of which are collinear, there is a point common to the stretches *AB*, *CD*, or to *AC*, *BD*, or to *AD*, *BC*. All these assumptions will be found implicit in I. 7, as may be seen by attempting a symbolic proof in which no figure is used.

Similar remarks apply to I. 16. In I. 12 it is assumed that a circle must meet a line in two points, if at all. But enough has been said to show that Euclid is not faultless, and that his explicit axioms are very insufficient. Let us, then, make an independent examination of metrical Geometry.

392. Metrical Geometry is usually said to be distinguished by the introduction of quantity. It is sufficient for the characterization of metrical Geometry to observe that it introduces, between every pair of points, a relation having certain properties in virtue of which it is numerically measurable—*i.e.* such that numbers can be given a one-one correspondence with the various relations of the class in question. The class of relations is called *distance*, and will be regarded, though this is not strictly necessary, as a class of magnitudes. Some of the properties of distance are as follows.

---

* See Pasch, *op. cit.* § 13, Grundsatz ix. The whole § is excellent.

(1) Every pair of points has one and only one distance.

(2) Distances are symmetrical relations.

(3) On a given straight line through a given point, there are two and only two points at a given distance from the given point.

(4) There is no maximum distance.

(5) The distance of a point from itself is zero*.

(6) There is no minimum to the distance between distinct points.

(7) If d, δ be two given distances, and A_0, A_1, A_2, ... A_n, ... be distinct points on a straight line, whose distances one from the next are all δ, then for some value of n, $A_0 A_n$ is greater than d.

(8) If A_0, A_n be any two points, there exist $n-1$ distinct points (whatever integer n may be) on the straight line $A_0 A_n$, such that the distances of each from the next, of A_0 from the first, and of A_n from the last, are all equal†.

393. It may be observed that, if we admit the axiom that the whole is greater than the part, the properties (1), (4), (5) and (6) belong to stretches, while (2) becomes admissible by abstracting from the sense of a stretch. With regard to the remaining properties, (3), (7) and (8), there is nothing in descriptive Geometry to show whether or not they belong to stretches. Hence we may, if we choose, regard these three properties as axioms regarding stretches, and drop the word *distance* altogether. I believe that this represents the simplest course, and, as regards actual space, the most correct. At the same time, there is no contradiction in regarding distances as new relations distinct from stretches‡. If we identify distance and stretch, what distinguishes metrical from descriptive Geometry is primarily the three additional axioms (3), (7) and (8), applied to a new indefinable, namely, the magnitude of divisibility of a stretch. This is not properly a notion of pure mathematics, since it cannot be derived from our original apparatus of logical notions. On the other hand, distance is not indefinable, being a class of one-one relations with certain assignable properties. On this point either course is logically permissible, but only distance can be introduced into pure mathematics in the strict sense in which the word is used in this work.

The above axioms are required for showing that all distances are numerically measurable in terms of any standard distance§. It is not necessary that distances should be magnitudes, or even relations; all that is essential is that distances should form a series with certain properties. If the points of a line form a continuous series, then

* See Part III, Chap. XXII.

† Further properties of distance will be added later on.

‡ Stretches are, of course, not properly relations; but this point is irrelevant in the present discussion.

§ See Part IV, Chap. XXXI.

distances do so also, in virtue of (3); thus all signless real numbers will be required for their measurement.

394. Assuming that distance and stretch are distinct, it may be asked whether distances do not suffice for generating order on the straight line, without the need of any asymmetrical transitive relation of points. This represents, I think, the usual view of philosophers; but it is by no means easy to decide whether it represents a tenable view. It might perhaps be thought that (2) might be dropped, and distance regarded as an asymmetrical relation. So long as we confine our attention to one line, this view seems unobjectionable. But as soon as we consider the fact that distances on different lines may be equal, we see that the difference of sense between AB and BA is not relevant to distance, since there is no such difference between distances on different lines. Thus if CD be a distance on another line, CD may be equal both to AB and BA, and hence AB and BA must be equal, not opposite, distances. And the same thing may be made evident by considering a sphere. For this certainly consists of points at a given distance from the centre; and thus points at opposite ends of a diameter must have the same distance from the centre. Distance, then, is symmetrical; but it does not follow that the order on a line cannot be generated by distance. Let A, B be given points on a line, and let C, C' be two points on AB whose distances from A are equal, and less than AB. If we now set up the axiom that either BC or BC' is less than AB, while the other, BC'' or BC, is greater than AB, we shall, I think, after some further axioms, be able to generate order without any other relation than distance. If A, B, C be three collinear points such that the distances AC, CB are both less than AB, then we shall say that C is between A and B. If A, B, C' be points such that AC', AB are both less than BC', then we shall say that A is between B and C'. If, finally, A, B, C'' be points such that AB, BC'' are both less than AC'', we shall say that B is between A and C'''. It remains to see whether, as the generation of a series requires, one of these always happens. Let A, B, C be any three collinear points. First suppose, if possible, that the distances AB, BC, CA are all equal. This case is not excluded by anything hitherto assumed; we require, therefore, the further axiom that, if AB, BC be equal, AC is not equal to either of them; and I think it will be prudent to assume that AC is greater than either. Thus the case of two equal distances and one less than either is excluded. Of the three distances AB, BC, AC, therefore, one must be the greatest: let this be AC. Then in virtue of the definition, B will be between A and C. But our difficulties are not at an end. For we require further that any point between A and B shall be between A and C; and that, if A be between D and C, B shall be between D and C. With regard to the first point, if E be between A and B, AE and EB are less than AB, and therefore less than AC. But nothing assures us that EC is less than

AC. For this purpose we need a new axiom, which will be just what we set out to prove, namely : If *AE*, *EB* be both less than *AB*, and *AB*, *BC* be both less than *AC*, then *EC* is less than *AC*. Finally, to prove that, if *A* be between *D* and *C*, and *B* between *A* and *C*, then *B* is between *D* and *C*. Here *DA*, *AC* are less than *DC*, and *AB*, *BC* are less than *AC*. Hence *BC* is less than *DC*; but nothing proves *BD* less than *DC*. For this we shall need a new axiom, and then at last our order will be definite. But the process, as is evident, is extremely complicated.

395. Moreover we still need a method of defining the straight line. Pieri has shown, in an admirable memoir*, how to deduce metrical geometry by taking *point* and *motion* as the only indefinables. In § 390, we objected to the introduction of motion, as usually effected, on the ground that its definition presupposes metrical properties; but Pieri escapes this objection by not defining motion at all, except through the postulates assumed concerning it. The straight line joining two points is the class of points that are unchanged by a motion which leaves the two points fixed. The sphere, the plane, perpendicularity, the order of points on a line, etc. are easily defined. This procedure is logically unimpeachable, and is probably the simplest possible for elementary geometry. But we must now return to the consideration of other suggested systems.

There is a method, invented by Leibniz† and revived by Frischauf‡ and Peano§, in which distance alone is fundamental, and the straight line is defined by its means. In this method distances are given to begin with as a class of relations which are the field of a certain transitive asymmetrical relation (greater and smaller); if we assume this relation to be continuous, distances will be . measurable; all distances have the same domain and the same converse domain, namely all the points of the space in question ; the locus of points equidistant from two fixed points is called a *plane*, and the intersection of two non-coincident planes, when it is not null, is called a *straight line*. (The definition of the straight line given by Peano‖ is as follows : The straight line *ab* is the class of points *x* such that any point *y*, whose distances from *a* and *b* are respectively equal to the distances of *x* from *a* and *b*, must be coincident with *x*.) Leibniz, who invented this method, failed, according to Couturat, to prove that there are straight lines, or that a straight line is determined by any two of its points. Peano has not, so far as I am aware, succeeded in proving either of these propositions, but it is of course possible to introduce them by means of axioms. Frischauf professes to demonstrate them, but his proofs are very informal, and it

* *Della geometria elementare como sistema ipotetico deduttivo*, Turin, 1899.

† Cf. Couturat, *La Logique de Leibniz*, Paris, 1901, Chap. ix, esp. p. 420.

‡ *Absolute Geometrie nach Johann Bolyai, Anhang.*

§ *Accademia Reale delle Scienze di Torino*, 1902-3, " La Geometria basata sulle idee di punto e distanza."　　　　　　　　　‖ *loc. cit.*

is difficult to know what axioms he is assuming. In any case, however, the definitions prove that, by a sufficient use of axioms, it is possible to construct a geometry in which distance is fundamental, and the straight line derivative. The method is so complicated as to be not practically desirable; but its logical possibility is nevertheless important.

396. It is thus plain that the straight line must be independent of distance, while distance *may* be independent of the straight line. Taking both as symmetrical relations, we *can*, by a very complicated series of axioms, succeed in generating order on the straight line and in explaining the addition and measurement of distances. But this complication, in most spaces *, is logically unnecessary, and is wholly avoided by deriving distances from stretches. We now start, as in descriptive Geometry, with an asymmetrical transitive relation by which the straight line is both defined and shown to be a series. We define as the *distance* of two points A and B the magnitude of divisibility of the stretch from A to B or B to A—for divisibility is a signless magnitude. Divisibility being a kind of magnitude, any two distances will be equal or unequal. As with all divisibilities, the sum of the divisibilities of AB and EF is the divisibility of the logical sum of the classes AB and EF, provided these classes have no common part. If they have a common part, we substitute for EF a stretch $E'F'$ equal to it and having no part in common with AB. The difference of the distances AB, EF (supposing AB the greater) is the divisibility of a stretch CD which, added logically to EF, and having no part in common with EF, produces a stretch equal to AB. It follows at once that, if A, B, C be collinear, and B be between A and C, $AB + BC = AC$ and $AC - AB = BC$. No further axiom is required for these propositions. For the proposition that, if $AB = A'B'$, and $CD = C'D'$, then $AB + CD = A'B' + C'D'$, we require only the general axiom, applicable to all divisibilities, that the sums of equals are equal. Thus by the help of the axioms (3), (7), (8) above, we have everything that is required for the numerical measurement (theoretically speaking) of all distances in terms of any given distance, and for the proof that change of unit involves multiplication throughout by a common factor.

397. With regard to magnitude of divisibility, in the sense in which this is relevant to metrical Geometry, it is important to realize that it is an ordinal notion, expressing a property of relations, not of their fields. We wish to say that a stretch of two inches has twice as much divisibility as a stretch of one inch, and that an area is infinitely more divisible than a stretch. Now, if we are dealing (as will be assumed in this discussion) with a continuous space, every stretch, area or volume is a class of 2^{a_0} terms; and considered as a class, it is the field of an infinite number of relations beside that (or those) belonging to it

---

* The only exceptions known to me are finite spaces of two dimensions. See Chap. xlix.

in respect of the space we are considering. The habit of allowing the imagination to dwell upon actual space has made the order of points appear in some way intrinsic or essential, and not merely relative to one of many possible ordering relations. But this point of view is not logical : it arises, in regard to actual space, only from the fact that the generating relations of actual space have a quite peculiar connection with our perceptions, and, through the continuity of motion, with time. From the standpoint of logic, no one of the relations having a given field has any preeminence, and the points of actual space, like any other class of 2^{a_0} terms, form, with regard to other sets of generating relations, other sorts of continuous spaces—indeed any other continuous space, having any finite number of dimensions, or even ω dimensions, can be formed of the points of a Euclidean space by attending to other generating relations.

From this it follows that magnitude of divisibility, if it is to distinguish a long stretch from a short one, or an area from a stretch, must be a property of the relations involved, not of the class of points composing the area or the stretch. It is not quite easy to define the exact property which is required ; for any two stretches are ordinally similar. We require some sense for the equality or inequality of the relations whose fields are the given stretches. Where coordinates (*i.e.* a correlation of the points of a line with the real numbers) have been already introduced, we may define the magnitude of a stretch as the difference of the coordinates of its end-points or its limits (according as the stretch has ends or not); but if this is done, the magnitudes of stretches will depend upon the necessarily more or less arbitrary plan upon which we have introduced our coordinates. This is the course adopted in the projective theory of distance—a course which has the merit of making metrical Geometry a logical development from projective axioms alone (see next chapter). The other course that may be adopted is, to assume that the generating relations of any two stretches have either a symmetrical transitive relation (equality), or an asymmetrical transitive relation or its converse (greater or less). Certain axioms will be required, as, for example, that if the points A, B, C, D are collinear, and AC is greater than AD, then BC is greater than BD^*. The relations of equal, greater and less may be regarded as defined by these axioms, and the common property of the generating relations of those stretches that are equal to a given stretch may be defined as the magnitude of divisibility of the said generating relations. The sense in which an area has infinitely more divisibility than a stretch is that, if n be any finite integer, and n stretches equal to a given stretch be removed from an area, there always remains an area, however great n may be. What is important to observe, in the above discussion, is that the logical parity

---

* Stretches are here regarded as having sign, so that, if AC is greater than AD, CA is less than DA.

of all the orders of which a class of terms is capable makes it necessary to regard the magnitudes with which metrical Geometry deals as belonging to relations or classes of relations, not, as is commonly supposed, to the class of points forming their fields.

398. In elliptic space, where the straight line is a closed series, the attempt to make distance independent of stretch leads to still further complications. We now no longer have the axiom that, if A, B, C be collinear, we cannot have $AB = BC = CA$; and we have to recognize two distances between every pair of points, which, when distance is taken as fundamental, becomes extremely awkward. We may however avoid admitting two distances by refusing to regard the greater of the two as properly a distance. This will then be only a stretch. If two distances are admitted, one is always greater than the other, except in a limiting case, when both are the lower limit of the greater distances and the upper limit of the lesser distances. Further if a, b, c, d be any four distinct points, the greater of the two distances ab is always greater than the lesser of the two distances cd. Thus the whole class of greater distances may be banished, and only greater stretches be admitted.

We must now proceed as follows. Distances are a class of symmetrical relations, which are magnitudes of one kind, having a maximum, which is a one-one relation whose field is all points, and a minimum, which is the distance of any point from itself. Every point on a given line has a given distance other than the maximum or minimum from two and only two other points on the line. If a, b, c, d be four distinct points on one line, we shall say that a and c are separated by b and d in the following four cases, of which (1) and (2) and also (3) and (4) are not mutually exclusive:

(1)　If $ab < ac \cdot bc < ac \cdot ad > ac$.

(2)　If $ab < ac \cdot bc < ac \cdot dc > ac$.

(3)　If $ab > ac \cdot ad < ac \cdot dc < ac$.

(4)　If $bc > ac \cdot ad < ac \cdot dc < ac$.

We then need Vailati's five axioms enumerated in Part IV, Chap. xxiv, in order to generate a closed series from the separation of couples so defined. Thus it is possible, though by a somewhat complicated process, to generate a closed series of points on a line by means of the symmetrical relation of distance.

I shall not work out in further detail the consequences of this hypothesis in elliptic space, but proceed at once to the hypothesis that distances are the magnitudes of stretches. When the number of dimensions exceeds two, the polar form of elliptic space is merely projective space together with the necessary metrical axioms; the antipodal form is a space in which two antipodal points together have the properties of a single projective point. Neglecting the latter, to which similar remarks will apply, I shall confine myself to the polar form. Since this is a projective space, every pair of points determines two segments on the line joining the points. The sum of these two segments, together with the two points, is the whole line, and therefore constant. It is an axiom that all complete straight lines have the same divisibility. The divisibility of either segment is *a* distance between the two points : when the two distances are equal, either may be called *the* distance ; when they are unequal, it will be convenient to call the smaller *the* distance, except in special problems. The whole theory then proceeds as in the case of descriptive space. But it is important to observe that, in elliptic space, the quadrilateral construction and the generation of order, being prior to stretches, are prior to distances, and are presupposed in metrical Geometry.

399. So far, therefore, metrical Geometry introduces three new axioms, and one new indefinable. The stretch in *every* series is a quantity, and metrical Geometry merely introduces such axioms as make all stretches of points measurable. A few words may be useful as to the sense in which, in a theoretical discussion, the word *measurement* is to be understood. The actual application of the foot-rule is here not in question, but only those properties of pure space which are presupposed in the use of the foot-rule. A set of magnitudes is *theoretically* measurable when there is a one-one relation between them and some or all numbers ; it is *practically* measurable when, given any magnitude, *we* can discover, with a certain margin of error, what the number is to which our magnitude has the relation in question. But how we are to discover this is a subsequent question, presupposing that there is such a proposition to be discovered, and soluble, if at all, by empirical means to be invented in the laboratory. With practical measurement, then, we are not at all concerned in the present discussion.

400. I come now to a more difficult question than distance, namely the question as to the definition of *angle*. Here, to begin with, we must deal with rays, not with whole straight lines. The ray may be taken either as an asymmetrical relation, or as the half-line on one side of a given point on a line. The latter usage is very convenient, and I shall frequently employ it. Elementary Geometry assumes that two rays starting from the same point determine a certain magnitude, called the angle between them. This magnitude may, however, be defined in various ways. In the first place, we must observe that, since the rays in a plane through a point form a closed series, every pair of rays through a point defines *two* stretches of rays. Of these, however, one stretch contains the opposites of both rays, while the other stretch contains the opposites of neither—except, indeed, in the one case where the two rays are each other's opposites. This case is met by Euclid's postulate that all right angles are equal—a postulate, however, which is now known to be demonstrable*. Omitting this case, the angle between two rays may be defined as that stretch of rays through their intersection which is bounded by the two rays and does not contain the opposite of either, *i.e.* if A, B be the rays, and $\breve{A}$, $\breve{B}$ their opposites, the angle is the class of rays C which are separated from $\breve{A}$ or $\breve{B}$ by A and B. We might also, but for an objection to be mentioned shortly, define the angle as all the points on such rays. A definition equivalent to this last, but simpler in form, and avoiding the mention of the opposite rays, is the following†. Let a, b be any two points of the rays A, B, and let c be any point of the stretch ab. Then the class of points c, for all possible positions of a and b on their respective rays, is the angle between A and B. That is, every pair of intersecting rays divides the plane of the rays into two parts: the part defined as above is the angle. Or rather, the part so defined is the angle as a quantity: the angle as a magnitude is the divisibility of this part. But to these latter definitions we shall find fatal objections, and we shall find it necessary to adhere to the definition as a stretch of rays.

401. Thus angle, like distance, is not a new indefinable, but like distance, it requires some new axioms. The angle between a ray A and its opposite A' cannot be defined as above, but may be defined as the logical sum of the angles between A and B, B and A' respectively. This limiting angle is greater than any other at the point, being in fact the whole half of the plane on one side of the straight line AA'. If the angles between A and B, B and A' are equal, each is called a right angle. (That there are such angles, can be proved if we assume continuity.) Two intersecting straight lines make four angles, which are equal in pairs. The order of a collection of rays through a point in a plane may be obtained by correlation with the points where these rays intersect a given straight line, provided there is any straight line which all of them intersect. But since rays through a point in a plane form a closed series, while the points on a line do not, we require a four-term relation for the former order. The following definition seems adequate. Given four rays OA, OB, OC, OD through a point O and in one plane, if these all meet a certain straight line in A, B, C, D respectively, and A and C are separated by B and D, then OA and OC are said to be separated by OB and OD. In projective space this suffices. But in descriptive space we must provide for other cases. Thus if OA, OB, OC meet the given line, and B is between OA and OC, while OD does not meet the given line, then OA and OC are again said to be separated by OB and OD. If, finally, OA' and OB' be the

---

* See *e.g.* Killing, *op. cit.* Vol. ii, p. 171. A strict proof will be found in Hilbert, *op. cit.* p. 16.

† Killing, *op. cit.* ii, p. 169.

opposites of OA and OB, then OA and OA' are separated by OB and OB'. In virtue of the descriptive axioms of the preceding chapter, the order among the rays so obtained will be unambiguous, *i.e.* independent of our choice of the line ABC, and will cover all cases.

But now we need axioms analogous to those which, in the case of distance, were numbered (3), (7) and (8). At any given point in a given ray, there must be, in a given plane, two and only two rays, on opposite sides of the given ray (*i.e.* separated from each other by the given ray and its opposite), which make a given angle with the given ray; and angles must obey the axioms of Archimedes and of linearity. But in addition to these axioms, which insure that angles shall be numerically measurable, we must have some method of connecting the measure of angles with that of distances, such as is required for the solution of triangles. Does this require a new axiom? Euclid appears to obtain this, by means of I. 47, II. 12, and II. 13, without any fresh axiom. For this result we depend upon the propositions on the congruence of triangles (I. 4, 8, 26), which demand only, as we saw, the axiom that, with one angle at a given point, and one side along a given ray through that point, there exist two and only two triangles in a given plane through the ray (one on each side of the given ray), which are equal in all respects to a given triangle. Thus it would seem that no fresh axioms are required for angles in a plane.

402. With regard to the definition of an angle as a portion of a plane, it is necessary (as in many other cases), if we retain this definition, somewhat to restrict the axiom that the whole is greater than the part. If a whole A has two parts B, C, which together constitute A, and if C be infinitesimal with respect to A, then B will be equal to A. This case occurs in a plane under the following circumstances. Let O, O' be any two points, OP, $O'P$ lines in one plane and making equal angles with the ray OO'*. Then in Euclidean or hyperbolic space these lines OP, $O'P'$ will not intersect; thus the angle between OO' and $O'P'$ will be part of the angle $O'OP$. Hence the above restriction is necessary as regards the axiom that the whole is greater than the part.

In Euclidean space this answer is sufficient, since, if OP makes with OO' a less angle than $O'P'$ does, OP and $O'P'$ will intersect. But in hyperbolic space, OP and $O'P'$ may not intersect even then. Hence if we adhere to the above definition of angle, we shall have to hold that the whole may be less than the part. This, however, is intolerable, and shows that the definition in question must be rejected. We may, however, still regard angle as the stretch of rays; for the rays in the angle at O' are not part of the rays in the angle at O. Hence it is only as a stretch of rays, or as the magnitude of such a stretch, that an angle can be properly defined.

* The angle between the rays OO', $O'P'$ is what Euclid would call the angle between OO' produced and $O'P'$.

As showing, in a curious manner, the increased power of deduction which results from the above axioms concerning distances and angles, we may remark that the uniqueness of the quadrilateral construction, which before could not be proved without three dimensions, can now be proved, as regards all constructions in one plane, without any assumption of points outside that plane. Nothing is easier than to prove this proposition by the methods of elementary coordinate Geometry. Thus although projective Geometry, as an independent science, requires three dimensions, any projective proposition concerning plane figures can be metrically proved, if the above axioms hold, for a two-dimensional space.

403. As regards figures of three dimensions, angles between planes and solid angles can be defined exactly as rectilinear angles were defined. Moreover fresh axioms will not be required, for the measurement of such angles can be deduced from the data we already possess.

With regard to areas and volumes some remarks seem necessary. Areas and volumes, like angles, are classes of points when taken as quantities, and divisibilities when taken as magnitudes. For areas and volumes we do not require afresh the axioms of Archimedes and of linearity, but we require one axiom apiece to give a criterion of equal areas and volumes, *i.e.* to connect their equality with that of distances and angles. Such an axiom is supplied, as regards areas, by the axiom that two congruent triangles have the same area, and as regards volumes, by the corresponding axiom concerning tetrahedra. But the existence of congruent tetrahedra, like that of congruent triangles, demands an axiom. For this purpose, Pasch* gives the following general axiom: If two figures are congruent, and a new point be added to one of them, a new point can be added to the other so that the two new figures are congruent. This axiom allows us to infer congruent tetrahedra from congruent triangles; and hence the measurement of volumes proceeds smoothly.

404. In three dimensions, a curious fact has to be taken account of, namely, the disjunction of right and left-handedness, or of clockwise and counter-clockwise. This fact is itself of a descriptive nature, and may be defined as follows. Between two non-coplanar rays, or between four non-coplanar points taken in an assigned order, there is always one of two opposite relations, which may be called right and left. The formal properties of these relations have been explained in Part IV (§ 222); for the present I am concerned with their geometrical consequences. In the first place, they cause volumes to become magnitudes with sign, in exactly the way in which distances on a straight line have sign when compounded with their sense. But in the case of distances, since not all are on one straight line, we could not thus compound distance and sense generally: we should require, for a compound, some more general

* *Op. cit.* p. 109.

notion than sense, such as vectors supply. Here, on the contrary, since, in a three-dimensional space, all volumes have one or other of two senses, the compound can be made for all volumes. Thus if the volume of the tetrahedron *abcd* has one sign, that of *bacd* will have the opposite sign. This is the familiar geometrical fact that the determinant giving the volume of a tetrahedron *abcd* has one or other sign according as the sense of *abcd* is the same as or different from that of *OXYZ*, where *O* is the origin and *X*, *Y*, *Z* any positive points on the axes. It is this fact, also, which gives signs to angular momentum in Dynamics. The importance of the fact (which itself seems to be an independent axiom) is this, that it makes a distinction between two figures whose metrical properties are all identical. It is this distinction which puzzled Kant, who, like most of his contemporaries, supposed all geometrical facts to be metrical. In itself, the fact would be no more puzzling than the distinction between the stretches *AB* and *BA*, which are metrically indistinguishable. But it becomes puzzling when metrical equality is supposed to result from motion and superposition. In our former definition of motion (§ 390) we omitted (as was then observed) a condition essential to its definition. Not only must two congruent figures be metrically equal, but there must be a continuous series of equal figures leading from the one to the other. Or, what amounts to the same thing, if *a*, *b*, *c*, *d* and *a′*, *b′*, *c′*, *d′* be homologous non-coplanar points in the two figures, the tetrahedra *abcd*, *a′b′c′d′* must have the same sense. In the case of equal and opposite tetrahedra, these conditions fail. For there is no gradual transition from clockwise to counter-clockwise; thus at some point in the series a sudden jump would be necessary. No motion will transform *abcd* into a tetrahedron metrically equal in all respects, but with the opposite sense. In this fact, however, there seems, to my mind, to be nothing mysterious, but merely a result of confining ourselves to three dimensions. In one dimension, the same would hold of distances with opposite senses; in two dimensions, of areas. It is only to those who regard motion as essential to the notion of metrical equality that right and left-handedness form a difficulty; in our theory, they are rather a confirmation than a stumbling-block.

With this we may end our brief review of metrical Geometry, leaving it to the next chapter to discuss its relation to projective Geometry and the projective theory of distance and angle.

CHAPTER XLVIII.

RELATION OF METRICAL TO PROJECTIVE AND DESCRIPTIVE GEOMETRY.

405. IN the present chapter I wish to discuss two questions. First, can projective and descriptive Geometry be established without any metrical presuppositions, or even without implying metrical properties? Secondly, can metrical Geometry be deduced from either of the others, or, if not, what unavoidable novelties does it introduce? The previous exposition has already dogmatically assumed certain answers to these questions, but we are now to examine critically the various possible answers.

The distinction between projective and descriptive Geometry is very recent, and is of an essentially ordinal nature. If we adopt the view—which, as we saw, is the simpler of two legitimate views—that the straight line is defined by a certain relation between any two of its points, then in projective Geometry this relation is symmetrical, while in descriptive Geometry it is asymmetrical. Beyond this we have the difference that, in projective Geometry, a line and a plane, two planes, or two lines in a plane, always intersect, while in descriptive Geometry the question whether this is the case or not is left open. But these differences are not very important for our present purpose, and it will therefore be convenient to speak of projective and descriptive Geometry together as non-quantitative Geometry.

The logical independence of non-quantitative Geometry is now scarcely open to question. We have seen, in Chapters XLV and XLVI, how it may be built up without any reference whatever to quantitative considerations. Quantity, in fact, though philosophers appear still to regard it as very essential to mathematics, does not occur in pure mathematics, and does occur in many cases not at present amenable to mathematical treatment. The notion which does occupy the place traditionally assigned to quantity is *order*; and this notion, we saw, is present in both kinds of non-quantitative Geometry. But the purity of the notion of order has been much obscured by the belief that all order depends upon distance—a belief which, though it is entertained by so excellent a writer as Meinong, we have seen to be false. Distance being essentially quantitative, to admit that series depend upon distance is to admit that order depends upon quantity. But this view leads at once to an endless regress, since distances have an order of magnitude, which would have to be derived from new distances of distances, and so on. And positively, an asymmetrical transitive relation suffices to generate a series, but does not imply distance. Hence the fact that the points of a line form a series does not show that Geometry must have metrical presuppositions, and no such presuppositions appear in the detail of projective or descriptive Geometry.

406. But although non-quantitative Geometry, as it now exists, is plainly independent of everything metrical, the historical development of the subject has tended greatly to obscure this independence. A brief historical review of the subject may be useful in showing the relation of the more modern to the more traditional methods.

In Euclid, and in Greek geometers generally, hardly any descriptive theorems are to be found. One of the earliest discoveries of an important descriptive theorem was the one named after Pascal*. Gradually it was found that propositions which assert points to be collinear or lines to be concurrent, or propositions concerning tangents, poles and polars, and similar matters, were unaltered by projection; that is, any such property belonging to a plane figure would belong also to the projection or shadow of this figure from any point on to any plane. All such properties (as, for instance, those common to all conics) were called projective or descriptive. Among these properties was anharmonic ratio, which was defined as follows. If A, B, C, D be four points on one straight line, their anharmonic ratio is $\dfrac{AB}{CB} \Big/ \dfrac{AD}{CD}$; if OA, OB, OC, OD be four lines through a point, their anharmonic ratio is $\dfrac{\sin AOB}{\sin COB} \Big/ \dfrac{\sin AOD}{\sin COD}$. In Chasles's great work on descriptive Geometry, and even in most recent works (such as Cremona's projective Geometry), this definition will be found at a very early stage in the development of the subject, together with a proof that anharmonic ratio is unaltered by projection. But such a definition is itself metrical, and cannot therefore be used to found a subject independent of metrical Geometry. With other portions of what used to be called descriptive or projective Geometry, the same lack of independence will be found. Consider, for example, the definition of a conic. To define it as a curve of the second degree would require projective coordinates, which there was no known method of introducing. To define it as a curve meeting any straight line in not more than two points would require the distinction of real and imaginary points, for if we confine ourselves to

---

* If a hexagon be inscribed in a conic, the three pairs of opposite sides intersect in collinear points.

real points there are innumerable curves other than conics which satisfy the definition. But imaginary points *are*, in ordinary metrical Geometry, imaginary coordinates, for which there is no purely geometrical interpretation; thus without projective coordinates, our definition again fails. To define a conic as the locus of points P for which the anharmonic ratio of PA, PB, PC, PD (where A, B, C, D are fixed points) is constant, again involves metrical considerations, so long as we have no projective definition of anharmonic ratio. And the same dependence upon metrical Geometry appears as regards any other projective or descriptive theorem, so long as the traditional order of ideas is adhered to.

The true founder of non-quantitative Geometry is von Staudt[*]. It was he who introduced the definition of a harmonic range by means of the quadrilateral construction, and who rendered it possible, by repetitions of this construction, to give projective definitions of all rational anharmonic ratios[†]. These definitions indicate the succession of quadrilateral constructions required in order to obtain a fourth point from three given points; thus, though they are essentially numerical, they have no reference whatever to quantity. But there remained one further step, before projective Geometry could be considered complete, and this step was taken by Pieri. In Klein's account, it remains doubtful whether *all* sets of four collinear points have an anharmonic ratio, and whether any meaning can be assigned to irrational anharmonic ratios. For this purpose, we require a method of generating order among *all* the points of a line. For, if there be no order but that obtained from Klein's method, there is no sense in which we can regard a point not obtained by that method as the limit of a series of points which are so obtained, since the limit and the series which it limits must always both belong to some one series. Hence there will be no way of assigning irrational coordinates to the points which do not have rational coordinates. There is, of course, no projective reason for supposing that there are such points; but there are metrical reasons, and in any case it is well, if possible, to be able to deal projectively with a continuous space. This is effected by Pieri, with the help of certain new axioms, but without any new indefinables. Thus at last the long process by which projective Geometry has purified itself from every metrical taint is completed.

407. Projective Geometry, having achieved its own independence, has, however, embarked upon a career of foreign aggrandisement; and in this we shall, I think, though on the whole favourable, be obliged to make some slight reservations. The so-called projective theory of distance aims at proving that metrical is merely a branch of projective

---

[*] *Geometrie der Lage*, Nürnberg, 1847; *Beiträge zur Geometrie der Lage, ib.* 1856, 1857, 1860.

[†] This step, I believe, is due to Klein. See *Math. Annalen*, Vols. IV, VI, XXXVII.

Geometry, and that distances are merely logarithms of certain anharmonic ratios. If this theory be correct, there is not a special subject of metrical Geometry, and the axioms by which, in the preceding chapter, we distinguished this subject, must be consequences of projective axioms. Let us examine the manner in which this result is obtained*.

We have already seen how to assign coordinates to every point of a line in projective space, and how to define the anharmonic ratio of any four points. We have seen also how to obtain a projective from a descriptive space. In a descriptive space, when an ideal point has a real correlative (*i.e.* when it is a sheaf of lines which has a vertex), we assign to the real point the coordinate which belongs to the ideal point considered as belonging to a projective space. In this way, the coordinate Geometry of the two spaces becomes very similar, the difference being that, in projective space, every real set of coordinates gives a real point, whereas, in descriptive space, this holds of each coordinate only within certain limits (both of which limits are excluded). In what follows, therefore, remarks concerning projective space will apply also to descriptive space except when the contrary is expressly stated.

Let us consider the anharmonic ratios of all ranges $axby$, where a, b are fixed points and x, y variable points on our line. Let α, ξ, β, η be the coordinates of these points. Then $\dfrac{\xi - a}{\xi - \beta} \Big/ \dfrac{\eta - a}{\eta - \beta}$ will be the anharmonic ratio of the four points, which, since α, β are constants, may be conveniently denoted by $(\xi\eta)$. If now ζ be the coordinate of any other point z, we have

$$(\zeta\eta)(\eta\zeta) = (\xi\zeta).$$

Hence $$\log(\xi\eta) + \log(\eta\zeta) = \log(\xi\zeta).$$

Thus the logarithm of the anharmonic ratio in question has one of the essential properties of distance, namely additiveness. If xy, yz, xz be the distances of x, y, z taken as having sign, we must have

$$xy + yz = xz.$$

We have also $\log(\xi\xi) = 0$ and $\log(\xi\eta) = -\log(\eta\xi)$, which are two further properties of distance. From these properties (of which the third follows from the other two) it is easy to show that all properties of distances which have no reference to the fixed points a, b belong to the logarithm in question. Hence, if the distances of points from a and b can also be made, by a suitable choice of a and b, to agree with those derived from the logarithm, we shall be able to identify distance with this logarithm. In this way—so it is contended—metrical Geometry may be wholly

---

* The projective theory of distance and angle is due to Cayley (*Sixth Memoir upon Quantics*, 1859) and to Klein (*Math. Annalen*, Vols. IV, VI, VII, XXXVII). A fuller discussion than the following will be found in my *Foundations of Geometry*, Cambridge, 1897, §§ 30–38.

brought under the projective sway; for a similar theory applies to angles between lines or planes.

408. Let us consider first the case where our projective points are the ideal points of a descriptive space. Let x be considered fixed, and distinct from a and b. Let y be moved so that η becomes more and more nearly equal to β. Then as η approaches β, $\log(\xi\eta)$ will be always finite, but will assume values exceeding any that may be assigned. This is mathematically expressed by saying that, if ξ be any number other than α and β, then $\log(\xi\beta)$ is infinite. (If ξ be equal to α or β, $\log(\xi\alpha)$ and $\log(\xi\beta)$ are indeterminate; this case will therefore be supposed excluded in what follows.) Hence a and b must be at an infinite distance from every point except each other; and their distance from each other is indeterminate. Again x and y must not be separated by a and b, *i.e.* y must belong to the segment axb, if we wish the distance to be real; for if $\xi - \alpha$ and $\xi - \beta$ have the same sign, $\eta - \alpha$ and $\eta - \beta$ must also have the same sign, but if $\xi - \alpha$ and $\xi - \beta$ have different signs, $\eta - \alpha$ and $\eta - \beta$ must also have different signs; and these conditions amount to the same as the condition that y must belong to the segment axb. Hence if we insist that any two real points (*i.e.* points which are not merely ideal) are to have a real distance (*i.e.* a distance measured by a number which is not complex or purely imaginary), we shall require a and b to fulfil the following conditions: (1) they must be ideal points to which no real ones correspond; (2) they must be the two limits of the series of those ideal points to which real points do correspond. These two conditions include all that has been said. For, in the first place, there is no real distance of any point from α or β; hence α and β must not be coordinates of real points. In the second place, on one of the two segments defined by a and b, there is a real distance xy however near ξ or η may approach to α or β; hence a and b are the limits of the ideal points to which real ones correspond. In the third place, it follows from the last proposition that all ideal points to which real ones correspond belong to one of the two segments ab, and all ideal points to which no real ones correspond (except a and b themselves) belong to the other of the two segments ab. When these conditions are satisfied, the function $\log(\xi\eta)$ will have all the properties which are required for a measure of distance.

The above theory is only applicable to descriptive space, for it is only there that we have a distinction between ideal and actual points. And in descriptive space we begin with an asymmetrical transitive relation by which order is generated on the straight line. Before developing a theory which is applicable to pure projective space, let us examine a little further the above theory, which may be called the *descriptive* theory of distance.

In the first place, the ideal points to which real ones correspond, which for shortness I shall call proper points, form part of the whole series of ideal points, which is closed. The proper points are a semi-continuous portion of this closed series, *i.e.* they have all the properties of a continuum except that of having two ends. It may happen that there is only one ideal point which is not proper, or it may happen that there are many. In the former case, the one purely ideal point will be the limit of the proper points in both directions. This is the case of Euclidean space, for in Euclidean space there is only one sheaf of lines to which a given line belongs and which has no vertex, namely the sheaf of lines parallel to the given line. Hence in this case the points a and b must be taken to be identical. The function $\log(\xi\eta)$ is then zero for all values of ξ and η, and is therefore useless as a measure of distance. But by a familiar process of proceeding to the limit, we can, in this case, obtain the value $\xi - \eta$ for the distance*. This is the usual measure of elementary Geometry; and for the distance of two points in a plane or in space we should similarly obtain the usual formula in this case. We see here the exact meaning of the common phrase that, in Euclidean space, $+\infty$ is the same as $-\infty$, or that the two ends of a line coincide. The fact is, of course, that the line has no ends, but that it determines only one ideal point which is not proper, and that this is the limit of proper ideal points in both directions: when it is added to the proper ideal points, we obtain a closed continuous series of sheaves to which the line in question belongs. In this way, a somewhat cryptic expression is found to have a very simple interpretation.

But it may happen also—and this is the case of hyperbolic space—that there are many improper ideal points on a line. In this case, the proper ideal points will have two different limits; these will be the sheaves of Lobatchewsky's parallels in the two directions. In this case, our function $\log(\xi\eta)$ requires no modification, but expresses distance as it stands. The ideal points a and b are distinct, which is commonly expressed by saying that our line has two real and distinct points at infinity.

Thus in descriptive space, in which our coordinates are obtained by correlation with those of the derived projective space, it is always possible to define a certain function of our projective coordinates which will fulfil the conditions required for a measure of distance. These conditions may be enumerated as follows†. (1) Every pair of real points is to have a distance whose measure is real and finite, and vanishes only when the two points coincide. (2) If x, y, z are collinear, and y lies between x and z, the sum of the measures of xy and yz is to be the measure of xz. (3) As the ideal point corresponding to y approaches

---

* See *e.g.* Klein, *Vorlesungen über nicht Euklidische Geometrie*, Göttingen, 1893, Vol. I, pp. 151 ff.

† Cf. Whitehead, *Universal Algebra*, Bk. VI, Chap. I. I confine myself in the text to distances on one straight line.

the ideal point which is the limit of proper ideal points, while x remains fixed, the absolute value of the measure of xy is to grow without limit.

It may well be asked, however, why we should desire to define a function of two variable points possessing these properties. If the mathematician replies that his only object is amusement, his procedure will be logically irreproachable, but extremely frivolous. He will, however, scarcely make this reply. We have, as a matter of fact, the notion of a stretch, and, in virtue of the general axiom that every class has some magnitude of divisibility, we know that the stretch has magnitude. But we do not know, without a special assumption to that effect, that the stretch fulfils the axioms of Archimedes and of linearity. When once these are assumed, the above properties of the measure of distance become properties which must belong to the measure of stretch. But if these two axioms are not assumed, there is no reason why there should be any magnitude having a measure possessing the above four characteristics. Thus the descriptive theory of distance, unless we regard it as purely frivolous, does not dispense with the need of the above axioms. What it does show—and this fact is extremely remarkable— is that, if stretches are numerically measurable, then they are measured by a constant multiple of the logarithm of the anharmonic ratio of the two ideal points associated with the ends of the stretch together with the two ideal points which limit the series of proper ideal points ; or, in case the latter pair are identical, the stretch is measured by a function obtained as the limit of the above when the said pair approach to identity and the constant factor increases without limit. This is a most curious result, but it does not obviate the need for the axioms which distinguish metrical Geometry. The same conclusion follows as regards metrical Geometry in a plane or in three dimensions ; but here new complications are introduced, which are irrelevant to the present issue, and will therefore not be discussed.

It is important to realize that the reference to two fixed ideal points, introduced by the descriptive theory of distance, has no analogue in the nature of distance or stretch itself. This reference is, in fact, a convenient device, but nothing more. The stretch, in descriptive space, is completely defined by its end-points, and in no way requires a reference to two further ideal points. And as descriptive Geometry starts with the stretch, it would be a needless complication to endeavour subsequently to obtain a definition of stretch in terms of four points. In short, even if we had a projective theory of distance in descriptive space, this would still be not purely projective, since the whole projective space composed of ideal elements is derived from axioms which do not hold in projective space.

409. It remains to examine the projective theory of distance in projective space. The theory we have hitherto examined, since it used the distinction of real and ideal elements, was descriptive, not projective; we have now to examine the corresponding theory for pure projective Geometry. Here there are no ideal elements of the above sort associated with our line; if, therefore, α and β be real and distinct numbers, they will be the coordinates of real and distinct points. Hence there will be real points x, y which will be separated by a and b, and will have an imaginary measure of distance. To this there could be no objection, but for the fact that we wish our measure to be the measure of a stretch. This is the reason why it is desired that any two real points should have a real measure of distance. In order to insure this result in a pure projective space, it is necessary that α and β should not be the co-ordinates of points at all, but should be conjugate complex numbers. It is further necessary that the constant multiple of the logarithm should be a pure imaginary. We then find that the distance of two real points always has a real measure, which is an inverse cosine*. In a projective space, the condition (2) of p. 424 introduces complications, since *between* has not, as in descriptive space, a simple meaning. The definition of *between* in this case is dealt with fully by Mr Whitehead in his *Universal Algebra* (§ 206).

410. But if such a function is to be properly geometrical, and to give a truly projective theory of distance, it will be necessary to find some geometrical entity to which our conjugate complex numbers α and β correspond. This can be done by means of involutions. Although, in a projective space, there are no ideal points, yet there are what may be called ideal point-pairs. In Chapter XLV we considered involutions with real double points: if a, b be two points on a line, all point-pairs x, x' such that x, x' are harmonic conjugates with respect to a, b form an involution. In this case, x and x' are said to be conjugate; a and b are each self-conjugate, and are called the double points of the involution. But there are also involutions without real double points. The general definition of an involution may be given as follows (substituting the relation of x to x' for the pair x, x'): An involution of points is a symmetrical one-one relation, other than identity, whose domain and converse domain are the same straight line, and which is such that any class of referents is projectively similar to the corresponding class of relata. Such a relation is either strictly an aliorelative, or is a self-relative as regards two and only two points, namely the double points of the involution. For every pair of distinct points on the line as double points there will be one and only one involution: *all* point-pairs (using this expression so as to exclude the identity of the two points of the pair) have a one-one correlation with *some* involutions. Thus involutions may be called ideal point-pairs: those that correspond

---

* This is the form originally given by Cayley in the Sixth Memoir upon Quantics. The simpler logarithmic form is due to Klein.

to an actual point-pair are called *hyperbolic*, the others *elliptic*. Thus an ideal point-pair is one and indivisible, being in fact a one-one relation. Two proper ideal point-pairs have an anharmonic ratio defined by their respective double points: two improper ideal point-pairs, or a proper and an improper ideal point-pair, have an analogous projective relation, which is measured by the function obtained as above from the supposition that α and β are conjugate complex numbers. This function may be called the anharmonic ratio of the two ideal point-pairs. If one be fixed and improper, the other variable and proper, an imaginary multiple of the logarithm of the resulting anharmonic ratio has the properties required for a measure of the distance of the actual point-pair corresponding to the proper ideal point-pair. This gives the pure projective theory of distance. But to this theory, as anything more than a technical development, there are the same objections as in the case of descriptive space; *i.e.* unless there be some magnitude determined by every actual point-pair, there is no reason for the process by which we obtain the above measure of distance; and if there is such a magnitude, then the above process gives merely the measure, not the definition, of the magnitude in question. Thus stretch or distance remains a fundamental entity, of which the properties are such that the above method gives a measure of it, but not a definition *.

411. There is however another and a simpler way of introducing metrical notions into a projective space, and in this way distance becomes a natural accompaniment of the introduction of coordinates. Let p, q, r be three fixed points, *abc* a line not passing through p or q or r but in the plane *pqr*. Let *qr* pass through a, *rp* through b, *pq* through c. Let R_1 be the relation which holds between x and y when these are points on *abc*, and *xr*, *yq* meet on *ap*; and let R_2, R_3 be similarly defined. Then a Möbius net may be regarded as constructed by repetitions of the relations R_1, R_2, R_3. We shall have, if xR_1y, yR_1z, then $xH_{ay}z$. We can define the square root of R_1, or any power of R_1 whose index is a positive or negative power of 2. Further, if s is any point of *qr*, and $xR_1'y$ means that x and y are on *abc* and *xr*, *ys* meet on *ap*, then $R_1R_1' = R_1'R_1$. From these propositions, which are proved by pure projective methods, it follows that if α and β be numbers, we may define $R_1^{\alpha+\beta}$ to mean $R_1^{\alpha}R_1^{\beta}$, provided R_1^{α} and R_1^{β} have been already defined; whence, since $R_1^{2^n}$ can be defined if n is a positive or negative integer, all rational powers of R_1 can be defined, and irrational powers can be defined as limits. Hence, if x be any real number, positive or negative, we can define R_1^x, for we may identify R_1^{-x} with $\tilde{R_1}^x$. We may now take this relation R_1^x as the *distance* of any two points between

* On the above method of introducing imaginaries in projective Geometry, see von Staudt, *Beiträge zur Geometrie der Lage*, 1, § 7.

which it holds, and regard x as the measure of the distance. We shall find that distances so defined have the usual properties of Euclidean distances, except that the distance of a from any other point is infinite. Thus on a projective line any two points do actually have a relation which may be called distance, and in this sense a projective theory of metrical properties can be justified. But I do not know whether this method can be extended to a plane or to space.

To sum up: Although the usual so-called projective theory of distance, both in descriptive and in projective space, is purely technical, yet such spaces do necessarily possess metrical properties, which can be defined and deduced without new indefinables or indemonstrables. But metrical Geometry, as an independent subject, requires the new idea of the magnitude of divisibility of a series, which is indefinable, and does not belong, properly speaking, to pure mathematics. This idea is applied to stretches, angles, areas, *etc.*, and it is assumed that all the magnitudes dealt with obey the axioms of Archimedes and linearity. Without these axioms, many of the usual metrical propositions cannot be proved in the usual metrical manner; with these axioms, the usual kind of elementary Geometry becomes possible, and such results as the uniqueness of the quadrilateral construction can be proved without three dimensions. Thus there is a genuinely distinct science of metrical Geometry, but, since it introduces a new indefinable, it does not belong to pure mathematics in the sense in which we have used the word in this work. It does not, as is often supposed, require distances and angles as new relations between points or lines or planes, but stretches and magnitudes of divisibility suffice throughout. On the other hand, projective and descriptive Geometry are both independent of all metrical assumptions, and allow the development of metrical properties out of themselves; hence, since these subjects belong to pure mathematics, the pure mathematician should adopt their theory of metrical matters. There is, it is true, another metrical Geometry, which does work with distances, defined as one-one relations having certain properties, and this subject is part of pure mathematics; but it is terribly complicated, and requires a bewildering number of axioms. Hence the deduction of metrical properties from the definition of a projective or descriptive space has real importance, and, in spite of appearances to the contrary, it affords, from the point of view of pure mathematics, a genuine simplification and unification of method.

CHAPTER XLIX.

DEFINITIONS OF VARIOUS SPACES.

412. In the preceding discussions of different Geometries, I have usually, for the sake of convenience, adhered to the distinction between definitions and indefinables on the one hand, and axioms or postulates on the other. But this distinction, in pure mathematics, has no validity except as regards the ideas and propositions of Logic. In pure mathematics, all the propositions state logical implications containing a variable. This is, in fact, the definition, or part of the definition, of pure mathematics. The implications stated must flow wholly from the propositions of Logic, which are prior to those of other branches of mathematics. Logic and the rest of pure mathematics are distinguished from applied mathematics by the fact that, in it, all the constants are definable in terms of some eight fundamental notions, which we agreed to call logical constants. What distinguishes other branches of mathematics from Logic is merely complication, which usually takes the form of a hypothesis that the variable belongs to some rather complicated class. Such a class will usually be denoted by a single symbol; and the statement that the class in question is to be represented by such and such a symbol is what mathematicians call a *definition*. That is to say, a definition is no part of mathematics at all, and does not make any statement concerning the entities dealt with by mathematics, but is simply and solely a statement of a symbolic abbreviation: it is a proposition concerning symbols, not concerning what is symbolized. I do not mean, of course, to affirm that the word *definition* has no other meaning, but only that this is its true mathematical meaning. All mathematics is built up by combinations of a certain number of primitive ideas, and all its propositions can, but for the length of the resulting formulae, be explicitly stated in terms of these primitive ideas; hence all definitions are theoretically superfluous. But further, when Logic is extended, as it should be, so as to include the general theory of relations, there are, I believe, no primitive ideas in mathematics except such as belong to the domain of Logic. In the previous chapters of this Part, I have spoken, as most authors do, of certain indefinables in Geometry.

But this was a concession, and must now be rectified. In mathematics, two classes of entities which have internal relations of the same logical type are equivalent. Hence we are never dealing with one particular class of entities, but with a whole class of classes, namely, with all classes having internal relations of some specified type. And by the *type* of a relation I mean its purely logical properties, such as are denoted by the words one-one, transitive, symmetrical, and so on. Thus for example we defined the class of classes called *progression* by certain logical characteristics of the internal relations of terms of any class which is a progression, and we found that finite Arithmetic, in so far as it deals with numbers, and not with the terms or classes of which numbers can be asserted, applies equally to all progressions. And when it is realized that all mathematical ideas, except those of Logic, can be defined, it is seen also that there are no primitive propositions in mathematics except those of Logic. The so-called axioms of Geometry, for example, when Geometry is considered as a branch of pure mathematics, are merely the protasis in the hypotheticals which constitute the science. They would be primitive propositions if, as in applied mathematics, they were themselves asserted; but so long as we only assert hypotheticals (*i.e.* propositions of the form " *A* implies *B* ") in which the supposed axioms appear as protasis, there is no reason to assert the protasis, nor, consequently, to admit genuine axioms. My object in the present chapter is to execute the purely formal task imposed by these considerations, and to set forth the strict definitions of various spaces, from which, without indefinables and without primitive propositions, the various Geometries will follow. I shall content myself with the definition of some of the more important spaces, since my object is chiefly to show that such definitions are possible.

413. (1) *Projective Space of three dimensions.* A projective space of three dimensions is any class of entities such that there are at least two members of the class; between any two distinct members there is one and only one symmetrical aliorelative, which is connected, and is transitive so far as its being an aliorelative will permit, and has further properties to be enumerated shortly; whatever such aliorelative may be taken, there is a term of the projective space not belonging to the field of the said aliorelative, which field is wholly contained in the projective space, and is called, for shortness, a *straight line*, and is denoted by ab, if a, b be any two of its terms; every straight line which contains two terms contains at least one other term; if a, b, c be any three terms of the projective space, such that c does not belong to the class ab, then there is at least one term of the projective space not belonging to any class cx, where x is any term of ab; under the same circumstances, if a' be a term of bc, b' a term of ac, the classes aa', bb' have a common part; if d be any term, other than a and b, of the class ab, and u, v any two terms such that d belongs to the class uv, but neither u nor v belongs to the class ab, and if y be the only term of the common part of au and bv, z the only term of the common part of av and bu, x the only term of the common part of yz and ab, then x is not identical with d (under these circumstances it may be proved that the term x is independent of u and v, and is uniquely determined by a, b, d; hence x and d have a symmetrical one-one relation which may be denoted, for brevity, by $xH_{ab}d$; if y, e be two further terms of the projective space, belonging to the class xd, and such that there are two terms g, h of the class xd for which we have $gH_{xd}h$ and $gH_{ye}h$, then we write for shortness $yQ_{xd}e$ to express this relation of the four terms x, d, y, e); a projective space is such that the relation Q_{xd}, whatever terms of the space x and d may be, is transitive; also that, if a, b, c, d be any four distinct terms of one straight line, two and only two of the propositions $aQ_{bc}d$, $aQ_{bd}c$, $aQ_{cd}b$ will hold; from these properties of projective space it results that the terms of a line form a series; this series is continuous in the sense defined in § 277; finally, if a, b, c, d, e be any five terms of a projective space, there will be in the class ae at least one term x, and in the class cd at least one term y, such that x belongs to the class by.

This is a formal definition of a projective space of three dimensions. Whatever class of entities fulfils this definition is a projective space. I have enclosed in brackets a passage in which no new properties of projective space are introduced, which serves only the purpose of convenience of language. There is a whole class of projective spaces, and this class has an infinite number of members. The existence-theorem may be proved to begin with, by constructing a projective space out of complex numbers in the purely arithmetical sense defined in § 360. We then know that the class of projective spaces has at least four members, since we know of four sub-classes contained under it, each of which has at least one member. In the first place, we have the above arithmetical space. In the second place, we have the projective space of descriptive Geometry, in which the terms of the projective space are sheaves of lines in the descriptive space. In the third place, we have the polar form of elliptic space, which is distinguished by the addition of certain metrical properties of stretches, consistent with, but not implied by, the definition of projective space; in the fourth place, we have the antipodal form of elliptic Geometry, in which the terms of the projective space are pairs of terms of the said elliptic space. And any number of varieties of projective space may be obtained by adding properties not inconsistent with the definition—for example, by insisting that all planes are to be red or blue. In fact, every class of 2^{a_0} terms (*i.e.* of the number of terms in a continuous series) is a projective space; for when two classes are similar, if one is the field of a certain relation, the other will be the field of a like relation. Hence by correlation with a projective space, any class of 2^{a_0} terms becomes itself a projective space. The fact is, that the standpoint of line-Geometry is more fundamental where definition is concerned : a projective space would be best defined as a class K of relations whose fields are straight lines satisfying the above conditions. This point is strictly analogous to the substitution of serial relations for series which we found desirable in Part IV. When a set of terms are to be regarded as the field of a class of relations, it is convenient to drop the terms and mention only the class of relations, since the latter involve the former, but not the former the latter.

It is important to observe that the definition of a space, as of most other entities of a certain complexity, is arbitrary within certain limits. For if there be any property which implies and is implied by one or more of the properties used in the definition, we may make a substitution of the new property in place of the one or more in question. For example, in place of defining the line by a relation between points, it is possible to define the line as a class having a certain relation to a couple of points. In such cases, we can only be guided by motives of simplicity.

It seems scarcely necessary to give a formal definition of descriptive or metrical space, since the above model serves to show how such a definition might be constructed. I shall instead give a definition of Euclidean space. This I shall give in a form which is inappropriate when Euclidean space is considered as the limit of certain non-Euclidean spaces, but is very appropriate to quaternions and the vector Calculus. This form has been adopted by Peano*, and leads to a very simple account of the Euclidean axioms. I shall not strictly follow Peano, but my account will be very similar to his.

414. (2) *Euclidean space of three dimensions.* A Euclidean space of three dimensions is a class of terms containing at least two members, and such that any two of them have one and only one asymmetrical one-one relation of a class, which will be called the class of vectors, defined by the following characteristics† : the converse of a vector, or the relative product of two vectors, is a vector; if a given vector holds between a and b, c and d, then the vector which holds between a and c is the same as that which holds between b and d; any term of the space has any assigned relation of the class to at least one term of the space ; if the nth power (where n is any integer) of any vector of the class is identity, then the vector itself is identity ; there is a vector whose nth power is a given vector ; any two vectors have one and only one symmetrical relation of a certain class having the following properties : the relation of any two vectors is measured by a real number, positive or negative, and is such

---

* "Analisi della Teoria dei vettori," Turin, 1898 (*Accademia Reale delle Scienze di Torino*).

† For the convenience of the reader, it may be well to observe that this relation corresponds to that of having a given distance in a given direction—direction being taken in the sense in which all parallel lines have the same direction.

that the relation of a vector to itself is always measured by a positive number, and that the measure of the relation of the relative product of two vectors to a third vector is the sum of the measures of their several relations to the third vector; there is a vector satisfying the definition of an irrational power of a vector given below; there are vectors which are not relative products of powers of two given vectors; if i, j, k be three vectors, no one of which is a relative product of powers of one or both of the others, then all vectors are relative products of powers of i, j, k.

The only points calling for explanation here are the notion of an irrational power of a vector and the measurable relation of two vectors. All rational powers are definite; for every vector has an nth root, and the nth root has an mth power, which is the m/nth power of the original vector   But it does not follow that real powers which are not rational can be defined. The definition of limits of classes of vectors given by Peano* is, when translated into relational language, the following. Let u be a class of real numbers, x_0 a number belonging to the derivative of u. Let some one-one relation subsist between all u's and some or all vectors; and let v be the class of vectors correlative to u. Then the vector a is said to be the limit of the class v as x approaches x_0 in the class u, when the limit of the measure of the relation to itself of the vector which, multiplied relatively into a, will give the correlate to x in the class v, is zero. The point of this definition is the use of the order obtained among vectors by means of the measurable relation which each has to itself. Thus suppose we have a progression x_1, x_2, ... x_n, ... of rational numbers, and suppose these to be respectively the measures of the relations to themselves of the vectors a_1, a_2, ... a_n, ... Then if x be the limit of x_1, x_2, ... x_n, ..., there is to be a vector whose relation to itself is measured by x, and this is to be the limit of the vectors a_1, a_2, ... a_n, ...; and thus irrational powers of a vector become definable. The other point to be examined is the measurable relation between two vectors. This relation measures, in terms of elementary Geometry, the product of the two stretches represented by the vectors into the cosine of the angle between them: it is, in the language of the calculus of extension, the internal product of the two vectors. To say that the relation is measurable in terms of real numbers means, in the sense in which this statement is employed, that all such relations have a one-one relation to some or all of the real numbers; hence, from the existence of irrational powers, it follows that all such relations form a continuous series; to say that the relation of a vector to itself is always measured by a positive number means that there exists a section (in Dedekind's sense) of the continuous series of relations, such that all those relations that vectors can have to themselves appear on one side of the section :

* *Op. cit.* p. 22.

while it can be proved that the relation which defines the section is that which the vector identity has to itself.

This definition is, of course, by no means the only one which can be given of Euclidean space, but it is, I think, the simplest. For this reason, and also because it belongs to an order of ideas which, being essentially Euclidean, is foreign to the methods of previous chapters, I have thought it worth while to insert it here.

415. As another example which may serve to enlarge our ideas, I shall take the space invented by Clifford, or rather the space which is formally analogous to his surface of zero curvature and finite extent *. I shall first briefly explain the nature of this space, and then proceed to a formal definition. Spaces of the type in question may have any number of dimensions, but for the sake of simplicity I shall confine myself to two dimensions. In this space, most of the usual Euclidean properties hold as regards figures not exceeding a certain size; that is to say, the sum of the angles of a triangle is two right angles, and there are motions, which may be called translations, in which all points travel along straight lines. But in other respects, the space is very different from Euclidean space. To begin with, the straight line is a closed series, and the whole space has a finite area. In the second place, every motion is a translation; a circular transformation (*i.e.* one which preserves distances from a certain fixed point unaltered) is never a motion, *i.e.* never leaves every distance unaltered; but all translations can, as in Euclidean space, be compounded out of translations in two fixed directions. In this space, as in Euclid, we have parallels, *i.e.* straight lines which remain at a constant distance apart, and can be simultaneously described in a motion; also straight lines can be represented by linear equations. But the formula for distance is quite unlike the Euclidean formula. Thus if πk be the length of the whole straight line, and (x, y), (x', y') be the coordinates of any two points (choosing a system in which the straight line has a linear equation), then if ω be the angle between the lines $x = 0$, $y = 0$, the distance of the two points in question is d, where

$$\cos \frac{d}{k} = \cos (x - x') \cos (y - y') - \cos \omega \sin (x - x') \sin (y - y'),$$

and the formula for the angle between two lines is similarly complicated. We may, in order to lead to these results, set up the following definition.

(3) *Clifford's space of two dimensions.* A Clifford's space of two dimensions is a class of at least two terms, between any two of which there are two relations of different classes, called respectively distance

---

* On the general subject of the spaces of which this is the simplest example, see Klein, *Math. Annalen* xxxvii, pp. 554–565, and Killing, *Grundlagen der Geometrie*, Vol. i, Chap. iv.

and direction, and possessing the following properties: a *direction* is a symmetrical aliorelative, transitive so far as its being an aliorelative will permit, but not connected; a term of the space together with all the terms to which the said term has a given relation of direction form what is called a straight line; no straight line contains all the terms of the space; every term of the space has any assigned relation of direction to some but not all other terms of the space; no pair of terms has more than one relation of direction; *distances* are a class of symmetrical relations forming a continuous series, having two ends, one of which is identity; all distances except identity are intransitive aliorelatives; every term of the space has any assigned relation of distance to some but not all of the terms of the space; any given term of the space has any given distance and direction from two and only two other terms of the space, unless the given distance be either end of the series of distances; in this case, if the given distance be identity, there is no term having this distance and also the given direction from the given term, but if the distance be the other end of the series, there is one and only one term having the given distance and the given direction from the given term; distances in one straight line have the properties, mentioned in Chapter XLVII, required for generating an order among the terms of one straight line; the only motions, *i.e.* one-one relations whose domain and converse domain are each the space in question and which leave all distances among the relata the same as those among the corresponding referents, are such as consist in combining a given distance, a given direction, and one of the two senses of the series constituting a straight line; and every such combination is equivalent to the relative product of some distance in one fixed direction with some distance in another fixed direction, both taken with a suitable sense; finally all possible directions form a single closed continuous series in virtue of mutual relations.

This completes, I think, the definition of a Clifford's space of two dimensions. It is to be observed that, in this space, distance cannot be identified with stretch, because (1) we have only two dimensions, so that we cannot generate a closed series of terms on a line by means of projective methods*, (2) the line is to be closed, so that we cannot generate order on the straight line by the descriptive method. It is for similar reasons that both directions and distances have to be taken as symmetrical relations; thus it is only after an order has been generated on a line that we can distinguish two senses, which may be associated with direction to render it asymmetrical, and with distances in a given direction to give them signs. It is important to observe that, when

---

* Mr W. E. Johnson has pointed out to me that this difficulty might be overcome by introducing the uniqueness of the quadrilateral construction by a special axiom—a method which would perhaps be simpler than the above.

distance is taken as independent of the straight line, it becomes necessary, in order to distinguish different spaces, to assign some property or properties of the one-one relations or transformations which leave distances unchanged. This method has been adopted by Lie in applying to Geometry the theory of continuous groups *, and has produced, in his hands and those of Klein, results of the greatest interest to non-Euclidean Geometry. But since, in most spaces, it is unnecessary to take distance as indefinable, I have been able, except in this instance of Clifford's space†, to adopt a simpler method of specifying spaces. For this reason, it was important to consider briefly some such space as Clifford's, in order to give an instance of the use of distance, and of what geometers call motion, in the definition of a space.

Enough has now been said, I hope, to show that the definition of a kind of space is always possible in purely logical terms, and that new indefinables are not required. Not only are the actual terms composing a space irrelevant, and only their relations important, but even the relations do not require individual determination, but only specification as members of certain logical classes of relations. These logical classes are the elements used in geometrical definitions, and these are definable in terms of the small collection of indefinables out of which the logical calculus (including that of relations) is built up. This result, which holds throughout pure mathematics, was the principal object of the present chapter.

---

* *Leipziger Berichte*, 1890.

† If I had defined an elliptic space of two dimensions, I should have had to take distance as distinct from stretch, because the projective generation of order fails in two dimensions.

CHAPTER L.

THE CONTINUITY OF SPACE.

416. It has been commonly supposed by philosophers that the continuity of space was something incapable of further analysis, to be regarded as a mystery, not critically inspected by the profane intellect. In Part V, I asserted that Cantor's continuity is all that we require in dealing with space. In the present chapter, I wish to make good this assertion, in so far as is possible without raising the question of absolute and relative position, which I reserve for the next chapter.

Let us begin with the continuity of projective space. We have seen that the points of descriptive space are ordinally similar to those·of a semi-continuous portion of a projective space, namely to the ideal points which have real correlatives. Hence the continuity of descriptive space is of the same kind as that of projective space, and need not, therefore, be separately considered. But metrical space will require a new discussion.

It is to be observed that Geometries, as they are treated now-a-days, do not begin by assuming spaces with an infinite number of points; in fact, space is, as Peano remarks*, a word with which Geometry can very easily dispense. Geometries begin by assuming a class-concept *point*, together with certain axioms from which conclusions can be drawn as to the number of points. So, in projective Geometry, we begin with the assumption that there are at least two points, and that any two points determine a class of points, the straight line, to which they and at least one other point belong. Hence we have three points. We now introduce the new assumption that there is at least one point not on any given straight line. This gives us a fourth point, and since there must be points on the lines joining it to our previous points, we obtain three more points—seven in all. Hence we can obtain an infinite denumerable series of points and lines, but we cannot, without a further assumption, prove that there are more than three points on any one line. Four points on a line result from the assumption that, if b and d

* *Riv. di Mat.* Vol. iv, p. 52.

be harmonic with respect to *a* and *c*, then *b* and *d* are distinct. But in order to obtain an infinite number of points on a line, we need the further assumptions from which the projective order results*. These assumptions necessitate a denumerable series of points on our line. With these, if we chose, we might be content. Such a series of points is obtained by successive quadrilateral constructions; and if we chose to define a space in which all points on a line could be obtained by successive quadrilateral constructions starting with any three points of the line, no contradiction would emerge. Such a space would have the ordinal type of the positive rationals and zero: the points on a line would form a compact denumerable series with one end. The extension, introduced by assuming that the series of points is continuous, is only necessary if our projective space is to possess the usual metrical properties—if, that is to say, there is to be a stretch, with one end and its straight line given, which is to be equal to any given stretch. With only rational points, this property (which is Euclid's postulate of the existence of the circle) cannot hold universally. But for pure projective purposes, it is irrelevant whether our space possesses or does not possess this property. The axiom of continuity itself may be stated in either of the two following forms. (1) All points on a line are limits of series of rational points, and all infinite series of rational points have limits; (2) if all points of a line be divided into two ,classes, of which one wholly precedes the other, then either the first class has a last term, or the last has a first term, but both do not happen. In the first of these ways, the continuity which results is exactly Cantor's, but the second, which is Dedekind's definition, is a necessary, not a sufficient, condition for Cantor's continuity. Adopting this first definition, the rational points, omitting their first term, form an endless compact denumerable series; all points form a perfect series; and between any two points there is a rational point, which is precisely the ordinal definition of continuity†. Thus if a projective space is to have continuity at all, it must have the kind of continuity which belongs to the real numbers.

417. Let us consider next the continuity of a metrical space; and, for the sake of definiteness, let us take Euclidean space. The question is here more difficult, for continuity is not usually introduced by an axiom *ad hoc*, but appears to result, in some sense, from the axioms of distance. It was already known to Plato that not all lengths are commensurable, and a strict proof of this fact is contained in the tenth book of Euclid. But this does not take us very far in the direction of Cantor's continuity. The gist of the assertion that not all lengths are commensurable, together with the postulate of the circle, may be expressed as follows. If *AB*, *AC* be two lengths along the same straight

---

* Cf. Pieri, *op. cit.* § 6, Prop. 1.   † See Part V, Chap. xxxvi.

line, it may happen that, if AB be divided into m equal parts, and AC into n equal parts, then, however m and n may be chosen, one of the parts of AB will not be equal to one of the parts of AC, but will be greater for some values of m and n, and less for others; also lengths equal to either may be taken along any given line and with any given end-points*. But this fact by no means proves that the points on a line are not denumerable, since all algebraic numbers are denumerable. Let us see, then, what our axioms allow us to infer.

In Greek Geometry there were two great sources of irrationals, namely, the diagonal of a square and the circumference of a circle. But there could be no knowledge that these are irrationals of different kinds, the one being measured by an algebraic number, the other by a transcendent number. No general method was known for constructing any assigned algebraic number†, still less for constructing an assigned transcendent number. And so far as I know, such methods, except by means of limits, are still wanting. Some algebraic and some transcendent numbers can be constructed geometrically without the use of limits, but the constructions are isolated, and do not follow any general plan. Hence, for the present, it cannot be inferred from Euclid's axioms that space has continuity in Cantor's sense, or that the points of space are not denumerable. Since the introduction of analytic Geometry, some equivalent assumption has been always tacitly made. For example, it has been assumed that any equation which is satisfied by real values of the variables will represent a figure in space ; and it seems even to be universally supposed that to every set of real Cartesian coordinates a point must correspond. These assumptions were made, until quite recent times, without any discussion at all, and apparently without any consciousness that they were assumptions.

When once these assumptions are recognized as such, it becomes apparent that, here as in projective space, continuity must be introduced by an axiom *ad hoc*. But as against the philosophers, we may make the following remark. Cantor's continuity is indubitably *sufficient* to satisfy all metrical axioms, and the only question is, whether existent space need have continuity of so high an order. In any case, if measurement is to be theoretically possible, space must not have a *greater* continuity than that of the real numbers.

The axiom that the points on a line form a continuous series may be put in the form which results from amending Dedekind, or in the form that a line is a perfect series. In the first form, every section of the line is definable by a single point, which is at one end of one of the parts produced by the section, while the other part has no end. In the second

---

* A length is not synonymous with a segment, since a length is regarded as essentially terminated. But a length is, for present purposes, synonymous with a stretch or a distance.

† For shortness, I shall identify numbers with the lengths which they measure.

form, which is preferable because, unlike the first, it completely defines
the ordinal type, every infinite series of points has a limit, and every point
is a limiting point. It is not necessary to add that the line has cohesion*,
for this results from the axioms of Archimedes and of linearity, which
are in any case essential to measurement. Whether the axiom of continuity be true as regards our actual space, is a question which I see no
means of deciding. For any such question must be empirical, and it
would be quite impossible to distinguish empirically what may be called
a rational space from a continuous space. But in any case there is
no reason to think that space has a higher power than that of the
continuum.

418. The axiom of continuity enables us to dispense with the
postulate of the circle, and to substitute for it the following pair.
(1) On any straight line there is a point whose distance from a given
point on the line is less than a given distance. (2) On any straight
line there is a point whose distance from a given point on or off the line
is greater than a given distance. From these two assumptions, together
with continuity, the existence of the circle can be proved. Since it is
not possible, conversely, to deduce continuity from the circle, and since
much of analytic Geometry might be false in a discontinuous space, it
seems a distinct advance to banish the circle from our initial assumptions,
and substitute continuity with the above pair of axioms.

419. There is thus no mystery in the continuity of space, and no
need of any notions not definable in Arithmetic. There is, however,
among most philosophers, a notion that, in space, the whole is prior to
the parts†; that although every length, area, or volume can be divided
into lengths, areas, or volumes, yet there are no indivisibles of which
such entities are composed. According to this view, points are mere
fictions, and only volumes are genuine entities. Volumes are not to be
regarded as classes of points, but as wholes containing parts which are
never simple. Some such view as this is, indeed, often put forward
as giving the very essence of what should be called continuity. This
question is distinct from the question of absolute and relative position,
which I shall discuss in the following chapter. For, if we regard
position as relative, our present question will arise again concerning
continuous portions of matter. This present question is, in fact,
essentially concerned with continuity, and may therefore be appro-
priately discussed here.

The series which arise in Arithmetic, whether continuous or not,
are essentially composed of terms—integers, rationals, real numbers, etc.
And where we come near to the continuity of space, as in the case of
the real numbers, each real number is a segment or infinite class

---

* See Part V, Chap. xxxv.

† Cf. Leibniz, *Phil. Werke* (Gerhardt), ii, p. 379 ; iv, p. 491 ; also my *Philosophy
of Leibniz*, Chap. ix.

of rationals, and no denial that a segment is composed of elements is possible. In this case, we start from the elements and gradually construct various infinite wholes. But in the case of space, we are told, it is infinite wholes that are given to begin with; the elements are only inferred, and the inference, we are assured, is very rash. This question is in the main one of Logic. Let us see how the above view is supported.

Those who deny indivisible points as constituents of space have had, in the past, two lines of argument by which to maintain their denial. They had the difficulties of continuity and infinity, and they had the way in which space is presented in what, according to their school, they called intuition or sensation or perception. The difficulties of continuity and infinity, as we saw in Part V, are a thing of the past; hence this line of argument is no longer open to those who deny points. As regards the other argument, it is extremely difficult to give it a precise form—indeed I suspect that it is impossible. We may take it as agreed that everything spatial, of whose existence we become immediately aware in sensation or intuition, is complex and divisible. Thus the empirical premiss, in the investigation of space, is the existence of divisible entities with certain properties. But here it may be well to make a little digression into the meaning of an empirical premiss.

420. An empirical premiss is a proposition which, for some reason or for no reason, I believe, and which, we may add, is existential. Having agreed to accept this proposition, we shall usually find, on examination, that it is complex, and that there are one or more sets of simpler propositions from which it may be deduced. If P be the empirical premiss, let A be the class of sets of propositions (in their simplest form) from which P may be deduced; and let two members of the class A be considered equivalent when they imply one another. From the truth of P we infer the truth of one set of the class A. If A has only one member, that member must be true. But if there are many members of the class A, not all equivalent, we endeavour to find some other empirical premiss P', implied by all sets of simple propositions of the class A'. If now it should happen that the classes A and A' have only one common member, and the other members of A are inconsistent with the other members of A', the common member must be true. If not, we seek a new empirical premiss P'', and so on. This is the essence of induction*. The empirical premiss is not in any essential sense a premiss, but is a proposition which we wish our deduction to arrive at. In choosing the premisses of our deduction, we are only guided by logical simplicity and the deducibility of our empirical premiss.

421. Applying these remarks to Geometry, we see that the common desire for self-evident axioms is entirely mistaken. This desire is due to

---

* Cf. Couturat, *La Logique de Leibniz*, Paris, 1901, p. 270.

the belief that the Geometry of our actual space is an *à priori* science, based on intuition. If this were the case, it would be properly deducible from self-evident axioms, as Kant believed. But if we place it along with other sciences concerning what exists, as an empirical study based upon observation, we see that all that can be legitimately demanded is that observed facts should follow from our premisses, and, if possible, from no set of premisses not equivalent to those which we assume. No one objects to the law of gravitation as being not self-evident, and similarly, when Geometry is taken as empirical, no one can legitimately object to the axiom of parallels—except, of course, on the ground that, like the law of gravitation, it need only be approximately true in order to yield observed facts. It cannot be maintained that no premisses except those of Euclidean Geometry will yield observed results; but others which are permissible must closely approximate to the Euclidean premisses. And so it is with continuity: we cannot prove that our actual space must be continuous, but we cannot prove that it is not so, and we can prove that a continuous space would not differ in any discoverable manner from that in which we live.

422. To return from this digression : we agreed that the *empirical* premisses, as regards the continuity of space, are concerned always with divisible entities which have divisible parts. The question before us is whether we are to infer from this that the *logical* premisses for the science of existing space (*i.e.* the definition of existing space) may or must be concerned with divisible entities. The question whether our premisses *must* be concerned with divisible entities is fully answered, in the negative, by actual Geometry, where, by means of indivisible points, a space empirically indistinguishable from that in which we live is constructed. The only reasons hitherto alleged by philosophers against regarding this answer as satisfactory, are either such as were derived from the difficulties of infinity and continuity, or such as were based upon a certain logical theory of relations. The former have been already disproved; the latter will be discussed in the next chapter. The question whether our premisses *may* be concerned with divisible entities is far more difficult, and can be answered only by means of the logical discussions of Part II. Whatever is complex, we then decided (§ 143), must be composed of simple elements ; and this conclusion carries us a long way towards the decision of our present question. But it does not quite end our doubts. We distinguished, in Part II, two kinds of wholes, namely *aggregates* and *unities*. The former may be identified, at any rate for present purposes, with classes, while the latter seem to be indistinguishable from propositions. Aggregates consist of units from whose addition (in the sense presupposed in Arithmetic) they result ; unities, on the contrary, are not reconstituted by the addition of their constituents. In all unities, one term at least is either a predicated predicate or a relating relation ; in aggregates, there is no such term.

Now what is really maintained by those who deny that space is composed of points is, I imagine, the view that space is a unity, whose constituents do not reconstitute it. I do not mean to say that this view is consciously held by all who make the denial in question, but that it seems the only view which renders the said denial reasonable.

Before discussing this opinion, it is necessary to make a distinction. An aggregate may be an aggregate of unities, and need by no means be an aggregate of simple terms. The question whether a space is an aggregate of unities or of simple terms is mathematically, though not philosophically, irrelevant; the difference of the two cases is illustrated by the difference between an independent projective space and the projective space defined in terms of the elements of a descriptive space. For the present, I do not wish to discuss whether points are unities or simple terms, but whether space is or is not an aggregate of points.

This question is one in which confusions are very liable to occur, and have, I think, actually occurred among those who have denied that a space is an aggregate. Relations are, of course, quite essential to a space, and this has led to the belief that a space *is*, not only its terms, but also the relations relating them. Here, however, it is easy to see that, if a space be the field of a certain class of relations, then a space is an aggregate; and if relations are essential to the definition of a space, there must be some class of relations having a field which is the space. The relations essential to Geometry will not hold between two spatially divisible terms: there is no straight line joining two volumes, and no distance between two surfaces. Thus, if a space is to be defined by means of a class of relations, it does not follow, as is suggested, that a space is a unity, but rather, on the contrary, that it is an aggregate, namely the field of the said class of relations. And against any view which starts from volumes or surfaces, or indeed anything except points and straight lines, we may urge, with Peano*, that the distinction between curves, surfaces, and volumes, is only to be effected by means of the straight line, and requires, even then, the most elaborate developments†. There is, therefore, no possibility of any definite Geometry without points, no logical reason against points, and strong logical reasons in their favour. We may therefore take it as proved that, if we are to construct any self-consistent theory of space, we must hold space to be an aggregate of points, and not a unity which is indefinable as a class. Space is, in fact, essentially a class, since it cannot be defined by enumeration of its terms, but only by means of its relation to the class-concept *point*. Space is nothing but the

---

* *Riv. di Mat.* iv, p. 53.

† Cf. Peano, "Sur une courbe qui remplit toute une aire plane," *Math. Annalen*, xxxvi, where it is shown that a continuous curve can be made to pass through all the points of the area of a square, or, for that matter, of the volume of a cube.

extension of the concept *point*, as the British army is the extension of the concept *British soldier*; only, since the number of points is infinite, Geometry is unable to imitate the Army-List by the issue of a Space-List.

Space, then, is composed of points; and if analytical Geometry is to be possible, the number of points must be either equal to, or less than, the number of the continuum. If the number be less, some propositions of the accepted Geometry will be false; but a space in which the number of points is equal to the number of finite numbers, and in which the points of a line form a series ordinally similar to the rationals, will, with suitable axioms, be empirically indistinguishable from a continuous space, and may be actual. Thus Arithmetic, as enlarged by Cantor, is undoubtedly adequate to deal with Geometry; the only question is, whether the more elaborate parts of its machinery are required. It is in number that we become certain of the continuum; among actual existents, so far as present evidence shows, continuity is possible, but cannot be rendered certain and indubitable.

CHAPTER LI.

LOGICAL ARGUMENTS AGAINST POINTS.

423. It has been an almost universal opinion among philosophers, ever since the time of Leibniz, that a space composed of points is logically impossible. It is maintained that the spatial relations with which we have been concerned do not hold between spatial points, which essentially and timelessly have the relations which they do have, but between material points, which are capable of motion, *i.e.* of a change in their spatial relations. This is called the theory of relative position, whereas the theory of spatial points is called the theory of absolute position. Those who advocate relative position usually also maintain that matter and spatial relations, on account of certain contradictions supposed to be found in them, are not real, but belong only to the world of appearance. This is, however, a further point, which need not be explicitly discussed in what follows. Apart from this point, the issue between the absolute and relative theories may be stated as follows: The absolute theory holds that there are true propositions in which spatial relations are asserted to hold timelessly between certain terms, which may be called spatial points; the relational theory holds that every true proposition asserting a spatial relation involves a time at which this relation holds between its terms, so that the simplest spatial propositions assert triangular relations of a time and two terms, which may be called material points.

The question as to which of these two theories applies to the actual world is, like all questions concerning the actual world, in itself irrelevant to pure mathematics*. But the argument against absolute position usually takes the form of maintaining that a space composed of points is logically inadmissible, and hence issues are raised which a philosophy of mathematics must discuss. In what follows, I am concerned only with the question: Is a space composed of points self-contradictory? It is true that, if this question be answered in the negative, the sole ground

---

* Some arguments on this point will be found in the earlier part of my paper, "Is position in Time and Space absolute or relative?" *Mind*, N.S., No. 39; the later portions of this paper are here reprinted.

for denying that such a space exists in the actual world is removed; but this is a further point, which, being irrelevant to our subject, will be left entirely to the sagacity of the reader.

424. The arguments against the absolute theory are, in my opinion, one and all fallacious. They are best collected in Lotze's *Metaphysic* (§ 108 ff.). They are there confused with arguments for the subjectivity of space—an entirely distinct question, as should have been evident from the fact that Kant, in the *Critique*, appears to have advocated the theory of absolute position *. Omitting arguments only bearing on this latter point, we have the following summary of Lotze's arguments against absolute space.

(1) Relations only are either (α) as presentations in a relating consciousness, or (β) as internal states in the real elements which are said to stand in these relations (§ 109).

(2) The being of empty space is neither the being which works effects (which belongs to a thing), nor the mere validity of a truth, nor the fact of being presented by us. What kind of being is it then? (§ 109).

(3) All points are exactly alike, yet every pair have a relation peculiar to themselves; but being exactly like every other pair, the relation should be the same for all pairs (§ 111).

(4) The being of every point must consist in the fact that it distinguishes itself from every other, and takes up an invariable position relatively to every other. Hence the being of space consists in an active mutual conditioning of its various points, which is really an interaction (§ 110).

(5) If the relations of points were a mere fact, they could be altered, at least in thought; but this is impossible: we cannot move points or imagine holes in space. This impossibility is easily explained by a subjective theory (§ 110).

(6) If there are real points, either (*a*) one point creates others in appropriate relations to itself, or (β) it brings already existing points into appropriate relations, which are indifferent to their natures (§ 111).

425. (1) All these arguments depend, at bottom, upon the first, the dogma concerning relations. As it is of the essence of the absolute theory to deny this dogma, I shall begin by examining it at some length†. "All relations," Lotze tells us, "only *are* as presentations in a relating consciousness, or as internal states in the real elements which, as we are wont to say, stand in these relations." This dogma Lotze regards as self-evident, as indeed he well may; for I doubt if there is one anterior philosopher, unless it be Plato, who does not, consciously or

---

* Cf. Vaihinger, *Commentar*, pp. 189–190.

† The logical opinions which follow are in the main due to Mr G. E. Moore, to whom I owe also my first perception of the difficulties in the relational theory of space and time.

unconsciously, employ the dogma as an essential part of his system. To deny it, therefore, is a somewhat hardy undertaking. Let us, nevertheless, examine the consequences to which the dogma leads us.

It would seem that, if we accept the dogma, we must distinguish two kind of relations, (a) those which are presentations in a relating consciousness, and (β) those which are internal states of the elements supposed to be related. These may be ultimately identical, but it will be safer in the mean time to treat them as different. Let us begin with those which are only presentations in a relating consciousness. These presentations, we must suppose, are beliefs in propositions asserting relations between the terms which appear related. For it must be allowed that there are beliefs in such propositions, and only such beliefs seem capable of being regarded as presentations in which relations have their being. But these beliefs, if the relations believed to hold have no being except in the beliefs themselves, are necessarily false. If I believe *A* to be *B*'s father, when this is not the case, my belief is erroneous ; and if I believe *A* to be west of *B*, when westerliness in fact exists only in my mind, I am again mistaken. Thus this first class of relations has no validity whatever, and consists merely in a collection of mistaken beliefs. The objects concerning which the beliefs are entertained are as a matter of fact wholly unrelated ; indeed there cannot even be *objects*, for the plural implies diversity, and all beliefs in the relation of diversity must be erroneous. There cannot even be one object distinct from myself, since this would have to have the relation of diversity to me, which is impossible. Thus we are committed, so far as this class of relations goes, to a rigid monism.

But now, what shall we say of the second class of relations, those namely which are reducible to internal states of the apparently related objects ? It must be observed that this class of relations presupposes a plurality of objects (two at least), and hence involves the relation of diversity. Now we have seen that, if there be diversity, it cannot be a relation of the first class ; hence it must itself be of the second class. That is, the mere fact that *A* is different from *B* must be reducible to internal states of *A* and *B*. But is it not evident that, before we can distinguish the internal states of *A* from those of *B*, we must first distinguish *A* from *B* ? i.e. *A* and *B* must *be* different, before they can have different states. If it be said that *A* and *B* are precisely similar, and are yet two, it follows even more evidently that their diversity is not due to difference of internal states, but is prior to it. Thus the mere admission that there are internal states of different things destroys the theory that the essence of relations is to be found in these states. We are thus brought back to the notion that the apparent relations of two things consist in the internal states of one thing, which leads us again to the rigid monism implied in the first type of relation.

Thus the theory of relations propounded by Lotze is, in fact, a theory that there are no relations. This has been recognised by the most logical adherents of the dogma—*e.g.* Spinoza and Mr Bradley—who have asserted that there is only one thing, God or the Absolute, and only one type of proposition, namely that ascribing predicates to the Absolute. In order to meet this development of the above theory of relations, it will be necessary to examine the doctrine of subject and predicate.

426. Every proposition, true or false—so the present theory contends—ascribes a predicate to a subject, and—what is a corollary from the above—there is only one subject. The consequences of this doctrine are so strange, that I cannot believe they have been realised by those who maintain it. The theory is in fact self-contradictory. For if the Absolute has predicates, then there are predicates; but the proposition "there are predicates" is not one which the present theory can admit. We cannot escape by saying that the predicates merely qualify the Absolute; for the Absolute cannot be qualified by nothing, so that the proposition "there are predicates" is logically prior to the proposition "the Absolute has predicates." Thus the theory itself demands, as its logical *prius*, a proposition without a subject and a predicate; moreover this proposition involves diversity, for even if there be only one predicate, this must be different from the one subject. Again, since there is a predicate, the predicate is an entity, and its predicability of the Absolute is a relation between it and the Absolute. Thus the very proposition which was to be non-relational turns out to be, after all, relational, and to express a relation which current philosophical language would describe as purely external. For both subject and predicate are simply what they are—neither is modified by its relation to the other. To be modified by the relation could only be to have some other predicate, and hence we should be led into an endless regress. In short, no relation ever modifies either of its terms. For if it holds between *A* and *B*, then it is between *A* and *B* that it holds, and to say that it modifies *A* and *B* is to say that it really holds between different terms *C* and *D*. To say that two terms which are related would be different if they were not related, is to say something perfectly barren; for if they were different, they would be other, and it would not be the terms in question, but a different pair, that would be unrelated. The notion that a term can be modified arises from neglect to observe the eternal self-identity of all terms and all logical concepts, which alone form the constituents of propositions*. What is called modification consists merely in having at one time, but not at another, some specific relation to some other specific term; but the term which sometimes has and sometimes has not the relation in question must be unchanged,

---

* See Mr G. E. Moore's paper on "The Nature of Judgment," *Mind*, N.S., Vol. VIII. Also *supra*, §§ 47, 48, where the view adopted differs somewhat from Mr Moore's.

otherwise it would not be *that* term which had ceased to have the relation.

The general objection to Lotze's theory of relations may be thus summed up. The theory implies that all propositions consist in the ascription of a predicate to a subject, and that this ascription is not a relation. The objection is, that the predicate is either something or nothing. If nothing, it cannot be predicated, and the pretended proposition collapses. If something, predication expresses a relation, and a relation of the very kind which the theory was designed to avoid. Thus in either case the theory stands condemned, and there is no reason for regarding relations as all reducible to the subject-predicate form.

427. (2) I come now to the second of Lotze's objections to empty space. This is again of a somewhat abstract logical character, but it is far easier to dispose of, since it depends upon a view more or less peculiar to Lotze. There are, it says, three and only three kinds of being, no one of which belongs to space. These are (α) the being of things, which consists in activity or the power to produce effects; (β) the validity of a truth; (γ) the being which belongs to the contents of our presentations.

The answer to this is, that there is only one kind of being, namely, being *simpliciter*, and only one kind of existence, namely, existence *simpliciter*. Both being and existence, I believe, belong to empty space; but being alone is relevant to the refutation of the relational theory— existence belongs to the question which Lotze confounds with the above, namely, as to the reality or subjectivity of space. It may be well first to explain the distinction of being and existence, and then to return to Lotze's three kinds of being.

Being is that which belongs to every conceivable term, to every possible object of thought—in short to everything that can possibly occur in any proposition, true or false, and to all such propositions themselves. Being belongs to whatever can be counted. If *A* be any term that can be counted as one, it is plain that *A* is something, and therefore that *A* is. "*A* is not" must always be either false or meaningless. For if *A* were nothing, it could not be said not to be; "*A* is not" implies that there is a term *A* whose being is denied, and hence that *A* is. Thus unless "*A* is not" be an empty sound, it must be false— whatever *A* may be, it certainly is. Numbers, the Homeric gods, relations, chimeras and four-dimensional spaces all have being, for if they were not entities of a kind, we could make no propositions about them. Thus being is a general attribute of everything, and to mention anything is to show that it is.

Existence, on the contrary, is the prerogative of some only amongst beings. To exist is to have a specific relation to existence—a relation, by the way, which existence itself does not have. This shows, incidentally, the weakness of the existential theory of judgment—the theory, that is, that every proposition is concerned with something that exists. For if this theory were true, it would still be true that existence itself is an entity, and it must be admitted that existence does not exist. Thus the consideration of existence itself leads to non-existential propositions, and so contradicts the theory. The theory seems, in fact, to have arisen from neglect of the distinction between existence and being. Yet this distinction is essential, if we are ever to deny the existence of anything. For what does not exist must be something, or it would be meaningless to deny its existence; and hence we need the concept of being, as that which belongs even to the non-existent.

Returning now to Lotze's three kinds of being, it is sufficiently evident that his views involve hopeless confusions.

(*a*) The being of things, Lotze thinks—following Leibniz here as elsewhere—consists in activity. Now activity is a highly complex notion, which Lotze falsely supposed unanalyzable. But at any rate it is plain that, if there be activity, what is active must both be and exist, in the senses explained above. It will also be conceded, I imagine, that existence is conceptually distinguishable from activity. Activity may be a universal mark of what exists, but can hardly be synonymous with existence. Hence Lotze requires the highly disputable proposition that whatever exists must be active. The true answer to this proposition lies (1) in disproving the grounds alleged in its favour, (2) in proving that activity implies the existence of time, which cannot be itself active. For the moment, however, it may suffice to point out that, since existence and activity are logically separable, the supposition that something which is not active exists cannot be logically absurd.

(*β*) The validity of a truth—which is Lotze's second kind of being—is in reality no kind of being at all. The phrase, in the first place, is ill-chosen—what is meant is the truth of a truth, or rather the truth of a proposition. Now the truth of a proposition consists in a certain relation to truth, and presupposes the being of the proposition. And as regards being, false propositions are on exactly the same level, since to be false a proposition must already be. Thus validity is not a kind of being, but being belongs to valid and invalid propositions alike.

(*γ*) The being which belongs to the contents of our presentations is a subject upon which there exists everywhere the greatest confusion. This kind is described by Lotze as " *ein Vorgestelltwerden durch uns.*" Lotze presumably holds that the mind is in some sense creative—that what it intuits acquires, in some sense, an existence which it would not have if it were not intuited. Some such theory is essential to every form of Kantianism—to the belief, that is, that propositions which are believed solely because the mind is so made that we cannot but believe them may yet be true in virtue of our belief. But the whole theory rests, if I am not mistaken, upon neglect of the fundamental distinction between an idea and its object. Misled by neglect of being, people have supposed that what does not exist is nothing. Seeing that numbers, relations, and many other objects of thought, do not exist outside the mind, they have supposed that the thoughts in which we think of these entities actually create their own objects. Every one except a philosopher can see the difference between a post and my idea of a post, but few see the difference between the number 2 and my idea of the number 2. Yet the distinction is as necessary in one case as in the other. The argument that 2 is mental requires that 2 should be essentially an existent. But in that case it would be particular, and it would be impossible for 2 to be in two minds, or in one mind at two times. Thus 2 must be in any case an entity, which will have being even if it is in no mind*. But further, there are reasons for denying that 2 is created by the thought which thinks it. For, in this case, there could never be two thoughts until some one thought so; hence what the person so thinking supposed to be two thoughts would not have been two, and the opinion, when it did arise, would be erroneous. And applying the same doctrine to 1; there cannot be one thought until some one thinks so. Hence Adam's first thought must have been concerned with the number 1; for not a single thought could precede this thought. In short, all knowledge must be recognition, on pain of being mere delusion; Arithmetic must be discovered in just the same sense in which Columbus discovered the West Indies, and we no more create numbers than he created the Indians. The number 2 is not purely mental, but is an entity which may be thought *of*. Whatever can be thought of has being, and its being is a precondition, not a result, of its being thought of. As regards the existence of an object of thought, however, nothing can be inferred from the fact of its being thought of, since it certainly does not exist in the thought which thinks of it. Hence, finally, no special kind of being belongs to the objects of our presentations as such. With this conclusion, Lotze's second argument is disposed of.

428. (3) Lotze's third argument has been a great favourite, ever since Leibniz introduced it. All points, we are told, are exactly alike, and therefore any two must have the same mutual relation as any other two; yet their mutual distances must differ, and even, according to Lotze (though in this, in the sense in which he seems to mean it, he is mistaken), the relation of every pair must be peculiar to that pair. This argument will be found to depend again upon the subject-predicate logic which we have already examined. To be exactly alike can only mean—as in Leibniz's Identity of Indiscernibles—not to have different predicates. But when once it is recognised that there is no essential distinction between subjects and predicates, it is seen that any two simple terms simply differ immediately—they are two, and this is the sum-total of their differences. Complex terms, it is true, have differences which can be revealed by analysis. The constituents

---

* Cf. Frege, *Grundgesetze der Arithmetik*, p. xviii.

of the one may be A, B, C, D, while those of the other are A, E, F, G. But the differences of B, C, D from E, F, G are still immediate differences, and immediate differences must be the source of all mediate differences. Indeed it is a sheer logical error to suppose that, if there were an ultimate distinction between subjects and predicates, subjects could be distinguished by differences of predicates. For before two subjects can differ as to predicates, they must already be two; and thus the immediate diversity is prior to that obtained from diversity of predicates. Again, two terms cannot be distinguished in the first instance by difference of relation to other terms; for difference of relation presupposes two distinct terms, and cannot therefore be the ground of their distinctness. Thus if there is to be any diversity at all, there must be immediate diversity, and this kind belongs to points.

Again, points have also the subsequent kind of diversity consisting in difference of relation. They differ not only, as Lotze urges, in their relations to each other, but also in their relations to the objects in them. Thus they seem to be in the same position as colours, sounds, or smells. Two colours, or two simple smells, have no intrinsic difference save immediate diversity, but have, like points, different relations to other terms.

Wherein, then, lies the plausibility of the notion that all points are exactly alike? This notion is, I believe, a psychological illusion, due to the fact that we cannot remember a point, so as to know it when we meet it again. Among simultaneously presented points it is easy to distinguish; but though we are perpetually moving, and thus being brought among new points, we are quite unable to detect this fact by our senses, and we recognise places only by the objects they contain. But this seems to be a mere blindness on our part—there is no difficulty, so far as I can see, in supposing an immediate difference between points, as between colours, but a difference which our senses are not constructed to be aware of. Let us take an analogy: Suppose a man with a very bad memory for faces: he would be able to know, at any moment, whether he saw one face or many, but he would not be aware whether he had ever seen any of the faces before. Thus he might be led to define people by the rooms in which he saw them, and to suppose it self-contradictory that new people should come to his lectures, or old people cease to do so. In the latter point, at least, it will be admitted by lecturers that he would be mistaken. And as with faces, so with points—inability to recognise them must be attributed, not to the absence of individuality, but merely to our incapacity.

429. (4) Lotze's fourth argument is an endeavour to effect a *reductio ad absurdum*, by proving that, on the absolute theory, points must interact. The being of every point, Lotze contends, must consist in the fact that it distinguishes itself from every other, and takes up an invariable position relatively to every other. Many fallacies are contained in this argument. In the first place, there is what may be called the ratiocinator's fallacy, which consists in supposing that everything has to be explained by showing that it is something else. Thus the being of a point, for Lotze, must be found in its difference from other points, while, as a matter of fact, its being is simply its being. So far from being explained by something else, the being of a point is presupposed in all other propositions about it, as *e.g.* in the proposition that the point differs from other points. Again, the phrase that the point distinguishes *itself* from all other points seems to be designed to imply some kind of self-assertion, as though the point would not be different unless it chose to differ. This suggestion helps out the conclusion, that the relations between points are in reality a form of interaction. Lotze, believing as he does that activity is essential to existence, is unable to imagine any other relation between existents than that of interaction. How hopelessly inapplicable such a view is, will appear from an analysis of interaction. Interaction is an enormously complex notion, presupposing a host of other relations, and involving, in its usual form, the distinction of a thing from its qualities—a distinction dependent on the subject-predicate logic already criticized. Interaction, to begin with, is either the simultaneous action of *A* on *B* and *B* on *A*, or the action of the present states of *A* and *B* conjointly on their states at the next instant. In either case it implies action. Action generally may be defined as a causal relation between one or more states of one or more things at the present instant and one or more states of the same or different things at a subsequent instant. When there is only one thing in both cases, the action is immanent if the thing be the same in cause and effect, transient if the cause be in one thing and the effect in another. In order to speak of action, rather than causality simply, it is necessary to suppose things enduring for a certain time, and having changing states. Thus the notion of interaction presupposes the following relations: (1) diversity between things; (2) diversity between the states of things; (3) simultaneity; (4) succession; (5) causality; (6) the relation of a thing to its states. This notion, involving, as a moment's inspection shows, six simpler relations in its analysis, is supposed to be *the* fundamental relation! No wonder absurdities are produced by such a supposition. But the absurdities belong to Lotze, not to space. To reduce the relations of points to interactions, on the ground that interaction is the type of all relations, is to display a complete incapacity in the simplest problems of analysis. The relations of points are not interactions, any more than before and after, or diversity, or greater and less, are interactions. They are eternal relations of entities, like the relation of 1 to 2 or of interaction itself to causality. Points do not *assign* positions to each other, as though they were each other's pew-openers: they eternally have the relations which they have, just like all other entities. The whole argument, indeed, rests upon an absurd dogma, supported by a false and scholastic logic.

430. (5) The fifth argument seems to be designed to prove the Kantian apriority of space. There are, it says, necessary propositions concerning space, which show that the nature of space is not a "mere fact." We are intended to infer that space is an *à priori* intuition, and a psychological reason is given why we cannot imagine holes in space. The impossibility of holes is apparently what is called a necessity of thought. This argument again involves much purely logical discussion. Concerning necessities of thought, the Kantian theory seems to lead to the curious result that whatever we cannot help believing must be false. What we cannot help believing, in this case, is something as to the nature of space, not as to the nature of our minds. The explanation offered is, that there is no space outside our minds; whence it is to be inferred that our unavoidable beliefs about space are all mistaken. Moreover we only push one stage farther back the region of "mere fact," for the constitution of our minds remains still a mere fact.

The theory of necessity urged by Kant, and adopted here by Lotze, appears radically vicious. Everything is in a sense a mere fact. A proposition is said to be proved when it is deduced from premisses; but the premisses, ultimately, and the rule of inference, have to be simply assumed. Thus any ultimate premiss is, in a certain sense, a mere fact. On the other hand, there seems to be no true proposition of which there is any sense in saying that it might have been false. One might as well say that redness might have been a taste and not a colour. What is true, is true; what is false, is false; and concerning fundamentals, there is nothing more to be said. The only logical meaning of necessity seems to be derived from implication. A proposition is more or less necessary according as the class of propositions for which it is a premiss is greater or smaller*. In this sense the propositions of logic have the greatest necessity, and those of geometry have a high degree of necessity. But this sense of necessity yields no valid argument from our inability to imagine holes in space to the conclusion that there cannot really be any space at all except in our imaginations.

431. (6) The last argument may be shortly disposed of. If points be independent entities, Lotze argues—so I interpret him—that we can imagine a new point coming into existence. This point, then, must have the appropriate relations to other points. Either it creates the other points with the relations, or it merely creates the relations to already existing points. Now it must be allowed that, if there be real points, it is not self-contradictory to suppose some of them non-existent. But strictly speaking, no single proposition whatever is self-contradictory. The nearest approach would be "No proposition is true," since this implies its own truth. But even here, it is not strictly self-contradictory

* Cf. G. E. Moore, "Necessity," *Mind*, N.S., No. 35.

to deny the implication. Everywhere we come upon propositions accepted because they are self-evident, and for no other reason: the law of contradiction itself is such a proposition. The mutual implication of all the points of space seems to be another; the denial of some only among points is rejected for the same reason as the assertion that such and such a proposition is both true and false, namely, because both are obviously untrue. But if, *per impossibile*, a point previously missing were to come into existence, it would not create new points, but would have the appropriate relations to already existing points. The point, in fact, would have already had being, and as an entity would have eternally had to other points the same relations as it has when it comes into existence. Thus Lotze's argument on this, as on other points, depends upon a faulty logic, and is easily met by more correct views as to the nature of judgment.

I conclude, from the above discussion, that absolute position is not logically inadmissible, and that a space composed of points is not self-contradictory. The difficulties which used to be found in the nature of infinity depended upon adherence to one definite axiom, namely, that a whole must have more terms than a part; those in the nature of space, on the other hand, seem to have been derived almost exclusively from general logic. With a subject-predicate theory of judgment, space necessarily appears to involve contradictions; but when once the irreducible nature of relational propositions is admitted, all the supposed difficulties vanish like smoke*. There is no reason, therefore, so far as I am able to perceive, to deny the ultimate and absolute philosophical validity of a theory of geometry which regards space as composed of points, and not as a mere assemblage of relations between non-spatial terms.

* Cf. my *Philosophy of Leibniz*, Cambridge, 1900, Chap. x.

CHAPTER LII.

KANT'S THEORY OF SPACE.

432. In the present chapter I do not propose to undertake a minute or textual examination of Kant's opinions; this has been done elsewhere, and notably in Vaihinger's monumental commentary, so well that it need not be done over again here. It is only the broad outlines of the Kantian doctrine that I wish to discuss. This doctrine, more or less modified, has held the field for over a century, and has won a nearly universal acceptance. As my views are, on almost every point of mathematical theory, diametrically opposed to those of Kant, it becomes necessary explicitly to defend the opinions in which I differ from him *. In this I shall pay special attention to what Kant calls the transcendental arguments, *i.e.* those derived from the nature of mathematics.

433. Broadly speaking, the way in which Kant seeks to deduce his theory of space from mathematics (especially in the Prolegomena) is as follows. Starting from the question: "How is pure mathematics possible?" Kant first points out that all the propositions of mathematics are synthetic. He infers hence that these propositions cannot, as Leibniz had hoped, be proved by means of a logical calculus; on the contrary, they require, he says, certain synthetic *à priori* propositions, which may be called axioms, and even then (it would seem) the reasoning employed in deductions from the axioms is different from that of pure logic. Now Kant was not willing to admit that knowledge of the external world could be obtained otherwise than by experience; hence he concluded that the propositions of mathematics all deal with something subjective, which he calls a form of intuition. Of these forms there are two, space and time; time is the source of Arithmetic, space of Geometry. It is only in the forms of time and space that objects can be experienced by a subject; and thus pure mathematics must be applicable to all experience. What is essential, from the logical point

---

* The theory of space which I shall discuss will be that of the Critique and the Prolegomena. Pre-critical works, and the *Metaphysische Anfangsgründe der Naturwissenschaft* (which differs from the Critique on this point), will not be considered.

of view, is, that the *à priori* intuitions supply methods of reasoning and inference which formal logic does not admit ; and these methods, we are told, make the figure (which may of course be merely imagined) essential to all geometrical proofs. The opinion that time and space are sub- jective is reinforced by the antinomies, where Kant endeavours to prove that, if they be anything more than forms of experience, they must be definitely self-contradictory.

In the above outline I have omitted everything not relevant to the philosophy of mathematics. The questions of chief importance to us, as regards the Kantian theory, are two, namely, (1) are the reasonings in mathematics in any way different from those of Formal Logic ? (2) are there any contradictions in the notions of time and space ? If these two pillars of the Kantian edifice can be pulled down, we shall have success- fully played the part of Samson towards his disciples.

434. The question of the nature of mathematical reasoning was obscured in Kant's day by several causes. In the first place, Kant never doubted for a moment that the propositions of logic are analytic, whereas he rightly perceived that those of mathematics are synthetic. It has since appeared that logic is just as synthetic as all other kinds of truth ; but this is a purely philosophical question, which I shall here pass by *. In the second place, formal logic was, in Kant's day, in a very much more backward state than at present. It was still possible to hold, as Kant did, that no great advance had been made since Aristotle, and that none, therefore, was likely to occur in the future. The syllogism still remained the one type of formally correct reasoning ; and the syl- logism was certainly inadequate for mathematics. But now, thanks mainly to the mathematical logicians, formal logic is enriched by several forms of reasoning not reducible to the syllogism †, and by means of these all mathematics can be, and large parts of mathematics actually have been, developed strictly according to the rules. In the third place, in Kant's day, mathematics itself was, logically, very inferior to what it is now. It is perfectly true, for example, that any one who attempts, without the use of the figure, to deduce Euclid's seventh proposition from Euclid's axioms, will find the task impossible ; and there probably did not exist, in the eighteenth century, any single logically correct piece of mathematical reasoning, that is to say, any reasoning which correctly deduced its result from the explicit premisses laid down by the author. Since the correctness of the result seemed indubitable, it was natural to suppose that mathematical proof was something different from logical proof. But the fact is, that the whole difference lay in the fact that mathematical proofs were simply unsound. On closer examination, it has been found that many of the propositions which,

---

* See my *Philosophy of Leibniz*, § 11.
† See Chap. ii *supra*, esp. § 18.

to Kant, were undoubted truths, are as a matter of fact demonstrably false*. A still larger class of propositions—for instance, Euclid's seventh proposition mentioned above—can be rigidly deduced from certain premisses, but it is quite doubtful whether the premisses themselves are true or false. Thus the supposed peculiarity of mathematical reasoning has disappeared.

The belief that the reasonings of Geometry are in any way peculiar has been, I hope, sufficiently refuted already by the detailed accounts which have been given of these reasonings, and especially by Chapter XLIX. We have seen that all geometrical results follow, by the mere rules of logic, from the definitions of the various spaces. And as regards the opinion that Arithmetic depends upon time, this too, I hope, has been answered by our accounts of the relation of Arithmetic to Logic. Indeed, apart from any detail, it seems to be refuted by the simple observation that time must have parts, and therefore plurality, whole and part, are prior to any theory of time. All mathematics, we may say—and in proof of our assertion we have the actual development of the subject—is deducible from the primitive propositions of formal logic: these being admitted, no further assumptions are required.

But admitting the *reasonings* of Geometry to be purely formal, a Kantian may still maintain that an *à priori* intuition assures him that the definition of three-dimensional Euclidean space, alone among the definitions of possible spaces, is the definition of an existent, or at any rate of an entity having some relation to existents which other spaces do not have. This opinion is, strictly speaking, irrelevant to the philosophy of mathematics, since mathematics is throughout indifferent to the question whether its entities exist. Kant thought that the actual *reasoning* of mathematics was different from that of logic; the suggested emendation drops this opinion, and maintains merely a new primitive proposition, to the effect that Euclidean space is that of the actual world. Thus, although I do not believe in any immediate intuition guaranteeing any such primitive proposition, I shall not undertake the refutation of this opinion. It is enough, for my purpose, to have shown that no such intuition is relevant in any strictly mathematical proposition.

435. It remains to discuss the mathematical antinomies. These are concerned with infinity and continuity, which Kant supposed to be specially spatio-temporal. We have already seen that this view is mistaken, since both occur in pure Arithmetic. We have seen also in Part V (especially in Chapter XLII) that the supposed antinomies of infinity and continuity, in their arithmetical form, are soluble; it remains to prove the same conclusion concerning Kant's spatio-temporal

---

* For example, the proposition that every continuous function can be differentiated.

form. The third and fourth antinomies are not relevant here, since they
involve causality ; only the first two, therefore, will be examined.

First Antinomy. *Thesis*: "The world has a beginning in time, and
as regards space also is enclosed within limits." This statement is not
concerned with pure time and pure space, but with the things in them.
The proof, such as it is, applies in the first instance to time only, and is
effected by *reductio ad absurdum.* "For assume," it says, "that the
world has no beginning in time, then an eternity has passed away
(*abgelaufen*) before every given point of time, and consequently an
infinite series of conditions of the things in the world has happened.
But the infinity of a series consists in this, that it can never be completed by successive synthesis. Consequently an infinite past series of
things in the world (*Weltreihe*) is impossible, and a beginning of the
world is a necessary condition of its existence, which was first to be
proved."

This argument is difficult to follow, and suggests a covert appeal to
causality and the supposed necessity for a first cause. Neglecting this
aspect of the argument, it would seem that, like most of the arguments
against infinity, it fails to understand the use of the class-concept and
the word *any*. It is supposed—so it would seem—that the events preceding a given event ought to be definable by extension, which, if their
number is infinite, is obviously not the case. "Completion by successive
synthesis" seems roughly equivalent to enumeration, and it is true that
enumeration of an infinite series is practically impossible. But the series
may be none the less perfectly definable, as the class of terms having a
specified relation to a specified term. It then remains a question, as
with all classes, whether the class is finite or infinite; and in the latter
alternative, as we saw in Part V, that there is nothing self-contradictory.
In fact, to elicit a contradiction, it would be necessary to state as an
axiom that every class must have a finite number of terms—an axiom
which can be refuted, and for which there are no grounds. It seems,
however, that previous events are regarded by Kant as *causes* of later
ones, and that the cause is supposed to be logically prior to the effect.
This, no doubt, is the reason for speaking of *conditions*, and for confining
the antinomy to events instead of moments. If the cause were logically
prior to the effect, this argument would, I think, be valid ; but we shall
find, in Part VII, that cause and effect are on the same logical level.
Thus the thesis of the first antinomy, in so far as it concerns time, must
be rejected as false, and the argument concerning space, since it depends
upon that regarding time, falls also.

Antithesis. "The world has no beginning, and no limits in space,
but is infinite both in respect of time and space." The proof of this
proposition assumes the infinity of pure time and space, and argues that
these imply events and things to fill them. This view was rejected, as
regards space, in the preceding chapter, and can be disproved, as regards

time, by precisely similar arguments ; it is in any case irrelevant to our
contention, since no proof is offered that time and space are themselves
infinite. This, in fact, seems incapable of proof, since it depends upon
the merely self-evident axiom that there is a moment before any given
moment, and a point beyond any given point. But as no converse proof
is valid, we may, in this instance, regard the self-evident as true.
Whether events had a beginning, and whether matter is bounded by
empty space, are questions which, if our philosophy of space and
time be sound, no argument independent of causality can decide.

Second Antinomy. *Thesis* : " Every complex substance in the world
consists of simple parts, and nothing exists anywhere except the simple,
or what is composed of simple parts." Here, again, the argument applies
to things *in* space and time, not to space and time themselves. We
may extend it to space and time, and to all collections, whether existent
or not. It is indeed obvious that the proposition, true or false, is
concerned purely with whole and part, and has no special relation to
space and time. Instead of a complex substance, we might consider the
numbers between 1 and 2, or any other definable collection. And with
this extension, the proof of the proposition must, I think, be admitted ;
only that *terms* or *concepts* should be substituted for *substances*, and
that, instead of the argument that relations between substances are
accidental (*zufällig*), we should content ourselves with saying that
relations imply terms, and complexity implies relations.

Antithesis. "No complex thing in the world consists of simple parts,
and nothing simple exists in it anywhere." The proof of this proposition, as of the first antithesis, assumes, what is alone really interesting to us, the corresponding property of space. "Space," Kant says,
"does not consist of simple parts, but of spaces." This dogma is
regarded as self-evident, though all employment of points shows that
it is not universally accepted. It appears to me that the argument of
the thesis, extended as I have just suggested, applies to pure space as to
any other collection, and demonstrates the existence of simple points
which compose space. As the dogma is not argued, we can only conjecture the grounds upon which it is held. The usual argument from
infinite division is probably what influenced Kant. However many parts
we divide a space into, these parts are still spaces, not points. But
however many parts we divide the stretch of ratios between 1 and 2
into, the parts are still stretches, not single numbers. Thus the argument
against points proves that there are no numbers, and will equally prove
that there are no colours or tones. All these absurdities involve a
covert use of the axiom of finitude, *i.e.* the axiom that, if a space does
consist of points, it must consist of some finite number of points. When
once this is denied, we may admit that no finite number of divisions of
a space will lead to points, while yet holding every space to be composed of points. A finite space is a whole consisting of simple parts,

but not of any finite number of simple parts. Exactly the same thing
is true of the stretch between 1 and 2. Thus the antinomy is not
specially spatial, and any answer which is applicable in Arithmetic is
applicable here also. The thesis, which is an essential postulate of Logic,
should be accepted, while the antithesis should be rejected.

Thus Kant's antinomies do not specially involve space and time: any
other continuous series, including that of real numbers, raises the same
problems. And what is more, the properties of space and time, to
which Kant appeals, are general properties of such series. Other
antinomies than Kant's—*e.g.* that concerning absolute and relative
position, or concerning the straight line as both a relation and a
collection of points—have been solved in the preceding chapters of
this Part. Kant's antinomies, which involve the difficulties of infinity,
are by far the most serious, and these being essentially arithmetical,
have been already solved in Part V.

436. Before proceeding to matter and motion, let us briefly recapitulate the results of this Part. Geometry, we said, is the study
of series having more than one dimension ; and such series arise wherever
we have a series whose terms are series. This subject is important in
pure mathematics, because it gives us new kinds of order and new
methods of generating order. It is important in applied mathematics,
because at least one series of several dimensions exists, namely, space.
We found that the abstract logical method, based upon the logic of
relations, which had served hitherto, was still adequate, and enabled us
to define all the classes of entities which mathematicians call spaces,
and to deduce from the definitions all the propositions of the corresponding Geometries. We found that the continuity and infinity
of a space can always be arithmetically defined, and that no new
indefinables occur in Geometry. We saw that the philosophical objections to points raised by most philosophers are all capable of being
answered by an amended logic, and that Kant's belief in the peculiarity
of geometrical reasoning, and in the existence of certain antinomies
peculiar to space and time, has been disproved by the modern realization
of Leibniz's universal characteristic. Thus, although we discussed no
problems specially concerned with what actually exists, we incidentally
answered all the arguments usually alleged against the existence of an
absolute space. Since common sense affirms this existence, there seems
therefore no longer any reason for denying it ; and this conclusion, we
shall find, will give us the greatest assistance in the philosophy of
Dynamics.

PART VII.

MATTER AND MOTION.

CHAPTER LIII.

MATTER.

437. THE nature of matter, even more than that of space, has always been regarded as a cardinal problem of philosophy. In the present work, however, we are not concerned with the question: What is the nature of the matter that actually exists? We are concerned merely with the analysis of rational Dynamics considered as a branch of pure mathematics, which introduces its subject-matter by definition, not by observation of the actual world. Thus we are not confined to laws of motion which are empirically verified: non-Newtonian Dynamics, like non-Euclidean Geometry, must be as interesting to us as the orthodox system. It is true that philosophical arguments against the reality of matter usually endeavour to raise logical objections to the notion of matter, and these objections, like the objections to absolute space, are relevant to a discussion of mathematical principles. But they need not greatly concern us at this stage, as they have mostly been dealt with incidentally in the vindication of space. Those who have agreed that a space composed of points is possible, will probably agree also that matter is possible. But the question of possibility is in any case subsequent to our immediate question, which is: What is matter? And here matter is to mean, matter as it occurs in rational Dynamics, quite independently of all questions as to its actual existence.

438. There is—so we decided in Part VI—no logical implication of other entities in space. It does not follow, merely because there is space, that therefore there are things in it. If we are to believe this, we must believe it on new grounds, or rather on what is called the evidence of the senses. Thus we are here taking an entirely new step. Among terms which appear to exist, there are, we may say, four great classes: (1) instants, (2) points, (3) terms which occupy instants but not points, (4) terms which occupy both points and instants. It seems to be the fact that there are no terms which occupy points but not instants. What is meant by *occupying* a point or an instant, analysis cannot explain; this is a fundamental relation, expressed by *in* or *at*, asymmetrical and intransitive, indefinable and simple. It is evident that bits of matter are among the terms of (4). Matter or materiality itself, the class-concept, is among the terms which do not exist, but bits of matter exist both in time and in space. They do not, however, form the whole of class (4): there are, besides, the so-called secondary qualities, at least colours, which exist in time and space, but are not matter. We are not called upon to decide as to the subjectivity of secondary qualities, but at least we must agree that they differ from matter. How, then, is matter to be defined?

439. There is a well-worn traditional answer to this question. Matter, we are told, is a substance, a thing, a subject, of which secondary qualities are the predicates. But this traditional answer cannot content us. The whole doctrine of subject and predicate, as we have already had occasion to argue, is radically false, and must be abandoned. It may be questioned whether, without it, any sense other than that of Chapter IV can be made of the notion of *thing*. We are sometimes told that things are organic unities, composed of many parts expressing the whole and expressed in the whole. This notion is apt to replace the older notion of substance, not, I think, to the advantage of precise thinking. The only kind of unity to which I can attach any precise sense—apart from the unity of the absolutely simple—is that of a whole composed of parts. But this form of unity cannot be what is called organic; for if the parts express the whole or the other parts, they must be complex, and therefore themselves contain parts; if the parts have been analyzed as far as possible, they must be simple terms, incapable of expressing anything except themselves. A distinction is made, in support of organic unities, between conceptual analysis and real division into parts. What is really indivisible, we are told, may be conceptually analyzable. This distinction, if the conceptual analysis be regarded as subjective, seems to me wholly inadmissible. All complexity is conceptual in the sense that it is due to a whole capable of logical analysis, but is real in the sense that it has no dependence upon the mind, but only upon the nature of the object. Where the mind can distinguish elements, there must *be* different elements to distinguish; though, alas! there are often different elements which the mind does not distinguish. The analysis of a finite space into points is no more objective than the analysis (say) of causality into time-sequence + ground and consequent, or of equality into sameness of relation to a given magnitude. In every case of analysis, there is a whole consisting of parts with relations; it is only the nature of the parts and the relations which distinguishes different cases. Thus the notion of an organic whole in the above sense must be attributed to defective analysis, and cannot be used to explain things.

It is also said that analysis is falsification, that the complex is not equivalent to the sum of its constituents and is changed when analyzed into these. In this doctrine, as we saw in Parts I and II, there is a measure of truth, when what is to be analyzed is a unity. A proposition has a certain indefinable unity, in virtue of which it is an assertion; and this is so completely lost by analysis that no enumeration of constituents will restore it, even though itself be mentioned as a constituent. There is, it must be confessed, a grave logical difficulty in this fact, for it is difficult not to believe that a whole must be constituted by its constituents. For us, however, it is sufficient to observe that all unities are propositions or propositional concepts, and that consequently nothing that exists is a unity. If, therefore, it is maintained that things are unities, we must reply that no things exist.

440. Thus no form of the notion of substance seems applicable to the definition of matter. The question remains: How and why is matter distinguished from the so-called secondary qualities? It cannot, I think, be distinguished as belonging to a different logical class of concepts; the only classes appear to be things, predicates, and relations, and both matter and the secondary qualities belong to the first class. Nevertheless the world of dynamics is sharply distinguished from that of the secondary qualities, and the elementary properties of matter are quite different from those of colours. Let us examine these properties with a view to definition.

The most fundamental characteristic of matter lies in the nature of its connection with space and time. Two pieces of matter cannot occupy the same place at the same moment, and the same piece cannot occupy two places at the same moment, though it may occupy two moments at the same place. That is, whatever, at a given moment, has extension, is not an indivisible piece of matter: division of space always implies division of any matter occupying the space, but division of time has no corresponding implication. (These properties are commonly attributed to matter: I do not wish to assert that they do actually belong to it.) By these properties, matter is distinguished from whatever else is in space. Consider colours for example: these possess impenetrability, so that no two colours can be in the same place at the same time, but they do not possess the other property of matter, since the same colour may be in many places at once. Other pairs of qualities, as colour and hardness, may also coexist in one place. On the view which regarded matter as the subject of which qualities were attributes, one piece of colour was distinguished from another by the matter whose attribute it was, even when the two colours were exactly similar. I should prefer to say that the colour is the same, and has no direct relation to the matter in the place. The relation is indirect, and consists in occupation of the same place. (I do not wish to decide any moot questions as to the secondary qualities, but merely to show the difference between the common-sense notions of these and of matter respectively.) Thus impenetrability and its converse seem to characterize matter sufficiently to distinguish it from whatever else exists in space. Two pieces of matter cannot occupy the same place and the same time, and one piece of matter cannot occupy two places at the same time. But the latter property must be understood of a simple piece of matter, one which is incapable of analysis or division.

Other properties of matter flow from the nature of motion. Every piece of matter persists through time: if it exists once, it would seem that it must always exist. It either retains its spatial position, or changes it continuously, so that its positions at various times form a continuous series in space. Both these properties require considerable discussion, which will follow at a later stage. They are purely kinematical, *i.e.* they involve none of the so-called laws of motion, but only the nature of motion itself.

A controversy has always existed, since early Greek times, as to the possibility of a vacuum. The question whether there is a vacuum cannot, I think, be decided on philosophical grounds, *i.e.* no decision is possible from the nature of matter or of motion. The answer belongs properly to Science, and therefore none will be suggested here.

We may sum up the nature of matter as follows. *Material unit* is a class-concept, applicable to whatever has the following characteristics: (1) A simple material unit occupies a spatial point at any moment; two units cannot occupy the same point at the same moment, and one cannot occupy two points at the same moment. (2) Every material unit persists through time; its positions in space at any two moments may be the same or different; but if different, the positions at times intermediate between the two chosen must form a continuous series. (3) Two material units differ in the same immediate manner as two points or two colours; they agree in having the relation of inclusion in a class to the general concept *matter*, or rather to the general concept *material unit*. Matter itself seems to be a collective name for all pieces of matter, as space for all points and time for all instants. It is thus the peculiar relation to space and time which distinguishes matter from other qualities, and not any logical difference such as that of subject and predicate, or substance and attribute.

441. We can now attempt an abstract logical statement of what rational Dynamics requires its matter to be. In the first place, time and space may be replaced by a one-dimensional and n-dimensional series respectively. Next, it is plain that the only relevant function of a material point is to establish a correlation between all moments of time and some points of space, and that this correlation is many-one. So soon as the correlation is given, the actual material point ceases to have any importance. Thus we may replace a material point by a many-one relation whose domain is a certain one-dimensional series, and whose converse domain is contained in a certain three-dimensional series. To obtain a material universe, so far as kinematical considerations go, we have only to consider a class of such relations subject to the condition that the logical product of any two relations of the class is to be null. This condition insures impenetrability. If we add that the one-dimensional and the three-dimensional series are to be both continuous, and that each many-one relation is to define a continuous function, we have all the kinematical conditions for a system of material particles, generalized and expressed in terms of logical constants.

CHAPTER LIV.

MOTION.

442. Much has been written concerning the laws of motion, the possibility of dispensing with Causality in Dynamics, the relativity of motion, and other kindred questions. But there are several preliminary questions, of great difficulty and importance, concerning which little has been said. Yet these questions, speaking logically, must be settled before the more complex problems usually discussed can be attacked with any hope of success. Most of the relevant modern philosophical literature will illustrate the truth of these remarks : the theories suggested usually repose on a common dogmatic basis, and can be easily seen to be unsatisfactory. So long as an author confines himself to demolishing his opponents, he is irrefutable ; when he constructs his own theory, he exposes himself, as a rule, to a similar demolition by the next author. Under these circumstances, we must seek some different path, whose by-ways remain unexplained. "Back to Newton" is the watchword of reform in this matter. Newton's scholium to the definitions contains arguments which are unrefuted, and so far as I know, irrefutable : they have been before the world two hundred years, and it is time they were refuted or accepted. Being unequal to the former, I have adopted the latter alternative.

The concept of motion is logically subsequent to that of occupying a place at a time, and also to that of change. Motion is the occupation, by one entity, of a continuous series of places at a continuous series of times. Change is the difference, in respect of truth or falsehood, between a proposition concerning an entity and a time T and a proposition concerning the same entity and another time T', provided that the two propositions differ only by the fact that T occurs in the one where T' occurs in the other. Change is continuous when the propositions of the above kind form a continuous series correlated with a continuous series of moments. Change thus always involves (1) a fixed entity, (2) a three-cornered relation between this entity, another entity, and some but not all, of the moments of time. This is its bare minimum. Mere existence at some but not all moments constitutes change on this definition. Consider pleasure, for example. This, we know, exists at some moments, and we may suppose that there are moments when it does not exist. Thus there is a relation between pleasure, existence, and some moments, which does not subsist between pleasure, existence, and other moments. According to the definition, therefore, pleasure changes in passing from existence to non-existence or *vice versâ*. This shows that the definition requires emendation, if it is to accord with usage. Usage does not permit us to speak of change except where what changes is an existent throughout, or is at least a class-concept one of whose particulars always exists. Thus we should say, in the case of pleasure, that my mind is what changes when the pleasure ceases to exist. On the other hand, if my pleasure is of different magnitudes at different times, we should say the pleasure changes its amount, though we agreed in Part III that not pleasure, but only particular amounts of pleasure, are capable of existence. Similarly we should say that colour changes, meaning that there are different colours at different times in some connection; though not colour, but only particular shades of colour, can exist. And generally, where both the class-concept and the particulars are simple, usage would allow us to say, if a series of particulars exists at a continuous series of times, that the class-concept changes. Indeed it seems better to regard this as the only kind of change, and to regard as unchanging a term which itself exists throughout a given period of time. But if we are to do this, we must say that wholes consisting of existent parts do not exist, or else that a whole cannot preserve its identity if any of its parts be changed. The latter is the correct alternative, but some subtlety is required to maintain it. Thus people say they change their minds: they say that the mind changes when pleasure ceases to exist in it. If this expression is to be correct, the mind must not be the sum of its constituents. For if it were the sum of *all* its constituents throughout time, it would be evidently unchanging; if it were the sum of its constituents at one time, it would lose its identity as soon as a former constituent ceased to exist or a new one began to exist. Thus if the mind is anything, and if it can change, it must be something persistent and constant, to which all constituents of a psychical state have one and the same relation. Personal identity could be constituted by the persistence of this term, to which all a person's states (and nothing else) would have a fixed relation. The change of mind would then consist merely in the fact that these states are not the same at all times.

Thus we may say that a term changes, when it has a fixed relation to a collection of other terms, each of which exists at some part of time, while all do not exist at exactly the same series of moments. Can we say, with this definition, that the universe changes? The universe is a somewhat ambiguous term: it may mean all the things that exist at a single moment, or all the things that ever have existed or will exist, or the common quality of whatever exists. In the two former senses it cannot change; in the last, if it be other than existence, it can change. Existence itself would not be held to change, though different terms exist at different times; for existence is involved in the notion of change as commonly employed, which applies only in virtue of the difference between the things that exist at different times. On the whole, then, we shall keep nearest to usage if we say that the fixed relation, mentioned at the beginning of this paragraph, must be that of a simple class-concept to simple particulars contained under it.

443. The notion of change has been much obscured by the doctrine of substance, by the distinction between a thing's nature and its external relations, and by the pre-eminence of subject-predicate propositions. It has been supposed that a thing could, in some way, be different and yet the same : that though predicates define a thing, yet it may have different predicates at different times. Hence the distinction of the essential and the accidental, and a number of other useless distinctions, which were (I hope) employed precisely and consciously by the scholastics, but are used vaguely and unconsciously by the moderns. Change, in this metaphysical sense, I do not at all admit. The so-called predicates of a term are mostly derived from relations to other terms ; change is due, ultimately, to the fact that many terms have relations to some parts of time which they do not have to others. But every term is eternal, timeless, and immutable ; the relations it may have to parts of time are equally immutable. It is merely the fact that different terms are related to different times that makes the difference between what exists at one time and what exists at another. And though a term may cease to exist, it cannot cease to be ; it is still an entity, which can be counted as *one*, and concerning which some propositions are true and others false.

444. Thus the important point is the relation of terms to the times they occupy, and to existence. Can a term occupy a time without existing? At first sight, one is tempted to say that it can. It is hard to deny that Waverley's adventures occupied the time of the '45, or that the stories in the 1,001 Nights occupy the period of Harun al Raschid. I should not say, with Mr Bradley, that these times are not parts of real time ; on the contrary, I should give them a definite position in the Christian Era. But I should say that the *events* are not real, in the sense that they never existed. Nevertheless, when a term exists at a time, there is an ultimate triangular relation, not reducible to a combination of separate relations to existence and the time respectively. This may be shown as follows. If "*A* exists now" can be analyzed into "*A* is now" and "*A* exists," where *exists* is used without any tense, we shall have to hold that "*A* is then" is logically possible even if *A* did not exist then ; for if occupation of a time be separable from existence, a term may occupy a time at which it does not exist, even if there are other times when it does exist. But, on the theory in question, "*A* is then" and "*A* exists" constitute the very meaning of "*A* existed then," and therefore, when these two propositions are true, *A* must have existed then. This can only be avoided by denying the possibility of analyzing "*A* exists now" into a combination of two-term relations; and hence non-existential occupation of a time, if possible at all, is radically different from the existential kind of occupation.

It should be observed, however, that the above discussion has a merely philosophical interest, and is strictly irrelevant to our theme. For existence, being a constant term, need not be mentioned, from a mathematical point of view, in defining the moments occupied by a term. From the mathematical point of view, change arises from the fact that there are propositional functions which are true of some but not all moments of time, and if these involve existence, that is a further point with which mathematics as such need not concern itself.

445. Before applying these remarks to motion, we must examine the difficult idea of occupying a place at a time. Here again we seem to have an irreducible triangular relation. If there is to be motion, we must not analyze the relation into occupation of a place and occupation of a time. For a moving particle occupies many places, and the essence of motion lies in the fact that they are occupied at different times. If "*A* is here now" were analyzable into "*A* is here" and "*A* is now," it would follow that "*A* is there then" is analyzable into "*A* is there" and "*A* is then." If all these propositions were independent, we could combine them differently: we could, from "*A* is now" and "*A* is there," infer "*A* is there now," which we know to be false, if *A* is a material point. The suggested analysis is therefore inadmissible. If we are determined to avoid a relation of three terms, we may reduce "*A* is here now" to "*A*'s occupation of this place is now." Thus we have a relation between *this time* and a complex concept, *A*'s occupation of this place. But this seems merely to substitute another equivalent proposition for the one which it professes to explain. But mathematically, the whole requisite conclusion is that, in relation to a given term which occupies a place, there is a correlation between a place and a time.

446. We can now consider the nature of motion, which need not, I think, cause any great difficulty. A simple unit of matter, we agreed, can only occupy one place at one time. Thus if *A* be a material point, "*A* is here now" excludes "*A* is there now," but not "*A* is here then." Thus any given moment has a unique relation, not direct, but *viâ A*, to a single place, whose occupation by *A* is at the given moment; but there need not be a unique relation of a given place to a given time, since the occupation of the place may fill several times. A moment such that an interval containing the given moment otherwise than as an end-point can be assigned, at any moment within which interval *A* is in the same place, is a moment when *A* is at rest. A moment when this cannot be done is a moment when A is in motion, provided A occupies *some* place at neighbouring moments on either side. A moment when there are such intervals, but all have the said moment as an end-term, is one of transition from rest to motion or *vice versâ*. Motion consists in the fact that, by the occupation of a place at a time, a correlation is established between places and times; when different times, throughout any period however short, are correlated with different places, there is motion; when different times, throughout some period however short, are all correlated with the same place, there is rest.

We may now proceed to state our doctrine of motion in abstract logical terms, remembering that material particles are replaced by many-one relations of all times to some places, or of all terms of a continuous one-dimensional series t to some terms of a continuous three-dimensional series s. Motion consists broadly in the correlation of different terms of t with different terms of s. A relation R which has a single term of s for its converse domain corresponds to a material particle which is at rest throughout all time. A relation R which correlates all the terms of t in a certain interval with a single term of s corresponds to a material particle which is at rest throughout the interval, with the possible exclusion of its end-terms (if any), which may be terms of transition between rest and motion. A time of momentary rest is given by any term for which the differential coefficient of the motion is zero. The motion is continuous if the correlating relation R defines a continuous function. It is to be taken as part of the definition of motion that it is continuous, and that further it has first and second differential coefficients. This is an entirely new assumption, having no kind of necessity, but serving merely the purpose of giving a subject akin to rational Dynamics.

447. It is to be observed that, in consequence of the denial of the infinitesimal, and in consequence of the allied purely technical view of the derivative of a function, we must entirely reject the notion of a *state* of motion. Motion consists *merely* in the occupation of different places at different times, subject to continuity as explained in Part V. There is no transition from place to place, no consecutive moment or consecutive position, no such thing as velocity except in the sense of a real number which is the limit of a certain set of quotients. The rejection of velocity and acceleration as physical facts (*i.e.* as properties belonging *at each instant* to a moving point, and not merely real numbers expressing limits of certain ratios) involves, as we shall see, some difficulties in the statement of the laws of motion; but the reform introduced by Weierstrass in the infinitesimal calculus has rendered this rejection imperative.

CHAPTER LV.

CAUSALITY.

448. A GREAT controversy has existed in recent times, among those who are interested in the principles of Dynamics, on the question whether the notion of causality occurs in the subject or not. Kirchoff* and Mach, and, in our own country, Karl Pearson, have upheld the view that Dynamics is purely descriptive, while those who adhere to the more traditional opinion maintain that it not merely registers sequences, but discovers causal connections. This controversy is discussed in a very interesting manner in Professor James Ward's *Naturalism and Agnosticism*, in which the descriptive theory is used to prove that Dynamics cannot give metaphysical truths about the real world. But I do not find, either in Professor Ward's book or elsewhere, a very clear statement of the issue between the two schools. The practical mathematical form of the question arises as regards *force*, and in this form, there can be no doubt that the descriptive school are in the right: the notion of force is one which ought not to be introduced into the principles of Dynamics. The reasons for this assertion are quite conclusive. Force is the supposed cause of acceleration: many forces are supposed to concur in producing a resultant acceleration. Now an acceleration, as was pointed out at the end of the preceding chapter, is a mere mathematical fiction, a number, not a physical fact; and a component acceleration is doubly a fiction, for, like the component of any other vector sum, it is not part of the resultant, which alone could be supposed to exist. Hence a force, if it be a cause, is the cause of an effect which never takes place. But this conclusion does not suffice to show that causality never occurs in Dynamics. If the descriptive theory were strictly correct, inferences from what occurs at some times to what occurs at others would be impossible. Such inferences must involve a relation of implication between events at different times, and any such relation is in a general sense causal. What does appear to be the case is, that the only causality occurring in Dynamics requires the whole configuration of the material world as a datum, and does not yield relations of particulars to par-

---

* *Vorlesungen über mathematische Physik*, Leipzig, 1883, Vorrede.

ticulars, such as are usually called causal. In this respect, there is
a difficulty in interpreting such seeming causation of particulars by
particulars as appears, for example, in the law of gravitation. On
account of this difficulty, it will be necessary to treat causation at some
length, examining first the meaning to be assigned to the causation of
particulars by particulars as commonly understood, then the meaning
of causality which is essential to rational Dynamics, and finally the
difficulty as regards component acceleration.

449. The first subject of the present chapter is the logical nature
of causal propositions. In this subject there is a considerable difficulty,
due to the fact that temporal succession is not a relation between events
directly, but only between moments*. If two events could be successive,
we could regard causation as a relation of succession holding between
two events without regard to the time at which they occur. If "*A*
precedes *B*" (where *A* and *B* are actual or possible temporal existents)
be a true proposition, involving no reference to any actual part of time,
but only to temporal succession, then we say *A causes B*. The law of
causality would then consist in asserting that, among the things which
actually precede a given particular existent *B* now, there is always one
series of events at successive moments which would necessarily have
preceded *B* then, just as well as *B* now; the temporal relations of *B*
to the terms of this series may then be abstracted from all particular
times, and asserted *per se*.

Such would have been the account of causality, if we had admitted
that events can be successive. But as we have denied this, we require
a different and more complicated theory. As a preliminary, let us
examine some characteristics of the causal relation.

A causal relation between two events, whatever its nature may be,
certainly involves no reference to constant particular parts of time. It
is impossible that we should have such a proposition as "*A* causes *B* now,
but not then." Such a proposition would merely mean that *A exists*
now but not then, and therefore *B* will exist at a slightly subsequent
moment, though it did not exist at a time slightly subsequent to the
former time. But the causal relation itself is eternal: if *A had* existed
at any other time, *B* would have existed at the subsequent moment.
Thus "*A* causes *B*" has no reference to constant particular parts of time.

Again, neither *A* nor *B* need ever exist, though if *A* should exist at
any moment, *B* must exist at a subsequent moment, and *vice versâ*. In
all Dynamics (as I shall prove later) we work with causal connections;
yet, except when applied to concrete cases, our terms are not existents.
Their non-existence is, in fact, the mark of what is called rational
Dynamics. To take another example: All deliberation and choice, all
decisions as to policies, demand the validity of causal series whose terms

---

* See my article in *Mind*, N.S., No. 39, "Is position in time and space absolute
or relative?"

do not and will not exist. For the rational choice depends upon the construction of two causal series, only one of which cån be made to exist. Unless both were valid, the choice could have no foundation. The rejected series consists of equally valid causal connections, but the events connected are not to be found among existents. Thus all statesmanship, and all rational conduct of life, is based upon the method of the frivolous historical game, in which we discuss what the world would be if Cleopatra's nose had been half an inch longer.

A causal relation, we have seen, has no essential reference to existence, as to particular parts of time. But it has, none the less, some kind of connection with both. If one of its terms is among existents, so is the other; if one is non-existent, the other is also non-existent. If one of the terms is at one moment, the other is at a later or earlier moment. Thus if *A* causes *B*, we have also " *A*'s existence implies *B*'s " and " *A*'s being at this moment implies *B*'s being at a subsequent moment." These two propositions are implied by " *A* causes *B* "; the second, at least, also implies " *A* causes *B*," so that we have here a mutual implication. Whether the first also implies " *A* causes *B*," is a difficult question. Some people would hold that two moments of time, or two points of space, imply each other's existence; yet the relation between these cannot be said to be causal.

It would seem that whatever exists at any part of time has causal relations. This is not a distinguishing characteristic of what exists, since we have seen that two non-existent terms may be cause and effect. But the absence of this characteristic distinguishes terms which cannot exist from terms which might exist. Excluding space and time, we may define as a *possible* existent any term which has a causal relation to some other term. This definition excludes numbers, and all so-called abstract ideas. But it admits the entities of rational Dynamics, which might exist, though we have no reason to suppose that they do.

If we admit (what seems undeniable) that whatever occupies any given time is both a cause and an effect, we obtain a reason for either the infinity or the circularity of time, and a proof that, if there are events at any part of time, there always have been and always will be events. If, moreover, we admit that a single existent *A* can be isolated as the cause of another single existent *B*, which in turn causes *C*, then the world consists of as many independent causal series as there are existents at any one time. This leads to an absolute Leibnizian monadism—a view which has always been held to be paradoxical, and to indicate an error in the theory from which it springs. Let us, then, return to the meaning of causality, and endeavour to avoid the paradox of independent causal series.

450. The proposition " *A* causes *B* " is, as it stands, incomplete. The only meaning of which it seems capable is " *A*'s existence at any time implies *B*'s existence at some future time." It has always been customary to suppose that cause and effect must occupy consecutive moments; but as time is assumed to be a compact series, there cannot be any consecutive moments, and the interval between any two moments will always be finite. Thus in order to obtain a more complete causal proposition, we must specify the interval between A and B. A causal connection then asserts that the existence of A at any one time implies the existence of B after an interval which is independent of the particular time at which A existed. In other words, we assert: "There is an interval t such that A's existence at any time t_1 implies B's existence at a time $t_1 + t$." This requires the measurement of time, and consequently involves either temporal distance, or magnitude of divisibility, which last we agreed to regard as not a motion of pure mathematics. Thus if our measure is effected by means of distance, our proposition is capable of the generalization which is required for a purely logical statement.

451. A very difficult question remains—the question which, when the problem is precisely stated, discriminates most clearly between monism and monadism. Can the causal relation hold between particular events, or does it hold only between the whole present state of the universe and the whole subsequent state? Or can we take a middle position, and regard one group of events now as causally connected with one group at another time, but not with any other events at that other time?

I will illustrate this difficulty by the case of gravitating particles. Let there be three particles A, B, C. We say that B and C both cause accelerations in A, and we compound these two accelerations by the parallelogram law. But this composition is not truly addition, for the components are not *parts* of the resultant. The resultant is a new term, as simple as its components, and not by any means their sum. Thus the effects attributed to B and C are never produced, but a third term different from either is produced. This, we may say, is produced by B and C together, taken as one whole. But the effect which they produce as a whole can only be discovered by supposing each to produce a separate effect: if this were not supposed, it would be impossible to obtain the two accelerations whose resultant is the actual acceleration. Thus we seem to reach an antinomy: the whole has no effect except what results from the effects of the parts, but the effects of the parts are non-existent.

The examination of this difficulty will rudely shake our cherished prejudices concerning causation. The laws of motion, we shall find, actually contradict the received view, and demand a quite different and far more complicated view. In Dynamics, we shall find (1) that the causal relation holds between events at three times, not at two; (2) that the whole state of the material universe at two of the three times is necessary to the statement of a causal relation. In order to provide for this conclusion, let us re-examine causality in a less conventional spirit.

452. Causality, generally, is the principle in virtue of which, from a sufficient number of events at a sufficient number of moments, one or more events at one or more new moments can be inferred. Let us suppose, for example, that, by means of the principle, if we are given e_1 events at a time t_1, e_2 at a time t_2,...e_n at a time t_n, then we can infer e_{n+1} events at a time t_{n+1}. If, then, $e_{r+1} \gtrless e_r$, and if the times t_r are arbitrary, except that t_{r+1} is after t_r, it follows that, from the original data, we can infer certain events at all future times. For we may choose e_1 of the events e_2,...e_n of the events e_{n+1}, and infer e_{n+1} events at a new time t_{n+2}. Hence by means of our supposed law, inference to future times is assured. And if, for any value of r, $e_{r+1} > e_r$, then more than e_{n+1} events can be inferred at the time t_{n+2}, since there are several ways of choosing e_r events out of e_{r+1} events. But if for any value of r, $e_{r+1} > e_r$, then inference to the past becomes in general impossible. In order that an *unambiguous* inference to the past may be possible, it is necessary that the implication should be reciprocal, *i.e.* that e_1 events at time t_1 should be implied by e_2 at t_2...e_{n+1} at t_{n+1}. But some inference to the past is possible without this condition, namely, that at time t_1 there were e_1 events implying, with the others up to t_n, the e_{n+1} events at time t_{n+1}. But even this inference soon fails if, for any value of r, $e_{r+1} > e_r$, since, after inferring e_1 events at time t_1, e_r for the next inference takes the place of e_{r+1}, but is too small to allow the inference. Thus if unambiguous inference to any part of time is to be possible, it is necessary and sufficient (1) that any one of the $n + 1$ groups of events should be implied by the other $\overline{n}$ groups; (2) that $e_r = e_{r+1}$ for all values of r. Since causality demands the possibility of such inference, we may take these two conditions as satisfied.

Another somewhat complicated point is the following. If $e_1 e_2...e_n$ cause e_{n+1}, and $e_2...e_{n+1}$, cause e_{n+2} and so on, we have an independent causal series, and a return to monadism, though the monad is now complex, being at each moment a group of events. But this result is not necessary. It may happen that only certain groups $e_1 e_2...e_n$ allow inference to e_{n+1}, and that $e_2 e_3...e_n$, e_{n+1} is not such a group. Thus suppose $e'_1 e'_2...e'_4$ simultaneous with $e_1...e_n$, and causing e'_{n+1}. It may be that $e_2 e_3...e_n e'_{n+1}$ and $e'_2 e'_3...e'_n e_{n+1}$ form the next causal groups, causing e_{n+2} and e'_{n+2} respectively. In this way no independent causal series will arise, in spite of particular causal sequences. This however remains a mere possibility, of which, so far as I know, no instance occurs.

Do the general remarks on the logical nature of causal propositions still hold good? Must we suppose the causal relation to hold directly between the *events* $e_1 e_2...e_{n+1}$, and merely to imply their temporal succession? There are difficulties in this view. For, having recognized that consecutive times are impossible, it has become necessary to assume finite intervals of time between e_1 and e_2, e_2 and e_3 etc. Hence the length of these intervals must be specified, and thus a mere reference to events, without regard to temporal position, becomes impossible. All we can say is, that only relative position is relevant. Given a causal relation in which the times are t_r, this relation will still be valid for times $T + t_r$. Thus the ultimate statement seems to be: given m events at any moment, m other events at a moment whose distance from the first is specified, and so on till we have n groups of events, then m new events can be inferred at any new moment whose distance from the first is specified, provided m and n have suitable values, and the groups of events be suitably chosen—where, however, the values to be assigned to m and n may depend upon the nature of the events in question. For example, in a material system consisting of N particles, we shall have $m = N$, $n = 2$. Here m depends upon the nature of the material system in question. What circumstances obtain in Psychology, it is as yet impossible to say, since psychologists have failed to establish any strict causal laws.

Thus rational Dynamics assume that, in an independent material system, the configurations at any two moments imply the configuration at any other moment. This statement is capable of translation into the language of pure mathematics, as we shall see in the next chapter. But it remains a question what we are to say concerning such causation of particulars by particulars as *appears* to be involved in such principles as the law of gravitation. But this discussion must be postponed until we have examined the so-called laws of motion.

CHAPTER LVI.

DEFINITION OF A DYNAMICAL WORLD.

453. BEFORE proceeding to the laws of motion, which introduce new complications of which some are difficult to express in terms of pure mathematics, I wish briefly to define in logical language the dynamical world as it results from previous chapters.

Let t be a one-dimensional continuous series, s a three-dimensional continuous series, which we will not assume to be Euclidean as yet. If R be a many-one relation whose domain is t and whose converse domain is contained in s, then R defines a motion of a material particle. The indestructibility and ingenerability of matter are expressed in the fact that R has the whole of t for its field. Let us assume further that R defines a continuous function in s.

In order to define the motions of a material system, it is only necessary to consider a class of relations having the properties assigned above to R, and such that the logical product of any two of them is null. This last condition expresses impenetrability. For it asserts that no two of our relations relate the same moment to the same point, *i.e.* no two particles can be at the same place at the same time. A set of relations fulfilling these conditions will be called a class of *kinematical motions*.

With these conditions, we have all that kinematics requires for the definition of matter; and if the descriptive school were wholly in the right, our definition would not add the new condition which takes us from kinematics to kinetics. Nevertheless this condition is essential to inference from events at one time to events at another, without which Dynamics would lose its distinctive feature.

454. A generalized form of the statement of causality which we require is the following: A class of *kinetic motions* is a class of kinematical motions such that, given the relata of the various component relations at n given times, the relata at all times are determinate. In ordinary Dynamics we have $n = 2$, and this assumption may be made without the loss of any interesting generality. Our assertion then amounts to saying that there is a certain specific many-one relation which holds between any two configurations and their times and any third time, as referent, and the configuration at the third time as relatum ; in ordinary language, given two configurations at two given times, the configuration at any other time is determinate. Formally, the principle of causality in this form may be stated as follows. If R be a relation which is any one of our motions, and t any time, let R_t be the relation holding only between t and the term to which t has the relation R. If K be the whole class of motions, let K_t be the whole class of such terms as R_t. Then K_t expresses the configuration of the system at the time t. Now let t', t'' be any other two times. Then K is a class of kinetic motions if there is a many-one relation S, the same for any three times, which holds between the class whose terms are t, t', t'', K_t, $K_{t'}$, as referent and the configuration $K_{t''}$ as relatum.

The particular causal laws of the particular universe considered are given when S is given, and *vice versâ* *. We may treat of a whole set of universes agreeing in having the same S, *i.e.* the same causal laws, and differing only in respect of the distribution of matter, *i.e.* the class K. This is the ordinary procedure of rational Dynamics, which commonly defines its S in the way believed to apply to the actual world, and uses its liberty only to imagine different material systems.

It will be observed that, owing to the rejection of the infinitesimal, it is necessary to give an integrated form to our general law of causality. We cannot introduce velocities and accelerations into statements of general principles, though they become necessary as soon as we descend to the laws of motion. A large part of Newton's laws, as we shall see in the next chapter, is contained in the above definition, but the third law introduces a radical novelty, and gives rise to the difficulty as to the causation of particulars by particulars, which we have mentioned but not yet examined.

* In the Dynamics applicable to the actual world, the specification of S requires the notion of mass.

CHAPTER LVII.

NEWTON'S LAWS OF MOTION.

455. THE present chapter will adopt, for the moment, a naïve attitude towards Newton's Laws. It will not examine whether they really hold, or whether there are other really ultimate laws applying to the ether; its problem is merely to give those laws a meaning.

The first thing to be remembered is—what physicists now-a-days will scarcely deny—that *force* is a mathematical fiction, not a physical entity. The second point is that, in virtue of the philosophy of the calculus, acceleration is a mere mathematical limit, and does not itself express a definite state of an accelerated particle. It may be remembered that, in discussing derivatives, we inquired whether it was possible to regard them otherwise than as limits—whether, in fact, they could be treated as themselves fractions. This we found impossible. In this conclusion there was nothing new, but its application in Dynamics will yield much that is distinctly new. It has been customary to regard velocity and acceleration as physical facts, and thus to regard the laws of motion as connecting configuration and acceleration. This, however, as an ultimate account, is forbidden to us. It becomes necessary to seek a more integrated form for the laws of motion, and this form, as is evident, must be one connecting three configurations.

456. The first law of motion is regarded sometimes as a definition of equal times. This view is radically absurd. In the first place, equal times have no definition except as times whose magnitude is the same. In the second place, unless the first law told us *when* there is no acceleration (which it does not do), it would not enable us to discover what motions are uniform. In the third place, if it is always significant to say that a given motion is uniform, there can be no motion by which uniformity is defined. In the fourth place, science holds that no motion occurring in nature is uniform; hence there must be a meaning of uniformity independent of all actual motions—and this definition is, the description of equal absolute distances in equal absolute times.

The first law, in Newton's form, asserts that velocity is unchanged in the absence of causal action from some other piece of matter. As it stands, this law is wholly confused. It tells us nothing as to how we are to discover causal action, or as to the circumstances under which causal action occurs. But an important meaning may be found for it, by remembering that velocity is a fiction, and that the only events that occur in any material system are the various positions of its various particles. If we then assume (as all the laws of motion ta itly do) that there is to be some relation between different configurations, the law tells us that such a relation can only hold between *three* configurations, not between two. For two configurations are required for velocity, and another for change of velocity, which is what the law asserts to be relevant. Thus in any dynamical system, when the special laws (other than the laws of motion) which regulate that system are specified, the configuration at any given time can be inferred when *two* configurations at *two* given times are known.

457. The second and third laws introduce the new idea of *mass* ; the third also gives one respect in which acceleration depends upon configuration.

The second law as it stands is worthless. For we know nothing about the impressed force except that it produces change of motion, and thus the law might seem to be a mere tautology. But by relating the impressed force to the configuration, an important law may be discovered, which is as follows. In any material system consisting of n particles, there are certain constant coefficients (masses) $m_1, m_2 \ldots m_n$ to be associated with these particles respectively; and when these coefficients are considered as forming part of the configuration, then m_1 multiplied by the corresponding acceleration is a certain function of the momentary configuration ; this is the same function for all times and all configurations. It is also a function dependent only upon the relative positions : the same configuration in another part of space will lead to the same accelerations. That is, if x_r, y_r, z_r be the coordinates of m_r at time t, we have $x_r = f_r(t)$ etc., and

$$ m_1\,\ddot{x}_1 = F\,(m_1,\ m_2,\ m_3,\ \ldots\ m_n,\ x_2 - x_1,\ x_3 - x_1 \ldots x_n - x_1,\ y_2 - y_1,\ \ldots). $$

This involves the assumption that $x_1 = f_1(r)$ is a function having a second differential coefficient $\ddot{x}_1$; the use of the equation involves the further assumption that $\ddot{x}_1$ has a first and second integral. The above, however, is a very specialized form of the second law ; in its general form, the function F may involve other coefficients than the masses, and velocities as well as positions.

458. The third law is very interesting, and allows the analysis of F into a vector sum of functions each depending only on m_1 and one other particle m_r and their relative position. It asserts that the acceleration of m_1 is made up of component accelerations having special reference respectively to $m_2, m_3 \ldots m_n$; and if these components be f_{12},

$f_{13}, \ldots f_{1n}$, it asserts that the acceleration of any other particle m_r has a corresponding component f_{r1} such that

$$m_r f_{r1} = - m_1 f_{1r}.$$

This law leads to the usual properties of the centre of mass. For if $\ddot{x}_{12}$ be the x-component of f_{12}, we have $m_1 \ddot{x}_{12} + m_2 \ddot{x}_{21} = 0$, and thus

$$\sum_r \sum_s m_r \ddot{x}_{rs} = 0.$$

Again, the special reference of f_{12} to m_2 can only be a reference to the mass m_2, the distance r_{12}, and the direction of the line 12; for these are the only intrinsic relations of the two particles. It is often specified as part of the third law that the acceleration is in the direction 12, and this seems worthy to be included, as specifying the dependence of f_{12} upon the line 12. Thus f_{12} is along 12, and

$$f_{12} = \phi \, (m_1, \, m_2, \, r_{12}),$$

$$f_{21} = \phi \, (m_2, \, m_1, \, - r_{12})$$

and

$$m_1 \, \phi \, (m_1, \, m_2, \, r_{12}) = - m_2 \, \phi \, (m_2, \, m_1, \, - r_{12}),$$

or, measuring f_{12} from 1 towards 2, and f_{21} from 2 towards 1, both will have the same sign, and

$$m_1 \, \phi \, (m_1, \, m_2, \, r_{12}) = m_2 \, \phi \, (m_2, \, m_1, \, r_{12}).$$

Hence $m_1 \, \phi \, (m_1, \, m_2, \, r_{12})$ is a symmetrical function of m_1 and m_2, say

$$\psi \, (m_1, \, m_2, \, r_{12}).$$

Thus

$$f_{12} = \frac{1}{m_1} \, \psi \, (m_1, \, m_2, \, r_{12}),$$

$$f_{21} = \frac{1}{m_2} \, \psi \, (m_1, \, m_2, \, r_{12}).$$

Thus the resultant acceleration of each particle is analyzable into components depending only upon itself and one other particle; but this analysis applies only to the statement in terms of acceleration. No such analysis is possible when we compare, not configuration and acceleration, but three configurations. At any moment, though the change of distance and straight line 12 is not due to m_1 and m_2 alone, yet the acceleration of m_1 consists of components each of which is the same it would be if there were only one other particle in the field. But where a finite time is in question this is no longer the case. The total change in the position of m_1 during a time t is not what it would have been if m_2 had first operated alone for a time t, then m_3 alone and so on. Thus we cannot speak of any total effect of m_2 or of m_3; and since momentary effects are fictions, there are really no independent effects of separate particles on m_1. The statement by means of accelerations is to be regarded as a mathematical device, not as though there really were an actual acceleration which is caused in one particle by one other. And thus we escape the very grave difficulty which we should otherwise

have to meet, namely, that the component accelerations, not being
(in general) parts of the resultant acceleration, would not be actual
even if we allowed that acceleration is an actual fact.

459. The first two laws are completely contained in the following
statement: In any independent system, the configuration at any time is
a function of that time and of the configurations at two given times,
provided we include in configuration the masses of the various particles
composing the system. The third law adds the further fact that the
configuration can be analyzed into distances and straight lines; the
function of the configuration which represents the acceleration of any
particle is a vector-sum of functions containing only one distance, one
straight line, and two masses each—moreover, if we accept the addition
to the third law spoken of above, each of these functions is a vector
along the join of the two particles which enter into it. But for this
law, it might happen that the acceleration of m_1 would involve the area
of the triangle 1 2 3, or the volume of the tetrahedron 1 2 3 4; and but
for this law, we should not have the usual properties of the centre of mass.

The three laws together, as now expounded, give the greater part of
the law of gravitation; this law merely tells us that, so far as gravitation
is concerned, the above function

$$\psi\,(m_1,\ m_2,\ r_{12}) = m_1\,m_2\,/r_{12}{}^2.$$

It should be remembered that nothing is known, from the laws of motion,
as to the form of ψ, and that we might have e.g. $\psi = 0$ if $r_{12} > R$. If
ψ had this form, provided R were small compared to sensible distances,
the world would seem as though there were no action at a distance.

It is to be observed that the first two laws, according to the above
analysis, merely state the general form of the law of causality explained
in Chapter IV. From this it results that we shall be able, with the
assumptions commonly made as to continuity and the existence of first
and second derivatives, to determine a motion completely when the
configuration and velocities at a given instant are given; and in particular, these data will enable us to determine the acceleration at the
given instant. The third law and the law of gravitation together add
the further properties that the momentary accelerations depend only
upon the momentary configuration, not upon the momentary velocities,
and that the resultant acceleration of any particle is the vector-sum
of components each dependent only on the masses and distances of the
given particle and one other.

The question whether Newtonian Dynamics applies in such problems
as those of the motion of the ether is an interesting and important one;
but in so far as it deals with the truth or falsehood of the laws of motion
in relation to the actual world, it is for us irrelevant. For us, as pure
mathematicians, the laws of motion and the law of gravitation are not
properly laws at all. but parts of the definition of a certain kind of matter.

460. By the above account the view of causality which has usually satisfied philosophers is contravened in two respects, (1) in that the relation embodied in a causal law holds between three events, not between two; (2) in that the causal law has the unity of a formula or function, *i.e.* of a constant relation, not merely that derived from repetition of the same cause. The first of these is necessitated by modern theories of the infinitesimal calculus; the second was always necessary, at least since Newton's time. Both demand some elucidation.

(1) The whole essence of dynamical causation is contained in the following equation: if t_1, t_2 be specified times, C_1, C_2 the corresponding configurations of any self-contained system, and C the configuration at any time t, then

$$C = F(C_1, t_1, C_2, t_2, t)$$

(a compressed form for as many equations as C has coordinates). The form of F depends only upon the number of particles and the dynamical laws of the system, not upon the choice of C_1 or C_2. The cause must be taken to be the *two* configurations C_1 and C_2, and the interval $t_2 - t_1$ may be any we please. Further t may fall between t_1 and t_2, or before both. The effect is any single one of the coordinates of the system at time t, or any collection of these coordinates; but it seems better to regard each coordinate as one effect, since each is given in one equation. Thus the language of cause and effect has to be greatly strained to meet the case, and seems scarcely worth preserving. The cause is two states of the whole system, at times as far apart as we please; the effect is one coordinate of the system at any time before, after, or between the times in the cause. Nothing could well be more unlike the views which it has pleased philosophers to advocate. Thus on the whole it is not worth while preserving the word *cause*: it is enough to say, what is far less misleading, that any two configurations allow us to infer any other.

(2) The causal law regulating any system is contained in the form of F. The law does not assert that one event A will always be followed by another B; if A be the configuration of the system at one time, nothing can be inferred as to that at another; the configuration might recur without a recurrence of any configuration that formerly followed it. If A be two configurations whose distance in time is given, then indeed our causal law does tell us what configurations will follow them, and if A recurred, so would its consequences. But if this were all that our causal law told us, it would afford cold comfort, since no configuration ever does actually recur. Moreover, we should need an infinite number of causal laws to meet the requirements of a system which has successively an infinite number of configurations. What our law does is to assert that an infinite class of effects have each the same functional relation to one of an infinite class of causes; and this is done by means of a formula. One formula connects *any* three configurations,

and but for this fact continuous motions would not be amenable to causal laws, which consist in specifications of the formula.

461. I have spoken hitherto of independent systems of n particles. It remains to examine whether any difficulties are introduced by the fact that, in the dynamical world, there are no independent systems short of the material universe. We have seen that no effect can be ascribed, within a material system, to any one part of the system ; the whole system is necessary for any inference as to what will happen to one particle. The only effect traditionally attributed to the action of a single particle on another is a component acceleration; but (a) this is not part of the resultant acceleration, (β) the resultant acceleration itself is not an event, or a physical fact, but a mere mathematical limit. Hence nothing can be attributed to particular particles. But it may be objected that we cannot know the whole material universe, and that, since no effect is attributable to any part as such, we cannot consequently know anything about the effect of the whole. For example, in calculating the motions of planets, we neglect the fixed stars ; we pretend that the solar system is the whole universe. By what right, then, do we assume that the effects of this feigned universe in any way resemble those of the actual universe?

The answer to this question is found in the law of gravitation. We can show that, if we compare the motions of a particle in a number of universes differing only as regards the matter at a greater distance than R, while much within this distance all of them contain much matter, then the motion of the particle in question relatively to the matter well within the distance R will be approximately the same in all the universes*. This is possible because, by the third law, a kind of fictitious analysis into partial effects is possible. Thus we can approximately calculate the effect of a universe of which part only is known. We must not say that the effect of the fixed stars is insensible, for we assume that they have no effect *per se* ; we must say that the effect of a universe in which they exist differs little from that of one in which they do not exist; and this we are able to prove in the case of gravitation. Speaking broadly, we require (recurring to our previous function ϕ) that, if ϵ be any number, however small, there should be some distance R such that, recurring to our previous function ϕ, if $\dfrac{d}{ds}$ denote differentiation in any direction, then

$$\frac{d}{ds}\int_r^\infty \phi(r)\,dr < \epsilon \text{ if } r > R.$$

When this condition is satisfied, the difference between the relative accelerations of two particles within a certain region, which results from assuming different distributions of matter at a distance greater than R from a certain point within the region, will have an assignable upper limit ; and hence there is an upper limit to the error incurred by pre-

* This is true only of *relative*, not of absolute motions.

tending that there is no matter outside the space of radius R. Hence approximation becomes possible in spite of the fact that the whole universe is involved in the exact determination of any motion.

The above leads to two observations of some interest. First, no law which does not satisfy the above inequality is capable of being practically applied or tested. The assumption that gravity varies as the direct distance, for example, could only be tested in a finite universe. And in all phenomena, such as those of electricity, we must assume, where the total effect is a sum or integral, or is calculated by means of a sum or integral, that the portion contributed to relative motions by large values of r is small. Secondly, the denial of any partial effect of a part is quite necessary if we are to apply our formulae to an infinite universe in the form of integrals. For an integral is not really an infinite sum, but the limit of a finite sum. Thus if each particle had a partial effect, the total effect of an infinite number of particles would *not* be an integral. But though an integral cannot represent an infinite sum, there seems no reason whatever why it should not represent the effect of a universe which has an infinite number of parts. If there are finite volumes containing an infinite number of particles, the notion of mass must be modified so as to apply no longer to single particles, but to infinite classes of particles. The density at a point will then be not the mass of that point, but the differential coefficient, at the point, of the mass with respect to the volume.

It should be observed that the impossibility of an independent system short of the whole universe does not result from the laws of motion, but from the special laws, such as that of gravitation, which the laws of motion lead us to seek.

462. The laws of motion, to conclude, have no vestige of self-evidence; on the contrary, they contradict the form of causality which has usually been considered evident. Whether they are ultimately valid, or are merely approximate generalizations, must remain doubtful: the more so as, in all their usual forms, they assume the truth of the axiom of parallels, of which we have so far no evidence. The laws of motion, like the axiom of parallels in regard to space, may be viewed either as parts of a definition of a class of possible material universes, or as empirically verified assertions concerning the actual material universe. But in no way can they be taken as *à priori* truths necessarily applicable to any possible material world. The *à priori* truths involved in Dynamics are only those of logic: as a system of deductive reasoning, Dynamics requires nothing further, while as a science of what exists, it requires experiment and observation. Those who have admitted a similar conclusion in Geometry are not likely to question it here; but it is important to establish separately every instance of the principle that knowledge as to what exists is never derivable from general philosophical considerations, but is always and wholly empirical.

CHAPTER LVIII.

ABSOLUTE AND RELATIVE MOTION.

463. In the justly famous scholium to the definitions, Newton has stated, with admirable precision, the doctrine of absolute space, time, and motion. Not being a skilled philosopher, he was unable to give grounds for his views, except an empirical argument derived from actual Dynamics. Leibniz, with an unrivalled philosophical equipment, controverted Newton's position in his letters against Clarke*; and the victory, in the opinion of subsequent philosophers, rested wholly with Leibniz. Although it would seem that Kant, in the Transcendental Aesthetic, inclines to absolute position in space, yet in the *Metaphysische Anfangsgründe der Naturwissenschaft* he quite definitely adopts the relational view. Not only other philosophers, but also men of science, have been nearly unanimous in rejecting absolute motion, the latter on the ground that it is not capable of being observed, and cannot therefore be a datum in an empirical study.

But a great difficulty has always remained as regards the argument from absolute rotation, adduced by Newton himself. This argument, in spite of a definite assertion that all motion is relative, is accepted and endorsed by Clerk Maxwell†. It has been revived and emphasized by Heymans‡, combated by Mach§, Karl Pearson||, and many others, and made part of the basis of a general attack on Dynamics in Professor Ward's *Naturalism and Agnosticism*. Let us first state the argument in various forms, and then examine some of the attempts to reply to it. For us, since absolute time and space have been admitted, there is no need to avoid absolute motion, and indeed no possibility of doing so. But if absolute motion is in any case unavoidable, this affords a new argument in favour of the justice of our logic, which, unlike the logic current among philosophers, admits and even urges its possibility.

---

* *Phil. Werke*, ed. Gerhardt, Vol. VII.
† *Matter and Motion*, Art. cv. Contrast Art. xxx.
‡ *Die Gesetze und Elemente des wissenschaftlichen Denkens*, Leyden, 1890.
§ *Die Mechanik in ihrer Entwickelung*, Leipzig, 1883. (Translated, London, 1902.)
|| *Grammar of Science*, London, 1892. (2nd edition, 1900.)

464. If a bucket containing water is rotated, Newton observes, the water will become concave and mount up the sides of the bucket. But if the bucket be left at rest in a rotating vessel, the water will remain level in spite of the relative rotation. Thus absolute rotation is involved in the phenomenon in question. Similarly, from Foucault's pendulum and other similar experiments, the rotation of the earth can be demonstrated, and could be demonstrated if there were no heavenly bodies in relation to which the rotation becomes sensible. But this requires us to admit that the earth's rotation is absolute. Simpler instances may be given, such as the case of two gravitating particles. If the motion dealt with in Dynamics were wholly relative, these particles, if they constituted the whole universe, could only move in the line joining them, and would therefore ultimately fall into one another. But Dynamics teaches that, if they have initially a relative velocity not in the line joining them, they will describe conics about their common centre of gravity as focus. And generally, if acceleration be expressed in polars, there are terms in the acceleration which, instead of containing several differentials, contain squares of angular velocities: these terms require absolute angular velocity, and are inexplicable so long as relative motion is adhered to.

If the law of gravitation be regarded as universal, the point may be stated as follows. The laws of motion require to be stated by reference to what have been called *kinetic* axes: these are in reality axes having no absolute acceleration and no absolute rotation. It is asserted, for example, when the third law is combined with the notion of mass, that, if m, m' be the masses of two particles between which there is a force, the component accelerations of the two particles due to this force are in the ratio $m_2 : m_1$. But this will only be true if the accelerations are measured relatively to axes which themselves have no acceleration. We cannot here introduce the centre of mass, for, according to the principle that dynamical facts must be, or be derived from, observable data, the masses, and therefore the centre of mass, must be obtained from the acceleration, and not *vice versâ*. Hence any dynamical motion, if it is to obey the laws of motion, must be referred to axes which are not subject to any forces. But, if the law of gravitation be accepted, no *material* axes will satisfy this condition. Hence we shall have to take *spatial* axes, and motions relative to these are of course absolute motions.

465. In order to avoid this conclusion, C. Neumann* assumes as an essential part of the laws of motion the existence, somewhere, of an absolutely rigid " Body *Alpha*," by reference to which all motions are to be estimated. This suggestion misses the essence of the discussion, which is (or should be) as to the logical *meaning* of dynamical pro-

---

* *Die Galilei-Newtonsche Theorie*, Leipzig, 1870, p. 15.

positions, not as to the way in which they are discovered. It seems sufficiently evident that, if it is necessary to invent a fixed body, purely hypothetical and serving no purpose except to be fixed, the reason is that what is really relevant is a fixed *place*, and that the body occupying it is irrelevant. It is true that Neumann does not incur the vicious circle which would be involved in saying that the Body *Alpha* is fixed, while all motions are relative to it; he asserts that it is rigid, but rightly avoids any statement as to its rest or motion, which, in his theory, would be wholly unmeaning. Nevertheless, it seems evident that the question whether one body is at rest or in motion must have as good a meaning as the same question concerning any other body; and this seems sufficient to condemn Neumann's suggested escape from absolute motion.

466. A development of Neumann's views is undertaken by Streintz[*], who refers motions to what he calls "fundamental bodies" and "fundamental axes." These are defined as bodies or axes which do not rotate and are independent of all outside influences. Streintz follows Kant's *Anfangsgründe* in regarding it as possible to admit absolute rotation while denying absolute translation. This is a view which I shall discuss shortly, and which, as we shall see, though fatal to what is desired of the relational theory, is yet logically tenable, though Streintz does not show that it is so. But apart from this question, two objections may be made to his theory. (1) If motion *means* motion relative to fundamental bodies (and if not, their introduction is no gain from a logical point of view), then the law of gravitation becomes strictly meaningless if taken to be universal—a view which seems impossible to defend. The theory requires that there should be matter not subject to any forces, and this is denied by the law of gravitation. The point is not so much that universal gravitation must be *true*, as that it must be significant— whether true or false is an irrelevant question. (2) We have already seen that absolute *accelerations* are required even as regards translations, and that the failure to perceive this is due to overlooking the fact that the centre of mass is not a piece of matter, but a spatial point which is only determined by means of accelerations.

467. Somewhat similar remarks apply to Mr W. H. Macaulay's article on "Newton's Theory of Kinetics[†]." Mr Macaulay asserts that the true way to state Newton's theory (omitting points irrelevant to the present issue) is as follows: "Axes of reference can be so chosen, and the assignment of masses so arranged, that a certain decomposition of the rates of change of momenta, relative to the axes, of all the particles of the universe is possible, namely one in which the components occur

---

[*] *Die physikalischen Grundlagen der Mechanik*, Leipzig, 1883; see esp. pp. 24, 25.

[†] Bulletin of the American Math. Soc., Vol. III. (1896–7). For a later statement of Mr Macaulay's views, see Art. *Motion, Laws of*, in the new volumes of the *Encycl. Brit.* (Vol. XXXI).

in pairs; the members of each pair belonging to two different particles, and being opposite in direction, in the line joining the particles, and equal in magnitude" (p. 368). Here again, a purely logical point remains. The above statement appears unobjectionable, but it does not show that absolute motion is unnecessary. The axes cannot be material, for all matter is or may be subject to forces, and therefore unsuitable for our purpose; they cannot even be defined by any fixed geometrical relation to matter. Thus our axes will really be spatial; and if there were no absolute space, the suggested axes could not exist. For apart from absolute space, any axes would have to be material or nothing. The axes can, in a sense, be defined by relation to matter, but not by a constant geometrical relation; and when we ask what property is changed by motion relative to such axes, the only possible answer is that the absolute position has changed. Thus absolute space and absolute motion are not avoided by Mr Macaulay's statement of Newton's laws.

468. If absolute rotation alone were in question, it would be possible, by abandoning all that recommends the relational theory to philosophers and men of science, to keep its logical essence intact. What is aimed at is, to state the principles of Dynamics in terms of sensible entities. Among these we find the metrical properties of space, but not straight lines and planes. Collinearity and coplanarity may be included, but if a set of collinear material points change their straight line, there is no sensible intrinsic change. Hence all advocates of the relational theory, when they are thorough, endeavour, like Leibniz*, to deduce the straight line from distance. For this there is also the reason that the field of a given distance is all space, whereas the field of the generating relation of a straight line is only that straight line, whence the latter, but not the former, makes an intrinsic distinction among the points of space, which the relational theory seeks to avoid. Still, we might regard straight lines as relations between *material* points, and absolute rotation would then appear as change in a relation between material points, which is logically compatible with a relational theory of space. We should have to admit, however, that the straight line was not a *sensible* property of two particles between which it was a relation; and in any case, the necessity for absolute translational accelerations remains fatal to any relational theory of motion.

469. Mach† has a very curious argument by which he attempts to refute the grounds in favour of absolute rotation. He remarks that, in the actual world, the earth rotates relating to the fixed stars, and that the universe is not given twice over in different shapes, but only once, and as we find it. Hence any argument that the rotation of the earth could be inferred *if* there were no heavenly bodies is futile. This argument contains the very essence of empiricism, in a sense in which

---

* See my article "Recent Work on Leibniz," in *Mind*, 1903.
† *Die Mechanik in ihrer Entwickelung*, 1st edition, p. 216.

empiricism is radically opposed to the philosophy advocated in the present work*. The logical basis of the argument is that all propositions are essentially concerned with actual existents, not with entities which may or may not exist. For if, as has been held throughout our previous discussions, the whole dynamical world with its laws can be considered without regard to existence, then it can be no part of the *meaning* of these laws to assert that the matter to which they apply exists, and therefore they can be applied to universes which do not exist. Apart from general arguments, it is evident that the laws are so applied throughout rational Dynamics, and that, in all exact calculations, the distribution of matter which is assumed is not that of the actual world. It seems impossible to deny significance to such calculations; and yet, if they have significance, if they contain propositions at all, whether true or false, then it can be no necessary part of their *meaning* to assert the existence of the matter to which they are applied. This being so, the universe is given, as an entity, not only twice, but as many times as there are possible distributions of matter, and Mach's argument falls to the ground. The point is important, as illustrating a respect in which the philosophy here advocated is to be reckoned with idealism and not with empiricism, in spite of the contention that what exists can only be known empirically.

Thus, to conclude: Absolute motion is essential to Dynamics, and involves absolute space. This fact, which is a difficulty in current philosophies, is for us a powerful confirmation of the logic upon which our discussions have been based.

* Cf. Art. "Nativism" in the *Dictionary of Philosophy and Psychology*, edited by Baldwin, Vol. ii, 1902.

CHAPTER LIX.

HERTZ'S DYNAMICS.

470. WE have seen that Newton's Laws are wholly lacking in self-evidence—so much so, indeed, that they contradict the law of causation in a form which has usually been held to be indubitable. We have seen also that these laws are specially suggestive of the law of gravitation. In order to eliminate what, in elementary Dynamics, is specially Newtonian, from what is really essential to the subject, we shall do well to examine some attempts to re-state the fundamental principles in a form more applicable to such sciences as Electricity. For this purpose the most suitable work seems to be that of Hertz*.

The fundamental principles of Hertz's theory are so simple and so admirable that it seems worth while to expound them briefly. His object, like that of most recent writers, is to construct a system in which there are only three fundamental concepts, space, time, and mass. The elimination of a fourth concept, such as force or energy, though evidently demanded by theory, is difficult to carry out mathematically. Hertz seems, however, to have overcome the difficulty in a satisfactory manner. There are, in his system, three stages in the specification of a motion. In the first stage, only the relations of space and time are considered: this stage is purely kinematical. Matter appears here merely as a means of establishing, through the motion of a particle, a one-one correlation between a series of points and a series of instants. At this stage a collection of n particles has $3n$ coordinates, all so far independent: the motions which result when all are regarded as independent are all the *thinkable* motions of the system. But before coming to kinetics, Hertz introduces an intermediate stage. Without introducing time, there are in any free material system direct relations between space and mass, which form the geometrical connections of the system. (These may introduce time in the sense of involving velocities, but they are independent of time in the sense that they are expressed at all times by the same equations, and that these do not contain the time explicitly.) Those among thinkable motions which satisfy the equations

* *Principien der Mechanik*, Leipzig, 1894.

of connection are called *possible* motions. The connections among the parts of a system are assumed further to be continuous in a certain well-defined sense (p. 89). It then follows that they can be expressed by homogeneous linear differential equations of the first order among the coordinates. But now a further principle is needed to discriminate among possible motions, and here Hertz introduces his only law of motion, which is as follows :

"Every free system persists in its state of rest or of uniform motion in a straightest path."

This law requires some explanation. In the first place, when there are in a system unequal particles, each is split into a number of particles proportional to its mass. By this means all particles become equal. If now there are *n* particles, their $3n$ coordinates are regarded as the coordinates of a point in space of $3n$ dimensions. The above law then asserts that, in a free system, the velocity of this representative point is constant, and its path from a given point to another neighbouring point in a given direction is that one, among the possible paths through these two points, which has the smallest curvature. Such a path is called a *natural* path, and motion in it is called a *natural* motion.

471. It will be seen that this system, though far simpler and more philosophical in form than Newton's, does not differ very greatly in regard to the problems discussed in the preceding chapter. We still have, what we found to be the essence of the law of inertia, the necessity for three configurations in order to obtain a causal relation. This broad fact must reappear in every system at all resembling ordinary Dynamics, and is exhibited in the necessity for differential equations of the second order, which pervades all Physics. But there is one very material difference between Hertz's system and Newton's—a difference which, as Hertz points out, renders an experimental decision between the two at least theoretically possible. The special laws, other than the laws of motion, which regulate any particular system, are for Newton laws concerning mutual accelerations, such as gravitation itself. For Hertz, these special laws are all contained in the geometrical connections of the system, and are expressed in equations involving only velocities (*v.* p. 48). This is a considerable simplification, and is shown by Hertz to be more conformable to phenomena in all departments except where gravitation is concerned. It is also a great simplification to have only one law of motion, instead of Newton's three. But for the philosopher, so long as this law involves second differentials (which are introduced through the curvature), it is a comparatively minor matter that the special laws of special systems should be of the first order.

The definition of mass as number of particles, it should be observed, is a mere mathematical device, and is not, I think, regarded by Hertz as anything more (*v.* p. 54). Not only must we allow the possibility of incommensurable masses, but even if this difficulty were overcome, it would still remain significant to assert that all our ultimate particles were equal. Mass would therefore still be a variety of magnitude, only that all particles would happen to be of the same magnitude as regards their mass. This would not effect any theoretical simplification, and we shall do well, therefore, to retain mass as an intensive quantity of which a certain magnitude belongs to a certain particle, without any implication that the particle is divisible. There is, in fact, no valid ground for denying ultimately different masses to different particles. The whole question is, indeed, purely empirical, and the philosopher should, in this matter, accept passively what the physicist finds requisite.

With regard to ether and its relations to matter, a similar remark seems to be applicable. Ether is, of course, matter in the philosophical sense; but beyond this the present state of Science will scarcely permit us to go. It should be observed, however, that in Electricity, as elsewhere, our equations are of the second order, thus indicating that the law of inertia, as interpreted in the preceding chapter, still holds good. This broad fact seems, indeed, to be the chief result, for philosophy, of our discussion of dynamical principles.

472. Thus to sum up, we have two principal results:

(1) In any independent system, there is a relation between the configurations at three given times, which is such that, given the configurations at two of the times, the configuration at the third time is determinate.

(2) There is no independent system in the actual world except the whole material universe; but if two universes which have the same causal laws as the actual universe differ only in regard to the matter at a great distance from a given region, the relative motions within this region will be approximately the same in the two universes—*i.e.* an upper limit can be found for the difference between the two sets of motions.

These two principles apply equally to the Dynamics of Newton and to that of Hertz. When these are abandoned, other principles will give a science having but little resemblance to received Dynamics.

473. One general principle, which is commonly stated as vital to Dynamics, deserves at least a passing mention. This is the principle that the cause and effect are equal. Owing to pre-occupation with quantity and ignorance of symbolic logic, it appears to have not been perceived that this statement is equivalent to the assertion that the implication between cause and effect is mutual. All equations, at bottom, are logical equations, *i.e.* mutual implications; quantitative equality between variables, such as cause and effect, involves a mutual formal implication. Thus the principle in question can only be maintained if cause and effect be placed on the same logical level, which, with the interpretation we were compelled to give to causality, it is no longer possible to do. Nevertheless, when one state of the universe is given, any two others have a mutual implication; and this is the source of the various laws of conservation which pervade Dynamics, and give the truth underlying the supposed equality of cause and effect.

474. We may now review the whole course of the arguments contained in the present work. In Part I, an attempt is made to analyze the nature of deduction, and of the logical concepts involved in it. Of these, the most puzzling is the notion of *class*, and from the contradiction discussed in Chapter x (though this is perhaps soluble by the doctrine of types*), it appeared that a tenable theory as to the nature of classes is very hard to obtain. In subsequent Parts, it was shown that existing pure mathematics (including Geometry and Rational Dynamics) can be derived wholly from the indefinables and indemonstrables of Part I. In this process, two points are specially important: the definitions and the existence-theorems. A definition is always either the definition of a class, or the definition of the single member of a unit class: this is a necessary result of the plain fact that a definition can only be effected by assigning a property of the object or objects to be defined, *i.e.* by stating a propositional function which they are to satisfy. A kind of grammar controls definitions, making it impossible *e.g.* to define Euclidean *Space*, but possible to define the class of Euclidean *spaces*. And wherever the principle of abstraction is employed, *i.e.* where the object to be defined is obtained from a transitive symmetrical relation, some class of classes will always be the object required. When symbolic expressions are used, the requirements of what may be called grammar become evident, and it is seen that the logical type of the entity defined is in no way optional.

The existence-theorems of mathematics—*i.e.* the proofs that the various classes defined are not null—are almost all obtained from Arithmetic. It may be well here to collect the more important of them. The existence of zero is derived from the fact that the null-class is a member of it; the existence of 1 from the fact that zero is a unit-class (for the null-class is its only member). Hence, from the fact that, if n be a finite number, $n + 1$ is the number of numbers from 0 to n (both inclusive), the existence-theorem follows for all finite numbers. Hence, from the class of the finite cardinal numbers themselves, follows the existence of α_0, the smallest of the infinite cardinal numbers; and from the series of finite cardinals in order of magnitude follows the existence of ω, the smallest of infinite ordinals. From the definition of the rational numbers and of their order of magnitude follows the existence of η, the type of endless compact denumerable series; thence, from the segments of the series of rationals, the existence of the real numbers, and of θ, the type of continuous series. The terms of the series of well-ordered types are proved to exist from the two facts: (1) that the number of well-ordered types from 0 to α is $\alpha + 1$, (2) that

---

* See Appendix B.

if u be a class of well-ordered types having no maximum, the series of all types not greater than every u is itself of a type greater than every u. From the existence of θ, by the definition of complex numbers (Chapter XLIV), we prove the existence of the class of Euclidean spaces of any number of dimensions; thence, by the process of Chapter XLVI, we prove the existence of the class of projective spaces, and thence, by removing the points outside a closed quadric, we prove the existence of the class of non-Euclidean descriptive (hyperbolic) spaces. By the methods of Chapter XLVIII, we prove the existence of spaces with various metrical properties. Lastly, by correlating some of the points of a space with all the terms of a continuous series in the ways explained in Chapter LVI, we prove the existence of the class of dynamical worlds. Throughout this process, no entities are employed but such as are definable in terms of the fundamental logical constants. Thus the chain of definitions and existence-theorems is complete, and the purely logical nature of mathematics is established throughout.

APPENDIXES

LIST OF ABBREVIATIONS.

Bs. *Begriffsschrift.* Eine der arithmetischen nachgebildete Formelsprache des reinen Denkens. Halle a/S, 1879.

Gl. *Grundlagen der Arithmetik.* Eine logisch-mathematische Untersuchung über den Begriff der Zahl. Breslau, 1884.

FT. *Ueber formale Theorien der Arithmetik.* Sitzungsberichte der Jenaischen Gesellschaft für Medicin und Naturwissenschaft, 1885.

FuB. *Function und Begriff.* Vortrag gehalten in der Sitzung vom 9. Januar, 1891, der Jenaischen Gesellschaft für Medicin und Naturwissenschaft. Jena, 1891.

BuG. *Ueber Begriff und Gegenstand.* Vierteljahrschrift für wiss. Phil., xvi 2 (1892).

SuB. *Ueber Sinn und Bedeutung.* Zeitschrift für Phil. und phil. Kritik, vol. 100 (1892).

KB. *Kritische Beleuchtung einiger Punkte in E. Schröder's Vorlesungen über die Algebra der Logik.* Archiv für syst. Phil., Vol. i (1895).

BP. *Ueber die Begriffsschrift des Herrn Peano und meine eigene.* Berichte der math.-physischen Classe der Königl. Sächs. Gesellschaft der Wissenschaften zu Leipzig (1896).

Gg. *Grundgesetze der Arithmetik.* Begriffsschriftlich abgeleitet. Vol. i. Jena, 1893. Vol. ii. 1903.

APPENDIX A.

THE LOGICAL AND ARITHMETICAL DOCTRINES OF FREGE.

475. THE work of Frege, which appears to be far less known than it deserves, contains many of the doctrines set forth in Parts I and II of the present work, and where it differs from the views which I have advocated, the differences demand discussion. Frege's work abounds in subtle distinctions, and avoids all the usual fallacies which beset writers on Logic. His symbolism, though unfortunately so cumbrous as to be very difficult to employ in practice, is based upon an analysis of logical notions much more profound than Peano's, and is philosophically very superior to its more convenient rival. In what follows, I shall try briefly to expound Frege's theories on the most important points, and to explain my grounds for differing where I do differ. But the points of disagreement are very few and slight compared to those of agreement. They all result from difference on three points: (1) Frege does not think that there is a contradiction in the notion of concepts which cannot be made logical subjects (see § 49 *supra*) ; (2) he thinks that, if a term *a* occurs in a proposition, the proposition can always be analysed into *a* and an assertion about *a* (see Chapter VII); (3) he is not aware of the contradiction discussed in Chapter x. These are very fundamental matters, and it will be well here to discuss them afresh, since the previous discussion was written in almost complete ignorance of Frege's work.

Frege is compelled, as I have been, to employ common words in technical senses which depart more or less from usage. As his departures are frequently different from mine, a difficulty arises as regards the translation of his terms. Some of these, to avoid confusion, I shall leave untranslated, since every English equivalent that I can think of has been already employed by me in a slightly different sense.

The principal heads under which Frege's doctrines may be discussed are the following: (1) meaning and indication ; (2) truth-values and judgment; (3) Begriff and Gegenstand; (4) classes; (5) implication and symbolic logic ; (6) the definition of integers and the principle of abstraction ; (7) mathematical induction and the theory of progressions. I shall deal successively with these topics.

476. *Meaning and indication.* The distinction between meaning (*Sinn*) and indication (*Bedeutung*)* is roughly, though not exactly, equivalent to my distinction between a concept as such and what the concept denotes (§ 96). Frege did not possess this distinction in the first two of the works under consideration (the *Begriffsschrift* and the *Grundlagen der Arithmetik*) ; it appears first in BuG. (cf. p. 198), and is specially dealt with in SuB. Before making the distinction, he thought that identity has to do with the names of objects (Bs. p. 13) : " A is identical with B " means, he says, that the sign A and the sign B have the same signification (Bs. p. 15)—a definition which, verbally at least, suffers from circularity. But later he explains identity in much the same way as it was explained in § 64. "Identity," he says, "calls for reflection owing to questions which attach to it and are not quite easy to answer. Is it a relation? A relation between Gegenstände? or between names or signs of Gegenstände?" (SuB. p. 25). We must distinguish, he says, the meaning, in which is contained.the way of being given, from what is indicated (from the *Bedeutung*). Thus "the evening star" and "the morning star" have the same indication, but not the same meaning. A word ordinarily stands for its indication; if we wish to speak of its meaning, we must use inverted commas or some such device (pp. 27–8). The indication of a proper name is the object which it indicates; the presentation which goes with it is quite subjective; between the two lies the meaning, which is not subjective and yet is not the object (p. 30). A proper name *expresses* its meaning, and *indicates* its indication (p. 31).

This theory of indication is more sweeping and general than mine, as appears from the fact that *every* proper name is supposed to have the two sides. It seems to me that only such proper names as are derived from concepts by means of *the* can be said to have meaning, and that such words as *John* merely indicate without meaning. If one allows, as I do, that concepts can be objects and have proper names, it seems fairly evident that their proper names, as a rule, will indicate them without having any distinct meaning; but the opposite view, though it leads to an endless regress, does not appear to be logically impossible. The further discussion of this point must be postponed until we come to Frege's theory of Begriffe.

477. *Truth-values and Judgment.* The problem to be discussed under this head is the same as the one raised in § 52†, concerning the difference between asserted and unasserted propositions. But Frege's position on this question is more subtle than mine, and involves a more radical analysis of judgment. His *Begriffsschrift*, owing to the absence of the distinction between meaning and indication, has a simpler theory than his later works. I shall therefore omit it from the discussions.

There are, we are told (Gg. p. x), three elements in judgment : (1) the recognition of truth, (2) the Gedanke, (3) the truth-value (*Wahrheitswerth*).

---

* I do not translate *Bedeutung* by *denotation*, because this word has a technical meaning different from Frege's, and also because *bedeuten*, for him, is not quite the same as *denoting* for me.

† This is the logical side of the problem of *Annahmen*, raised by Meinong in his able work on the subject, Leipzig, 1902. The logical, though not the psychological, part of Meinong's work appears to have been completely anticipated by Frege.

Here the Gedanke is what I have called an unasserted proposition—or rather, what I called by this name covers both the Gedanke alone and the Gedanke together with its truth-value. It will be well to have names for these two distinct notions; I shall call the Gedanke alone a *propositional concept*; the truth-value of a Gedanke I shall call an *assumption**. Formally at least, an assumption does not require that its content should be a propositional concept: whatever x may be, "the truth of x" is a definite notion. This means the true if x is true, and if x is false or not a proposition it means the false (FuB. p. 21). In like manner, according to Frege, there is "the falsehood of x"; these are not assertions and negations of propositions, but only assertions of truth or of falsity, *i.e.* negation belongs to what is asserted, and is not the opposite of assertion†. Thus we have first a propositional concept, next its truth or falsity as the case may be, and finally the assertion of its truth or falsity. Thus in a hypothetical judgment, we have a relation, not of two judgments, but of two propositional concepts (SuB. p. 43).

This theory is connected in a very curious way with the theory of meaning and indication. It is held that every assumption indicates the true or the false (which are called truth-values), while it means the corresponding propositional concept. The assumption "$2^2 = 4$" indicates the true, we are told, just as "2^2" indicates 4‡ (FuB. p. 13; SuB. p. 32). In a dependent clause, or where a name occurs (such as Odysseus) which indicates nothing, a sentence may have no indication. But when a sentence has a truth-value, this is its indication. Thus every assertive sentence (*Behauptungssatz*) is a proper name, which indicates the true or the false (SuB. pp. 32—4; Gg. p. 7). The sign of judgment (*Urtheilstrich*) does not combine with other signs to denote an object; a judgment indicates nothing, but asserts something. Frege has a special symbol for judgment, which is something distinct from and additional to the truth-value of a propositional concept (Gg. pp. 9—10).

478. There are some difficulties in the above theory which it will be well to discuss. In the first place, it seems doubtful whether the introduction of truth-values marks any real analysis. If we consider, say, "Caesar died," it would seem that what is asserted is the propositional concept "the death of Caesar," not "the truth of the death of Caesar." This latter seems to be merely another propositional concept, asserted in "the death of Caesar is true," which is not, I think, the same proposition as "Caesar died." There is great difficulty in avoiding psychological elements here, and it would seem that Frege has allowed them to intrude in describing judgment as the recognition of truth (Gg. p. x). The difficulty is due to the fact that there is a psychological sense of assertion, which is what is lacking to Meinong's *Annahmen*, and that this does not run parallel with the logical sense. Psychologically, any proposition, whether true or false, may be merely thought of, or may be actually asserted: but for this possibility, error would be impossible. But logically, true propositions only are asserted,

---

* Frege, like Meinong, calls this an *Annahme*: FuB. p. 21.

† Gg. p. 10. Cf. also Bs. p. 4.

‡ When a term which indicates is itself to be spoken of, as opposed to what it indicates, Frege uses inverted commas. Cf. § 56.

though they may occur in an unasserted form as parts of other propositions. In "p implies q," either or both of the propositions p, q may be true, yet each, in this proposition, is unasserted in a logical, and not merely in a psychological, sense. Thus assertion has a definite place among logical notions, though there is a psychological notion of assertion to which nothing logical corresponds. But assertion does not seem to be a constituent of an asserted proposition, although it is, in some sense, contained in an asserted proposition. If p is a proposition, "p's truth" is a concept which has being even if p is false, and thus "p's truth" is not the same as p asserted. Thus no concept can be found which is equivalent to p asserted, and therefore assertion is not a constituent in p asserted. Yet assertion is not a term to which p, when asserted, has an external relation; for any such relation would need to be itself asserted in order to yield what we want. Also a difficulty arises owing to the apparent fact, which may however be doubted, that an asserted proposition can never be part of another proposition : thus, if this be a fact, where any statement is made about p asserted, it is not really about p asserted, but only about the assertion of p. This difficulty becomes serious in the case of Frege's one and only principle of inference (Bs. p. 9): "p is true and p implies q; therefore q is true*." Here it is quite essential that there should be three actual assertions, otherwise the assertion of propositions deduced from asserted premisses would be impossible; yet the three assertions together form one proposition, whose unity is shown by the word *therefore*, without which q would not have been deduced, but would have been asserted as a fresh premiss.

It is also almost impossible, at least to me, to divorce assertion from truth, as Frege does. An asserted proposition, it would seem, must be the same as a true proposition. We may allow that negation belongs to the content of a proposition (Bs. p. 4), and regard every assertion as asserting something to be true. We shall then correlate p and not-p as unasserted propositions, and regard "p is false" as meaning "not-p is true." But to divorce assertion from truth seems only possible by taking assertion in a psychological sense.

479. Frege's theory that assumptions are proper names for the true or the false, as the case may be, appears to me also untenable. Direct inspection seems to show that the relation of a proposition to the true or the false is quite different from that of (say), "the present King of England" to Edward VII. Moreover, if Frege's view were correct on this point, we should have to hold that in an asserted proposition it is the meaning, not the indication, that is asserted, for otherwise, all asserted propositions would assert the very same thing, namely the true, (for false propositions are not asserted). Thus asserted propositions would not differ from one another in any way, but would be all strictly and simply identical. Asserted propositions have no indication (FuB. p. 21), and can only differ, if at all, in some way analogous to meaning. Thus the meaning of the unasserted proposition together with its truth-value must be what is asserted,

---

* Cf. *supra*, § 18, (4) and § 38.

if the meaning simply is rejected. But there seems no purpose in introducing the truth-value here : it seems quite sufficient to say that an asserted proposition is one whose meaning is true, and that to say the meaning is true is the same as to say the meaning is asserted. We might then conclude that true propositions, even when they occur as parts of others, are always and essentially asserted, while false propositions are always unasserted, thus escaping the difficulty about *therefore* discussed above. It may also be objected to Frege that " the true " and " the false," as opposed to truth and falsehood, do not denote single definite things, but rather the classes of true and false propositions respectively. This objection, however, would be met by his theory of ranges, which correspond approximately to my classes ; these, he says, are things, and the true and the false are ranges (*v. inf.*).

480. *Begriff and Gegenstand. Functions.* I come now to a point in which Frege's work is very important, and requires careful examination. His use of the word *Begriff* does not correspond exactly to any notion in my vocabulary, though it comes very near to the notion of an assertion as defined in § 43, and discussed in Chapter VII. On the other hand, his *Gegenstand* seems to correspond exactly to what I have called a *thing* (§ 48). I shall therefore translate *Gegenstand* by *thing*. The meaning of *proper name* seems to be the same for him as for me, but he regards the range of proper names as confined to things, because they alone, in his opinion, can be logical subjects.

Frege's theory of functions and *Begriffe* is set forth simply in FuB. and defended against the criticisms of Kerry* in BuG. He regards functions— and in this I agree with him—as more fundamental than predicates and relations ; but he adopts concerning functions the theory of subject and assertion which we discussed and rejected in Chapter VII. The acceptance of this view gives a simplicity to his exposition which I have been unable to attain ; but I do not find anything in his work to persuade me of the legitimacy of his analysis.

An arithmetical function, *e.g.* $2x^3 + x$, does not denote, Frege says, the result of an arithmetical operation, for that is merely a number, which would be nothing new (FuB. p. 5). The essence of a function is what is left when the x is taken away, *i.e.*, in the above instance, $2(\quad)^3 + (\quad)$. The argument x does not belong to the function, but the two together make a whole (*ib.* p. 6). A function may be a proposition for every value of the variable ; its value is then always a truth-value (p. 13). A proposition may be divided into two parts, as " Caesar " and " conquered Gaul." The former Frege calls the *argument*, the latter the *function*. Any thing whatever is a possible argument for a function (p. 17). (This division of propositions corresponds exactly to my *subject* and *assertion* as explained in § 43, but Frege does not restrict this method of analysis as I do in Chapter VII.) A thing is anything which is not a function, *i.e.* whose expression leaves no empty place. The two following accounts of the nature of a function are quoted from the earliest and one of the latest of Frege's works respectively.

(1) "If in an expression, whose content need not be propositional

* Vierteljahrschrift für wiss. Phil., vol. XI, pp. 249–307.

(*beurtheilbar*), a simple or composite sign occurs in one or more places, and we regard it as replaceable, in one or more of these places, by something else, but by the same everywhere, then we call the part of the expression which remains invariable in this process a *function*, and the replaceable part we call its argument" (Bs. p. 16).

(2) "If from a proper name we exclude a proper name, which is part or the whole of the first, in some or all of the places where it occurs, but in such a way that these places remain recognizable as to be filled by one and the same arbitrary proper name (as argument positions of the first kind), I call what we thereby obtain the name of a function of the first order with one argument. Such a name, together with a proper name which fills the argument-places, forms a proper name" (Gg. p. 44).

The latter definition may become plainer by the help of some examples. "The present king of England" is, according to Frege, a proper name, and "England" is a proper name which is part of it. Thus here we may regard England as the argument, and "the present king of" as function. Thus we are led to "the present king of x." This expression will always have a meaning, but it will not have an indication except for those values of x which at present are monarchies. The above function is not propositional. But "Caesar conquered Gaul" leads to "x conquered Gaul"; here we have a propositional function. There is here a minor point to be noticed: the *asserted* proposition is not a proper name, but only the assumption is a proper name for the true or the false (*v. supra*); thus it is not "Caesar conquered Gaul" as asserted, but only the corresponding assumption, that is involved in the genesis of a propositional function. This is indeed sufficiently obvious, since we wish x to be able to be any thing in "x conquered Gaul," whereas there is no such asserted proposition except when x did actually perform this feat. Again consider "Socrates is a man implies Socrates is a mortal." This (unasserted) is, according to Frege, a proper name for the true. By varying the proper name "Socrates," we can obtain three propositional functions, namely "x is a man implies Socrates is a mortal," "Socrates is a man implies x is a mortal," "x is a man implies x is a mortal." Of these the first and third are true for all values of x, the second is true when and only when x is a mortal.

By suppressing in like manner a proper name in the name of a function of the first order with one argument, we obtain the name of a function of the first order with two arguments (Gg. p. 44). Thus *e.g.* starting from "$1 < 2$," we get first "$x < 2$," which is the name of a function of the first order with one argument, and thence "$x < y$," which is the name of a function of the first order with two arguments. By suppressing a function in like manner, Frege says, we obtain the name of a function of the second order (Gg. p. 44). Thus *e.g.* the assertion of existence in the mathematical sense is a function of the second order : "There is at least one value of x satisfying ϕx" is not a function of x, but may be regarded as a function of ϕ. Here ϕ must on no account be a thing, but may be any function. Thus this proposition, considered as a function of ϕ, is quite different from functions of the first order, by the fact that the possible arguments are different. Thus given any proposition, say $f(a)$, we may consider either $f(x)$, the function of the first order resulting from varying a and keeping f constant, or $\phi(a)$, the function of the second order got by varying f and keeping a fixed; or, finally, we may consider $\phi(x)$, in which both f and a are separately varied. (It is to be observed that such notions as $\phi(a)$, in which we consider any proposition concerning a, are involved in the identity of indiscernibles as stated in § 43.) Functions of the first order with two variables, Frege points out, express relations (Bs. p. 17); the referent and the relatum are both subjects in a relational proposition (Gl. p. 82). Relations, just as much as predicates, belong, Frege rightly says, to pure logic (*ib.* p. 83).

481. The word *Begriff* is used by Frege to mean nearly the same thing as *propositional function* (*e.g.* FuB. p. 28)*; when there are two variables, the Begriff is a relation. A thing is anything not a function, *i.e.* anything whose expression leaves no empty place (*ib.* p. 18). To Frege's theory of the essential cleavage between things and Begriffe, Kerry objects (*loc. cit.* p. 272 ff.) that Begriffe also can occur as subjects. To this Frege makes two replies. In the first place, it is, he says, an important distinction that some terms can only occur as subjects, while others can occur also as concepts, even if Begriffe can also occur as subjects (BuG. p. 195). In this I agree with him entirely; the distinction is the one employed in §§ 48, 49. But he goes on to a second point which appears to me mistaken. We can, he says, have a concept falling under a higher one (as Socrates falls under man, he means, not as Greek falls under man); but in such cases, it is not the concept itself, but its name, that is in question (BuG. p. 195). "The concept horse," he says, is not a concept, but a thing; the peculiar use is indicated by inverted commas (*ib.* p. 196). But a few pages later he makes statements which seem to involve a different view. A concept, he says, is essentially predicative even when something is asserted of it: an assertion which can be made of a concept does not fit an object. When a thing is said to fall under a concept, and when a concept is said to fall under a higher concept, the two relations involved, though similar, are not the same (*ib.* p. 201). It is difficult to me to reconcile these remarks with those of p. 195; but I shall return to this point shortly.

Frege recognizes the unity of a proposition: of the parts of a propositional concept, he says, not all can be complete, but one at least must be incomplete (*ungesättigt*) or predicative, otherwise the parts would not cohere (*ib.* p. 205). He recognizes also, though he does not discuss, the oddities resulting from *any* and *every* and such words: thus he remarks that every positive integer is the sum of four squares, but "every positive integer" is not a possible value of x in "x is the sum of four squares." The meaning of "every positive integer," he says, depends upon the context (Bs. p. 17)—a remark which is doubtless correct, but does not exhaust the subject. Self-contradictory notions are admitted as concepts: F is a concept if "a falls under the concept F" is a proposition whatever thing a may be (Gl. p. 87). A concept is the indication of a predicate; a thing is what can never be

---

* "We have here a function whose value is always a truth-value. Such functions with one argument we have called Begriffe; with two, we call them relations." Cf. Gl. pp. 82—3.

the whole indication of a predicate, though it may be that of a subject (BuG. p. 198).

482. The above theory, in spite of close resemblance, differs in some important points from the theory set forth in Part I above. Before examining the differences, I shall briefly recapitulate my own theory.

Given any propositional concept, or any unity (see § 136), which may in the limit be simple, its constituents are in general of two sorts : (1) those which may be replaced by anything else whatever without destroying the unity of the whole ; (2) those which have not this property. Thus in "the death of Caesar," anything else may be substituted for Caesar, but a proper name must not be substituted for *death*, and hardly anything can be substituted for *of*. Of the unity in question, the former class of constituents will be called *terms*, the latter *concepts*. We have then, in regard to any unity, to consider the following objects :

(1) What remains of the said unity when one of its terms is simply removed, or, if the term occurs several times, when it is removed from one or more of the places in which it occurs, or, if the unity has more than one term, when two or more of its terms are removed from some or all of the places where they occur. This is what Frege calls a function.

(2) The class of unities differing from the said unity, if at all, only by the fact that one of its terms has been replaced, in one or more of the places where it occurs, by some other terms, or by the fact that two or more of its terms have been thus replaced by other terms.

(3) Any member of the class (2).

(4) The assertion that every member of the class (2) is true.

(5) The assertion that some member of the class (2) is true.

(6) The relation of a member of the class (2) to the value which the variable has in that member.

The fundamental case is that where our unity is a propositional concept. From this is derived the usual mathematical notion of function, which might at first sight seem simpler. If $f(x)$ is not a propositional function, its value for a given value of x ($f(x)$ being assumed to be one-valued) is the term y satisfying the propositional function $y = f(x)$, *i.e.* satisfying, for the given value of x, some relational proposition ; this relational proposition is involved in the definition of $f(x)$, and some such propositional function is required in the definition of any function which is not propositional.

As regards (1), confining ourselves to one variable, it was maintained in Chapter VII that, except where the proposition from which we start is predicative or else asserts a fixed relation to a fixed term, there is no such entity : the analysis into argument and assertion cannot be performed in the manner required. Thus what Frege calls a function, if our conclusion was sound, is in general a non-entity. Another point of difference from Frege, in which, however, he appears to be in the right, lies in the fact that I place no restriction upon the variation of the variable, whereas Frege, according to the nature of the function, confines the variable to things, functions of the first order with one variable, functions of the first order with two variables, functions of the second order with one variable, and so on. There are thus for him an infinite number of different kinds of variability. This arises from the fact that he regards as distinct the concept occurring as such and the concept occurring as term, which I (§ 49) have identified. For me, the functions, which cannot be values of variables in functions of the first order, are non-entities and false abstractions. Instead of the rump of a proposition considered in (1), I substitute (2) or (3) or (4) according to circumstances. The ground for regarding the analysis into argument and function as not always possible is that, when one term is removed from a propositional concept, the remainder is apt to have no sort of unity, but to fall apart into a set of disjointed terms. Thus what is fundamental in such a case is (2). Frege's general definition of a function, which is intended to cover also functions which are not propositional, may be shown to be inadequate by considering what may be called the identical function, *i.e.* x as a function of x. If we follow Frege's advice, and remove x in hopes of having the function left, we find that nothing is left at all; yet nothing is not the meaning of the identical function. Frege wishes to have the empty places where the argument is to be inserted indicated in some way; thus he says that in $2x^3 + x$ the function is $2(\quad)^3 + (\quad)$. But here his requirement that the two empty places are to be filled by the same letter cannot be indicated: there is no way of distinguishing what we mean from the function involved in $2x^3 + y$. The fact seems to be that we want the notion of any term of a certain class, and that this is what our empty places really stand for. The function, as a single entity, is the relation (6) above; we can then consider any relatum of this relation, or the assertion of all or some of the relata, and any relation can be expressed in terms of the corresponding referent, as "Socrates is a man" is expressed in terms of Socrates. But the usual formal apparatus of the calculus of relations cannot be employed, because it presupposes propositional functions. We may say that a propositional function is a many-one relation which has all terms for the class of its referents, and has its relata contained among propositions*: or, if we prefer, we may call the class of relata of such a relation a propositional function. But the air of formal definition about these statements is fallacious, since propositional functions are presupposed in defining the class of referents and relata of a relation.

Thus by means of propositional functions, propositions are collected into classes. (These classes are not mutually exclusive.) But we may also collect them into classes by the terms which occur in them : all propositions containing a given term a will form a class. In this way we obtain propositions concerning variable propositional functions. In the notation $\phi(x)$, the ϕ is essentially variable ; if we wish it not to be so, we must take some particular proposition about x, such as "x is a class" or "x implies x." Thus $\phi(x)$ essentially contains two variables. But, if we have decided that ϕ is not a separable entity, we cannot regard ϕ itself as the second variable. It will be necessary to take as our variable either the relation of x to $\phi(x)$, or else the class of propositions $\phi(y)$ for different values of y but for constant ϕ. This does not matter formally, but it is important for logic to be clear as to

---

* Not all relations having this property are propositional functions; *v. inf.*

the meaning of what appears as the variation of ϕ. We obtain in this way another division of propositions into classes, but again these classes are not mutually exclusive.

In the above manner, it would seem, we can make use of propositional functions without having to introduce the objects which Frege calls functions. It is to be observed, however, that the kind of relation by which propositional functions are defined is less general than the class of many-one relations having their domain coextensive with terms and their converse domain contained in propositions. For in this way any proposition would, for a suitable relation, be relatum to any term, whereas the term which is referent must, for a propositional function, be a constituent of the proposition which is its relatum*. This point illustrates again that the class of relations involved is fundamental and incapable of definition. But it would seem also to show that Frege's different kinds of variability are unavoidable, for in considering (say) $\phi(2)$, where ϕ is variable, the variable would have to have as its range the above class of relations, which we may call *propositional relations*. Otherwise, $\phi(2)$ is not a proposition, and is indeed meaningless, for we are dealing with an indefinable, which demands that $\phi(2)$ should be the relatum of 2 with regard to some propositional relation. The contradiction discussed in Chapter x seems to show that some mystery lurks in the variation of propositional functions; but for the present, Frege's theory of different kinds of variables must, I think, be accepted.

483. It remains to discuss afresh the question whether concepts can be made into logical subjects without change of meaning. Frege's theory, that when this appears to be done it is really the name of the concept that is involved, will not, I think, bear investigation. In the first place, the mere assertion "not the concept, but its name, is involved," has already made the concept a subject. In the second place, it seems always legitimate to ask : "what is it that is named by this name?" If there were no answer, the name could not be a name; but if there is an answer, the concept, as opposed to its name, can be made a subject. (Frege, it may be observed, does not seem to have clearly disentangled the logical and linguistic elements of naming : the former depend upon denoting, and have, I think, a much more restricted range than Frege allows them.) It is true that we found difficulties in the doctrine that everything can be a logical subject : as regards "any a," for example, and also as regards plurals. But in the case of "any a," there is ambiguity, which introduces a new class of problems; and as regards plurals, there are propositions in which the many behave like a logical subject in every respect except that they are many subjects and not one only (see §§ 127, 128). In the case of concepts, however, no such escapes are possible. The case of asserted propositions is difficult, but is met, I think, by holding that an asserted proposition is merely a true proposition, and is therefore asserted wherever it occurs, even when grammar would lead to the opposite conclusion. Thus, on the whole, the doctrine of concepts which cannot be made subjects seems untenable.

484. *Classes.* Frege's theory of classes is very difficult, and I am not

---

* The notion of a constituent of a proposition appears to be a logical indefinable.

sure that I have thoroughly understood it. He gives the name *Werthver-lauf* * to an entity which appears to be nearly the same as what I call the class as one. The concept of the class, and the class as many, do not appear in his exposition. He differs from the theory set forth in Chapter VI chiefly by the fact that he adopts a more intensional view of classes than I have done, being led thereto mainly by the desirability of admitting the null-class and of distinguishing a term from a class whose only member it is. I agree entirely that these two objects cannot be attained by an extensional theory, though I have tried to show how to satisfy the requirements of formalism (§§ 69, 73).

The extension of a *Begriff*, Frege says, is the range of a function whose value for every argument is a truth-value (FuB. p. 16). Ranges are things, whereas functions are not (*ib.* p. 19). There would be no null-class, if classes were taken in extension; for the null-class is only possible if a class is not a collection of terms (KB. pp. 436–7). If x be a term, we cannot identify x, as the extensional view requires, with the class whose only member is x; for suppose x to be a class having more than one member, and let y, z be two different members of x; then if x is identical with the class whose only member is x, y and z will both be members of this class, and will therefore be identical with x and with each other, contrary to the hypothesis†. The extension of a *Begriff* has its being in the *Begriff* itself, not in the individuals falling under the *Begriff* (*ib.* p. 451). When I say something about all men, I say nothing about some wretch in the centre of Africa, who is in no way indicated, and does not belong to the indication of *man* (p. 454). *Begriffe* are prior to their extension, and it is a mistake to attempt, as Schröder does, to base extension on individuals; this leads to the calculus of regions (*Gebiete*), not to Logic (p. 455).

What Frege understands by a range, and in what way it is to be conceived without reference to objects, he endeavours to explain in his *Grundgesetze der Arithmetik*. He begins by deciding that two propositional functions are to have the same range when they have the same value for every value of x, *i.e.* for every value of x both are true or both false (pp. 7, 14) This is laid down as a primitive proposition. But this only determines the equality of ranges, not what they are in themselves. If $X(\xi)$ be a function which never has the same value for different values of ξ and if we denote by ϕ' the range of ϕx, we shall have $X(\phi') = X(\psi')$ when and only when ϕ' and ψ' are equal, *i.e.* when and only when ϕx and ψx always have the same value. Thus the conditions for the equality of ranges do not of themselves decide what ranges are to be (p. 16). Let us decide arbitrarily —since the notion of a range is not yet fixed—that the true is to be the range of the function " x is true " (as an assumption, not an asserted proposition), and the false is to be the range of the function " $x = $ not every term is identical with itself." It follows that the range of ϕx is the true when and only when the true and nothing else falls under the *Begriff* ϕx; the range of ϕx is the false when and only when the false and nothing else falls under the *Begriff* ϕx; in other cases, the range is neither the true nor

---

* I shall translate this as *range*.   † *Ib.* p. 444. Cf. *supra*, § 74.

the false (pp. 17—18). If only one thing falls under a concept, this one thing is distinct from the range of the concept in question (p. 18, note)— the reason is the same as that mentioned above.

There is an argument (p. 49) to prove that the name of the range of a function always has an indication, *i.e.* that the symbol employed for it is never meaningless. In view of the contradiction discussed in Chapter x, I should be inclined to deny a meaning to a range when we have a proposition of the form $\phi[f(\phi)]$, where f is constant and ϕ variable, or of the form $f_x(x)$, where x is variable and f_x is a propositional function which is determinate when x is given, but varies from one value of x to another— provided, when f_x is analyzed into things and concepts, the part dependent on x does not consist only of things, but contains also at least one concept. This is a very complicated case, in which, I should say, there is no class as one, my only reason for saying so being that we can thus escape the contradiction.

485. By means of variable propositional functions, Frege obtains a definition of the relation which Peano calls ϵ, namely the relation of a term to a class of which it is a member*. The definition is as follows : "$a\epsilon u$" is to mean the term (or the range of terms if there be none or many) x such that there is a propositional function ϕ which is such that u is the range of ϕ and ϕa is identical with x (p. 53). It is observed that this defines $a\epsilon u$ whatever things u and u may be. In the first place, suppose u to be a range. Then there is at least one ϕ whose range is u, and any two whose range is u are regarded by Frege as identical. Thus we may speak of *the* function ϕ whose range is u. In this case, $a\epsilon u$ is the proposition ϕa, which is true when a is a member of u, and is false otherwise. If, in the second place, u is not a range, then there is no such propositional function as ϕ, and therefore $a\epsilon u$ is the range of a propositional function which is always false, *i.e.* the null-range. Thus $a\epsilon u$ indicates the true when u is a range and a is a member of u ; $a\epsilon u$ indicates the false when u is a range and a is not a member of u ; in other cases, $a\epsilon u$ indicates the null-range.

It is to be observed that from the equivalence of $x\epsilon u$ and $x\epsilon v$ for all values of x we can only infer the identity of u and v when u and v are ranges. When they are not ranges, the equivalence will always hold, since $x\epsilon u$ and $x\epsilon v$ are the null-range for all values of x ; thus if we allowed the inference in this case, any two objects which are not ranges would be identical, which is absurd. One might be tempted to doubt whether u and v must be identical even when they are ranges : with an intensional view of classes, this becomes open to question.

Frege proceeds (p. 55) to an analogous definition of the propositional function of three variables which I have symbolised as $x\,R\,y$, and here again he gives a definition which does not place any restrictions on the variability of R. This is done by introducing a *double range*, defined by a propositional function of two variables ; we may regard this as a class of couples with sense†. If then R is such a class of couples, and if $(x\,;\,y)$ is a member of this

---

* Cf. §§ 21, 76, *supra*.

† Neglecting, for the present, our doubts as to there being any such entity as a couple with sense, cf. § 98.

class, $x \, R \, y$ is to hold; in other cases it is to be false or null as before. On this basis, Frege successfully erects as much of the logic of relations as is required for his Arithmetic; and he is free from the restrictions on the variability of R which arise from the intensional view of relations adopted in the present work (cf. § 83).

486. The chief difficulty which arises in the above theory of classes is as to the kind of entity that a range is to be. The reason which led me, against my inclination, to adopt an extensional view of classes, was the necessity of discovering some entity determinate for a given propositional function, and the same for any equivalent propositional function. Thus "x is a man" is equivalent (we will suppose) to "x is a featherless biped," and we wish to discover some one entity which is determined in the same way by both these propositional functions. The only single entity I have been able to discover is the class as one—except the derivative class (also as one) of propositional functions equivalent to either of the given propositional functions. This latter class is plainly a more complex notion, which will not enable us to dispense with the general notion of *class*; but this more complex notion (so we agreed in § 73) must be substituted for the class of terms in the symbolic treatment, if there is to be any null-class and if the class whose only member is a given term is to be distinguished from that term. It would certainly be a very great simplification to admit, as Frege does, a range which is something other than the whole composed of the terms satisfying the propositional function in question ; but for my part, inspection reveals to me no such entity. On this ground, and also on account of the contradiction, I feel compelled to adhere to the extensional theory of classes, though not quite as set forth in Chapter VI.

487. That some modification in that doctrine is necessary, is proved by the argument of KB. p. 444. This argument appears capable of proving that a class, even as one, cannot be identified with the class of which it is the only member. In § 74, I contended that the argument was met by the distinction between the class as one and the class as many, but this contention now appears to me mistaken. For this reason, it is necessary to re-examine the whole doctrine of classes.

Frege's argument is as follows. If a is a class of more than one term, and if a is identical with the class whose only term is a, then to be a term of a is the same thing as to be a term of the class whose only term is a, whence a is the only term of a. This argument *appears* to prove not merely that the extensional view of classes is inadequate, but rather that it is wholly inadmissible. For suppose a to be a collection, and suppose that a collection of one term is identical with that one term. Then, if a can be regarded as one collection, the above argument proves that a is the only term of a. We cannot escape by saying that ϵ is to be a relation to the class-concept or the concept of the class or the class as many, for if there is any such entity as the class as one, there will be a relation, which we may call ϵ, between terms and their classes as one. Thus the above argument leads to the conclusion that either (a) a collection of more than one term is not identical with the collection whose only term it is, or (β) there is no collection as one term at all in the case of a collection of many terms, but the collection is strictly and only many. One or other of these must be admitted in virtue of the above argument.

488. (α) To either of these views there are grave objections. The former is the view of Frege and Peano. To realize the paradoxical nature of this view, it must be clearly grasped that it is not only the collection as many, but the collection as one, that is distinct from the collection whose only term it is. (I speak of collections, because it is important to examine the bearing of Frege's argument upon the possibility of an extensional standpoint.) This view, in spite of its paradox, is certainly the one which seems to be required by the symbolism. It is quite essential that we should be able to regard a class as a single object, that there should be a null-class, and that a term should not (in general, at any rate) be identical with the class of which it is the only member. It is subject to these conditions that the *symbolic* meaning of *class* has to be interpreted. Frege's notion of a range may be identified with the collection as one, and all will then go well. But it is very hard to see any entity such as Frege's range, and the argument that there must be such an entity gives us little help. Moreover, in virtue of the contradiction, there certainly are cases where we have a collection as many, but no collection as one (§ 104). Let us then examine (β), and see whether this offers a better solution.

(β) Let us suppose that a collection of one term is that one term, and that a collection of many terms is (or rather are) those many terms, so that there is not a single term at all which is the collection of the many terms in question. In this view there is, at first sight at any rate, nothing paradoxical, and it has the merit of admitting universally what the Contradiction shows to be sometimes the case. In this case, unless we abandon one of our fundamental dogmas, ε will have to be a relation of a term to its class-concept, not to its class; if *a* is a class-concept, what appears symbolically as the class whose only term is *a* will (one might suppose) be the class-concept under which falls only the concept *a*, which is of course (in general, if not always) different from *a*. We shall maintain, on account of the contradiction, that there is not always a class-concept for a given propositional function φ*x*, *i.e.* that there is not always, for every φ, some class-concept *a* such that *x*ε*a* is equivalent to φ*x* for all values of *x*; and the cases where there is no such class-concept will be cases in which φ is a quadratic form.

So far, all goes well. But now we no longer have one definite entity which is determined equally by any one of a set of equivalent propositional functions, *i.e.* there is, it might be urged, no meaning of *class* left which is determined by the extension alone. Thus, to take a case where this leads to confusion, if *a* and *b* be different class-concepts such that *x*ε*a* and *x*ε*b* are equivalent for all values of *x*, the class-concept under which *a* falls and nothing else will not be identical with that under which falls *b* and nothing else. Thus we cannot get any way of denoting what should symbolically correspond to the class as one. Or again, if *u* and *v* be similar but different classes, "similar to *u*" is a different concept from "similar to *v*"; thus, unless we can find some extensional meaning for *class*, we shall not be able to say that the number of *u* is the same as that of *v*. And all the usual elementary problems as to combinations (*i.e.* as to the number of classes of specified kinds contained in a given class) will have become impossible and even meaningless. For these various reasons, an objector might contend, something like the class as one must be maintained; and Frege's range fulfils the conditions required. It would seem necessary therefore to accept ranges by an act of faith, without waiting to see whether there are such things.

Nevertheless, the non-identification of the class with the class as one, whether in my form or in the form of Frege's range, appears unavoidable, and by a process of exclusion the class as many is left as the only object which can play the part of a class. By a modification of the logic hitherto advocated in the present work, we shall, I think, be able at once to satisfy the requirements of the Contradiction and to keep in harmony with common sense*.

489. Let us begin by recapitulating the possible theories of classes which have presented themselves. A class may be identified with (a) the predicate, (β) the class concept, (γ) the concept of the class, (δ) Frege's range, (ϵ) the numerical conjunction of the terms of the class, (ζ) the whole composed of the terms of the class.

Of these theories, the first three, which are intensional, have the defect that they do not render a class determinate when its terms are given. The other three do not have this defect, but they have others. (δ) suffers from a doubt as to there being such an entity, and also from the fact that, if ranges are terms, the contradiction is inevitable. (ϵ) is logically unobjectionable, but is not a single entity, except when the class has only one member. (ζ) cannot always exist as a term, for the same reason as applies against (δ); also it cannot be identified with the class on account of Frege's argument†.

Nevertheless, without a single object‡ to represent an extension, Mathematics crumbles. Two propositional functions which are equivalent for all values of the variable may not be identical, but it is necessary that there should be some object determined by both. Any object that may be proposed, however, presupposes the notion of *class*. We may define *class* optatively as follows: A class is an object uniquely determined by a propositional function, and determined equally by any equivalent propositional function. Now we cannot take as this object (as in other cases of symmetrical transitive relations) the class of propositional functions equivalent to a given propositional function, unless we already have the notion of *class*. Again, equivalent relations, considered intensionally, may be distinct: we want therefore to find some one object determined equally by any one of a set of equivalent relations. But the only objects that suggest themselves are the class of relations or the class of couples forming their common range; and these both presuppose *class*. And without the notion of class, elementary problems, such as "how many combinations can be formed of m objects n at a time?" become meaningless. Moreover, it appears immediately evident that there is some sense in saying that two class-concepts have the *same*

* The doctrine to be advocated in what follows is the direct denial of the dogma stated in § 70, note.

† *Archiv* I. p. 444.

‡ For the use of the word *object* in the following discussion, see § 58, note.

extension, and this requires that there should be some object which can be called the extension of a class-concept. But it is exceedingly difficult to discover any such object, and the contradiction proves conclusively that, even if there be such an object sometimes, there are propositional functions for which the extension is not one term.

The class as many, which we numbered (ϵ) in the above enumeration, is unobjectionable, but is many and not one. We may, if we choose, represent this by a single symbol: thus $x\epsilon u$ will mean "x is one of the u's." This must not be taken as a relation of two terms, x and u, because u as the numerical conjunction is not a single term, and we wish to have a meaning for $x\epsilon u$ which would be the same if for u we substituted an equal class v, which prevents us from interpreting u intensionally. Thus we may regard "x is one of the u's" as expressing a relation of x to many terms, among which x is included. The main objection to this view, if only single terms can be subjects, is that, if u is a symbol standing essentially for many terms, we cannot make u a logical subject without risk of error. We can no longer speak, one might suppose, of a class of classes; for what should be the terms of such a class are not single terms, but are each many terms*. We cannot assert a predicate of many, one would suppose, except in the sense of asserting it of each of the many; but what is required here is the assertion of a predicate concerning the many as many, not concerning each nor yet concerning the whole (if any) which all compose. Thus a class of classes will be many many's; its constituents will each be only many, and cannot therefore in any sense, one might suppose, be single constituents. Now I find myself forced to maintain, in spite of the apparent logical difficulty, that this is precisely what is required for the assertion of number. If we have a class of classes, each of whose members has two terms, it is necessary that the members should each be genuinely two-fold, and should not be each one. Or again, "Brown and Jones are two" requires that we should not combine Brown and Jones into a single whole, and yet it has the form of a subject-predicate proposition. But now a difficulty arises as to the number of members of a class of classes. In what sense can we speak of two couples? This seems to require that each couple should be a single entity; yet if it were, we should have two units, not two couples. We require a sense for diversity of collections, meaning thereby, apparently, if u and v are the collections in question, that $x\epsilon u$ and $x\epsilon v$ are not equivalent for all values of x.

490. The logical doctrine which is thus forced upon us is this: The subject of a proposition may be not a single term, but essentially many terms; this is the case with all propositions asserting numbers other than 0 and 1. But the predicates or class-concepts or relations which can occur in propositions having plural subjects are different (with some exceptions) from those that can occur in propositions having single terms as subjects. Although a class is many and not one, yet there is identity and diversity among classes, and thus classes can be counted as though each were a genuine unity; and in this sense we can speak of *one* class and of the classes which are members of a

---

* Wherever the context requires it, the reader is to add "provided the class in question (or all the classes in question) do not consist of a single term."

class of classes. *One* must be held, however, to be somewhat different when asserted of a class from what it is when asserted of a term; that is, there is a meaning of *one* which is applicable in speaking of *one term*, and another which is applicable in speaking of *one class*, but there is also a general meaning applicable to both cases. The fundamental doctrine upon which all rests is the doctrine that the subject of a proposition may be plural, and that such plural subjects are what is meant by classes which have more than one term*.

It will now be necessary to distinguish (1) terms, (2) classes, (3) classes of classes, and so on *ad infinitum* ; we shall have to hold that no member of one set is a member of any other set, and that $x\epsilon u$ requires that x should be of a set of a degree lower by one than the set to which u belongs. Thus $x\epsilon x$ will become a meaningless proposition ; and in this way the contradiction is avoided.

491. But we must now consider the problem of classes which have one member or none. The case of the null-class might be met by a bare denial— this is only inconvenient, not self-contradictory. But in the case of classes having only one term, it is still necessary to distinguish them from their sole members. This results from Frege's argument, which we may repeat as follows. Let u be a class having more than one term ; let ιu be the class of classes whose only member is u. Then ιu has one member, u has many ; hence u and ιu are not identical. It may be doubted, at first sight, whether this argument is valid. The relation of x to u expressed by $x\epsilon u$ is a relation of a single term to many terms ; the relation of u to ιu expressed by $u\epsilon\iota u$ is a relation of many terms (as subject) to many terms (as predicate)†. This is, so an objector might contend, a different relation from the previous one ; and thus the argument breaks down. It is in different senses that x is a member of u and that u is a member of ιu ; thus u and ιu may be identical in spite of the argument.

This attempt, however, to escape from Frege's argument, is capable of refutation. For all the purposes of Arithmetic, to begin with, and for many of the purposes of logic, it is necessary to have a meaning for ϵ which is equally applicable to the relation of a term to a class, of a class to a class of classes, and so on. But the chief point is that, if every single term is a class, the proposition $x\epsilon x$, which gives rise to the Contradiction, must be admissible. It is only by distinguishing x and ιx, and insisting that in $x\epsilon u$ the u must always be of a type higher by one than x, that the contradiction can be avoided. Thus, although we may identify the class with the numerical conjunction of its terms, wherever there are many terms, yet where there is only one term we shall have to accept Frege's range as an object distinct from its only term. And having done this, we may of course also admit a range in the case of a null propositional function. We shall differ from Frege only in regarding a range as in no case a term, but an object of a different logical type, in the sense that a propositional function $\phi(x)$, in which x may be any term, is in general meaningless if for x we substitute a

---

* Cf. §§ 128, 132 *supra*.
† The word *predicate* is here used loosely, not in the precise sense defined in § 48.

range; and if x may be any range of terms, $\phi(x)$ will in general be meaning-less if for x we substitute either a term or a range of ranges of terms. Ranges, finally, are what are properly to be called *classes*, and it is of them that cardinal numbers are asserted.

492. According to the view here advocated, it will be necessary, with every variable, to indicate whether its field of significance is terms, classes, classes of classes, or so on*. A variable will not be able, except in special cases, to extend from one of these sets into another; and in $x\epsilon u$, the x and the u must always belong to different types; ϵ will not be a relation between objects of the same type, but $\epsilon\acute{\epsilon}$ or $\epsilon\acute{R}\acute{\epsilon}$ † will be, provided R is so. We shall have to distinguish also among relations according to the types to which their domains and converse domains belong; also variables whose fields include relations, these being understood as classes of couples, will not as a rule include anything else, and relations between relations will be different in type from relations between terms. This seems to give the truth—though in a thoroughly extensional form—underlying Frege's distinction between terms and the various kinds of functions. Moreover the opinion here advocated seems to adhere very closely indeed to common sense.

Thus the final conclusion is, that the correct theory of classes is even more extensional than that of Chapter vi; that the class as many is the only object always defined by a propositional function, and that this is adequate for formal purposes; that the class as one, or the whole composed of the terms of the class, is probably a genuine entity except where the class is defined by a quadratic function (see § 103), but that in these cases, and in other cases possibly, the class as many is the only object uniquely defined.

The theory that there are different kinds of variables demands a reform in the doctrine of formal implication. In a formal implication, the variable does not, in general, take all the values of which variables are susceptible, but only all those that make the propositional function in question a proposition. For other values of the variable, it must be held that any given propositional function becomes meaningless. Thus in $x\epsilon u$, u must be a class, or a class of classes, or etc., and x must be a term if u is a class, a class if u is a class of classes, and so on; in every propositional function there will be some range permissible to the variable, but in general there will be possible values for other variables which are not admissible in the given case. This fact will require a certain modification of the principles of Symbolic Logic; but it remains true that, in a formal implication, all propositions belonging to a given propositional function are asserted.

With this we come to the end of the more philosophical part of Frege's work. It remains to deal briefly with his Symbolic Logic and Arithmetic; but here I find myself in such complete agreement with him that it is hardly necessary to do more than acknowledge his discovery of propositions which, when I wrote, I believed to have been new.

493. *Implication and Symbolic Logic.* The relation which Frege employs as fundamental in the logic of propositions is not exactly the same as what I have called implication: it is a relation which holds between

---

* See Appendix B. † On this notation, see §§ 28, 97.

p and q whenever q is true or p is not true, whereas the relation which I employ holds whenever p and q are propositions, and q is true or p is false. That is to say, Frege's relation holds when p is not a proposition at all, whatever q may be; mine does not hold unless p and q are propositions. His definition has the formal advantage that it avoids the necessity for hypotheses of the form "p and q are propositions"; but it has the disadvantage that it does not lead to a definition of *proposition* and of negation. In fact, negation is taken by Frege as indefinable; *proposition* is introduced by means of the indefinable notion of a truth-value. Whatever x may be, "the truth-value of x" is to indicate the true if x is true, and the false in all other cases. Frege's notation has certain advantages over Peano's, in spite of the fact that it is exceedingly cumbrous and difficult to use. He invariably defines expressions for all values of the variable, whereas Peano's definitions are often-preceded by a hypothesis. He has a special symbol for assertion, and he is able to assert for all values of x a propositional function not stating an implication, which Peano's symbolism will not do. He also distinguishes, by the use of Latin and German letters respectively, between *any* proposition of a certain propositional function and *all* such propositions. By always using implications, Frege avoids the logical product of two propositions, and therefore has no axioms corresponding to Importation and Exportation *. Thus the joint assertion of p and q is the denial of "p implies not-q."

494. *Arithmetic.* Frege gives exactly the same definition of cardinal numbers as I have given, at least if we identify his *range* with my *class*†. But following his intensional theory of classes, he regards the number as a property of the class-concept, not of the class in extension. If u be a range, the number of u is the range of the concept "range similar to u." In the *Grundlagen der Arithmetik*, other possible theories of number are discusssed and dismissed. Numbers cannot be asserted of objects, because the same set of objects may have different numbers assigned to them (Gl. p. 29); for example, one army is so many regiments and such another number of soldiers. This view seems to me to involve too physical a view of objects: I do not consider the army to be the same object as the regiments. A stronger argument for the same view is that 0 will not apply to objects, but only to concepts (p. 59). This argument is, I think, conclusive up to a certain point; but it is satisfied by the view of the symbolic meaning of classes set forth in §73. Numbers themselves, like other ranges, are things (p. 67). For defining numbers as ranges, Frege gives the same general ground as I have given, namely what I call the principle of abstraction‡. In the *Grundgesetze der Arithmetik*, various theorems in the foundations of cardinal Arithmetic are proved with great elaboration, so great that it is often very difficult to discover the difference between successive steps in a demonstration. In view of the contradiction of Chapter x, it is plain that some emendation is required in Frege's principles; but it is hard to believe that it can do more than introduce some general limitation which leaves the details unaffected.

---

* See § 18, (7), (8).   † See Gl. pp. 79, 85; Gg. p. 57, Df. Z.   ‡ Gl. p. 79; cf. § 111 *supra.*

495. In addition to his work on cardinal numbers, Frege has, already in the *Begriffsschrift*, a very admirable theory of progressions, or rather of all series that can be generated by many-one relations. Frege does not confine himself to one-one relations: as long as we move in only one direction, a many-one relation also will generate a series. In some parts of his theory, he even deals with general relations. He begins by considering, for any relation $f(x, y)$, functions F which are such that, if $f(x, y)$ holds, then $F(x)$ implies $F(y)$. If this condition holds, Frege says that the property F is inherited in the f-series (Bs. pp. 55—58). From this he goes on to define, without the use of numbers, a relation which is equivalent to "some positive power of the given relation." This is defined as follows. The relation in question holds between x and y if every property F, which is inherited in the f-series and is such that $f(x, z)$ implies $F(z)$ for all values of z, belongs to y (Bs. p. 60). On this basis, a non-numerical theory of series is very successfully erected, and is applied in Gg. to the proof of propositions concerning the number of finite numbers and kindred topics. This is, so far as I know, the best method of treating such questions, and Frege's definition just quoted gives, apparently, the best form of mathematical induction. But as no controversy is involved, I shall not pursue this subject any further.

Frege's works contain much admirable criticism of the psychological standpoint in logic, and also of the formalist theory of mathematics, which believes that the actual symbols are the subject-matter dealt with, and that their properties can be arbitrarily assigned by definition. In both these points, I find myself in complete agreement with him.

496. Kerry (*loc. cit.*) has criticized Frege very severely, and professes to have proved that a purely logical theory of Arithmetic is impossible (p. 304). On the question whether concepts can be made logical subjects, I find myself in agreement with his criticisms ; on other points, they seem to rest on mere misunderstandings. As these are such as would naturally occur to any one unfamiliar with symbolic logic, I shall briefly discuss them.

The definition of numbers as classes is, Kerry asserts, a ὕστερον πρότερον. We must know that every concept has only *one* extension, and we must know what *one* object is; Frege's numbers, in fact, are merely convenient symbols for what are commonly called numbers (p. 277). It must be admitted, I think, that the notion of *a term* is indefinable (cf. § 132 *supra*), and is presupposed in the definition of the number 1. But Frege argues— and his argument at least deserves discussion—that *one* is not a predicate, attaching to every imaginable term, but has a less general meaning, and attaches to concepts (Gl. p. 40). Thus *a term* is not to be analyzed into *one* and *term*, and does not presuppose the notion of *one* (cf. § 72 *supra*). As to the assumption that every concept has only one extension, it is not necessary to be able to state this in language which employs the number 1: all we need is, that if ϕx and ψx are equivalent propositions for all values of x, then they have the same extension—a primitive proposition whose symbolic expression in no way presupposes the number 1. From this it follows that if a and b are both extensions of ϕx, a and b are identical, which again does not formally involve the number 1. In like manner, other objections to Frege's definition can be met.

Kerry is misled by a certain passage (Gl. p. 80, note) into the belief that Frege identifies a concept with its extension. The passage in question appears to assert that the number of u might be defined as the concept "similar to u" and not as the range of this concept; but it does not say that the two definitions are equivalent.

There is a long criticism of Frege's proof that 0 is a number, which reveals fundamental errors as to the existential import of universal propositions. The point is to prove that, if u and v are null-classes, they are similar. Frege defines similarity to mean that there is a one-one relation R such that "x is a u" implies "there is a v to which x stands in the relation R," and vice versa. (I have altered the expressions into conformity with my usual language.) This, he says, is equivalent to "there is a one-one relation R such that 'x is a u' and 'there is no term of v to which x stands in the relation R' cannot both be true, whatever value x may have, and vice versa"; and this proposition is true if "x is a u" and "y is a v" are always false. This strikes Kerry as absurd (pp. 287—9). Similarity of classes, he thinks, implies that they have terms. He affirms that Frege's assertion above is contradicted by a later one (Gl. p. 89): "If a is a u, and nothing is a v, then 'a is a u' and 'no term is a v which has the relation R to a' are both true for all values of R." I do not quite know where Kerry finds the contradiction; but he evidently does not realize that false propositions imply all propositions and that universal propositions have no existential import, so that "all a is b" and "no a is b" will both be true if a is the null-class.

Kerry objects (p. 290, note) to the generality of Frege's notion of relation. Frege asserts that any proposition containing a and b affirms a relation between a and b (Gl. p. 83); hence Kerry (rightly) concludes that it is self-contradictory to deny that a and b are related. So general a notion, he says, can have neither sense nor purpose. As for sense, that a and b should both be constituents of one proposition seems a perfectly intelligible sense; as for purpose, the whole logic of relations, indeed the whole of mathematics, may be adduced in answer. There is, however, what seems at first sight to be a formal disproof of Frege's view. Consider the propositional function "R and S are relations which are identical, and the relation R does not hold between R and S." This contains two variables, R and S; let us suppose that it is equivalent to "R has the relation T to S." Then substituting T for both R and S, we find, since T is identical with T, that "T does not have the relation T to T" is equivalent to "T has the relation T to T." This is a contradiction, showing that there is no such relation as T. Frege might object to this instance, on the ground that it treats relations as terms; but his double ranges, which, like single ranges, he holds to be things, will bring out the same result. The point involved is closely analogous to that involved in the Contradiction: it was there shown that some propositional functions with one variable are not equivalent to any propositional function asserting membership of a fixed class, while here it is shown that some containing two variables are not equivalent to the assertion of any fixed relation. But the refutation is the same in the case of relations as it was in the previous case. There is a hierarchy of relations according to the type of objects constituting their fields. Thus relations between terms are distinct from those between classes, and these again are distinct from relations between relations. Thus no relation can have itself both as referent and as relatum, for if it be of the same order as the one, it must be of a higher order than the other; the proposed propositional function is therefore meaningless for all values of the variables R and S.

It is affirmed (p. 291) that only the concepts of 0 and 1, not the objects themselves, are defined by Frege. But if we allow that the range of a Begriff is an object, this cannot be maintained; for the assigning of a concept will carry with it the assigning of its range. Kerry does not perceive that the uniqueness of 1 has been proved (*ib.*): he thinks that, with Frege's definition, there might be several 1's. I do not understand how this can be supposed: the proof of uniqueness is precise and formal.

The definition of immediate sequence in the series of natural numbers is also severely criticized (p. 292 ff.). This depends upon the general theory of series set forth in Bs. Kerry objects that Frege has defined "F is inherited in the f-series," but has not defined "the f-series" nor "F is inherited." The latter essentially ought not to be defined, having no precise sense; the former is easily defined, if necessary, as the field of the relation f. This objection is therefore trivial. Again, there is an attack on the definition: "y follows x in the f-series if y has all the properties inherited in the f-series and belonging to all terms to which x has the relation f*." This criterion, we are told, is of doubtful value, because no catalogue of such properties exists, and further because, as Frege himself proves, following x is itself one of these properties, whence a vicious circle. This argument, to my mind, radically misconceives the nature of deduction. In deduction, a proposition is proved to hold concerning *every* member of a class, and may then be asserted of a particular member: but the proposition concerning *every* does not necessarily result from enumeration of the entries in a catalogue. Kerry's position involves acceptance of Mill's objection to Barbara, that the mortality of Socrates is a necessary premiss for the mortality of all men. The fact is, of course, that general propositions can often be established where no means exist of cataloguing the terms of the class for which they hold; and even, as we have abundantly seen, general propositions fully stated hold of *all* terms, or, as in the above case, of *all* functions, of which no catalogue can be conceived. Kerry's argument, therefore, is answered by a correct theory of deduction; and the logical theory of Arithmetic is vindicated against its critics.

Note. The second volume of *Gg.*, which appeared too late to be noticed in the Appendix, contains an interesting discussion of the contradiction (pp. 253—265), suggesting that the solution is to be found by denying that two propositional functions which determine equal classes must be equivalent. As it seems very likely that this is the true solution, the reader is strongly recommended to examine Frege's argument on the point.

---

* Kerry omits the last clause, wrongly ; for not all properties inherited in the f-series belong to all its terms ; for example, the property of being greater than 100 is inherited in the number-series.

APPENDIX B.

THE DOCTRINE OF TYPES.

497. THE doctrine of types is here put forward tentatively, as affording a possible solution of the contradiction; but it requires, in all probability, to be transformed into some subtler shape before it can answer all difficulties. In case, however, it should be found to be a first step towards the truth, I shall endeavour in this Appendix to set forth its main outlines, as well as some problems which it fails to solve.

Every propositional function $\phi(x)$—so it is contended—has, in addition to its range of truth, a range of significance, *i.e.* a range within which x must lie if $\phi(x)$ is to be a proposition at all, whether true or false. This is the first point in the theory of types; the second point is that ranges of significance form *types*, *i.e.* if x belongs to the range of significance of $\phi(x)$, then there is a class of objects, the *type* of x, all of which must also belong to the range of significance of $\phi(x)$, however ϕ may be varied; and the range of significance is always either a single type or a sum of several whole types. The second point is less precise than the first, and the case of numbers introduces difficulties; but in what follows its importance and meaning will, I hope, become plainer.

A *term* or *individual* is any object which is not a range. This is the lowest type of object. If such an object—say a certain point in space—occurs in a proposition, any other individual may *always* be substituted without loss of significance. What we called, in Chapter VI, the class as one, is an individual, provided its members are individuals: the objects of daily life, persons, tables, chairs, apples, etc., are classes as one. (A person is a class of psychical existents, the others are classes of material points, with perhaps some reference to secondary qualities.) These objects, therefore, are of the same type as simple individuals. It would seem that all objects designated by single words, whether things or concepts, are of this type. Thus *e.g.* the relations that occur in actual relational propositions are of the same type as things, though relations in extension, which are what Symbolic Logic employs, are of a different type. (The intensional relations which occur in ordinary relational propositions are not determinate when their extensions are given, but the extensional relations of Symbolic Logic are classes of couples.) Individuals are the only objects of which numbers cannot be significantly asserted.

The next type consists of ranges or classes of individuals. (No ordinal ideas are to be associated with the word *range*.) Thus "Brown and Jones" is an object of this type, and will in general not yield a significant proposition if substituted for "Brown" in any true or false proposition of which Brown is a constituent. (This constitutes, in a kind of way, a justification for the grammatical distinction of singular and plural; but the analogy is not close, since a range may have one term or more, and where it has many, it may yet appear as singular in certain propositions.) If u be a range determined by a propositional function $\phi(x)$, not-u will consist of all objects for which $\phi(x)$ is false, so that not-u is contained in the range of significance of $\phi(x)$, and contains only objects of the same type as the members of u. There is a difficulty in this connection, arising from the fact that two propositional functions $\phi(x)$, $\psi(x)$ may have the same range of truth u, while their ranges of significance may be different; thus not-u becomes ambiguous. There will always be a minimum type within which u is contained, and not-u may be defined as the rest of this type. (The sum of two or more types is a type; a minimum type is one which is not such a sum.) In view of the Contradiction, this view seems the best; for not-u must be the range of falsehood of "x is a u," and "x is an x" must be in general meaningless; consequently "x is a u" must require that x and u should be of different types. It is doubtful whether this result can be insured except by confining ourselves, in this connection, to minimum types.

There is an unavoidable conflict with common sense in the necessity for denying that a mixed class (*i.e.* one whose members are not all of the same minimum type) can ever be of the same type as one of its members. Consider, for example, such phrases as "Heine and the French." If this is to be a class consisting of two individuals, "the French" must be understood as "the French nation," *i.e.*, as the class as one. If we are speaking of the French as many, we get a class consisting not of two members, but of one more than there are Frenchmen. Whether it is possible to form a class of which one member is Heine, while the other is the French as many, is a point to which I shall return later; for the present it is enough to remark that, if there be such a class, it must, if the Contradiction is to be avoided, be of a different type both from classes of individuals and from classes of classes of individuals.

The next type after classes of individuals consists of classes of classes of individuals. Such are, for example, associations of clubs; the members of such associations, the clubs, are themselves classes of individuals. It will be convenient to speak of *classes* only where we have classes of individuals, of *classes of classes* only where we have classes of classes of individuals, and so on. For the general notion, I shall use the word *range*. There is a progression of such types, since a range may be formed of objects of any given type, and the result is a range of higher type than its members.

A new series of types begins with the couple with sense. A range of such types is what Symbolic Logic treats as a relation : this is the extensional view of relations. We may then form ranges of relations, or relations of relations, or relations of couples (such as separation in Projective Geometry*),

* Cf. § 203.

or relations of individuals to couples, and so on; and in this way we get, not merely a single progression, but a whole infinite series of progressions. We have also the types formed of trios, which are the members of triple relations taken in extension as ranges; but of trios there are several kinds that are reducible to previous types. Thus if $\phi(x, y, z)$ be a propositional function, it may be a product of propositions $\phi_1(x) . \phi_2(y) . \phi_3(z)$ or a product $\phi_1(x) . \phi_2(y, z)$, or a proposition about x and the couple (y, z), or it may be analyzable in other analogous ways. In such cases, a new type does not arise. But if our proposition is not so analyzable—and there seems no à *priori* reason why it should always be so—then we obtain a new type, namely the trio. We can form ranges of trios, couples of trios, trios of trios, couples of a trio and an individual, and so on. All these yield new types. Thus we obtain an immense hierarchy of types, and it is difficult to be sure how many there may be; but the method of obtaining new types suggests that the total number is only a_0 (the number of finite integers), since the series obtained more or less resembles the series of rationals in the order $1, 2, ..., n, ..., 1/2$, $1/3, ..., 1/n, ..., 2/3, ..., 2/5, ...2/(2n+1), ...$ This, however, is only a conjecture.

Each of the types above enumerated is a *minimum* type; *i.e.*, if $\phi(x)$ be a propositional function which is significant for one value of x belonging to one of the above types, then $\phi(x)$ is significant for every value of x belonging to the said type. But it would seem—though of this I am doubtful—that the sum of any number of minimum types is a type, *i.e.* is a range of significance for certain propositional functions. Whether or not this is universally true, *all ranges* certainly form a type, since every range has a number; and so do all objects, since every object is identical with itself.

Outside the above series of types lies the type *proposition*; and from this as starting-point a new hierarchy, one might suppose, could be started; but there are certain difficulties in the way of such a view, which render it doubtful whether propositions can be treated like other objects.

498. Numbers, also, are a type lying outside the above series, and presenting certain difficulties, owing to the fact that every number selects certain objects out of every other type of ranges, namely those ranges which have the given number of members. This renders the obvious definition of 0 erroneous; for every type of range will have its own null-range, which will be a member of 0 considered as a range of ranges, so that we cannot say that 0 is the range whose only member is *the* null-range. Also numbers require a consideration of the totality of types and ranges; and in this consideration there may be difficulties.

Since all ranges have numbers, ranges are a range; consequently $x \epsilon x$ is sometimes significant, and in these cases its denial is also significant. Consequently there is a range w of ranges for which $x \epsilon x$ is false: thus the Contradiction proves that this range w does not belong to the range of significance of $x \epsilon x$. We may observe that $x \epsilon x$ can only be significant when x is of a type of infinite order, since, in $x \epsilon u$, u must always be of a type higher by one than x; but the range of all ranges is of course of a type of infinite order.

Since numbers are a type, the propositional function "x is not a u,"

where u is a range of numbers, must mean "x is a number which is not a u";
unless, indeed, to escape this somewhat paradoxical result, we say that,
although numbers are a type in regard to certain propositions, they are not
a type in regard to such propositions as "u is contained in v" or "x is a u."
Such a view is perfectly tenable, though it leads to complications of which it
is hard to see the end.

That propositions are a type results from the fact—if it be a fact—that
only propositions can significantly be said to be true or false. Certainly
true propositions appear to form a type, since they alone are asserted (cf.
Appendix A. § 479). But if so, the number of propositions is as great as
that of all objects absolutely, since every object is identical with itself, and
"x is identical with x" has a one-one relation to x. In this there are,
however, two difficulties. First, what we called the propositional concept
appears to be always an individual; consequently there should be no more
propositions than individuals. Secondly, if it is possible, as it seems to be, to
form ranges of propositions, there must be more such ranges than there are
propositions, although such ranges are only some among objects (cf. § 343).
These two difficulties are very serious, and demand a full discussion.

499. The first point may be illustrated by somewhat simpler ones.
There are, we know, more classes than individuals; but predicates are
individuals. Consequently not all classes have defining predicates. This
result, which is also deducible from the Contradiction, shows how necessary
it is to distinguish classes from predicates, and to adhere to the extensional
view of classes. Similarly there are more ranges of couples than there are
couples, and therefore more than there are individuals; but verbs, which
express relations intensionally, are individuals. Consequently not every
range of couples forms the extension of some verb, although every such
range forms the extension of some propositional function containing two
variables. Although, therefore, verbs are essential in the logical genesis of
such propositional functions, the intensional standpoint is inadequate to give
all the objects which Symbolic Logic regards as relations.

In the case of propositions, it seems as though there were always an
associated verbal noun which is an individual. We have "x is identical
with x" and "the self-identity of x," "x differs from y" and "the difference
of x and y"; and so on. The verbal noun, which is what we called the
propositional concept, appears on inspection to be an individual; but this is
impossible, for "the self-identity of x" has as many values as there are
objects, and therefore more values than there are individuals. This results
from the fact that there are propositions concerning every conceivable object,
and the definition of identity shows (§ 26) that every object concerning which
there are propositions, is identical with itself. The only method of evading
this difficulty is to deny that propositional concepts are individuals; and
this seems to be the course to which we are driven. It is undeniable,
however, that a propositional concept and a colour are two objects; hence
we shall have to admit that it is possible to form mixed ranges, whose
members are not all of the same type, but such ranges will be always of
a different type from what we may call pure ranges, $i.e.$ such as have only
members of one type. The propositional concept seems, in fact, to be nothing

other than the proposition itself, the difference being merely the psychological one that *we* do not assert the proposition in the one case, and do assert it in the other.

500. The second point presents greater difficulties. We cannot deny that there are ranges of propositions, for we often wish to assert the logical product of such ranges; yet we cannot admit that there are more ranges than propositions. At first sight, the difficulty might be thought to be solved by the fact that there is a proposition associated with every range of propositions which is not null, namely the logical product of the propositions of the range*; but this does not destroy Cantor's proof that a range has more sub-ranges than members. Let us apply the proof by assuming a particular one-one relation, which associates every proposition p which is not a logical product with the range whose only member is p, while it associates the product of all propositions with the null-range of propositions, and associates every other logical product of propositions with the range of its own factors. Then the range w which, by the general principle of Cantor's proof, is not correlated with any proposition, is the range of propositions which are logical products, but are not themselves factors of themselves. But, by the definition of the correlating relation, w ought to be correlated with the logical product of w. It will be found that the old contradiction breaks out afresh; for we can prove that the logical product of w both is and is not a member of w. This seems to show that there is no such range as w; but the doctrine of types does not show why there is no such range. It seems to follow that the Contradiction requires further subtleties for its solution; but what these are, I am at a loss to imagine.

Let us state this new contradiction more fully. If m be a class of propositions, the proposition "every m is true" may or may not be itself an m. But there is a one-one relation of this proposition to m: if n be different from m, "every n is true" is not the same proposition as "every m is true." Consider now the whole class of propositions of the form "every m is true," and having the property of not being members of their respective m's. Let this class be w, and let p be the proposition "every w is true." If p is a w, it must possess the defining property of w; but this property demands that p should not be a w. On the other hand, if p be not a w, then p does possess the defining property of w, and therefore is a w. Thus the contradiction appears unavoidable.

In order to deal with this contradiction, it is desirable to reopen the question of the identity of equivalent propositional functions and of the nature of the logical product of two propositions. These questions arise as follows. If m be a class of propositions, their logical product is the proposition "every m is true," which I shall denote by $\wedge^{\prime}m$. If we now consider the logical product of the class of propositions composed of m

---

* It might be doubted whether the relation of ranges of propositions to their logical products is one-one or many-one. For example, does the logical product of p and q and r differ from that of pq and r? A reference to the definition of the logical product (p. 21) will set this doubt at rest; for the two logical products in question, though equivalent, are by no means identical. Consequently there is a one-one relation of all ranges of propositions to some propositions, which is directly contradictory to Cantor's theorem.

together with ∧'*m*, this is equivalent to "Every *m* is true and every *m* is true,' *i.e.* to "every *m* is true" *i.e.* to ∧'*m*. Thus the logical product of the new class of propositions is equivalent to a member of the new class, which is the same as the logical product of *m*. Thus if we identify equivalent propositional functions (∧'*m* being a propositional function of *m*), the proof of the above contradiction fails, since every proposition of the form ∧'*m* is the logical product both of a class of which it is a member and of a class of which it is not a member.

But such an escape is, in reality, impracticable, for it is quite self-evident that equivalent propositional functions are often not identical. Who will maintain, for example, that "*x* is an even prime other than 2" is identical with "*x* is one of Charles II.'s wise deeds or foolish sayings"? Yet these are equivalent, if a well-known epitaph is to be credited. The logical product of all the propositions of the class composed of *m* and ∧'*m* is "Every proposition which either is an *m* or asserts that every *m* is true, is true"; and this is not identical with "every *m* is true," although the two are equivalent. Thus there seems no simple method of avoiding the contradiction in question.

The close analogy of this contradiction with the one discussed in Chapter x strongly suggests that the two must have the same solution, or at least very similar solutions. It is possible, of course, to hold that propositions themselves are of various types, and that logical products must have propositions of only one type as factors. But this suggestion seems harsh and highly artificial.

To sum up: it appears that the special contradiction of Chapter x is solved by the doctrine of types, but that there is at least one closely analogous contradiction which is probably not soluble by this doctrine. The totality of all logical objects, or of all propositions, involves, it would seem, a fundamental logical difficulty. What the complete solution of the difficulty may be, I have not succeeded in discovering; but as it affects the very foundations of reasoning, I earnestly commend the study of it to the attention of all students of logic.

INDEX

*The references are to pages. References in black type are to passages
where a technical term is defined or explained.*

Absolute, 226, 448
Abstraction, principle of, ix, 166, **219**, 242,
285, 305, 314, 497, 519
Acceleration, 474, 483;
absolute, 490, 491
Achilles and the tortoise, 350, 358
Action and Reaction, 483
Activity, 450
Addition, arithmetical, 118, 307; of indi-
viduals, 71, 133-135; logical, **17**, 21,
116; ordinal, 318; of quantities, 179,
180; relational, 182, 254; of relations,
321; relative, **26**, 387 *n.*; of vectors,
477
Adjectives, 20 *n.*, 42
Aggregates, 67, **139**, 442;
and classes as one, 141; infinite, 143
Algebra, universal, 376
Aliorelative, **203** *n.*, 320 *n.*
All, 72, 105, 113, 305
Analysis, how far falsification, 141, 466;
conceptual and real, 466
And, 67, 69, **71**, 130
Angles, 205, 414;
axioms of, 415, 416
Anharmonic ratio, **390**, 391, 420
Antinomies, of infinity, 188, 190-193;
Kant's, 259, 458-461
Any, 45, 46, **57**, 105, 263, 305, 351;
and kindred words, 55, **56**, 59, 89, 91
Archimedes, axiom of, 181, 252, **254**, 288,
332, 333, 337, 408
Area, 333, 417
Arithmetic, has no indemonstrables, 127;
and progressions, 240; relation-, 321
Arrow, Zeno's argument of, 350
Assertion, **34-35**, 48, 100, 502 ff.
Assertions, 39, 44, 82, 83, 98, 106, 505
Associative law, 307
Assumptions, 503
Axioms, in Geometry, 373, 441

Being, 43, 49, 71, 446, 449
Bernouilli, 329 *n.*
Bernstein, 306 *n.*, 367 *n.*
Bettazzi, 181 *n.*, 185
Between, 200, 201, 205, 207, **214**;
three theories of, 208; is a relation

between its terms? 210; and difference
of sense, 211; indefinable? 213; in
projective Geometry, 391, 393, 426; in
descriptive Geometry, 393
Bolyai, 373
Bolzano, 70, 201 *n.*, 307, 357 *n.*
Boole, 10, 24, 376
Borel, 306 *n.*, 367 *n.*
Bradley, 41, 43 *n.*, 47, 90, 99, 161 *n.*, 221,
224, 448, 471
Burali-Forti, 112 *n.*, 323, 364 *n.*

Calculus, propositional, 13-18; of classes,
18-23; of relations, 23-26; logical, 142;
infinitesimal, 259, 276, 304, 325-330,
338 ff.; principles of a, 376
Cantor, Georg, viii, 101, 111, 112, 119, 120,
121 *n.*, 144, 157, 161, 177, 199, 239 *n.*,
245, 259 ff., 267, 270 ff., 282, 331, 334,
347, 350, 353, 371, 375, 381, 390,
437 ff., 444, 527;
on irrationals, 283; on continuity, 287 ff.;
on transfinite cardinals, 304-311; on
transfinite ordinals, 312-324; on infinitesimal segments, 335; on orders of
infinity, 336; against greatest number,
363 ff.
Carroll, Lewis, 18 *n.*, 35
Cassirer, 287 *n.*
Cauchy, 329 *n.*
Causal laws, 481, 486
Causality, 474-479, 481;
in rational dynamics, 479
Causation, of particulars by particulars,
vii, 475, 477, 481, 487
Cause, equal to effect? 496
Cayley, 422 *n.*
Chain, 245, **246**;
of an element, 245, **246**
Change, 347, 469 ff.
Chasles, 420
Circle, postulate of, 438, 440
Class, v, ix, 18 ff., 40, 66-81, 349, 356,
497, 510 ff.;
extensional view of, 20, 67, 69, 131 ff.,
513, 526; intensional genesis of, 67, 515;
concept of, **67**; as many, 68, 76, 104,
106, 132; as one, 76, 103, 104, 106,

132, 513, 523; always definable by a
predicate? 98, 526; when a member of
itself, 102; defined by relation, 97, 98;
of terms not having a given relation to
themselves, 102; multiplicative, 308; infinite, 72, 106, 260, 306, 356, 357; denumerable, 309; and well-ordered series,
322; of one term, see *Individual*
Class-concept, 19, 20, 54, 56, 58, 67, 101,
113;
distinct from class, 68, 116, 131, 514
Clifford, 434
Cohen, 276 *n*., 326, 338–345
Collections, 69, 133, 140, 513, 514
Colours, 466, 467
Commutative law, 118, 240, 307; 312
Composition, 17, 31
Concepts, 44, 211, 508;
as such and as terms, 45; variation of,
86; propositional, 503, 526; can they
be subjects? 46, 507, 510
Congruent figures, 417
Conjunction, numerical, 57, 67, 72, 113,
131 ff.; propositional, 57; variable, 57
Connection, 202, 239
Consecutive, 201
Constants, logical, 3, 7, 8, 11, 106, 429;
and parameters, 6
Constituent, of a proposition, 356, 510;
of a whole, 143, 144
Continuity, 188, 193, 259, 286 ff., 368;
Dedekind's axiom of, 279, 294; ordinal,
296–303; philosophy of, 346–354; antinomies of, 347 ff.; in projective Geometry, 387, 390, 437; of Euclidean
space, 438 ff.
Continuum, in philosophical sense, 146,
440; in mathematical sense, 297, 299 *n*.,
310; composed of elements, 344, 347,
353, 440 ff.; primarily arithmetical, 444
Contradiction, the, vi, ix, 20, 66, 79, 97,
101–107, 305, 362, 513, 515, 517, 523,
524, 525;
Frege's solution of, 522; law of, 455
Coordinates, 439; projective, 385, 388, 390,
422, 427
Correlation, 260; of classes, 261; of series,
261, 321
Counting, 114, 133, 309
Couples, are relations classes of? 24, 99,
524; with sense, 99, 512, 524
Couples, separation of, 200, 205, 214, 237;
and transitive asymmetrical relations,
215, 238; in projective geometry, 386,
387
Couturat, 66, 194 *n*., 267 *n*., 291 *n*., 296 *n*.,
310 *n*., 326 *n*., 410 *n*., 441 *n*.
Cremona, 384 *n*., 420

Dedekind, 90, 111, 157, 199, 239 *n*., 245–
251, 294, 307, 315, 357 *n*., 381, 387, 438;
on irrationals, 278 ff.
Deduction, 522; principles of, 4, 15, 16
Definition, 15, 27, 111, 429, 497;
and *the*, 62; always nominal, 112; by
abstraction, 114, 219, 249

De Morgan, 23, 64 *n*., 218 *n*., 219 *n*., 326, 376
Denoting, 45, 47, 53, 106, 131;
and predicates, 54; and *any*, etc., 55, 62;
are there different kinds of? 56, 61; and
identity, 63; and infinite classes, 72, 73,
145, 350
Derivatives, of a series, 290 ff., 323; of
functions, 328
Descartes, 157
Dichotomy, Zeno's argument of, 348
Differential coefficients, 173, 328
Dimensions, 372, 374; definable logically,
376; axiom of three, 388, 399
Dini, 324 *n*., 327, 328 *n*., 329 *n*.
Direction, 435
Disjunction, 15 *n*., 17, 31; variable and
constant, 22, 58
Distance, 171, 179, 182 *n*., 195, 252–256,
288, 353;
measurement of, 180, 181, 254, 408; and
order, 204, 409, 419; and relative position, 252; not implied by order, 252,
254; definition of, 253; and limits, 254;
and stretch, 254, 342, 352, 408 ff., 435;
in Arithmetic, 254; axioms of, 407 ff.,
413, 424; and straight line, 410; projective theory of, 422, 425, 427; descriptive theory of, 423–5
Distributive law, 240, 307
Diversity, 23; conceptual, 46
Divisibility, infinite, 460
Divisibility, magnitude of, 149, 151, 153,
173, 230, 333, 345, 411, 425, 428; and
measurement, 178; not a property of
wholes as such, 179, 412
Domain, see *Relation*
Duality, logical, 26; geometrical, 375, 392
Du Bois Reymond, 181 *n*., 254, 336
Dynamics, vi; as pure mathematics, 465;
two principles of, 496

Economics, mathematical, 233 *n*.
Electricity, 494, 496
Empiricism, 373, 492
Epistemology, 339
Equality, 219, 339; of classes, 21; of re-
lations, 24; of quantities, 159
Equivalence, of propositions, 15, 527
Ether, 485, 496
Euclid, 157, 287, 373, 404, 420, 438;
his errors, 405–407
Euler, 329 *n*.
Evellin, 352
Existence, vii, 449, 458, 472;
of a class, 21, 32
Existence-theorems, ix, 322, 431, 497;
and Euclid's problems, 404
Exponentiation, 120, 308
Exportation, 16
Extension and Intension, 66

Fano, 385 *n*.
Field, see *Relation*
Finite, 121, 192, 371
Finitude, axiom of, 188, 191, 460; abso-
lute and relative, 332